The So-Called Eighth *Stromateus* by Clement of Alexandria

Philosophia Antiqua

A SERIES OF STUDIES ON ANCIENT PHILOSOPHY

VOLUME 144

The titles published in this series are listed at *brill.com/pha*

The So-Called Eighth *Stromateus* by Clement of Alexandria

Early Christian Reception of Greek Scientific Methodology

By

Matyáš Havrda

BRILL

LEIDEN | BOSTON

Library of Congress Cataloging-in-Publication Data

Names: Havrda, Matyáš, 1972- | Clement, of Alexandria, Saint, approximately 150-approximately 215. Stromata. Liber 8. | Clement, of Alexandria, Saint, approximately 150-approximately 215. Stromata. Liber 8. English.
Title: The so-called eighth Stromateus by Clement of Alexandria : early Christian reception of Greek scientific methodology / by Matyáš Havrda.
Description: Leiden ; Boston : Brill, [2017] | Series: Philosophia antiqua, ISSN 0079-1687 ; volume 44 | Includes bibliographical references and index.
Identifiers: LCCN 2016030309 (print) | LCCN 2016035973 (ebook) | ISBN 9789004310087 (hardback : acid-free paper) | ISBN 9789004325289 (e-book)
Subjects: LCSH: Clement, of Alexandria, Saint, approximately 150-approximately 215. | Philosophy, Ancient. | Science–Philosophy–Early works to 1800. | Science–Methodology–Early works to 1800. | Clement, of Alexandria, Saint, approximately 150-approximately 215. Stromata.
Classification: LCC B666.S63 S63 2017 (print) | LCC B666.S63 (ebook) | DDC 189/.2–dc23
LC record available at https://lccn.loc.gov/2016030309

Typeface for the Latin, Greek, and Cyrillic scripts: "Brill". See and download: brill.com/brill-typeface.

ISSN 0079-1687
ISBN 978-90-04-31008-7 (hardback)
ISBN 978-90-04-32528-9 (e-book)

This book is printed on acid-free paper and produced in a sustainable manner.

Contents

Preface

This book is a study of the so-called eighth *Stromateus* by Clement of Alexandria (d. before 221 C.E.). Its core is a lemmatic commentary, accompanied by the Greek text, translation, and an introduction, which deals with the perplexing question of the origin and purpose of this mysterious text.

The 'eighth book' is not a standard book of the *Stromateis*. To all appearances, it is a collection of excerpts, which, for the most part, are of a purely philosophical nature. For various reasons (mainly, I suspect, because it falls in-between different fields of historical expertise), the text has been rather neglected by scholars, both those working in ancient philosophy and the experts on Clement. The last monograph dedicated to it (one of the last German dissertations written in Latin) was published more than a century ago, and the number of scholars who have explored it in detail (apart from a few passages, immortalized by their inclusion in von Arnim's collection of Stoic fragments) would easily fit into a medium-sized seminar room.

This obscurity contrasts with the fact that the text deals, in an informed and intelligent manner, with a number of topics that scholars, especially those working on ancient philosophy, usually find interesting: demonstration, dialectic, division and definition, categories, causation, and scepticism. A closer look shows that these topics are treated in a didactic and introductory style, but not in a way that could be described as naïve or conventional. Here is a text standing in the tradition of Aristotle's *Organon*, combining freely but rigorously Aristotelian elements with those coming from later sources (Stoic, Platonist, sceptic, and medical); chiefly, as it seems, in order to provide methodological instruction.

One might wonder, then, if the 'eighth *Stromateus*' has anything to do with the Christian apologist and biblical exegete under whose name it is preserved. However, there is no doubt that it was written by Clement: First of all, there are several parallels, some of them almost verbatim, between the 'eighth book' and the regular books of the *Stromateis*. Second, the first two pages clearly conform to Clement's style and preoccupations. Third, the hand of a Christian writer can be occasionally recognized even in more densely philosophical parts. Most likely, the text consists of Clement's excerpts from a philosophical source, to which he sometimes added his own gloss or comment. Nonetheless, as I will argue in this book, it seems unlikely that he ever intended it to be the continuation of the *Stromateis*.

As a matter of fact, the very title of our text is problematic: In the Byzantine manuscript which contains the *Stromateis*, the text known as the eighth book

is placed at the beginning of a larger section, which also includes other fragmentary material of a more theological nature (the *Excerpta ex Theodoto* and *Eclogae propheticae*). However, the title 'eighth *Stromateus*', written at the head of the first page of that section, might well pertain to this material as a whole, not just to its first, philosophical part. The convention of using the title with reference to the first part only goes back to 1550 (the first printed edition of the *Stromateis*) and does not seem to have much support either in the manuscript or in the ancient testimonies on Clement's works. Thus, when speaking of the 'eighth *Stromateus*' in this study, we follow the modern convention; however, by adding the "so-called" (or the inverted commas) to it, we signal our reluctance to accept this convention as true. If one could choose to replace one convention with another, a better candidate for a title would be 'liber logicus', a description used by Daniel Heinse in 1616.

The scholarly debate about the 'eighth *Stromateus*' spans four centuries. It is concerned chiefly with the following questions: How to explain the incoherent and fragmentary character of this text? What is its place within Clement's *œuvre*? How is it relevant to Clement's thought? And what are its philosophical sources? Many conflicting answers to these questions have been produced—most of them in the 19th and early 20th century—but rarely has one study attempted to deal with them all. Today, the state of the art is hard to define. Some of the questions are occasionally addressed on the margins of other topics. Others are dealt with in connection with broader issues regarding the 'meta-Stromatic' material, without sufficient attention being paid to the peculiar contents of 'liber logicus'. Selected passages have attracted scholars working in ancient philosophy, but few have explored them in their broader context. The present study is an attempt to re-visit these questions: In a critical dialogue with the four centuries of scholarship, it proposes an argument whose aim is to accommodate all the major concerns which have arisen in the course of the debate.

The most intriguing aspect of the riddle is the question of Clement's philosophical sources. The tradition of the source-critical investigation of 'liber logicus', started by Hans von Arnim and including, in our own days, Jaap Mansfeld and Teun Tieleman, has produced a wealth of comparative material, which enables us to localize the main source of the text in the philosophical landscape of the 2nd century A.D. with reasonable assurance. I have argued in earlier studies, and continue to argue in this book, that the source, or the main one, is the lost treatise *On Demonstration*, written by Clement's older contemporary, the great doctor-*cum*-philosopher Galen of Pergamum. No doubt, readers will approach this proposal with caution and not everyone will be convinced. In the part of the introduction called 'The Galen Hypothesis', I have assembled

the reasons that, in the course of my work on the commentary, have convinced *me*. I am aware how bold the proposal is. However, other attempts at explaining the origin of the text and its relation to Galen's works, such as those submitted by Solmsen or Tieleman, strike me as less plausible. Whether or not everyone else will find the same reasons equally compelling is another matter:

> … ἢ γὰρ εὐτυχεῖς
> σὺν τῷ θεῷ φανούμεθ', ἢ πεπτωκότες.[1]

Like Zeus, this book was conceived in Crete. I started my investigation of the sources of '*Stromateis* VIII' while a visiting fellow at the University of Crete, Rethymno, in 2009. As an outcome of this 6-month visit, generously funded by the Alexander S. Onassis Foundation, I produced two articles arguing in favour of the Galenic provenance of the text.[2] During this happy time, right before the outbreak of the Greek financial crisis, my main hosts at the Department of Philosophy and Social Studies were Andrei Lebedev and George Karamanolis. With George, we met on a regular basis, reading and discussing the 'eighth book' together. His interest and encouragement was important for me at this early stage, when I felt a little bit lonely with a text that seemed to interest no one else and that I failed to understand.

The bulk of the book was written in the years 2012–2015, when I was a research fellow at the Institute for Classical Philology at Humboldt University Berlin. My research was part of the programme 'Medicine of the Mind, Philosophy of the Body', funded by the Alexander von Humboldt Foundation and directed by Philip van der Eijk. I am eternally grateful to Philip for accepting me into his research group, whose rigorous work ethic combined with selfless curiosity affected my work daily. In Berlin, I had many opportunities to present parts of my work in front of a learned audience, whether at the Institute for Classical Philology, the Institute for Philosophy at HU, or in seminars organized within the 'Philosophy, Science, and the Sciences' programme. My warm thanks to the organizers and participants of these events, especially Stephen Menn, whose feedback I found particularly useful. Of colleagues and friends whose advice had a direct impact on the book, I would additionally like to mention Fabio Acerbi, István Bodnár, Riccardo Chiaradonna, Sean Coughlin, Philip van der Eijk, Katerina Ierodiakonou, Ricardo Julião, Matteo Martelli,

1 Sophocles, *Oed. Tyr.* 145 f.

2 "Galenus Christianus? The Doctrine of Demonstration in *Stromata* VIII and the Question of Its Source," *Vigiliae Christianae* 65/4 (2011), pp. 343–375; "Categories in *Stromata* VIII," *Elenchos* 33/2 (2012), pp. 197–225.

Vladimír Mikeš, Oliver Overwien, Marwan Rashed, Christina Savino, Teun Tieleman, and Máté Veres.

Several scholars have kindly provided written comments on parts of my commentary relevant to their expertise. They include Alain Le Boulluec, Jonathan Barnes, Luca Castagnoli, and David Sedley. All these attentive readers have contributed more than I could acknowledge in each instance. I would particularly like to express my deep gratitude to Jonathan Barnes, whose detailed and penetrating notes on the sections on demonstration saved me from many errors and helped me raise my own standards of what a good commentary should look like. Barnes has also supplied a number of textual notes, most of which are mentioned in the critical apparatus to the Greek text. Of course, all responsibility for the remaining (or new) blunders is mine.

Further thanks are due to Brill's anonymous readers, who made helpful suggestions for improvement, and to the editors of the Philosophia Antiqua series for accepting my manuscript for publication. Katharina Heß helped me compile the bibliography and index locorum. Early stage of my research (2009–2011) and the final production phase of the manuscript was co-funded by Palacký University Olomouc.

My last thanks go to my first teacher of philosophy, Zdeněk Kratochvíl, who brought me to the study of Clement some twenty years ago; and to my wife Mai, who has accompanied me all along and shared my enthusiasm for my work. τὰ δ' ἄλλα σιγῶ.

Abbreviations

CAG	Commentaria in Aristotelem Graeca
CCTC	Cambridge Classical Texts and Commentaries
CMG	Corpus medicorum Graecorum
CML	Corpus medicorum Latinorum
DK	*Die Fragmente der Vorsokratiker*, ed. H. Diels and W. Kranz, Vols. I–II, Dublin/Zurich: Weidman, 1972
DL	Diogenes Laertius, *Vitae philosophorum*, ed. T. Dorandi, Cambridge: Cambridge University Press, 2013
Dox. Gr.	*Doxographi Graeci*, ed. H. Diels, Berlin: Reimer, 1879
EK	Posidonius, ed. L. Edelstein and I.G. Kidd, Cambridge: Cambridge University Press, 1972–1999
FDS	*Fragmente zur Dialektik der Stoiker*, ed. K. Hülser, Vols. I–IV, Stuttgart/Bad Cannstatt: frommann-holzboog, 1987–1988
GCS	Die griechischen christlichen Schriftsteller der ersten drei Jahrhunderte
K.	K.G. Kühn, ed., *Claudii Galeni opera omnia*, Vols. I–XX; Leipzig: Knobloch, 1821–1833
L	Laurentianus Pluteus 5.3
LCL	Loeb Classical Library
LS	A.A. Long and D.N. Sedley, *The Hellenistic Philosophers*, Vols. I–II, Cambridge: Cambridge University Press, 1987
LSJ	H.G. Liddle and R. Scott, *The Greek-English Lexicon*, revised and augmented by H.S. Jones, 9th edition with revised supplement, Oxford: Clarendon Press, 1996
MGO	Medicorum Graecorum opera quae exstant, ed. K.G. Kühn
OCT	Oxford Classical Texts
PhA	Philosophia Antiqua (Brill)
PL	Patrologiae cursus completus, series Latina, ed. J.-P. Migne
SC	Sources Chrétiennes (Cerf)
Suppl. Vig. Chr.	Supplements to Vigiliae Christianae (Brill)
SVF	*Stoicorum veterum fragmenta*, ed. H. von Arnim, Vols. I–III, Leipzig: Teubner, 1903–1905

INTRODUCTION

The Riddle of the 'Eighth *Stromateus*': Questions and Solutions

Ancient and Byzantine Testimonies

In the fourth century A.D., in his account of church history under Septimius Severus, Eusebius, a bishop of Palestinian Caesarea, says that Clement of Alexandria's *Stromateis* comprised eight books.[1] This information is likely based on the bishop's first-hand acquaintance with a manuscript of the *Stromateis* available in the Caesarean library.[2] Eusebius often quotes from Clement, including the *Stromateis*, but never from the eighth book; nor does he say anything about its contents. However, Acacius 'the One-Eyed', Eusebius' pupil and successor, does quote a passage he introduces as coming from the eighth book of the *Stromateis*.[3] Surprisingly, the reader of modern editions of Clement's works will not find this quotation in the eighth book, but in a text that is usually considered to be a different piece of writing, a text never mentioned by Eusebius (or anyone else in antiquity), namely the *Eclogae propheticae*. This indicates that in fourth century Caesarea there was a manuscript of Clement's works in which the material known as *Eclogae propheticae*, or some part of it, was placed and designated in such a way as to suggest that it belonged to the eighth book of the *Stromateis*.

There has always been something odd about the eighth book. In a Syriac collection of *testimonia patrum* dated between the 8th and 9th century, whose Greek original is believed to go back to the 6th century, there is a quotation of the definition of *hairesis* found in *Strom.* VIII (5) 16, 2. The definition is introduced by the words "Clement, *Stromateis*, end of book eight". However, in modern editions of Clement's works it is not placed at the end of the eighth

1 Eusebius, *Eccl. Hist.* VI 13, 1 (GCS 9/2: 541,9–12): τοῦ δὲ Κλήμεντος Στρωματεῖς, οἱ πάντες ὀκτώ, παρ' ἡμῖν σῴζονται, οὓς καὶ τοιαύτης ἠξίωσεν προγραφῆς Τίτου Φλαυίου Κλήμεντος τῶν κατὰ τὴν ἀληθῆ φιλοσοφίαν γνωστικῶν ὑπομνημάτων στρωματεῖς κτλ.

2 Cf. Andrew Carriker, *The Library of Eusebius of Caesarea* (Suppl. Vig. Chr. 67; Leiden/Boston: Brill, 2003), pp. 196f.

3 Cf. Theodor Zahn, *Forschungen zur Geschichte des neutestamentichen Kanon und der altkirchlichen* Literatur III: *Supplementum Clementinum* (Erlangen: Deichert, 1884), p. 29.

 | DOI: 10.1163/9789004325289_002

book, but in its first half.[4] In the ninth century, Photius, the learned bishop of Constantinople, inspected several manuscripts of the *Stromateis* and made the following observation: Whereas in the case of the first seven books, each book has the same title and contents in all the codices, when it comes to the eighth book, there is disagreement. Some codices provide it with the superscription τίς ὁ σῳζόμενος πλούσιος; ('Which rich man shall be saved?'), followed by the first lines of a text we know under that title (henceforth referred to as *QDS*), Clement's famous homily on Mark 10:17–27, preserved independently of the *Stromateis*. Other codices, according to Photius, have the title στρωματεὺς ὄγδοος ('The eighth *Stromateus*'), the same sort of title by which the other books of the *Stromateis* are introduced, followed by the words, "But the most ancient philosophers, etc.," i.e. the words we now identify as the beginning of the eighth book.[5] It is not clear why Photius prefers to say that in some codices *QDS* assumes the role of the eighth book of the *Stromateis*, rather than that some codices only contain seven books of the *Stromateis*, followed by *QDS*. Perhaps the reason is that the codex as a whole ran under the title στρωματεῖς, thus suggesting that everything contained in it belonged to this work. Photius also says that in one ancient codex (ἔν τινι παλαιῷ βιβλίῳ) he found not only the title στρωματεῖς, but a more complete one: 'Titus Flavius Clemens, presbyter of Alexandria, *Stromateis* of gnostic notes according to the true philosophy, books I, II, III, IV, V, VI, VII, and VIII.'[6] Photius does not reveal which of the two versions of the *Stromateis* he discovered in that codex; but since this codex explicitly claimed to contain *eight* books of the *Stromateis*, it seems likely that it also contained a text described as στρωματεὺς ὄγδοος.

We shall probably never be sure on what grounds Acacius included a passage from the *Eclogae propheticae* in the eighth book of the *Stromateis*, or why a compiler of patristic testimonies two centuries later thought that the eighth book ended after mere sixteen paragraphs of Stählin's edition. Neither do we know why some of the codices seen by Photius did not contain the eighth book at all, nor why these codices placed *QDS* after the seventh book instead.

---

4 Cf. below, commentary on 89,24–90,6.

5 Photius, *Bibliotheca* 111 (89b27–35 Bekker): ἀλλ' ὁ μὲν πρῶτος μέχρι τοῦ ἑβδόμου τὴν αὐτὴν ἔχουσιν ἐπιγραφὴν καὶ ἐνιαῖοι τυγχάνουσιν ἐν ἅπασι τοῖς βιβλίοις. ὁ μέντοι ὄγδοος διάφορός τέ ἐστι καὶ τῇ ἐπιγραφῇ καὶ τῷ ἐδάφει. ἔν τισι μὲν γάρ "Τίς ὁ σῳζόμενος πλούσιος;" ἐπιγράφεται, καὶ ἄρχεται οὕτως· "οἱ μὲν τοὺς ἐγκωμιαστικοὺς λόγους" καὶ ἑξῆς· ἔν τισι δὲ Στρωματεὺς ὄγδοος, ὥσπερ καὶ οἱ πρὸ αὐτοῦ ἑπτά, ἐπιγράφεται, καὶ ἀπάρχεται· "ἀλλ' οὐδὲ οἱ παλαίτατοι τῶν φιλοσόφων" καὶ ἑξῆς.

6 Ibid. 89b22–27: εὗρον μέντοι γε ἔν τινι παλαιῷ βιβλίῳ τὴν αὐτὴν πραγματείαν οὐχὶ Στρωματεῖς μόνον ἐπιγραφομένην ἀλλ' ὁλοκλήρως οὕτως· "Τίτου Φλαβίου Κλήμεντος, πρεσβυτέρου Ἀλεξανδρείας, τῶν κατὰ τὴν ἀληθῆ φιλοσοφίαν γνωστικῶν ὑπομνημάτων στρωματέων α, β, γ, δ, ε, ς, ζ καὶ η".

But the simplest and, I believe, the most plausible explanation may be found when we turn our attention to another important witness, namely the only manuscript on which our knowledge of the *Stromateis* depends,[7] the celebrated Codex Pluteus 5.3 from the Medicean Library in Florence, dated to the eleventh century.[8]

First of all, a brief description is due. The Florentine codex contains 388 folios (775 pages) with the text of Clement's *Stromateis*. On folio 346 *verso*, after the last page of the seventh book, a new section begins, introduced by the superscription στρωματεὺς ὄγδοος (see plate 1). What follows until the end of the manuscript (the text ends abruptly after 84 pages, about one ninth of the whole manuscript) is divided into four parts: The first part (fol. 346ᵛ–353ʳ) follows after the superscription mentioned above, covering less than fourteen pages. This section corresponds to *Strom.* VIII (1) 1, 1–(5) 16, 3 in Stählin's edition. The second part, starting on fol. 353ʳ, is introduced by the superscription αἱ τῶν ζητήσεων ἔφοδοι καὶ ἀρχαὶ περὶ ταῦτα καὶ ἐν τούτοις εἰσίν ('Procedures and principles of inquiries involve and consist in these matters'). (See plate 2). Almost exactly the same words are repeated after another sixteen pages (fol. 361ʳ), where they clearly function as a *subscriptio*. Here they are followed by a simply decorated line with floral motifs on each side, the same kind of line that marks the end of most books of the preceding *Stromateis*.[9] (See plate 3). This is the end of the second part. Afterwards, on the same page, there is another superscription ('Epitomes from the writings of Theodotus and the so-called eastern teaching from the time of Valentinus'),[10] and thirty-four pages later (378ʳ) yet another one ('Selections from the prophetic writings').[11] These

7 The text of the *Stromateis* is also preserved in Parisinus Suppl. Graecus 250 from the 16th century. But this manuscript depends on Plut. 5.3 (11th century). Cf. Otto Stählin, *Clemens Alexandrinus* I, GCS 12 (Leipzig 1905), p. XXXIX.

8 The codex (often referred to simply as 'Laurentianus' or L) is described in detail by Otto Stählin in the introduction to his GCS edition of Clement's works (GCS 12, pp. XXXIX–XLII). It is accessible online through the Bibliotheca Laurenziana website.

9 Cf. Plut. 5.3, fols. 65ʳ, 111ᵛ, 144ʳ, 190ᵛ, and 296ʳ. The end of books V (237ᵛ) and VII (346ʳ) is not marked by that line; book V, moreover, lacks a *subscriptio*.

10 ἐκ τῶν θεοδότου καὶ τῆς ἀνατολικῆς καλουμένης διδασκαλίας κατὰ τοὺς οὐαλεντίνου χρόνους ἐπιτομαί.

11 ἐκ τῶν προφητικῶν ἐκλογαί. Curiously, this title is written at the very bottom of the page, i.e. virtually as a *subscriptio* of the preceding text. Cf. Alain Le Boulluec, "Extraits d' œuvres de Clement d' Alexandrie: La transmission et le sens de leur titres," in idem, *Alexandrie antique et chrétienne: Clément et Origène* (Paris: Institut d' Études Augustiniennes, 2006), pp. 113 f., pointing out that the scribe did not understand the title as the beginning of a new work.

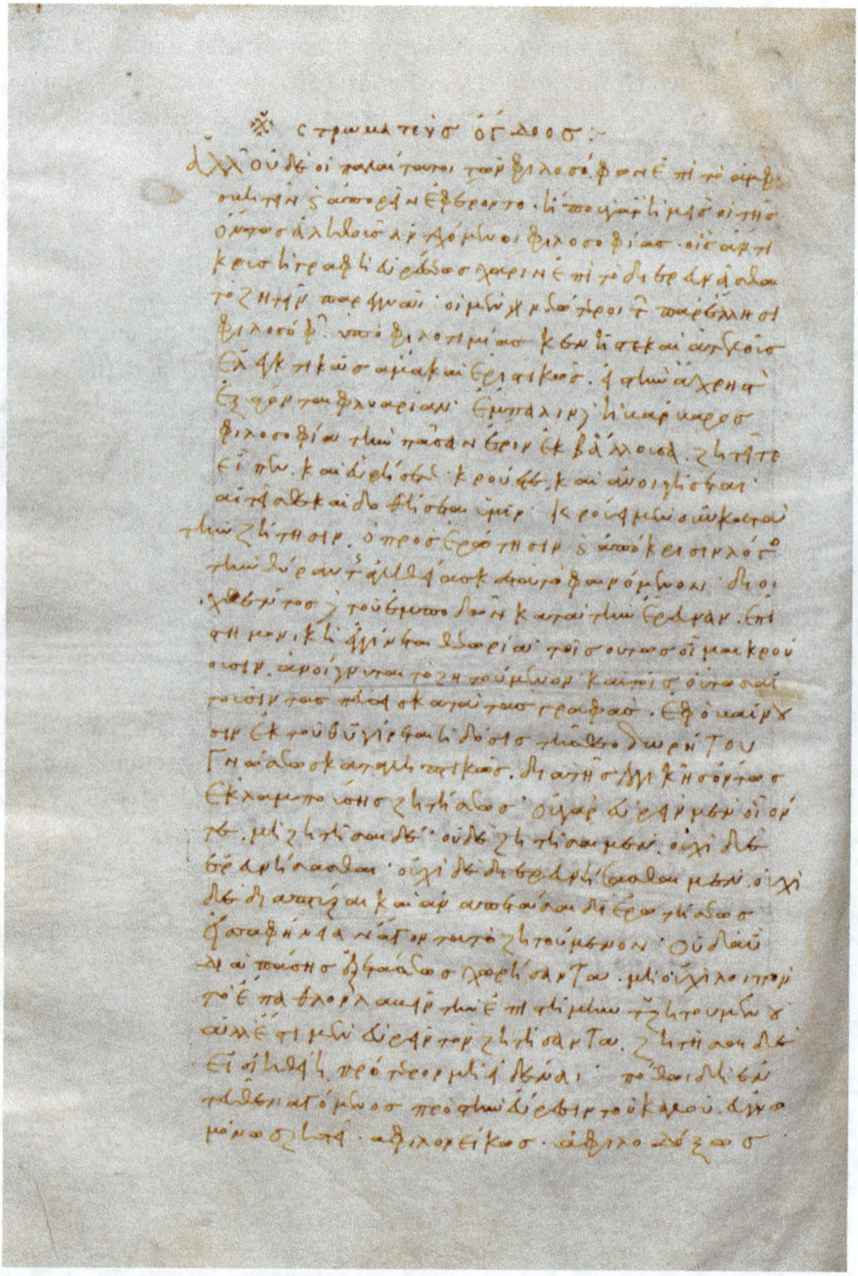

PLATE 1 *Florence, The Bibliotheca Medicea Laurenziana, ms. Plut. 5.3, fol. 346ᵛ.*
REPRODUCED WITH PERMISSION OF MIBACT

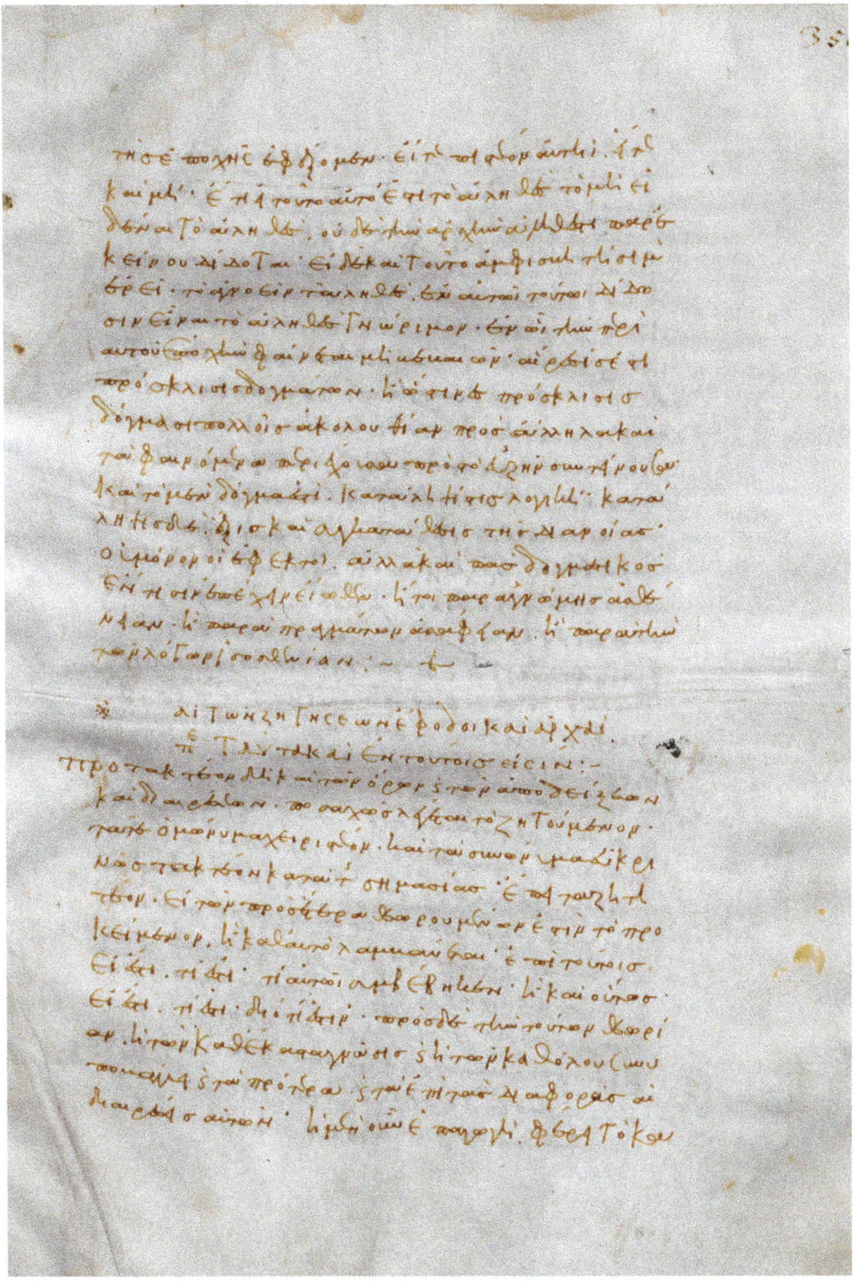

PLATE 2 *Florence, The Bibliotheca Medicea Laurenziana, ms. Plut. 5.3, fol. 353*[r]*.*

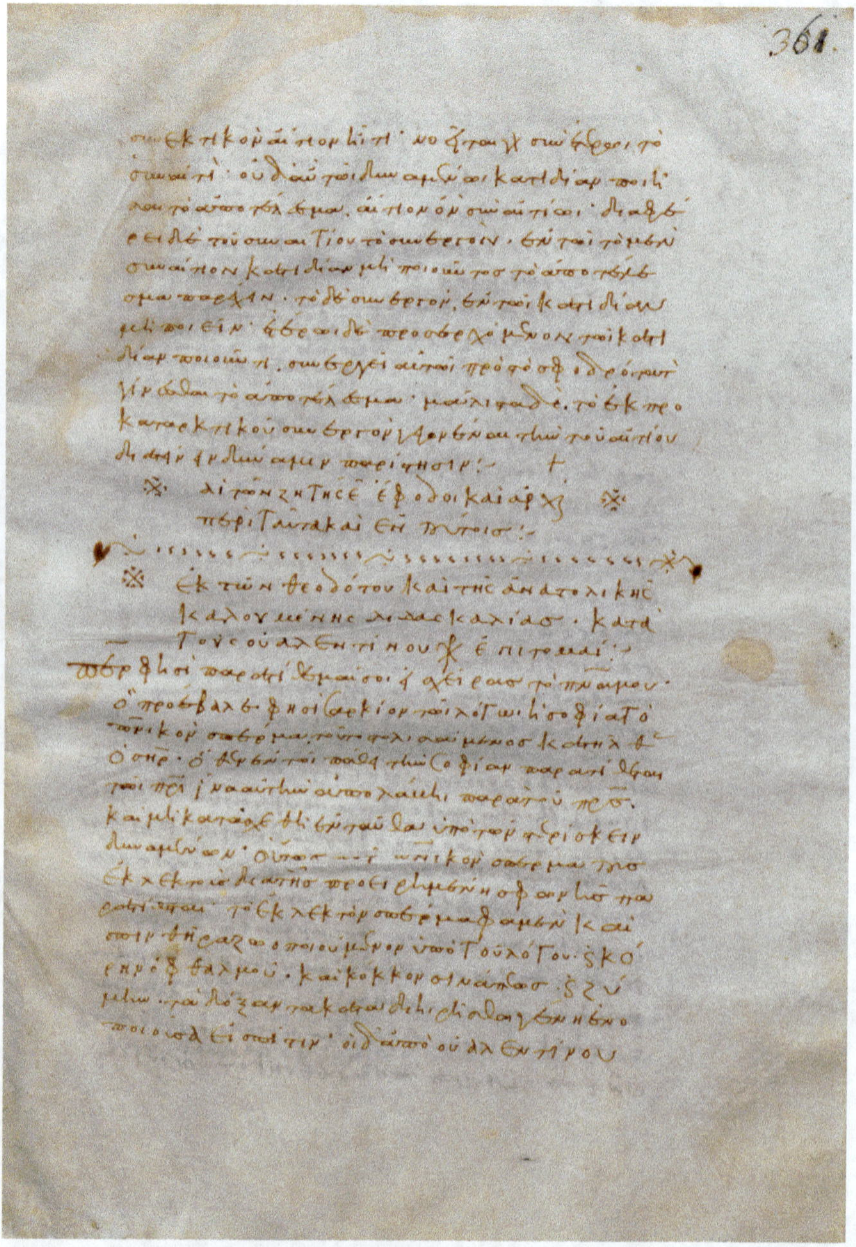

PLATE 3 *Florence, The Bibliotheca Medicea Laurenziana, ms. Plut. 5.3, fol. 361*[r]*.*
REPRODUCED WITH PERMISSION OF MIBACT

two superscriptions designate the beginning of sections nowadays known by their Latin titles *Excerpta ex Theodoto* and *Eclogae propheticae*, respectively.[12] There is no *subscriptio* at the end of the manuscript.[13]

Any reader of Plut. 5.3 will find it hard to decide where the text called στρωματεὺς ὄγδοος is supposed to end. There are three options: (1) It might appear that the extent of the eighth book is most naturally circumscribed by the first superscription on the one hand (στρωματεὺς ὄγδοος) and the second superscription on the other (αἱ τῶν ζητήσεων ἔφοδοι κτλ.). According to this interpretation, the eighth book would consist of a mere fourteen pages of the manuscript ('the first part' mentioned above), followed by another, slightly longer text concerned with the methods and principles of inquiry ('the second part'). Despite the arrangement of his own edition, this is how Otto Stählin understands the matter, citing the above-mentioned Syriac quotation in support of this view.[14] (2) Alternatively, one might think that the decorated line on fol. 361ʳ marks the end of the eighth book, just as it marks the end of most of the preceding books of the *Stromateis*. Starting from the first printed edition of the *Stromateis*, published in Florence in 1550 by Piero Vettori (Petrus Victorius), this has been the traditional view, reflected in our habitual way of speaking about Clement's works. Accordingly, we refer to the first and the second parts collectively as *Stromateis* VIII, distinguishing it from the next two parts, which are treated as independent writings.[15] This interpretation is strongly supported by the contents: there is clear thematic continuity between the first and the second part

12 For the origin of the Latin titles, cf. Le Boulluec, "Extraits," pp. 110 and 113 f.

13 The manuscript ends abruptly and we know that something was lost at the end, as more of the *Eclogae* is preserved in several manuscripts containing excerpts from Clement's works (cf. Stählin in Clemens Alexandrinus I, pp. XLVII–LI, on the 'Excerpthandschriften'). However, whatever was lost, it must have been lost earlier than in our manuscript, whose text ends at the *recto* side of the last folio, leaving the final page empty (*pace* Stählin, ibid., arguing that the 'Excerpthandschriften' are based on Plut. 5.3).

14 Cf. Stählin, GCS 12, p. XLI, who suggests that in Codex Laurentianus the eighth book ends with § 16.

15 Vettori faithfully reproduces the four titles mentioned above, not indicating any hierarchy among them (or rather among the first three ones, as the fourth, ἐκ τῶν προφητικῶν ἐκλογαί, is written in the minuscules for some reason). But his interpretation of the texts above the line as the eighth book of the *Stromateis* is indicated by the headers of the pages on which they are written: Pages 329–338 of his edition, covering fols. 346ᵛ–361ʳ, consist of two separate sections, each introduced by a title of the same size (στρωματεὺς ὄγδοος and αἱ τῶν ζητήσεων ἔφοδοι κτλ., respectively), but both sharing the same header, viz. στρωματέων λόγος ὄγδοος. Each of the subsequent sections (*Exc.* and *Ecl.*) has its own header in turn. Vettori does not reproduce the *subscriptio* on fol. 361ʳ.

(both being concerned with aspects of scientific methodology), whereas their topics and style are markedly different from those of parts three and four, the former dealing with Valentinian writings, the latter with selected biblical passages. (3) Finally, we might believe that the title στρωματεὺς ὄγδοος relates to the subsequent material as a whole. The following consideration speaks in favour of this view. As already mentioned, there is no indication in the manuscript that the eighth book of the *Stromateis* ends either on fol. 353ʳ or 361ʳ: On fol. 353ʳ, there is neither a line separating the first part from the second, nor any *subscriptio*. On fol. 361ʳ, there is both a line and a *subscriptio*; nevertheless, the *subscriptio* does not mark the end of *Stromateis* VIII, but of a section starting on fol. 353ʳ (αἱ τῶν ζητήσεων ἔφοδοι κτλ.). If the scribe had thought that *Stromateis* VIII ended in either of these places, presumably he would have shown it by means he uses elsewhere to this effect.[16]

Putting aside for the moment the question which of these options is more plausible than others, it is worth noting that their variety could help us explain the inconsistent testimonies concerning *Stromateis* VIII. The simplest way to account for the fact that Acacius includes a passage from the *Eclogae* in the eighth book of the *Stromateis* is to assume that he already held a similar text to the one we possess, with the same title of an unspecified extension, and chose to include all of its contents under the heading στρωματεὺς ὄγδοος. Likewise, whoever assigned a quotation from *Strom.* VIII (5) 16, 2–3 to the end of the eighth book could have done so on the basis of a manuscript similar to ours, while applying the superscription to the first part only.[17] The decision of Clement's Renaissance editor to locate the end of *Stromateis* VIII before the start of the *Excerpta* does not seem to have any precedent among earlier authors. Nevertheless, there are some grounds for believing that the

---

16 As mentioned above, the only instance in our manuscript where a book of the *Stromateis* is concluded neither by a decorated line nor by a *subscriptio* is the end of book V (fol. 237ᵛ). Interestingly, this omission is not due to a lack of space on the page where the fifth book ends: there is enough space for approximately six more lines. Perhaps the copyist did not wish to introduce any element he had not found in his antigraph, where the omission of the *subscriptio* could have been due to a lack of space. In any case, in this instance, no doubt could arise as to where the end is located, as the fifth book ends with a concluding formula and the sixth book starts (with a superscription) on another page. Had the scribe of Plut. 5.3 wished to suggest that *Strom.* VIII ended either on fol. 353ʳ or 361ʳ, he could have at least started to copy the following treatise on another page.

17 Cf. Le Boulluec, "Extraits," pp. 116 f. Zahn, *Forschungen*, pp. 116 f., thinks that there were different versions of the eighth book in circulation.

manuscript on which he based that decision, namely Plut. 5.3,[18] was similar to the text of *Stromateis* VIII known to Photius. Stählin observes that the scribe of the manuscript also copied a part of the Arethas codex, the famous collection of patristic texts transcribed under the supervision of the Cappadocian bishop Arethas in the ninth century, containing, among other things, Clement's *Protrepticus* and *Paedagogus*. According to Stählin, this suggests that the antigraph of our manuscript could have belonged to Arethas' library as well.[19] But if it did belong to that library, it seems plausible that it was based on some of the manuscripts known to Photius (who was a teacher of Arethas), the most likely candidate being the 'ancient codex' mentioned in the *Bibliotheca*. In fact, Photius' information about the superscription and the first line of the eighth book exactly corresponds to our manuscript and there is no reason to think that the rest of the text was significantly different.[20] If, like Eusebius before him, Photius never mentions such writings as the *Excerpta* or *Eclogae*, it is presumably because he did not distinguish them from *Stromateis* VIII. On the other hand, the fact that some Byzantine editors of Clement's work discarded the eighth book, placing *QDS* in its stead, seems to signal an embarrassment on their part; this, again, can be explained by the shape in which the text called στρωματεὺς ὄγδοος finds itself in our manuscript: a barely coherent collection of notes of questionable theological value, divided into thematically disconnected sections. No wonder someone chose to replace it with an essay as readable and pious as *QDS*.

Composition and Contents

Towards the end of the seventh book, Clement characterizes his writing as follows:

> Now, it seems that what are known as *Stromateis* are not to be compared to ornamental parks with rows of ordered plantations to please the eye, but rather to some thickly wooded hill, overgrown with cypresses and planes and bay-tree and ivy, and at the same time planted with apple trees and olives and figs, the cultivation of fruit-bearing and of woodland trees

18 Cf. Stählin, GCS 12, p. LXVI.

19 GCS 12, p. XL n. 1.

20 Photius' quotation from the end of the seventh book in *Bibl.* 111 (89b15–20 Bekker) shows that his text was different from (inferior to) the text of Plut. 5.3. On the other hand, we need not assume that he quoted this passage from the 'ancient codex'.

> being intentionally mingled together, since Scripture desires to withdraw from observation on account of those who venture secretly to steal its fruits. It is by transplanting the suckers and trees from these preserves that the gardener will furnish a beautiful park and pleasure-ground. Our *Stromateis* therefore make no pretense of order or of choice diction, seeing that in this kind of composition the Greeks purposely object to over-sweetness of style, and sow their doctrines secretly and not in a plain, unmistakable manner, seeking to exercise the diligence and ingenuity of the readers, if there should be such.[21]

This splendid description-*cum*-justification of Clement's chosen genre (based, as Clement points out, on established models) did not escape the attention of Photius, who, while paraphrasing Clement's own words, as he read them in one of his copies, says that in the *Stromateis* different thematic sections (κεφαλαία) are juxtaposed "disconnectedly and as if without order" (σποράδην ... καὶ ὥσπερ οὐκ ἐν τάξει).[22] Whether or not this is a fair characterisation of the *Stromateis* as a whole, it certainly fits everything that follows after the seventh book in our manuscript. If the composition of the first seven books appears to be loose at times and if the continuity of thought is not always transparent, the text that follows often disintegrates into a heap of disconnected segments without any structure or continuity whatsoever. Those of the Byzantine readers who replaced the eighth book with another treatise may have shared this impression. But, as the example of Photius shows, the contrast need not have appeared so stark to everyone, especially in the light of Clement's own description of his manner of writing.

On closer inspection, however, the difference between the seven books and what comes after them is so looming that it turns out to be extremely unlikely that Clement ever intended this, at least in the form preserved for us, as another book of the same work. After all, the *Stromateis* is a fairly coherent work which, despite its indirect and allusive way of speaking and despite its many digressions, evolves according to a plan. This plan, it is true, also evolves in the course of the writing, but it never completely loses the track of its earlier phases. Most importantly, every new phase is articulated by Clement, usually at the beginning of a book, in such a way as to assure the reader of continuity in the narrative.[23] There is no trace of anything similar after *Stromateis* VII. What

21 *Strom.* VII (18) 111, 1–3; trans. Hort and Mayor, slightly modified.

22 *Bibl.* 111 (89b12–22 Bekker).

23 On Clement's plans cf. below, pp. 50–54.

follows after the title στρωματεὺς ὄγδοος starts abruptly, without introduction; it consists of a number of thematic sections, some of them even having their own titles, without any indication how they fit together.

As far as the contents are concerned, there is a major breaking point before the start of the *Excerpta*. This is the place where the Byzantine scribe drew his decorated line and where the Renaissance editor marked the end of *Stromateis* VIII. The texts below this line, up to the end of the manuscript, may be described as fragments of biblical exegesis, at first (in the *Excerpta*) coming mainly from Valentinian sources (with Clement's occasional comments) and then (in the *Eclogae*) presented mainly in Clement's own voice. The texts above the line do not form a coherent whole, but are unified, at the very least, by their predominantly philosophical character.

Modern Interpretations

a *Early Modern Readers: From Heinse to Bunsen*

The incoherent arrangement and peculiar contents of the 'meta-Stromatic material', as I venture to call these texts, have long drawn the attention of scholars, calling for an explanation. At first, the debate focussed on the texts *above the line*, thought to be identified as the 'eighth book' in the manuscript, and on the question whether they really belonged to the *Stromateis*. The first scholar to doubt this was the Dutch humanist Daniel Heinse (Heinsius), who, in his 1616 edition of Clement's works, points out that the 'eighth book' lacks integrity, is much shorter than the other books of the *Stromateis* and, above all, does not share their overall theological bent: "Most certainly, this book has nothing in common with the others ... Even though in the other books [Clement] uses various digressions, while dealing mainly with the perfection of the gnostic, nevertheless the overall argument at least is theological and aims towards this goal. But this whole book is in fact about logic."[24] Heinse describes the book as a 'fragment', suggesting that it was taken from a different treatise and replaced the real eighth book of the *Stromateis*, which is now lost. Heinse's candidate for the original source is Clement's lost *Hypotyposeis*,

24 "Annotationes in Clementem Alexandrinum," attached to D. Heinse, ed., *Clementis Alexandrini opera graece et latine quae extant* (Leiden: Ex bibliopolio Commeliniano, 1616), p. 49b (*ad* p. 551): *Certissimus est hunc librum nihil cum reliquis commune habere ... Quanquam enim variis in reliquis utatur digressionibus, cum προηγουμένως de perfectione agat gnostici, tamen totum saltem argumentum est theologicum, et ad eumdem tendit finem. At vero totus his liber est logicus.*

from which several fragments (none of them containing anything philosophical) have been preserved.[25]

The view that the so-called 'eighth book' is largely irrelevant to the goals and concerns permeating the whole of the *Stromateis* and that it is a fragment rather than a coherent exposition was shared by other scholars of the 17th and 18th century. A question was even raised to what extent it conforms to Clement's style of writing.[26] Summarising these qualms, French Benedictine scholar Nicolas Le Nourry concludes that the inscription στρωματεὺς ὄγδοος was not added to this material by Clement himself, but by "other, ignorant people" (*ab aliis indoctis hominibus*). Describing the text as a "treatise on logic or dialectic" (*logicae seu dialecticae tractatus*), which "pertains to Christians no more than to pagans or anyone whatsoever", Le Nourry leaves the question of its source and authorship open.[27]

However, it would be rash to say that the texts above the line are completely devoid of a Christian or (broadly speaking) theological perspective. Particularly the first two pages of the manuscript, referred to as 'chapter one' since

---

25 Fragments of the *Hypotyposeis* (including a Latin text called *Adumbrationes*) are published in GCS 17, pp. 195–215. Since Stählin's edition, the collection has been significantly extended. Cf. Bogdan Bucur, *Angelomorphic Pneumatology: Clement of Alexandria and Other Early Christian Witnesses* (Suppl. Vig. Chr. 95; Leiden/Boston: Brill, 2009), pp. 6f.; Jana Plátová, "Bemerkungen zu den Hypotyposen-Fragmenten des Clemens Alexandrinus," in *Studia Patristica* 46, ed. J. Baun, A. Cameron, M. Edwards, and M. Vinzent (Leuven/Paris/Walpole, MA: Peeters, 2010), pp. 181–187; eadem, "Die Fragmente des Clemens Alexandrinus in den griechischen und arabischen Katenen," in *Studia Patristica* 54, ed. M. Vinzent, L. Mellerin, and H.A.G. Houghton (Leuven/Paris/Walpole, MA: Peeters, 2013), pp. 3–9. Plátová has recently published a new edition of the fragments in Klement Alexandrijský, *Exegetické zlomky. Eclogae Propheticae—Hypotyposes* (Prague: OIKOYMENH, 2014), pp. 117–185. Heinse thought that *Strom.* VIII was originally part of a logical propaedeutic to Christian doctrine; his belief that the *Hypotyposeis* must have included some philosophical material was suggested to him by the title of Sextus' *Hypotyposeis*, which he regarded as a 'sceptic' counterpart to the 'dogmatic' contents of our 'eighth book' ("Annotationes," p. 50ab).

26 Cf. Louis-Sébastien Le Nain de Tillemont, *Mémoires pour servir à l'histoire ecclesiastique des six premiers siècles*, III (Paris: Robustel, 1695), p. 652, Note 5 ('Sur le huittieme livre des Stromates'): "Cet écrit de dialectique est visiblement imparfait (il faudroit voir s' il a le style et l' obscurité de S. Clement)."

27 Nicolas Le Nourry, 'De omnibus Clementis Alexandrini operibus. Dissertatio tertia,' in idem, *Apparatus ad bibliothecam maximam veterum patrum et antiquorum scriptorum ecclesiasticorum*, III (Paris: Anisson, 1703), chapter 1, article 2, cols. 1290–1292. Le Nourry's three treatises on Clement's works are reprinted in Migne's Patrologia Graeca, vol. IX.

the 1715 Oxford edition,[28] have been cited as evidence against the opinion that the so-called eighth book has "nothing in common" with the other *Stromateis*. In a penetrating and enjoyable study of Clement's thought, published in 1851, Hubert Joseph Reinkens, a Catholic theologian and classicist from Wrocław, points out that these initial pages betray the same theological interests we find in the other books.[29] And this is indeed true: the first chapter is both Clementine and Christian, speaking as it does on behalf of the followers of the "barbarian philosophy" and dealing with a method of inquiry "concerning Scriptures" (κατὰ τὰς γραφάς) in view of the biblical command "Seek and you will find, knock and it will open, ask and it will be given to you" (Matt 7:7).[30] Reinkens also notes a number of parallels between the 'eighth book' and the other books of the *Stromateis*, citing them as another sign of continuity.[31] These observations (and other traces of Clement's Christian interests not mentioned by Reinkens) can be explained in different ways, some of which will be discussed below.[32] Reinkens explains them to the effect that the text as a whole can be reconciled with the aim and argument of the previous *Stromateis*.[33] Conceding that, for the most part (viz. from chapter two onwards), it deals with no recognizably Christian themes, Reinkens ascribes this peculiarity to the fact that the text is a fragment. Unlike Heinse, he suggests that it is a fragment of the eighth book of the *Stromateis*, separated from its original doctrinal setting.[34]

As far as the rest of the meta-Stromatic material is concerned, it had been barely noticed by scholars before 1854, when the Prussian theologian and diplomat Christian Karl Josias von Bunsen published the first volume of his *Analecta Ante-Nicaena*, containing the fragments of Clement's *Hypotyposeis*. Following Heinse's suggestion about the part known as the 'eighth book' and applying it to the texts below the line as well, von Bunsen included all this material among the fragments of the *Hypotyposeis*, giving the 'eighth book' the noble title *Isagoge dialectica*. Whatever plausibility there is to von Bunsen's view of the source, his arrangement highlighted the question of the origin and 'genre' of the meta-Stromatic material as a whole, a question obscured since

28 See below, p. 26.

29 Joseph Hubert Reinkens, *De Clemente presbytero Alexandrino, homine, scriptore, philosopho, theologo liber* (Wrocław: Aderholz, 1851), pp. 237 and 242.

30 *Strom.* VIII (1) 1, 1–2, 1.

31 Cf. *De Clemente*, pp. 242–245 and 250.

32 Cf. below, pp. 73–76.

33 Cf. esp. *De Clemente*, pp. 248–251.

34 *De Clemente*, p. 238.

John Potter's 1715 edition (later reprinted in Migne's *Patrologia Graeca*), where the 'eighth book' was followed by *QDS*.

b *The Shaping of Contemporary Views: From Zahn to Stählin*

As a matter of fact, it cannot be denied that, despite their differences, all three texts after *Stromateis* VII have something in common: All contain disconnected chunks of heterogeneous material, without even a pretended unity of argument or purpose;[35] all start and end abruptly; their syntax is sometimes garbled and their message muddled. No part of the material is preserved in a form that could ever be intended for publication by its author, especially an author as conscious of Greek literary conventions and concerned about his audience as Clement. We have seen that early modern scholars, noticing this peculiarity in the case of the 'eighth book', speak of its lack of integrity or perfection and describe it as a fragment. As regards the other two texts, their state could be explained by the titles attached to them in the manuscript: the first one is called ἐπιτομαί and the second ἐκλογαί. Both of these titles seem to suggest that the texts are derived from other sources, though perhaps not in the same way. ἐκ τῶν προφητικῶν ἐκλογαί renders the view that the third text consists of passages selected from the "prophetic writings", i.e. from Scripture. Whoever added it to the text might have wished to point out its affinity with an early Christian genre of 'selections' of biblical passages pertaining to Christian faith.[36] The title ἐκ τῶν Θεοδότου … ἐπιτομαί is intriguing. It seems to indicate that what follows consists of epitomes or 'summaries' from the writings of Theodotus concerning Valentinian teaching. Since there is no doubt that the text, in one way or another, goes back to Clement (containing as it does a number of parallels with the *Stromateis*), the most obvious interpretation of the title is that it is a collection of excerpts made by Clement himself, which was left to posterity more or less in the way it had been produced. But if so, then perhaps something similar is true of the first text as well. Could this explain the peculiar character of the meta-Stromatic material as a whole?

---

35 This is not to say that no argument or purpose is discernable *within* these disconnected chunks (whose length varies greatly) or that the chunks, within the limits of each collection, have nothing in common.

36 Cf. Eusebius, *Eccl. Hist.* IV 26, 13, referring to Ἐκλογαί by Melito of Sardes, a piece of writing described by its author as "selections from the Law and the Prophets concerning the Saviour and the whole of our faith" (ἐκλογὰς ἔκ τε τοῦ νόμου καὶ τῶν προφητῶν περὶ τοῦ σωτῆρος καὶ πάσης τῆς πίστεως ἡμῶν). Eusebius himself wrote a treatise called αἱ περὶ τοῦ Χριστοῦ προφητικαὶ ἐκλογαί ("Selected prophetic passages about Christ"); cf. Le Boulluec, "Extraits," pp. 112 f.

One of the first scholars to broach the possibility that the *Excerpta* and the *Eclogae* reached their present shape by the hand of Clement himself, "recorded perhaps as collections of material for the purpose of later elaboration," is Theodor Zahn.[37] However, Zahn dismisses the idea, favouring the notion that both texts are excerpts made by somebody else from Clement's finished work.[38] Surprisingly, Zahn's argument partly relies on the testimony of the title ἐκ τῶν Θεοδότου … ἐπιτομαί, mentioned above. Zahn argues that the author of this title could not have thought that these excerpts derived from the writings of the Valentinian Theodotus, for the views of the Valentinians and of Theodotus himself are criticized there. And since the excerpts follow a text described as the eighth book of Clement's *Stromateis*, and since no other author is mentioned in their title, Zahn concludes that they must be excerpts *from* Clement and the eighth book of the *Stromateis*. The same conclusion, according to Zahn, applies to the title and the contents of the *Eclogae*.[39]

But this is not a convincing solution. First of all, if the texts are excerpts from Clement's finished work, they cannot derive from one and the same book of the *Stromateis*. The lengths of individual books of the *Stromateis* range between 63 (book III) and 129 (book I) manuscript pages, the median length (represented by book V) being 92 pages. The total length of the meta-Stromatic material is 82 pages. It is inconceivable that material as divergent and scrappy would derive from a single book even remotely approaching the size and coherence of the other books of the *Stromateis*. Second, the phrase ἐκ τῶν Θεοδότου … ἐπιτομαί hardly provides the sense required by Zahn's interpretation, as it refers to τὰ Θεοδότου (rather than τὰ Κλήμεντος) as that from which the summaries are made.[40]

---

37 Zahn, *Forschungen*, p. 118.

38 It is not clear if the rejected idea is Zahn's own; with regard to the *Excerpta*, cf. already Ernst Renan, *Marc Aurèle et la fin du monde antique* (Paris: Calmann-Lévy, 1882), p. 118: "Les écrits de Théodote étaient aussi habituellement entre les mains de Clément, et des extraits paraissent nous en être parvenus dans la grande masse de notes que s'était faite le laborieux stromatiste." Cf. Stählin, GCS 12, pp. XLI–XLII.

39 Zahn, *Forschungen*, p. 117, suggests that these two sets of excerpts follow the initial (more or less complete) part of the eighth book similarly as excerpts from the seven books of the *Stromateis* follow the complete text of the *Paedagogus* in certain manuscripts, viz. Codex Neapolitanus II. AA 14 (fol. 106–166) and Codex Ottobonensis 94 (fol. 243–296); for these and other 'Excerpthandschriften', which all seem to go back to Ottobonensis 98 (fol. 1–64), cf. Stählin, GCS 12, pp. XLVII–LI. Almost a century after Zahn, a similar interpretation of the words ἐπιτομαί and ἐκλογαί is proposed by Pierre Nautin; cf. below, p. 22.

40 On this point, see further below, p. 22.

On the other hand, if the titles should mean that both texts are *merely* collections of material derived from other sources, they would be misleading. Far from being mere 'selections' of biblical passages, the *Eclogae* are rather snippets of biblical commentaries;[41] and though the *Excerpta* do contain quotations from Theodotus and summaries of the Valentinian teaching, Clement often attaches his views and comments.[42] But there is no reason to think that selections or summaries from other sources should contain no contribution on the part of the person who makes them. Eusebius' treatise αἱ περὶ τοῦ Χριστοῦ προφητικαὶ ἐκλογαί was not only a collection of biblical passages, but also of commentaries to these passages. Even more pertinent to our case is the attested practice of ancient authors to take excerpts from their sources and to add their own annotations or comments in the phase of writing that precedes the composition of the final work.[43] Could the *Excerpta* and perhaps other parts of the meta-Stromatic material be explained as witnesses to this preparatory phase of the process of writing? Over the course of twenty years since the publication of Zahn's study, this has actually become an accepted view of the matter, emerging gradually from detailed work focussing on the *Excerpta* and especially on the 'eighth book'. This is how a new consensus took shape.

In 1892, Paul Ruben publishes a doctoral dissertation on the *Excerpta*, in which he addresses the question of the origin and purpose of this text.[44] With regard to the rest of the meta-Stromatic material, Ruben accepts Zahn's conclusion that its 'dialectical' sections (i.e. our '*Strom.* VIII') belong to the introductory part of the original eighth book. In addition, he suggests that some material in the *Eclogae* also comes from this book (whereas the rest

---

41 Cf. Michel Cambe, *Avenir solaire et angélique des justes: le psaume 19 (18) commenté par Clément d'Alexandrie* (Cahiers de la Biblia patristica 10; Strasbourg: Brepols, 2009).

42 Cf. the edition and notes by François Sagnard, *Extraits de Théodote*, SC 23; Paris: Éditions du Cerf, 1948.

43 For a full discussion of the evidence, cf. Tiziano Dorandi, *Nell' officina dei classici: Come lavoravano gli autori antichi* (Rome: Carocci, 2007), pp. 29–46. Clement's technique of note-taking has been studied by Annewies van den Hoek, "Techniques of Quotation in Clement of Alexandria: A View of Ancient Literary Working Methods," *Vigiliae Christianae* 50/3 (1996), pp. 223–243.

44 *Clementis Alexandrini Excerpta ex Theodoto*, Leipzig: Teubner, 1892. The dissertation, written under the direction of Franz Bücheler and Hermann Usener, was defended at the University of Bonn. Ruben (1866–1943) was a German Jewish scholar who, after his promotion, was mainly active in the field of Old Testament textual criticism. Recently, his biography was published by Björn Biester, *Der innere Beruf zur Wissenschaft: Paul Ruben (1866–1943). Studien zur deutsch-jüdischen Wissenschaftsgeschichte*, Berlin: Dietrich Reimer Verlag, 2001.

he regards as fragments of the *Hypotyposeis*).[45] Nevertheless, Ruben rejects Zahn's proposal that the *Excerpta* also come from the same source; in his view their contents do not correspond to Clement's plans for the eighth book, as outlined in the *Stromateis*, and the majority of doctrines presented there are not Clement's own. According to Ruben, these ἐπιτομαί do not draw from Clement's finished piece of writing at all. Instead, they are excerpts from other writings made by Clement himself in preparation for a future work.[46]

Ruben's dissertation is a starting-point of a study presented by Hans von Arnim as an inaugural professorial lecture at the University of Rostock in 1894.[47] As a classicist, von Arnim is mainly interested in the philosophical part of the meta-Stromatic material, i.e. in our 'eighth book'. But he explores the question of its origin in view of the material as a whole. While accepting Ruben's proposal with regard to the *Excerpta*, von Arnim extends it to the *Eclogae* as well, suggesting that this text too is a collection of excerpts from Valentinian writings (*sic*!) accompanied by Clement's critical comments and exegetical notes.[48] As far as the 'eight book' is concerned, von Arnim divides the text into two parts. Whereas the first part is elaborated, the second part consists of excerpts prepared by Clement in view of a planned continuation.[49] When it comes to the origin and purpose of the first part, however, von Arnim is not as clear as one might wish. Sometimes he suggests that what is left of the elaborated part could be somebody's excerpts from Clement's book, but he does not specify whether this book is the eighth book of the *Stromateis* or some other treatise. Moreover, in his comments on specific passages (the 'elaborated part' included), von Arnim mostly derives their contents from one and the same philosophical source. It is therefore understandable that Adolf von Harnack, when paraphrasing von Arnim's thesis with approval, disregards the distinction between the two parts altogether and describes the contents of the 'eighth book' as "excerpts from pagan philosophical works".[50] The same

45 Ruben, *Excerpta*, p. 9.

46 *Excerpta*, p. 17. According to Ruben, the future work Clement had in mind was the treatise *On the Principles and Theology* mentioned in QDS 26, 8; cf. below, p. 53.

47 Hans von Arnim, *De octavo Clementis Stromateorum libro* (Rostock: Adler, 1894).

48 Von Arnim, *De octavo*, pp. 6–8.

49 In this connection von Arnim recalls the above-mentioned testimonies quoting *Strom.* VIII (5) 16, 3 as the end of the eighth book and interprets them to the effect that in some manuscripts the 'eighth book' ended with these words. This seems to indicate that von Arnim regards these words as a borderline between the elaborated part and the excerpts.

50 Adolf von Harnack, *Geschichte der altchristlichen Litteratur bis Eusebius*, II/2 (Leipzig: Hinrich, 1904), p. 17.

interpretation was adopted by Otto Stählin in his GCS edition of Clement's works and became a standard view of Clementine scholarship for most of the 20th century.[51]

c *Other Solutions: Ernst, Bousset, Nautin*

However, the 'standard view' has not gone unchallenged. One intriguing modification was already proposed in a 1910 Göttingen dissertation by Wilhelm Ernst and further developed by Wilhelm Bousset in a monograph published in 1915.[52] Ernst's dissertation is, to a large extent, a source-critical study of the 'eighth book', picking up on the insights of von Arnim's pioneering work and its splendid sequel, a 1905 Berlin dissertation by Christiane von Wedel.[53] Apart from noting a number of valuable parallels in philosophical literature, Ernst also provides a full list of correspondences between the 'eighth book' and other books of the *Stromateis*, quoting them in parallel columns. As we have seen, some of these parallels had already been noted by Reinkens, who took them as indications of continuity between the *Stromateis* and 'the eighth book'. But Ernst interprets them differently. On the one hand he endorses the view that the 'eighth book' is a collection of excerpts Clement made for his own use. On the other hand he disputes von Arnim's opinion that its contents correspond to Clement's plans for the continuation of the *Stromateis*: Rather than compiling the excerpts in view of this planned continuation, Clement made this collection *before* he composed the *Stromateis*, making use of the material

---

51 Cf. Stählin, GCS 12, pp. XLI–XLII.

52 Wilhelm Ernst, *De Clementis Alexandrini Stromatum libro VIII. qui fertur* (Göttingen: Officina Hubertiana, 1910); Wilhelm Bousset, *Jüdisch-christlicher Schulbetrieb in Alexandria und Rom: Literarische Untersuchungen zu Philo und Clemens von Alexandria, Justin und Irenäus* (Göttingen: Vandenhoeck & Ruprecht, 1915), pp. 248–263. I have been unable to find any information about Wilhelm Ernst, apart from the biographical profile in his dissertation: Born in 1885, he studied classical philology and history in Göttingen. His dissertation was supervised by Max Pohlenz and Paul Wendland.

53 *Symbola ad Clementis Alexandrini stromatum librum VIII. interpretandum* (Weimar: R. Wagner iun., 1905). Christiane von Wedel (her married name was Mewaldt von Wedel; 1881–1965) was one of the first female graduates at the Friedrich Wilhelm University in Berlin. She studied classical philology with Diels and Wilamowitz (who both supervised her dissertation) and theology with Harnack. After her graduation in 1905, she worked for several years for the Kirchenväterkommission (KVK), collating manuscripts of Eusebius' works; cf. Petra Hoffmann, *Weibliche Arbeitswelten in der Wissenschaft: Frauen an der Preußischen Akademie der Wissenschaften zu Berlin 1890–1945* (Berlin: Humboldt-Universität, diss., 2008), pp. 261 f. She was the wife of the classicist Johannes Mewaldt.

within the *Stromateis* itself.[54] In the course of his well argued and richly documented study, Ernst makes a number of interesting observations about the source of these excerpts and its philosophical background. Unfortunately, he concludes the argument with an attempt to identify Clement's source as one of his teachers at the Christian catechetical school in Alexandria and the 'eighth book' as lecture notes which Clement made as a student.[55] This bold idea is taken up by Wilhelm Bousset, who applies it to the interpretation of the meta-Stromatic material as a whole, suggesting that it derives from the lectures of Clement's teacher Pantaenus.[56]

In 1976, the French patrologist Pierre Nautin published an article in which he rejected both the 'lecture notes' hypothesis and the 'standard view', offering an alternative that may be described as a refurbished version of Zahn's solution.[57] In opposition to Ernst, Nautin points out that parallels between the 'eighth book' and the regular books of the *Stromateis* do not always support the conclusion that Clement drew on the former when composing the latter: for the parallels are sometimes less complete in the 'eighth book'.[58] Nautin also shows that passages quoted by Bousset as evidence for his thesis (especially in the *Eclogae* and the *Stromateis*) do not corroborate his argument at all.[59] As far as the 'standard view' is concerned, Nautin finds it implausible that either Clement or his friends would ever have allowed for a publication of private notes or mere drafts, thereby putting the author's reputation at risk. Nautin claims that there is no example of such a practice in antiquity.[60] In contrast, there is an example of excerpts made by a scribe at the end of a manuscript; it appears in one of the papyrus codices discovered in Tura (Egypt) in 1941, which contains a collection of extracts from Origen's writings. According to Nautin, these extracts, partly published by Jean Scherer in 1956, "make exactly the same impression" as the texts following *Stromateis* VII. This indicates that the "morceaux juxtaposés" at the end of Plut. 5.3 came into being in precisely

54 This thesis, elaborated by Ernst in detail, had already been proposed by von Wedel, *Symbola*, pp. 44 f.

55 *De Clementis*, pp. 52–58.

56 Bousset's hypothesis did not take root in Clementine scholarship; most 20th century scholars have followed the 'standard view' or a slightly modified version of it; cf. the overview in Nautin's article cited in the next note, p. 281, n. 36.

57 "La fin des *Stromates* et les *Hypotyposes* de Clément d'Alexandrie," *Vigiliae Christianae* 30/4 (1976), pp. 268–302.

58 "La fin," pp. 273–278.

59 "La fin," pp. 278–281.

60 "La fin," pp. 271 f.

the same way as the extracts in the Tura papyrus, i.e. as excerpts made by a copyist from Clement's finished works.[61] Analyzing Clement's literary plans, as outlined in the *Stromateis*, Nautin identifies these finished works partly as the original eighth book of the *Stromateis*,[62] and partly as the *Hypotyposeis*.[63]

Nautin's hypothesis is the most recent attempt to solve the riddle of the meta-Stromatic material. It has been endorsed (with some qualifications) by distinguished Clementine scholars and accepted beyond the field of patristic studies.[64] Not everyone has been convinced, but little argument has been produced to substantiate the qualms.[65] Yet there are several reasons for doubt.

First of all, it should be noted that the perceived similarity between the meta-Stromatic material and the excerpts in the Tura papyrus seems to be due mainly to the fact that both are collections of excerpts. But this is not the point at issue: the main difference between the 'standard view' and the Nautin hypothesis concerns the question of whether the author of these excerpts is Clement or someone else. As mentioned above, there is nothing improbable about the idea of Clement making excerpts from other sources or drafts of his

---

61 "La fin," pp. 269f., 273, 282, 298, 301f.

62 According to Nautin, extracts from this book end before the chapter on causes, i.e. after *Strom.* VIII (8) 24, 9; cf. the dissenting remarks by Le Boulluec, "Extraits," pp. 115–117.

63 "La fin," pp. 282–298, 301f.

64 Cf. esp. Le Boulluec, "Extraits," *passim*; Jaap Mansfeld, *Heresiography in Context: Hippolytus'* Elenchos *as a Source for Greek Philosophy* (PhA 56; Leiden/New York/Cologne: Brill, 1992), p. 62 n. 14. Other scholars sympathetic to Nautin's conclusions (chiefly in view of the *Excerpta* and/or *Eclogae* and their supposed connection with the *Hypotyposeis*) include Patrick Descourtieux (*Clément d'Alexandrie: Les Stromates. Stromate* VI, SC 446, Paris: Éditions du Cerf, 1999, p. 399 n. 4); Bogdan Bucur (*Angelomorphic Pneumatology*, pp. 6–21 *et passim*); and Michel Cambe (*Avenir solaire*, pp. 12–17 and 165–179).

65 Bucur complains that Nautin's critics "dismiss [his] proposal as simply a 'personal theory' (Annewies van den Hoek), and a 'perplexing' one at that (Nardi), while focusing only on his discussion of the Clementine program, without any objection to the first half of his study, which discusses the state of the manuscript, weighs various proposals to explain the situation, and draws the comparison with the Tura Codex II of Origen" (*Angelomorphic Pneumatology*, p. 10). Here, Bucur refers to remarks made about the Nautin hypothesis by van den Hoek in her introduction to *Clement d'Alexandrie: Les Stromates. Stromate* IV (SC 463; Paris: Éditions du Cerf, 2001, p. 13 and n. 7) and by Carlo Nardi, *Clemente Alessandrino: Estratti profetici* (Florence: Nardini, 1985), p. 11. It is true that Nautin's critics tend to dismiss his conclusions rather than refute his arguments. The most extensive attempt to take a "considered look" at Nautin's article is Andrew C. Itter, *Esoteric Teaching in the Stromateis of Clement of Alexandria* (Suppl. Vig. Chr. 97; Leiden/Boston: Brill, 2009), pp. 59–74. Itter believes that Nautin's arguments are inconclusive.

own commentaries during the preparatory phase of his work.[66] But if he ever made such excerpts or drafts, they too would presumably look as "condensed and discontinuous" as the material on the Tura papyrus.

Second, we may doubt whether the comparison between the two texts renders "precisely the same impression". According to Scherer, the length of excerpts from Origen's writings on the Tura papyrus ranges from one word to several pages; some expressions are noted simply for being eye-catching or funny, with little regard for preserving the continuity of thought.[67] But no such thoughtless scraps are found in the meta-Stromatic material; even where the text disintegrates into loosely connected segments (as for instance in *Strom.* VIII [5] 16, 2–3), they still consist of meaningful sentences.[68] More importantly, Origen's excerptor announces Origen's views in his polemic against Celsus with the word Ὠριγένης, just as he announces the views of Celsus with the word Κέλσος.[69] There are no such marks of distance between Clement and the excerptor in the meta-Stromatic material. Even when Clement contrasts his views with those of Theodotus and other Valentinians, as he occasionally does in the *Excerpta*, he speaks in the first person plural.[70] Another difference is in the titles. The only titles in the Tura papyrus are those which refer to the source text and its author.[71] But if we follow Nautin's interpretation of the meta-Stromatic material, we must believe that whoever made these excerpts from *Stromateis* VIII and the *Hypotyposeis* failed to indicate the latter source while adding, of his own accord, other titles characterizing the contents of the excerpted passages.[72]

---

66 Cf. van den Hoek, "Techniques of Quotation," p. 242 n. 78, pointing out that "the character of these writings [sc. *Strom.* VIII, *the Excerpta*, and the *Eclogae*] seems to correspond with [Clement's] technique of note-taking". Cf. also above, p. 16.

67 Jean Scherer, "Introduction" in idem, *Extraits des livres I et II du* Contre Celse *d'Origène d'après la papyrus no. 88747 du Musée du Caire* (Cairo: Institute français d'archéologie orientale, 1956), pp. 26–28.

68 One might quote, as a counter-example, a passage in *Strom.* VIII (9) 30, 4, where a phrase consisting of two biblical quotations (τὸ "ὀφθαλμὸν ἀντὶ ὀφθαλμοῦ" καὶ "ψυχὴν ἀντὶ ψυχῆς") stands in isolation from its immediate context. But this is clearly a gloss made on the margin of an argument starting at 30, 1 and ending at 30, 5; cf. below, *ad loc.* Very brief phrases are also found in *Ecl. Proph.* 43, 1–2, but these, again, are meaningful exegetical notes.

69 Cf. e.g. *Exc. C. Cels.* 2–3 (63,6; 64,7.12; 65.8 Scherer); cf. also Scherer's "Introduction," p. 27.

70 Cf. e.g. *Exc. Th.* 1, 1; 8, 1. Cf. also *Ecl. Proph.* 56, 3: ὁ Πάνταινος δὲ ἡμῶν.

71 Cf. "Introduction," pp. 5–8.

72 Cf. Antonio Servino, "Clemente Alessandrino: il problema di Stromata VIII," *Quaderni del Dipartimento di filologia, linguistica e tradizione classica «Augusto Rostagni»* 17 (2001),

Incidentally, as already mentioned in connection with the Zahn hypothesis, these titles constitute a major obstacle to the view that the relevant texts are excerpts from Clement's finished works. Neither the title of the *Excerpta* nor that of the *Eclogae* suggests this. Moreover, the *Excerpta* claim to be ἐκ τῶν Θεοδότου. Yet, like Zahn, Nautin quotes these titles in his favour, suggesting that they were added by Clement's excerptor. He argues, namely, that the words ἐκ τῶν Θεοδότου ... ἐπιτομαί do not refer to the writings of Theodotus, but to the *teachings* of Theodotus, as reported by Clement, just as the word διδασκαλία, used in the same case afterwards (καὶ τῆς ἀνατολικῆς καλουμένης διδασκαλίας κατὰ τοὺς Οὐαλεντίνου χρόνους), refers to a particular teaching.[73] But this is a forced interpretation. Although the word ἐπιτομή could refer to a summary of doctrines in antiquity (as exemplified by Theophrastus' αἱ τῶν φυσικῶν δοξῶν ἐπιτομαί), there seems to be no evidence that it was ever used with the preposition ἐκ in this sense. As far as I can see, the expression 'ἐκ τῶν X [gen.] ἐπιτομή' always refers to a summary from a work written by X.[74] It is true that the title also speaks about summaries "from the so-called eastern teaching" etc., but this may be explained as an epexegetic reference to the *contents* of the excerpted work.[75]

Should we believe that anyone would publish mere excerpts and sketches written by Clement for his private use? Nautin thinks we should not, citing the concerns of potential editors about Clement's reputation and the lack of comparative evidence. But we know nothing about Clement's editors and their

---

pp. 97–104, here 102, who finds it difficult to explain why any copyist would proceed in this way.

73 Nautin, "La fin," pp. 268, 273 and note 17. Cf. also Le Boulluec, "Extraits," pp. 110 f., proposing that "[l]e neuter τὰ Θεοδότου ne désigne pas 'les œuvres de Théodote', mais 'les [idées, ou les thèses] de Théodote', sur le même plan que la διδασκαλία, l' 'enseignement', la 'doctrine', dans le deuxième member du titre".

74 Thus, for example, Stobaeus introduces two excerpts from a diatribe written by Teles of Megara ('A Comparison of Poverty and Wealth') with the title 'Ἐκ τῶν Τέλητος ἐπιτομή (*Anth.* IV 32a, 21 and 33, 31: Wachsmuth—Hense V,785,1 and 808,12). A summary of *Church History* by Philostorgius bears the title 'Ἐκ τῶν ἐκκλησιαστικῶν ἱστοριῶν Φιλοστοργίου ἐπιτομή (GCS 21: 4,1–3).

75 According to this interpretation, then, the title would indicate that a particular writing of Theodotus, dealing with the 'eastern' Valentinian teaching, is the main source of Clement's excerpts. For a different view, cf. Einar Thomassen, *The Spiritual Seed: The Church of the 'Valentinians'* (Nag Hammadi & Manichaean Studies 60; Leiden/Boston: Brill, 2006), p. 28 n. 3, who takes the title to mean that "the excerpts come from Theodotus and from *other* documents of the eastern school" (italics mine). In either case, of course, the question of how many sources Clement *actually* used is a different matter.

motivations. It is just as reasonable to speculate about their concerns as it is to imagine that this intriguing material, discovered perhaps after Clement's death, was deemed worthy of being taken into some ecclesiastical library, where it was copied and kept alongside other scrolls containing Clement's completed works. As far as comparable cases are concerned, there is indeed evidence that an unfinished work or notes unintended for publication could have circulated by the initiative of the author's friends or followers without his consent or after his death. Aristotle's 'esoteric' writings are one case in point, but more pertinent to Clement are Galen's complaints about the spread of unfinished drafts of his works or private commentaries entrusted by him to friends or stolen by servants.[76]

Moreover, we know far too little about the circumstances under which the material was 'published'. The earliest report about the eighth book of the *Stromateis* comes from Eusebius, who was active roughly a century after Clement's death. If it is true, as I have proposed above, that Acacius, Eusebius' disciple and successor in Caesarea, already had a manuscript of the *Stromateis* similar to ours and took the whole of the meta-Stromatic material for the eighth book of that work, it is reasonable to extend this conjecture to Eusebius, too. So between Clement and Eusebius someone must have decided that these texts together constituted the eighth book of the *Stromateis*. Presumably, when making this decision, he was not thinking of them as mere drafts, rather per-

76 Cf. Galen, *Lib. Prop.* prooem. 6 (135,18–20 Boudon-Millot/XIX,10 K.): "… my writings were given to friends and pupils without proper ascription, because they were not intended for publication, but merely for those who asked to have the records of what they had heard …" (φίλοις γὰρ ἢ μαθηταῖς ἐδίδοτο χωρὶς ἐπιγραφῆς ὡς ἂν οὐδὲν πρὸς ἔκδοσιν ἀλλ' αὐτοῖς ἐκείνοις γεγονότα δεηθεῖσιν ὧν ἤκουσαν ἔχειν ὑπομνήματα); ibid. 14, 10 (166,5–8 Boudon-Millot/XIX,41 K.): "In fact, some of the notes I had written I later acquired from others, as I had given some to friends, and some had been stolen and published by servants" (καὶ μέντοι καὶ τῶν ὑπομνημάτων ὧν ἔγραψα τὰ μὲν ὑπ' ἐμοῦ δοθέντα φίλοις, τὰ δ' ὑπὸ τῶν οἰκετῶν κλεψάντων ἐκδοθέντα παρ' ἄλλων ἔλαβον ὕστερον); cf. also *idem*, AA I 1 (II,217,11–13 K.). Cf. Ann Ellis Hanson, "Galen: Author and Critic," in G.W. Most (ed.), *Editing Texts/Texte Edieren. Aporemata: Kritische Studien zur Philologiegeschichte*, II (Göttingen: Vandenhoeck & Ruprecht, 1998), pp. 22–53, here 28–35. A similar testimony (pupils publishing lecture notes without the consent of their master) had previously appeared in Quintilian, *Inst. Or.* I, praef. 7. Cf. Harry Y. Gamble, *Books and Readers in the Early Church: A History of Early Christian Texts* (New Haven/London: Yale University Press, 1995), p. 118: "The currency of unauthorized, excerpted, or corrupted texts is so frequently given as the reason for publication that it is almost a *topos* in the literature of the period." As a matter of fact, as Alain Le Boulluec points out to me, Clement himself was aware of the possibility that his writings might be published without his knowledge or against his will; cf. *Strom.* I (1) 14, 4.

haps supposing that their disorderly character suited the genre of the work as a whole.[77] Three things could have helped this identification: (1) At the end of the seventh book, Clement clearly indicates that another book of the *Stromateis* is to follow.[78] (2) The first couple of pages of the meta-Stromatic material (the beginning of 'book eight') look as if they were an introduction to Clement's own exposition. (3) The length of the material as a whole roughly corresponds to the length of a regular book of the *Stromateis*.[79] Does this reflect the length of the scroll on which this material was written, either by Clement himself, or by someone who was in charge of his literary estate? Was it attached to other scrolls, some of which, at least, were of a similar length, containing the seven books of the *Stromateis*?[80] Was it, then, identified as the eighth book by someone who read the beginning but only glimpsed through the rest of the scroll?

---

77 Cf. Photius, *Bibl.* 111, quoted above, p. 10.

78 *Strom.* VII (15) 89, 1: εἰς τὸν ἑξῆς προϊέναι Στρωματέα. VII (18) 111, 4: μετὰ τὸν ἕβδομον τοῦτον ἡμῖν Στρωματέα τῶν ἑξῆς ἀπ' ἄλλης ἀρχῆς ποιησόμεθα τὸν λόγον.

79 The numbers of the manuscript pages of individual books are as follows: Book I: 129 pp. (fols. 1–65ʳ, at least one folio missing); II: 90+ pp. (fols. 65ᵛ–111ʳ + 5 lines on fol. 111ᵛ); III: 63+ pp. (fols. 111ᵛ–143ᵛ + 13 lines on fol. 144ʳ); IV: 91 pp. (fols. 144ʳ–190ᵛ, the book starts and ends roughly in the middle of a page); V: 92 pp. (fols. 191ʳ–237ᵛ); VI: 115 pp. (fols. 238ʳ–296ʳ); VII: 98 pp. (296ᵛ–346ʳ); '*Strom.* VIII' + *Exc. Th.* + *Ecl. Proph.*: 82 pp. (346ᵛ–388ʳ, at least one folio lost in the antigraph of Plut. 5.3 or earlier; cf. above, p. 7 n. 13).

80 In a recent article, Annewies van den Hoek interestingly suggests that the original text of *Stromateis* I–VII could have been written on four scrolls, one containing book I, one containing books II and III, one books IV and V, and one books VI and VII. Van den Hoek draws attention to the relative sizes of the books, but especially to the programmatic preambles found at the beginning of books II, IV, and VI ("*Stromateis* Book VII in Light of Recent Scholarship: Approaches and Perspectives," in *The Seventh Book of the Stromateis*, Suppl. Vig. Chr. 117, Leiden/Boston: Brill, 2012, pp. 3–36, here 5 and n. 8). Preambles apart, it is indeed striking that book III (exceptionally short) is almost half the size of book I (exceptionally long). Van den Hoek's proposal also seems to be supported by the fact that, according to her count (based on the GCS edition), the sum total of books IV and V is equal to the sum total of books VI and VII (173 pages). But the matter may be more intricate. First of all, a page count based on the GCS edition must be taken with caution, as each page includes a section containing critical apparatus and references, whose length varies. In fact, when counted according to the manuscript, the results are rather different: books IV+V 183 pp., books VI + VII 213 pp. (thus the sum total of books VI and VII is over 80 pages more than the total of book I). Moreover, book II ends in the middle of an argument (the discussion about marriage, continued in book III), which seems to suggest that it has reached the end of the scroll; cf. *Strom.* II (23) 147, 5. Finally, it is worth noting that three books are almost equal in length (book II: 90+ pp., IV: 91 pp.; V: 92 pp.): is it because each covered a scroll of the same size?

The answer is: we don't know. It is at best a pleasing conjecture, another story submitted to the festival of reasoned imagination. But at least it shows that the process through which something like an edition of Clement's works took shape need not have involved the dilemma of whether mere drafts and notes of a respected author ought to be published or not.

These remarks, I believe, cast doubt on the strength of Nautin's arguments based on the comparison with the Tura papyrus and the perceived implausibility of the 'standard view'. But they do not vindicate the 'standard view' nor exclude the possibility that some version of the Zahn/Nautin hypothesis might be right after all. No firm judgement about the origin and purpose of the meta-Stromatic material can be made on the basis of external evidence alone. But a correct interpretation of the titles does not suffice either, for we do not know who attached them and why. If there is a solution to the riddle, it can only be found through a detailed analysis of the material itself and its comparison with Clement's plans for the continuation of the *Stromateis*. Nautin is aware of this and drafts such an analysis and comparison in his article, but it is far too sketchy and selective to yield convincing results. The present study does not aspire at correcting Nautin's conclusions. Its goal is less ambitious: To focus solely on the first part of the meta-Stromatic material—the 'liber logicus', as Heinse describes it, or the 'eighth book', as it is traditionally called—in order to find out which interpretive options as to its origin and purpose withstand close and sustained scrutiny of its contents.

Liber logicus

a *Intellectual Setting*(s)

The 'eighth book' is a difficult text. Not only because it deals with difficult topics in a reader-unfriendly manner, but also because it is straddled between fundamentally different intellectual traditions: on the one hand the tradition of Christian apologetic and exegetic discourse (the forerunner of Christian 'theology'), and on the other the tradition of Greek philosophy. Both of these traditions make use of philosophical vocabulary and patterns of thought, but each in a different way, governed by different concerns and presuppositions. The *Stromateis*, of course, just as the rest of Clement's *œuvre*, belongs with, and represents, the *former* tradition. We have seen that scholars are divided as to whether the 'eighth book' is continuous with the *Stromateis*, or whether it has "nothing in common" with it. This dilemma goes back to the distinction just drawn: the contrary interpretations reflect the conflicting nature of the evidence itself.

This two-faced character of the 'eighth book' also seems to be responsible for the fact that the text has seldom been studied as a whole. For it falls within the domain of not one but two historical disciplines, patristics and ancient philosophy, each of which is aware of its own limitations and hesitant to overstep the borderlines of its expertise. A patristic scholar will read with pleasure the introductory pages, where philosophical (mainly Platonic) allusions are cleverly intertwined in the texture of biblical exegesis, as they are all over the place in the *Stromateis*.[81] He will be ready to believe, with Gentien Hervet, that "the goal and plan of the eighth book is to teach how one ought to proceed towards the 'Barbarian' or 'Hebrew', which is the same thing as Christian, philosophy".[82] But as he reads on, things start looking strange: the discussion becomes very technical, ever more difficult to follow, and its relevance to anything Christian vanishes out of sight. In contrast, a historian of ancient philosophy will find more of interest in this text than anywhere else in Clement's writings. But, knowing that it is written by a Christian writer, he will be discouraged from a fuller treatment by the perceived incapacity to contextualize it properly.

Nevertheless, it is telling that most studies which do resolve to investigate the text as a whole primarily address the question of Clement's philosophical sources, rather than their use in the context of his thought. This shows something important, namely that the purely philosophical material, that is to say, the material with no (or marginal) traces of Clement's Christian interests, dominates the scene. Reinkens correctly notes that the only part of the 'eighth book' to deal with Christian teaching is the first two pages, marked off as the first chapter ever since 1715, when William Lowth, a canon of Winchester Cathedral, proposed a chapter division of the *Stromateis* for the Potter edition.[83] In these two pages, every single sentence breathes the air of a Christian thinker like Clement, a thinker as deeply versed in Greek philosophical rhetoric as he is concerned with advocating biblical exegesis as the best philosophical method

81 Cf. Carlo Nardi, "Socratismo evangelico nell' *Ottavo Stromateus* (cap. I) di Clemente Alessandrino," *Annali del Dipartimento di Filosofia* 5 (1989), pp. 23–36.

82 *Videtur, inquit, huius octavi libri esse scopus et institutum, ut doceat, quomodo ad barbarae, seu Hebraeae, quae eadem est, quae christiana, philosophiae cognitionem sit procedendum.* Hervetus' note, made in the commentary to his Latin translation of Clement's works (revised edition, 1590) and reprinted as part of Potter's edition of Clement (see the next note), is quoted approvingly by Reinkens, *De Clemente*, p. 241.

83 Reinkens, *De Clemente*, p. 238. The Oxonian bishop (later, the Archibishop of Canterbury) John Potter acknowledges the contribution of William Lowth in the preface to Potter's edition *Clementis Alexandrini opera quae extant* (Oxford: E theatro Sheldoniano, 1715). Cf. Stählin, GCS 12, p. LXX.

of discovering the truth. Reinkens is also correct in observing that the chief goal of the following chapters is to explain "the rules of demonstration" (*demonstrationis regulae*), a topic which seems to be of fundamental importance to Clement's project of Christian philosophy.[84] But precisely because demonstration has such a prominent role in Clement's project of *Christian* philosophy, that is to say, philosophy based on the authority of Scripture and the tradition of the church, it is surprising that the very text in which he treats the topic with thoroughness and expertise unparalleled anywhere in his writings is also the one most completely devoid of anything Christian. No wonder that scholars who ventured to study this material in detail have been compelled to inquire about Clement's philosophical sources, rather than about his own projects and views.

b *Thematic Division*

Before turning to the question of Clement's sources, let us briefly summarize the contents of the 'eighth book'. They fall within several thematic sections, some of which overlap with Lowth's chapter divisions. Lowth does not always manage to grasp the boundaries between themes, however, and a more precise division of the material might look like this:[85]

α. 'Seek and you will find' (1, 1–5 = chap. 1, L 346ᵛ–347ʳ)
β. Teaching on demonstration (3, 1–8, 3 = chap. 2 and part of chap. 3, L 347ʳ–349ᵛ)
γ. Method of discovery (8, 4–15, 1 = the rest of chap. 3, chap. 4 and part of chap. 5, L 349ᵛ–352ᵛ)
δ. Suspension of judgement I: negative (15, 2–16, 3 = the rest of chap. 5, L 352ᵛ–353ʳ)
ε. Division and definition (17, 1–21, 6 = chap. 6, L 353ʳ–355ʳ)
ζ. Suspension of judgement II: positive (22, 1–4 = chap. 7, L 355ʳ–355ᵛ)
η. Categories (23, 1–24, 9 = chap. 8, L 355ᵛ–357ʳ)
θ. Causes (25, 1–33, 9 = chap. 9, L 357ʳ–361ʳ)

In the manuscript, the text extends over thirty pages (fols. 346ᵛ–361ʳ). The first section (1, 1–5), covering barely one page and a half, discusses the question of 'seeking' or inquiry (ζήτησις) on the basis of Matthew 7:7. This is the

84 Cf. Reinkens, *De Clemente*, pp. 238 f. and 243 f.

85 Numbers in the brackets refer to paragraphs introduced by Reinhold Klotz in his 1832 edition and to their subdivisions added by Otto Stählin in 1909 (GCS 17).

only undoubtedly Christian part of the text. Sections β and γ, which together comprise twelve pages, are a fairly continuous exposition on the theory of demonstration. The first (3, 1–8, 3) is an introduction whose aim is to arrive at a definition of demonstration, starting from an agreed meaning of its name and proceeding towards an account of its essence (demonstration being understood as a valid argument based on premisses that are evident to the intellect or to sense-perception). The second (8, 4–15, 1) applies the theory of demonstration to the solution of problems, taking as an example the problem: 'Is the embryo an animal or not?' Attached to this exposition is a syntactically unrelated and thematically distinct text dealing, in a polemical manner, with scepticism, more specifically with the claim ascribed to the sceptic opponents (referred to as 'the suspenders [of judgement]') that "nothing is firm"; this brief, but dense and sparkling polemic is followed by loosely attached definitions of concepts relevant to the discussion (15, 2–16, 3). In the manuscript, it is introduced by the words "Against the Pyrrhonians" (incorrectly intertwined in the syntax of the next sentence), which seem to have entered the text either as a title or as a marginal gloss.

On the last line of section δ, as we have reached the fourteenth page of the manuscript, we arrive at a mark indicating the end of something and then, following a blank line, an obtrusive title appears, written in the same form as the title at the head of the book, indicating the beginning of something else. As mentioned above, this title is repeated with a minor modification at the end of the 'eighth book', thus indicating that everything between the fourteenth page and the end of the book somehow belongs together.[86] But in fact the texts following the title are even less integrated into a continuous whole than those preceding it. Moreover, one of these texts (22, 1–4) seems to be closely linked to section δ, since it also addresses the notion of the suspension of judgement, even though this time the perspective is clearly sympathetic with the sceptic position.

Of the remaining three sections, two more (ε and η) are concerned with logical issues: one with division and definition (17, 1–21, 6) and the other with the (Aristotelian) categories (23, 1–24, 9). Both are brief and terse, covering complex issues in a drastically abbreviated manner, with no regard whatsoever for the philosophically untrained reader. The last section (25, 1–33, 9) is also the longest (unless we take sections β and γ as one), extending over nine pages. It introduces several classifications of causes and deals with various aspects of causation.

---

86 Cf. above, p. 3.

c *Sources: Review of Scholarship*

Defining one's research objective as a source-critical inquiry, i.e. as an inquiry whose aim is to distinguish, characterize, or even identify the source or sources of a given text, is problematic. Unless we are lucky enough to have it preserved independently, it is a risky procedure to try to separate a source from the context of its reception. This is especially true in the case of a writer like Clement, who on the one hand does clearly use a large number of diverse sources, but on the other tends to adapt them, often in a subtle and ingenious manner, to his own (apologetic, pedagogical, or exegetical) goals.[87] However, as we have seen, the 'eighth book' is unlike any other. Although qualms about 'source criticism' should not be forgotten (after all, it is always Clement who speaks), the peculiar form and content of this text seem to call for the old-fashioned approach as the most promising line of research.

The main source-critical studies of the 'eighth book' have already been mentioned: all were produced at German universities at the turn of the 19th and 20th century. Hans von Arnim's 1894 Rostock lecture opened up this vista of inquiry, pointing out, for the first time, the importance of Clement's text to the historian of ancient philosophy. Despite his opinion that different parts of the text have different origins (one part coming from the finished eighth book, the rest from Clement's notes), von Arnim postulates a common philosophical source of sections α, β, γ, ε, and η, i.e. of all sections dealing with logical issues. Describing the source as *liber dialecticus* or *introductio dialectica*, he compares its genre to the textbooks of Apuleius, Pseudo-Galen, and Albinus.[88] However, von Arnim is more interested in the chapter on causes, which, in his

87 For Clement's (broadly speaking) apologetic concerns, cf. Annewies van den Hoek, "Apologetic and Protreptic Discourse in Clement of Alexandria," in *L'apologétique chrétienne gréco-latine à l'époque prénicénienne*, ed. A. Wlosok and F. Paschoud (Entretiens Hardt 51; Genève: Vandœuvre, 2005), pp. 69–102, here 92: "In a broad sense apologetic themes and discourse permeate everything that Clement wants to include in his vision of the 'true philosophy'." For his strategies in adapting motifs from Greek philosophy cf. Alain Le Boulluec, "Comment Clément applique t-il dans le Stromate VII, à l'intention des philosophes, la méthode définie dans le prologue (1–3)?" in *The Seventh Book of the Stromateis*, ed. M. Havrda, V. Hušek, and J. Plátová (Suppl. Vig. Chr. 117; Leiden/Boston: Brill, 2012), pp. 39–62. An excellent collection of 'case studies' is provided by Dietmar Wyrwa, *Die christliche Platonaneignung in den* Stromateis *des Clemens von Alexandrien* (Berlin/New York: de Gruyter, 1983), *passim*. With regard to exegesis, particularly instructive is Clement's appropriation of Philo the Jew; cf. Annewies van den Hoek, *Clement of Alexandria and His Use of Philo in the* Stromateis: *An Early Christian Reshaping of a Jewish Model*, Suppl. Vig. Chr. 3; Leiden: Brill, 1988.

88 Von Arnim, *De octavo*, p. 16.

view, pertains more to physics than to dialectics. He believes that this chapter consists of excerpts from two different books, a Stoic and a Peripatetic one (large chunks of this section later appear in von Arnim's collection of Stoic fragments).[89] Another passage from the 'eighth book', which, in von Arnim's view, qualifies as a Stoic fragment, is the polemic against ἐποχή in section δ. At the background to its counterpart section ζ, von Arnim detects yet another source, which he calls *liber scepticus*.[90]

Christiane von Wedel's dissertation, defended in Berlin in 1905, mainly deals with the question of Clement's sceptic sources and makes new observations on the chapter on causes. According to von Wedel, the chapter on causes is not compiled by Clement from two distinct sources, Stoic and Aristotelian, but rather from one source in which the doctrines of the two schools are compared and integrated.[91] Also of note is von Wedel's comment on the anti-sceptic polemic in section δ, according to which the arguments against scepticism probably derive from the same source as the material dealing with demonstration and related issues.[92]

The 1910 Göttingen dissertation by Wilhelm Ernst complements von Wedel's work and takes her conclusions on the philosophical background and number of Clement's sources a step further. Focussing mainly on sections dealing with demonstration, the categories, and causes, Ernst lists a number of parallels in philosophical literature that, in his view, point towards a common source for all the material collected in the 'eighth book'.[93] Many of the passages discussed by Ernst have parallels in authors rooted in the Aristotelian tradition, such as Aspasius, Galen, and Alexander of Aphrodisias, as well as in Neo-Platonist commentators on Aristotle. At the same time, there are elements of Stoic origin in some of these passages, often, as Ernst points out, accommodated to the Aristotelian framework. According to Ernst, the philosophy of Clement's source is of the same kind throughout the whole book, heavily drawing on the Aristotelian tradition while incorporating some Stoic motifs.[94]

89 *De octavo*, pp. 12 f. and 16.

90 *De octavo*, pp. 11 f. and 15 f.

91 Von Wedel, *Symbola*, pp. 22–44, here esp. 26.

92 *Symbola*, p. 21.

93 Ernst, *De Clementis*, here esp. 51: *Res eo deducta est, ut dubium esse non possit, quin Clemens materiam quam liber* VIII. *exhibet ab uno auctore petierit. Nam per totum librum elementa doctrinae et principia quaedam philosophiae referuntur.*

94 Cf. *De Clementis*, pp. 15–30 (on demonstration), esp. 16 f., 23 f. and 26 f., 31–41 (on causes), and 51 f.

Since 1910, no monograph has been dedicated to the 'eighth book', but some of its sections have been treated on the margins of other topics. In particular, the book has attracted the attention of scholars dealing with Middle Platonist philosophy and doxography. In his Cambridge dissertation published in 1937, Reginald E. Witt discusses the text in connection with his attempt to identify the sources of Middle Platonist eclecticism. Witt traces various motifs of the book to Antiochus of Ascalon and suggests that the source is some Peripatetic authority of the 2nd century influenced by Antiochus.[95] Middle Platonist parallels with the sections on demonstration, method of discovery, division and definition are further explored by Jaap Mansfeld and Teun Tieleman in connection with their research on doxographic manuals and their influence on Middle Platonist writers such as Alcinous and Galen. Neither Mansfeld nor Tieleman make an attempt to identify the source of the 'eighth book'; they rather use the text as an indicator of traditional elements in the Middle Platonist scholastic literature.[96]

As far as the section on causes is concerned, new observations were made by Jean-Joël Duhot in his 1989 study on the Stoic notion of causality. Duhot argues that most of the passages taken from this chapter in von Arnim's collection of Stoic fragments are in fact not Stoic at all. Although they do contain Stoic elements, these are mixed with elements of the Aristotelian and medical origin. As for the doctrinal background to the chapter as a whole, Duhot says that this "mélange des sources" cannot be clearly attributed to a specific school of thought. Nevertheless, he describes it as a document of scholastic Aristotelianism strongly marked by Stoic influence and elaborated in such a way as to be useful for medicine.[97]

---

95 Reginald E. Witt, *Albinus and the History of Middle Platonism* (Cambridge: Cambridge University Press, 1937), pp. 31–41. Describing the character of *Strom.* VIII as "unmistakably Antiochean" Witt suggests that Aristocles of Messene "or a similar Peripatetic authority" could be Clement's source (p. 41). The Antiochean trace is notably followed by Salvatore R.C. Lilla, *Clement of Alexandria: A Study in Christian Platonism and Gnosticism* (Oxford: Oxford University Press, 1971), pp. 120–127 and 134f.

96 Cf. Jaap Mansfeld, *Heresiography*, pp. 62–64 and 80–84; cf. also idem, "Doxography and Dialectic. The *Sitz im Leben* of the 'Placita'," *ANRW* II 36.4 (1990), pp. 3056–3229, here 3184–3193; Teun Tieleman, *Galen and Chrysippus On the Soul: Argument and Refutation in* De placitis, *Books II–III* (PhA 68; Leiden/New York/Cologne: Brill, 1996), esp. pp. 20, 24f., 30, 104, and 127.

97 Jean-Joël Duhot, *La conception stoicienne de la causalité* (Paris: Vrin, 1989), pp. 211–235.

d *How Many Sources?*

Looking at the material through the lens of these studies, we can see that the number of sources used by Clement for the composition of the 'eighth book' is probably not large. The first section apart, the number seems not to exceed four:

(1) There is little doubt that sections β and γ, comprising more than one third of the text, come from the same source: a treatise which deals, in an introductory manner, with demonstration, and shows how the demonstrative method is applied to questions proposed for inquiry. Between the two sections, there is continuity in topic, vocabulary, and style.
(2) It is also very likely that the compressed sections ε (division/definition) and η (the categories) have their origin in one and the same text, which discussed aspects of scientific inquiry against the backdrop of Aristotle's *Organon*. The theme of section η is ushered in the part of section ε, where the categories are mentioned (20, 2).
(3) Both sections dealing with scepticism seem to belong together: they are both concerned with the same topic, the suspension of judgement, describing it from two different perspectives. There is clear continuity between the last part of section δ (16, 3) and section ζ (22, 1–3).
(4) Finally, the section on causes, although containing doctrinal elements of different schools, also probably originates from one (philosophical) source, which sets out to confront different causal models while exploring various aspects of causation. The point of view from which these different models are confronted does not always transpire, but sometimes it does (cf. e.g. 26, 1–3; 26, 4–27, 2; 30, 1–5; 31, 1–5), and there is nothing to indicate that it belongs to a Christian writer.[98]

Could this number be further reduced? All scholars working on this issue have thought that the source of ε and η is the same as the source of β and γ. Did the *introductio dialectica* dealing with demonstration also discuss division, definition, and the categories? We can hardly be sure. The styles of the two pairs of sections are different: sections β and γ are more discursive, sometimes repetitive, more generous with examples and elucidations, more continuous, containing even traces of literary arrangement, such as introductions and summaries. By contrast, sections ε and η are dry and compendious, sometimes

98 It is true that this section contains some Christian elements (cf. 28, 5; 29, 3–5; 30, 4), but they appear to be marginal to the points at issue.

incoherent, apparently without any literary pretensions. But there are also similarities. All four sections deal with aspects of Aristotelian logic, organising Aristotelian views in such a way as to make them useful for inquiry.[99] Moreover, section ε is explicitly concerned with demonstration.[100] Could it be that the difference in style between β and γ on the one hand and ε and η on the other is due to different degrees of conciseness on the part of the excerptor? It is a tempting conclusion, especially since it would explain why Clement included excerpts about such issues as division, definition, and the categories in his collection at all.[101]

As mentioned above, von Wedel thinks that the polemic against ἐποχή in section δ also comes from the same source as the preceding and following texts. In her view, if the polemic were Clement's own or if it came from a different source, it would be hard to account for its position between fragments dealing with logical issues. On the other hand, according to von Wedel, it makes sense to believe that a source dealing with demonstration, problems, inquiry, and related topics also included a refutation of those who undermine inquiry by rejecting the possibility of demonstrating anything.[102] We may add, approvingly, that section β includes an implicit reference to a view that demonstration does not exist (3, 1–2). Von Wedel is right to point out that the claim rejected in section δ, viz. that 'nothing is firm' (a claim curiously described as something that ἐποχή "says"), undermines the possibility of demonstration (cf. esp. 4, 1; 6, 7–7, 2).[103] It is also true that the location of this material in the 'eighth book' is better explained if we suppose that it comes from the same source as the preceding and following sections. On the other hand, unlike in ε and η, there is little in the text itself to support this conjecture.

As far as the chapter on causes is concerned, Ernst argues that it also derives from the same source as the rest of the material. This is by no means impossible. In our text's view, demonstration produces scientific knowledge (ἐπιστήμη) in the souls of students (7, 7; cf. 5, 3). At the same time, it reminds us that "we know scientifically when we know the cause" (18, 1). Thus it makes sense that a treatise

99 Cf. the notion of τὸ ζητούμενον in *Strom.* VIII 3, 1; 4, 1; 6, 6; 8, 4; 9, 1–5; 14, 2 on the one hand and 17, 1; 18, 6; 23, 3 on the other.

100 Cf. *Strom.* VIII (6) 17, 1.8.

101 More on this point below, p. 42–43.

102 *Symbola*, pp. 15 f.

103 There is also a reference to the sceptics in 1, 1; however, its bearing on the present question depends on the solution to another difficulty, namely, the origin of the first chapter.

on demonstration would include a discussion of causes.[104] In addition to a similar theoretical background, pointed out by Ernst, there are some specific indications of continuity with the section on demonstration that do (mildly) support the view of a common source.[105] Perhaps the strongest indication is the presence of medical material throughout the chapter, as noted by Duhot. But to bring out the relevance of Duhot's observation, we must turn to the results of the most recent inquiry into Clement's sources, made by the present author.

e *The Galen Hypothesis*

In an article published in 2011, I proposed the hypothesis that the source of sections β and γ (3, 1–15, 1) is the Roman doctor-*cum*-philosopher Galen of Pergamum (129–ca. 215), perhaps his lost treatise *On Demonstration* (*De demonstratione*, henceforth DD).[106] Scholars have long noticed similarities between these sections and the writings of Galen, considering their number and coherence "striking".[107] Friedrich Solmsen even broaches the possibility that Galen might be Clement's source, but rejects it as "dramatization".[108] However, my

---

104 Cf. Le Boulluec, "Extraits," p. 116.

105 Cf. below, p. 263.

106 "Galenus Christianus? The Doctrine of Demonstration in *Stromata* VIII and the Question of Its Source," *Vigiliae Christianae* 65/4 (2011), pp. 343–375.

107 Cf. Tieleman, *Galen and Chrysippus*, p. 127. Ernst, *De Clementis*, pp. 11–24, already indicates the following parallels: (a) *Strom.* VIII (3) 6, 1 and Galen, *Inst. log.* 11, 1: the definition of ἔνδειξις; (b) *Strom.* VIII (3) 6, 2 and Galen, *Inst. log.* 11, 2: the description of demonstration as a conclusion drawn from true premisses; (c) *Strom.* VIII (3) 7, 2 and Galen, *Inst. log.* 1, 5; 8, 3; 16, 6–7; 17, 7: the description of the first premisses as that which is credible by itself; (d) *Strom.* VIII (3) 7, 3; 14, 3 and Galen, *MM* I 5 (X,39,7–9 K.); *PHP* IX 7, 4 (CMG V 4,1,2: 586,19 f./V,778 K.): the description of the first premisses as "plain to sense-perception and intellection"; (e) *Strom.* VIII (3) 8, 2 and Galen, *PHP* II 3, 12 (CMG V 4,1,2: 112,4 f./V,222 K.): indifference concerning names by which premisses are called; (f) *Strom.* VIII (3) 8, 4 and Galen, *Inst. log.* 1, 2: on the necessity of having premisses "appropriate" to the demonstrated matter. Friedrich Solmsen, "Early Christian Interest in the Theory of Demonstration," in W. den Boer et al. (eds.), *Romanitas et Christianitas: Studia J.H. Waszink Oblata* (Amsterdam: North-Holland, 1973), pp. 281–291, remarks that if we examine the fragments of Galen's treatise *On Demonstration* with Aristotle's *Analytics* as well as *Stromateis* VIII in mind, "we are struck by the number of tenets they have in common" (p. 285). Apart from the parallels noted by Ernst, Solmsen refers to "verbatim agreement" between *MM* I 5 (X,39,9 f. K.) and *Strom.* VIII (2) 4, 2. The list is further extended by Tieleman, *Galen and Chrysippus*, pp. 20, 24 f., 30, and 104.

108 "Early Christian Interest," p. 286. More optimistic is Robert M. Grant, "Paul, Galen, and Origen," *Journal of Theological Studies* 34/2 (1983), pp. 533–536, here 534 f.: "Conceivably

own examination of the evidence, which includes correspondences unknown to earlier scholars, has brought me to the conclusion that the Galenic trace is plausible, indeed more plausible than any other explanation on the table.

Out of the numerous parallels between the 'eighth book' and Galen, two, I believe, are especially revealing. As already mentioned, in section β Clement introduces a teaching on demonstration. He does so by pursuing a certain method applied to 'everything sought', i.e. to every subject matter of inquiry:

> Indeed, it is true of everything sought that if one wants to argue correctly about it, he will refer the argument to an agreed starting-point, which is no other than what people of the same nation and language agree to be the meaning of the noun. Then, having started from there, one must inquire whether the signified item, with which the argument is concerned, exists or not. Next, if it is shown to exist, one must precisely investigate its nature, what it is like, and never transgress the given order.[109]

Following these guidelines, Clement suggests that the starting point of the teaching in question is to "explain the proposed name [i.e. the name 'demon-

---

there is a reflection of Galen's own attitude in the writings of Clement of Alexandria. Clement was much concerned with 'proof' in *Stromateis* VIII. Whatever the precise dates of his works may be, he certainly did not write *Strom.* I (144) before 193, well after Galen's major works appeared. And while the collection of materials that constitutes *Stromateis* VIII was almost certainly used earlier, he may have been concerned with all of it just because of Galen's comments. Certainly, Clement is much more concerned with the subject than his predecessors were." In my 2011 article (noted above) I overlooked this fine paragraph. Unfortunately, Grant does not pursue his idea any further, considering Origen "a better witness" to the acquaintance of early Christian writers with Galen. For Origen's acquaintance with Galen, see further Ilaria Ramelli, "Origen, Greek Philosophy, and the Birth of the Trinitarian Meaning of Hypostasis," *Harvard Theological Review* 105/3 (2012), pp. 302–350, here 322. Parallels between Clement and Galen are also noted by Silke-Petra Bergjan, "Logic and Theology in Clement of Alexandria: The Purpose of the 8th Book of the *Stromata*," *Zeitschrift für antikes Christentum* 12 (2009), pp. 396–413, here esp. 397 f., 400, and 403. However, Bergjan emphasizes the contrast between Clement and Galen, namely the former's apparent lack of interest in the syllogistic theory (cf. pp. 400 and 404–407). For a sceptic assessment of efforts to find Galenic influence on several Christian writers (Athenagoras, Clement, and Origen) cf. Jonathan Barnes, "Galen, Christians, Logic," in *Logical Matters: Essays in Ancient Philosophy* II (Oxford: Clarendon Press, 2012), pp. 1–21, here 12 and n. 22.

109 *Strom.* VIII (2) 3, 3–4.

stration'] with an account so clearly that all who speak the same language will follow".[110] Pointing out situations in which the word 'demonstration' is used, he suggests that its users (a philosopher, a rhetorician, or a judge) use it as a word with meaning and that the item it signifies is understood to exist.[111] After explaining why every statement about "anything sought" must be referred to something agreed by all (this being an agreed account of the proposed name), Clement suggests the following account of the name 'demonstration': "... all human beings will agree that demonstration is an argument that furnishes credence from something agreed to something disputed."[112] This account is a starting point for the subsequent passage, in which two kinds of demonstration are distinguished: demonstration "in the proper sense", also called epistemic, and a deficient kind of demonstration based on reputable premisses. The conditions of demonstration in the proper sense are then further explored to show that demonstration is a valid argument based on something plainly apparent to sense perception and intellection.[113]

In *DD*, Galen must have described the method of inquiry in similar terms. In the first book of his *Method of Healing* (*De methodo medendi*, henceforth *MM*), he sets himself the task of discovering the art of healing in such a way as to guide other people towards it. While pursuing this task, he claims to follow methods he had established in *DD*:

> I will go through the rest of the discourse in order for you, making use of the methods I established in the treatise *On Demonstration*. In that work it was shown that the origins of all demonstration are those things which are plainly apparent to sense perception and intellection, and how in every inquiry into something it is necessary to replace its name with an account.

Calling this "the starting point of true teaching", he goes on to explain that his goal is to discover therapies for all diseases. In order to reach this goal, he must first explore how many diseases there are, i.e. to make a division of the genus 'disease'. But this, again, can only be done when we have reached firm knowledge of the thing which is to be divided, in other words, when we have "explained by an account what disease is". Galen continues as follows:

---

110 *Strom.* VIII (2) 3, 1.
111 *Strom.* VIII (2) 3, 2.
112 *Strom.* VIII (3) 5, 1.
113 *Strom.* VIII (3) 5, 2–8, 3.

> How then do we find this out correctly by method? How else than in the way described in *On Demonstration*? First of all, a notion must be agreed upon, without which it is impossible to discover the essence of the thing proposed. We have said that that we must take this notion as something agreed by all, for otherwise it would not be rightly called a starting point. What then is the notion of being ill agreed by all human beings? And to what underlying thing do they particularly refer with this verb, 'being ill'?[114]

Having proposed an account of the notions of 'health' and 'disease', as the words are used in ordinary language, Galen goes on to explore the essence of the matter, pointing out that "discoveries, inquiries, and demonstrations" about it can no longer be derived from the opinions of the multitude, but "from scientific premisses, the manner of whose discovery was spoken of in [*On Demonstration*]".

There are obvious correspondences between the method described by Galen with regard to the teaching on disease and therapy and the one described by Clement with regard to the teaching on demonstration. In both cases, one has to start by establishing the agreed meaning of the proposed name and then proceed to the investigation of the essence of the subject matter. In Clement's rendering of the method, there is yet another stage of inquiry between these two, at which we ask "whether the signified item exists". Galen does not address this question in the passage quoted above. However, later in the treatise, in a polemic against those who believe that names always refer to particulars, Galen asks a rhetorical question that does imply a distinction between the meaning of the proposed name and the existence of the signified item: "Do the words 'animal' and 'disease' seem to you to signify nothing, but have a sense similar to that of 'blituri' and 'scindapsus'? Or is it the case that they signify, but yet there is no object underlying the words, as is the case with 'Scylla' and 'centaur'?"[115]

It is surely remarkable that Clement and Galen describe the same method of inquiry. But more remarkable is the fact that while describing it, they use the same words and sometimes even the same syntax. Consider the following synopsis of passages mentioned in the preceding paragraphs:

---

114 *MM* I 5 (X,39–41 K.).

115 *MM* II 7 (X,144,9–14 K.), trans. Hankinson.

Galen	Clement
Galen, *MM* I 5 (X,39,5–10) Καί σοι τὸν ἑξῆς λόγον ἤδη ἅπαντα ποιήσομαι, χρώμενος ταῖς μεθόδοις ἃς ἐν τοῖς περὶ τῆς ἀποδείξεως ὑπομνήμασι κατεστησάμην.	
ὅτι τε γὰρ *ἀρχαὶ πάσης ἀποδείξεώς εἰσι τὰ πρὸς αἴσθησίν τε καὶ νόησιν ἐναργῶς φαινόμενα*	***Strom.* VIII (3) 7, 3** εἶεν δ' ἂν καὶ ἄλλαι *τῶν ἀποδείξεων ἀρχαὶ* (…), *τὰ πρὸς αἴσθησίν τε καὶ νόησιν ἐναργῶς φαινόμενα.*
καὶ ὡς ἐπὶ πάντων *τῶν ζητουμένων εἰς λόγον χρὴ μεταλαμβάνεσθαι τοὔνομα*, δι' ἐκείνων ἀποδέδεικται·	***Strom.* VIII (2) 4, 2** πᾶν οὖν τὸ προβληθὲν ὄνομα *μεταλαμβάνειν χρὴ εἰς λόγον* ὁμολογούμενόν τε καὶ σαφῆ τοῖς κοινωνοῦσι τῆς σκέψεως, ἀρχὴν μὲν τῆς διδασκαλίας ἐσόμενον, ἐξηγησόμενον δὲ τὴν *τῶν ζητουμένων* εὕρεσιν.
MM I 5 (X,40,2–41,1) *ἥδ' οὖν ἡμῖν ἀρχὴ τῆς ἀληθοῦς ἔστω διδασκαλίας*, ἣν ἂν εὐθὺς ἀπ' ἀρχῆς ἐποιησάμεθα, ζητοῦντές τε τὴν τέχνην αὐτοὶ καὶ ποδηγοῦντες ἑτέρους μηδέπω διεστραμμένους οὕτως ὥστ' ἐπεὶ πρόκειται θεραπείας εὑρεῖν ἁπάντων τῶν νοσημάτων, ἀναγκαῖον ἐπίστασθαι πρότερον ὁπόσα τὰ σύμπαντά ἐστιν· ἀλλ' ἐπεὶ μήτε διαφορὰς μήτε εἴδη δυνατὸν ἐξευρεῖν γένους μηδενὸς ἄνευ τοῦ βεβαίως αὐτὸ τὸ τεμνόμενον ἐπίστασθαι, χρὴ δήπου καὶ νῦν *ὅ τί ποτέ ἐστι νόσημα τῷ λόγῳ*	***Strom.* VIII (2) 3, 1** Τίς ἂν οὖν ἄλλη βελτίων ἢ ἐναργεστέρα *μέθοδος εἰς ἀρχὴν τῆς τοιᾶσδε εἴη ἂν διδασκαλίας*

<table>
<tr><td>διελθεῖν, ἵν' οὕτως ἐπιχειρήσωμεν ὀρθῶς αὐτοῦ τῇ διαιρέσει. πῶς οὖν ἐξεύρωμεν αὐτὸ ὀρθῶς μεθόδῳ; πῶς δ' ἄλλως ἢ ὡς ἐν τοῖς περὶ ἀποδείξεως ἐλέγετο;</td><td>ἢ τὸ προταθὲν ὄνομα λόγῳ διελθεῖν οὕτω σαφῶς ὡς πάντας ἀκολουθῆσαι τοὺς ὁμοφώνους; ἆρ' οὖν τοιοῦτόν ἐστι ⟨τὸ⟩ ὄνομα τῆς ἀποδείξεως,</td><td></td></tr>
<tr><td></td><td>οἷόν περ τὸ βλίτυρι, φωνὴ μόνον οὐδὲν σημαίνουσα;</td><td>Galen MM II 7 (X,144,9–14 K.)
ἆρά γε οὐδὲν αἱ φωναὶ δοκοῦσί σοι σημαίνειν, ἥ τε ζῷον καὶ ἡ νόσος, ἀλλ' ὁμοίως ἐκφωνεῖσθαι τῷ βλίτυρι καὶ σκινδαψός;</td></tr>
<tr><td>τῆς ἐννοίας πρότερον ὁμολογηθείσης, ἧς χωρὶς οὐχ οἷόν τέ ἐστιν εὑρεθῆναι τὴν οὐσίαν τοῦ προκειμένου πράγματος· αὐτὴν δὲ τὴν ἔννοιαν ὁμολογουμένην ἅπασιν ἐλέγομεν χρῆναι λαμβάνειν, ἢ οὐδ' ἂν ἀρχὴν δεόντως ὀνομάζεσθαι.</td><td>Strom. VIII (2) 3, 3–4
περὶ παντὸς τοίνυν τοῦ ζητουμένου εἴ τις ὀρθῶς διαλαμβάνοι, οὐκ ἂν ἐφ' ἑτέραν ἀρχὴν ὁμολογουμένην μᾶλλον ἀναγάγοι τὸν λόγον ἢ τὸ πᾶσι τοῖς ὁμοεθνέσι τε καὶ ὁμοφώνοις ἐκ τῆς προσηγορίας ὁμολογούμενον σημαίνεσθαι.</td><td></td></tr>
<tr><td></td><td>εἶτα ἐντεῦθεν ὁρμηθέντα ζητεῖν ἀνάγκη, εἰ ὑπάρχει τὸ σημαινόμενον τοῦτο π⟨ερὶ⟩ οὗ ὁ λόγος εἴτε καὶ μή·</td><td>ἢ σημαίνουσι μέν, οὐδὲν δ' ὑπόκειται πρᾶγμα ταῖς φωναῖς, ὡς ἐν τῷ Σκύλλα καὶ Κένταυρος, ἢ καὶ σημαίνουσι καὶ τὸ σημαινόμενον ἓν ὑπάρχει;</td></tr>
<tr><td>τίς οὖν ὑπὸ πάντων ἐστὶν ἀνθρώπων ὁμολογουμένη περὶ τοῦ νοσεῖν ἔννοια; καὶ κατὰ τίνος μάλιστα φέρουσιν ὑποκειμένου πράγματος τουτὶ τὸ ῥῆμα τὸ νοσεῖν;</td><td>ἐφεξῆς δέ, εἴπερ ὑπάρχειν δειχθείη, ζητητέον τούτου τὴν φύσιν ἀκριβῶς, ὁποία τίς ἐστιν, καὶ μή ποτε ὑπερβαίνει⟨ν⟩ τὴν δοθεῖσαν τάξιν.</td><td></td></tr>
</table>

Another intriguing parallel is found in Galen's treatise *On the Usefulness of the Parts* (*De usu partium*, henceforth *UP*). Here Galen reminds his readers that in *DD* he dealt with the question of whether the embryo (literally, "the thing in the womb") is an animal. This corresponds to the contents of our section γ, where the same question is used as an example through which "the method of discovery" (that is to say, the discovery of premisses appropriate to

a given problem) is explained. Again, not only the topic and the context of the discussion (namely, the theory of demonstration), but also Galen's choice of words and their combination closely resonate with Clement:

Galen, *UP* XV 5 (II,357,24–28 Helmreich/IV,238 f. K.)	***Strom.* VIII (4) 12, 2–3**
περὶ δὲ τοῦ ζῷον ἤδη τὸ κατὰ γαστρὸς ὑπάρχειν,	Ἀλλ' εἰ μὲν ἀποκρίνεσθαι ἕλοιτο (*sc.* ὁ ζητήσας, *εἰ ζῷον τὸ κατὰ γαστρός*), πάντων πρῶτον ἐρωτητέον αὐτὸν ἐφ' ὅ τι φέρει πρᾶγμα τὸ ζῷον ὄνομα, κἀπειδὰν τοῦτο ἀποκρίνηται, πάλιν ἐρωτητέον ὅ τι ποτὲ αὐτῷ σημαίνει τὸ κυούμενον ἢ *τὸ κατὰ γαστρός*, εἰ [τὰ διαπεπλασμένα ἤδη καὶ τὰ ζῷα] καὶ τὸ σπέρμα αὐτὸ ⟨τὸ⟩ καταβεβλημένον τὸ κατὰ
ὅταν γε διαπεπλασμένον ἅπασιν ᾖ τοῖς μορίοις, ἔν τε τοῖς Περὶ ἀποδείξεως ὑπομνήμασιν εἴρηται κἀν τοῖς Περὶ τῶν Ἱπποκράτους τε καὶ Πλάτωνος δογμάτων.	γαστρὸς αὐτῷ σημαίνειν βούλεται ἢ *μόνα τὰ διηρθρωμένα τε καὶ ἤδη διαπεπλασμένα*, τὰ ἔμβρυα καλούμενα.

These are by no means the only doctrinal, lexical, or syntactic correspondences between the 'eighth book' and the writings of Galen: many more are noted in the commentary.[116] But they are particularly interesting in that they pertain specifically to Galen's *DD*. Taking these parallels as a point of departure, I have considered three options of how they might be explained: (1) They are mutually unconnected representations of the same tradition or "trend", as Solmsen puts it; (2) Clement draws from a source that was also used by Galen; (3) Clement draws from Galen. Exploring these options, I have reached the conclusion that the third is the most likely. To make a long story short, the parallels are too numerous, too close, and too specific to choose the first, and there is no evidence to support the second.[117] On the other hand, there are reasons to

116 For sections β and γ, the reader may consult the commentary on the following passages: 81,9 f.; 81,11; 82,2 f.; 82,4 f.; 82,14–16; 82,16–18; 82,18–21; 82,22 f.; 82,24; 82,27 f.; 83,1; 83,3 f.; 83,7 f.; 83,24 f.; 83,26 f.; 84,9–13; 84,14 f.; 84,16 f.; 84,27–30; 84,32; 85,1b–13; 85,4; 85,7 f.; 85,15 f.; 85,17b; 85,23 f.; 86,1–5; 86,11 f.; 86,32 f.; 87,1 f.; 87,4–7; 87,7 f.; 88,15; and 88,29 f.

117 For the full argument, cf. "Galenus Christianus," pp. 368–372. Particularly noteworthy is the presence of medical notions like ἔνδειξις or πρόγνωσις in section β (5, 2 and 6, 1; cf. below, on 82,14–16 and 82,27 f.) and the interest in the powers of stones, plants, and animals in section γ, which seems to betray pharmacological concerns (9, 1/85,4). Apart from the embryo

believe that Clement, Galen's younger contemporary, could have heard about *DD* and, had he heard about it, would have been interested in reading it.[118] The relative chronology of Clement, who wrote the first book of the *Stromateis* after 192, and Galen, who wrote *DD* before 162, makes it possible.[119]

In the course of writing the commentary, I have obviously kept an eye on the Galen hypothesis. My aim was then twofold: I tried to point out parallels in Galen's works that struck me as especially close to Clement's text and/or especially helpful in understanding its doctrinal background. But I also tried to look for other than Galenic parallels and, generally, for any indications that might speak against the hypothesis. As a matter of fact, I did find a number of non-Galenic parallels to various motifs and formulations in sections β and γ. Most of them come from Aristotle or Aristotle's commentators, e.g. Clement's contemporary Alexander of Aphrodisias. But none of these parallels is as close and specific as the Galenic ones. Generally, when exploring sections β and γ, I failed to find anything that contradicts or mitigates the argument presented in my 2011 article.

How about the rest of the material? The optimistic scenario is that most, or even all, of it comes from the same source, thus giving us unprecedented access

---

question, Clement mentions another problem for demonstration, namely, 'In which part of the body is the ruling faculty of the soul?' (*Strom.* VIII [4] 14, 4: ἐν τίνι τοῦ σώματος μορίῳ τὸ ἡγεμονικόν ἐστι τῆς ψυχῆς). Galen discusses this problem, by way of a demonstrative method, in *PHP*. Cf. Tieleman, *Galen and Chrysippus*, p. 30. Clement also mentions the typically Galenic (not Clementine) view that the substance of the ruling faculty of the soul is unknown (14, 4/88,29: ἀγνοοῦμεν τὴν οὐσίαν). If Clement drew from a source of Galen, then it must have been a source that had already combined medical and philosophical concerns in a characteristically Galenic manner.

118 Cf. "Galenus Christianus," pp. 373–375.

119 Cf. ibid., p. 373. For Clement, cf. *Strom.* I (21) 144, 3–5, reflecting the death of Commodus. That Galen composed *DD* before his first journey to Rome, when employed as a physician to the gladiators in Pergamum, is forcefully argued by Iwan von Müller, "Über Galens Werk vom wissenschaftlichen Beweis," in *Abhandlungen der königlich-bayerischen Akademie der Wissenschaften, phil.-philol. Kl.* 20 (1897), pp. 403–478, here 411–414. According to this dating, Galen would have been about thirty years old when composing the treatise. This, now conventional, view has been recently questioned by Jacques Jouanna, "Médecine et philosophie: la reception de la science aristotélicienne chez Galien," in Y. Lehman (ed.), *Aristoteles Romanus. La réception de la science aristotélicienne dans l'Empire gréco-romain* (Turnhout: Brepols, 2013), pp. 159–181, here 171f., who finds it hard to believe that Galen would have composed a treatise to which he later attributed so much wisdom and importance at such a young age. But Jouanna's conclusion is weakened by the fact that he leaves von Müller's arguments intact and unmentioned.

to the contents of a lost treatise of Galen, most likely *DD*. A sceptic will retort that even if we get convinced of the Galenic background to sections β and γ, it will tell us little about the origin of the other sections.

Scepticism is a prerequisite of sound judgement, while optimism borders upon vice in historical scholarship. But one has to admit that, in this case, there is much to be said in favour of the optimistic view. As we have seen, signs of continuity can be detected between sections β and γ on the one hand, and ε and η on the other.[120] Moreover, the chapter on the categories (section η) contains views and expressions adopted by Galen, and other views and expressions whose closest parallels are found in Aspasius, a second-century Peripatetic, whose introductory treatise on the categories was known to Galen and recommended by him to his pupils.[121] The chapter on division and definition (section ε) also includes elements that, apart from other sources, can be found in Galen.[122] In addition, both chapters deal with topics that were certainly or almost certainly discussed in *DD*.[123] By contrast, if we suppose that Clement took them from a different source than the material on demonstration, it will be hard to explain why he was interested in these topics at all. Did he find them noteworthy for their own sake, or because he thought they were relevant to the topics that occupied him in the preceding sections, i.e. demonstration and inquiry? Probably the latter. But shall we suppose, then, that he recognized this relevance on his own, independently of the source he excerpted in sections β and γ, and that he turned to a different source to learn more about these issues? Perhaps, if Clement was a reader of Aristotle's works on logic, whose main concern was to give a unified account of the Aristotelian methodology of science, it

---

120 Cf. above, p. 33.

121 Cf. my "Categories in *Stromata* VIII," *Elenchos* 33/2 (2012), pp. 197–225 and commentary below, esp. on 94,7 f.; 95,1–3; 95,4 f.; and 95,24–26.

122 Cf. below, on 90,9 f.; 90,18 f.; 90,25; 91,7; 91,9 f.; 91,10–12; 91,16 f.; 92,1–3; 92,3 f.; 92,4 f.; 92,21 f.; 93,5 f.; 93,14 f.; and 93,15–17.

123 For division and definition, cf. Jonathan Barnes, "Galen on Logic and Therapy," in F. Kudlien—R.J. Durling (eds.), *Galen's Method of Healing* (Studies in Ancient Medicine 1; Leiden: Brill, 1991), pp. 50–102, here esp. 68 and 72–76; Riccardo Chiaradonna, "Le traité de Galien *Sur la démonstration* et sa postérité tardo-antique," in R. Chiaradonna and F. Trabattoni (eds.), *Physics and Philosophy of Nature in Greek Neoplatonism* (PhA 115; Leiden/Boston: Brill, 2008), pp. 43–77, here 45. Cf. also Teun Tieleman, "Methodology," in *The Cambridge Companion to Galen*, ed. R.J. Hankinson (Cambridge: Cambridge University Press, 2009), pp. 49–65, here 59 f.; R.J. Hankinson, "Epistemology," in *The Cambridge Companion to Galen*, pp. 157–183, here 167 f.; Jane Hood, "Galen's Aristotelian Definitions," in D. Charles (ed.), *Definition in Greek Philosophy* (Oxford: Oxford University Press, 2010), pp. 450–466. For the categories, see my "Categories," pp. 223 f.

would be a reasonable conjecture: For in the *Posterior Analytics*, Aristotle does discuss the relation between demonstration and definition and the value of the method of division in that regard. And it makes sense for anyone thinking about Aristotle's methodology to include his distinction of the ten categories as the elementary kinds of predicates under which any subject matter of inquiry will fall. But nothing in Clement's other writings, including passages where demonstration is mentioned, allows us to believe that he aspired at anything nearly as ambitious as integrating different aspects of Aristotle's logical theory, indeed that he knew much about the *Organon* at all. Thus the relevance of such issues as division, definition, and the categories to the theory of demonstration was probably suggested to him by someone who did possess this knowledge and aspiration—the most obvious candidate being the writer excerpted in sections β and γ.

But if we conclude that sections β, γ, ε, and η, comprising about two thirds of the 'eighth book', draw from the same source, and that this source is Galen, what shall we think of the chapter on causes (section θ)? It would be a marvellous discovery indeed if this poorly organized, but well-informed and sometimes philosophically penetrating chapter, which contains invaluable testimonies about ancient theories of causation, could be traced back to this author. Again, suspicious as we might be about marvellous things of this kind, it is true that some indications do point in that direction. To begin with, as Duhot observes, some elements in this chapter are likely of medical origin. Duhot singles out the following passages:

> P1. *Str.* VIII (9) 28,7: τὸ συνεκτικὸν αἴτιον οὐ δεῖται χρόνου· τὸ γὰρ καυτήριον ἅμα τῷ κατὰ τῆς σαρκὸς ἐπερεισθῆναι τὴν ἀλγηδόνα παρέχει. τῶν προκαταρκτικῶν τὰ μὲν χρόνου δεῖται, ἄχρις ἂν γένηται τὸ ἀποτέλεσμα, τὰ δὲ οὐ δεῖται, ὡς ἡ πτῶσις τοῦ κατάγματος. μή τι οὖν οὐ κατὰ στέρησιν χρόνου ἄχρονα λέγεται ταῦτα, ἀλλὰ κατὰ μείωσιν, ὡς καὶ τὸ ἐξαίφνης, μηδὲ αὐτὸ χωρὶς χρόνου γενόμενον.
>
> Synectic causes do not need time. For a cautery produces pain immediately as it is pressed onto flesh. Of the procatarctic causes, some need time before the effect comes about, whereas some do not, as when a fall causes a fracture. Of course, when saying that they do not need time, we do not mean it in the sense of a negation of time, but of its diminution, like with a sudden event, which also does not come about without time.
>
> P2. *Str.* VIII (9) 30, 1: ἀλλήλων οὐκ ἔστι τὰ αἴτια, ἀλλήλοις δὲ αἴτια. ἡ γὰρ σπληνικὴ διάθεσις προϋποκειμένη οὐ πυρετοῦ αἴτιος, ἀλλὰ τοῦ γίνεσθαι τὸν

πυρετόν· καὶ ὁ πυρετὸς προϋποκείμενος οὐ σπληνός, ἀλλὰ τοῦ αὔξεσθαι τὴν διάθεσιν.

Causes are not causes of each other, but there are causes to each other. For the pre-existing condition of the spleen is the cause, not of fever, but of the fever's coming about; and the pre-existing fever is the cause, not of the spleen, but of its condition's being intensified.

P3. *Str.* VIII (9) 31,4–5: τῷ ὄντι δὲ προκαταρκτικὰ μὲν αἴτια ἑνὸς γίνεται πλείονα κατὰ γένος καὶ κατ' εἶδος, καὶ κατὰ γένος μὲν τοῦ νοσεῖν ὁπωσοῦν, οἷον ψῦξις, ἔγκαυσις, κόπος, ἀπεψία, μέθη, κατ' εἶδος δὲ τοῦ πυρετοῦ. τὰ δὲ συνεκτικὰ αἴτια κατὰ γένος μόνον, οὐκέτι δὲ καὶ κατ' εἶδος. τοῦ γὰρ εὐωδιάζεσθαι κατὰ γένος ἑνὸς ὄντος πολλὰ τὰ αἴτια κατ' εἶδος, οἷον λιβανωτός, ῥόδον, κρόκος, στύραξ, σμύρνα, μύρον· τὸ γὰρ ῥόδον οὐκ ἂν οὕτως εὐῶδες εἴη ὡς ἡ σμύρνα.

But in fact, in the case of procatarctic causes, there can be many [causes] of one effect according to genus and according to species. [Many of one] according to genus are the causes of becoming ill in any way whatsoever, for example cold, heat, fatigue, indigestion, drunkenness; [many of one] according to species are the causes of fever. 5 (A)s far as synectic causes are concerned, there are only [many of one effect] according to genus, but not according to species. Thus smelling good is one effect according to genus, but it has many causes according to species, for example frankiscence, rose, saffron, storax, myrrh, or sweet oil. For rose certainly does not smell the same as myrrh.

P4. *Str.* VIII (9) 32,7: καὶ τὰ μὲν προκαταρκτικά, τὰ δὲ συνεκτικά, τὰ δὲ συναίτια, τὰ δὲ συνεργά. καὶ τὰ μὲν τοῦ κατὰ φύσιν, τὰ δὲ τοῦ παρὰ φύσιν, τὰ δὲ νόσου, καὶ κατὰ τὸ ὑποβεβηκὸς τὰ μὲν παθῶν, τὰ δὲ τοῦ μεγέθους αὐτῶν, τὰ δὲ χρόνων καὶ καιρῶν.

And some are procatarctic, some synectic, some joint-causes, some auxiliary. And some are causes of effects that are natural, some of those that are unnatural, some of disease, and, on a lower level, some of affections, some of their magnitude, some of temporal intervals.

P5. *Str.* VIII (9) 33, 1: Τῶν μὲν οὖν προκαταρκτικῶν αἰρομένων μένει τὸ ἀποτέλεσμα, συνεκτικὸν δέ ἐστιν αἴτιον, οὗ παρόντος μένει τὸ ἀποτέλεσμα καὶ αἰρομένου αἴρεται.

> When procatarctic causes are removed, the effect remains, whereas a synectic cause is one during whose presence the effect remains and on whose removal the effect is removed.

At least one more passage can be added to Duhot's list:

> P6. *Str.* VIII (9) 32, 4: Τῶν οὖν αἰτίων τὰ μέν ἐστι πρόδηλα, τὰ ἐπιλογισμῷ λαμβανόμενα, τὰ δὲ ἄδηλα, τὰ ἀναλογισμῷ.

> Of causes, some are clear, viz. those grasped by epilogism, whereas some are unclear, viz. those grasped by analogism.

As the reader may check in the commentary, all these passages contain conceptual distinctions typical of medical literature, most of which can only be found in medical sources. This in itself, of course, does not constitute a serious argument about authorship. The borderline between philosophy and medicine is narrow in antiquity and it is perfectly conceivable that a writer whose primary concern is philosophical would include medical theories when discussing other people's views. A case in point is Sextus Empiricus, who mentions the views of Herophilus, Erasistratus, and Asclepiades alongside those of the 'dogmatist' philosophers.[124] In fact, one of the above-mentioned passages (P6 and the following) has a parallel in Sextus' classification of signs in *M.* VIII 145–155, which seems to go back to medical sources as well.[125] On the other hand, Sextus' interest in medicine is probably due to the circumstance that he was a doctor himself and part of his agenda was to refute medical 'dogmatism'. As far as I am aware, there is just one other writer of the same period whose works contain a comparable blend of medical and philosophical material: Galen. Moreover, there is abundant evidence of Galen's interest in causal theories and many of the terms and distinctions discussed in our chapter are used by him in his works.[126] Finally, this chapter too contains correspondences in thought, vocabulary, or syntax with Galen's writings. Consider the following instances:

---

124 Cf. e.g. *M.* VII 91; 201–202; 323; 380; VIII 220.

125 In *M.* VIII 156, Sextus lumps 'dogmatist philosophers' and 'rationalist doctors' together. The same theory is reported in *PH* II 97–101, where the notion of the indicative sign is (probably incorrectly) blended with the Stoic definition of sign in general (cf. 101 and 104). For the medical background to Sextus' classification, cf. James Allen, *Inference from Signs: Ancient Debates about the Nature of Evidence* (Oxford: Oxford University Press, 2001), pp. 87–146.

126 For Galen's interest in causation, cf. esp. the pioneering works of R.J. Hankinson: "Galen's Theory of Causation," *ANRW* II 37.2 (1994), pp. 1757–1774; Galen: *On Antecedent Causes*, ed.,

Strom. VIII (9) 25, 4/96,5: ὁ δὲ χρόνος τῶν ὧν οὐκ ἄνευ λόγον ἐπέχει	Galen, *MM* III 10 (X,224,5 K.), on time and other items "having the status of prerequisites (τὸν ὧν οὐκ ἄνευ λόγον ἔχουσαι)"; *Symp. Diff.* 1, 12 (CMG V,5,1: 206,14/VII,48 K.), on that which "has the status of prerequisites (τὸν ὧν οὐκ ἄνευ λόγον ἐπέχει)"; *CP* 7, 89 (CMG Suppl. II: 21,32 f.), on things that "have the status of prerequisites (*rationem eorum sine quibus non … habere*)"
25, 5/96,6: *αἴτιον δὲ κυρίως λέγεται τὸ παρεκτικόν τινος ἐνεργητικῶς*, ἐπεὶ καὶ τὸν σίδηρον τμητικόν φαμεν εἶναι οὐ μόνον ἐν τῷ τέμνειν, ἀλλὰ καὶ ἐν τῷ μὴ τέμνειν· οὕτως οὖν καὶ τὸ παρεκτικὸν ἄμφω σημαίνει *καὶ τὸ ἤδη ἐνεργοῦν καὶ τὸ μηδέπω μέν, δυνάμει δὲ κεχρημένον τοῦ ἐνεργῆσαι.*	*CAM* 14, 3 (CMG V,1,3: 98,5–8/I,272 K.): *τὰ μὲν οὖν αἴτια κυριώτατα μέν, ὅταν ἐνεργῇ, ταύτης τῆς προσηγορίας ἀξιοῦται*, λέγεται μέντοι γε πολλάκις, *εἰ καὶ μηδέπω μηδὲν ἐνεργεῖ, κατ' αὐτὸ τὸ δύνασθαι μόνον*· οἷον καὶ ἡ ἀπεψία νόσων αἰτία λέγεται, κἂν μηδέπω νοσάζῃ.
26, 1/96,11: οἱ μὲν οὖν σωμάτων, οἱ δ' ἀσωμάτων φασὶν εἶναι τὰ αἴτια	*CP* 16, 199 (CMG Suppl. II: 54,1–4), quoting Herophilus: … ⟨*aut corporea existens causa corporis alucuius est factiva*⟩ *aut corporis factiva alicuius existit incorporea ens* etc.
27,1–2/97,8–11: καὶ τὸν οἰκοδόμον οἰκοδομεῖν λέγομεν κατὰ τὴν ἐπὶ τὸ γενησόμενον ἀναφοράν· οὕτως φαμὲν χλαμύδα ὑφαίνεσθαι· τὸ γὰρ ποιοῦν ἐνεργείας δηλωτικὸν ὑπάρχει, οὐκ ἔστι δὲ ἑτέρου μὲν τὸ ποιοῦν, ἑτέρου δὲ τὸ αἴτιον, ἀλλὰ ταὐτοῦ, τῆς χλαμύδος καὶ τῆς οἰκίας.	*Thras.* 27 (67,24–68,4 Helmreich /V,854,13–17 K.): οὐ γὰρ οἰκοδομεῖν οἰκίαν αὐτὸ δὴ τοῦτο τέλος ἐστὶ τῆς οἰκοδομικῆς, ἀλλ' οἰκία, καθάπερ οὐδ' ὑφαίνειν ἐσθῆτα καὶ ναῦν συμπήττειν καὶ σκίμποδα καὶ τῶν ἄλλων ἕκαστον, ἀλλ' αὐτὸ τὸ δημιουργηθέν.
28,4/98,9 f.: *πᾶν τὸ ἐνεργοῦν παρέχει τὸ* ἀποτέλεσμα *μετ' ἐπιτηδειότητος τοῦ πάσχοντος.*	*Diff. Feb.* I 6 (VII,290,11–14 K.): χρὴ γὰρ μεμνῆσθαι … *ὡς οὐδὲν τῶν αἰτίων ἄνευ τῆς τοῦ πάσχοντος ἐπιτηδειότητος ἐνεργεῖν πέφυκεν.*

---

intro., trans. and comm. (Cambridge: Cambridge University Press, 1998); *Cause and Explanation in Ancient Greek Thought* (Oxford: Clarendon Press, 1998), pp. 373–402; "Causation in Galen," in *Galien et la philosophie* (Entr. Hardt 49; Genève: Vandœuvre, 2002), pp. 31–72.

28, 7/98,19–22: τὸ συνεκτικὸν αἴτιον οὐ δεῖται χρόνου ... τῶν προκαταρκτικῶν τὰ μὲν χρόνου δεῖται ... τὰ δὲ οὐ δεῖται κτλ.	*Caus. Morb.* 2 (VII,8,18–9,1 K.), re. procatarctic causes: χρόνου δεῖται ... τὸ μέλλον ποιήσειν ὁπωσοῦν.
29, 1/98,28–30 τὸ αἴτιον τῶν "πρός τι".	*Ars. Med.* 23 (344,18 Boudon/I,366 K.): πᾶν αἴτιον ἐν τῷ πρός τι.
29, 4/99,4 f. *οὐ δύναται δὲ τὸ αὐτὸ πρὸς ἑαυτῷ λαμβανόμενον ἐνεργεῖν* ἅμα καὶ διατίθεσθαι.	*Plen.* 3 (VII,525,7 f. K.): ποιεῖν δ' εἰς ἑαυτὸ λέγειν ὁτιοῦν, ἢ *ἐνεργεῖν εἰς ἑαυτό, παρὰ τὴν ἔννοιάν ἐστιν.*
30, 1/99,13–16: ἡ γὰρ σπληνικὴ διάθεσις προϋποκειμένη οὐ πυρετοῦ αἴτιος, ἀλλὰ τοῦ γίνεσθαι τὸν πυρετόν.	*Diff. Feb.* II 6 (VII,346,6 f.): πολλάκις δὲ καὶ ἡ κατὰ τὸ ἧπαρ ἢ τὸν σπλῆνα διάθεσις ἤνεγκε πυρετόν.
30, 2/99,17 f.: οἱ ἐπὶ τῆς *ψαλίδος λίθοι ἀλλήλοις εἰσὶν* αἴτιοι τοῦ μένειν κατηγορήματος	*Symp. Diff.* 1, 13 (CMG V,5,1: 206,15–208,1/VII,48 K.): ... ἐνδέχεταί τινα *στοῖχον αἰτιῶν* γενέσθαι πολλάκις ἄλληλα διαδεχομένων, ὡς εἰ καὶ *ψηφίδων ἐφεξῆς ἀλλήλαις κειμένων* πλειόνων κινήσειέ τις τὴν πρώτην, αὕτη δὲ τὴν δευτέραν κἀκείνη τὴν τρίτην καὶ οὕτω κατὰ τὸ ἑξῆς ἑκάστη τὴν μετ' αὐτήν.
30, 4/99,26–28: ὁ μὲν γὰρ πλήξας τινὰ θανασίμως αἴτιός ἐστιν αὐτῷ τοῦ θανάτου ἢ τοῦ γίνεσθαι τὸν θάνατον, ἀντιπληγεὶς δὲ ὑπ' αὐτοῦ θανασίμως ἔσχεν αὐτὸν ἀνταίτιον κτλ.	CP XI 146 f.: ... *hunc hominem vidistis in stadiis milies pugnantem et percucientem maximis plagis antagonistas sicut et ipsum ab illis vicissim hoc patientem, sed tamen neque iste ab illorum aliquo neque illorum aliquis ab isto percussus mortuus est. Non enim est causa mortis plagatio, etc.*
31, 4/100,13–16 *προκαταρκτικὰ μὲν αἴτια* ἑνὸς γίνεται πλείονα κατὰ γένος καὶ κατ' εἶδος, καὶ κατὰ γένος μὲν τοῦ νοσεῖν ὁπωσοῦν, οἷον *ψῦξις, ἔγκαυσις, κόπος, ἀπεψία, μέθη*, κατ' εἶδος δὲ τοῦ πυρετοῦ.	*Sect. Int.* 7 (16,12–20 Helmreich/I,84,18–85,4 K.): φέρε δὴ πρῶτον ἐπισκεψώμεθα *περὶ τῶν προκαταρκτικῶν καλουμένων αἰτίων* ... τί δή ποτε *ψύξεις τε καὶ ἐγκαύσεις καὶ μέθας* καὶ ἀπεψίας καὶ πλησμονὰς καὶ ἐνδείας *καὶ κόπους* καὶ ἀργίας καὶ ἐδεσμάτων ποιότητας καὶ ἐθῶν ὑπαλλαγάς ... μάτην πολυπραγμονεῖτε;

(cont.)

31, 5/100,17–19: τοῦ γὰρ εὐωδιάζεσθαι *κατὰ γένος ἑνὸς ὄντος* πολλὰ τὰ αἴτια κατ' εἶδος, οἷον *λιβανωτός, ῥόδον, κρόκος, στύραξ, σμύρνα, μύρον.*	*Comp. Med. Loc.* IV 5 (XII,719,1–7 K.): ὁποῖα φάρμακά ἐστι *ῥόδα τε καὶ λιβανωτός … ἐκ τοῦ αὐτοῦ δὲ γένους αὐτῷ… τόν τε κρόκον καὶ τὴν σμύρναν* κτλ. *Ibid.* II 1 (XII,561,11–14 K.): σμύρνα, λίβανος … κρόκος … *στύραξ.*
32, 2/100,20–22 παρὰ *τὸ μέγεθος* τοῦ αἰτίου … παρὰ *τὴν ἐπιτηδειότητα* τοῦ πάσχοντος.	*Caus. Morb.* 2 (VII,8,11–14 K.): πῶς οὖν ἐπὶ τοῦ πυρὸς οὐδὲν τούτων θαυμάζων ἐπὶ τοῦ κόπου θαυμάζεις, εἰ καὶ *μεγέθους* δεῖται καὶ χρόνου καὶ *τοῦ σώματος ἐπιτηδείως ἔχοντος* ἐξάπτεσθαι;
32, 4–5/101,4 f. τὰ μέν ἐστι πρόδηλα, τὰ *ἐπιλογισμῷ* λαμβανόμενα, τὰ δὲ ἄδηλα, τὰ *ἀναλογισμῷ*, καὶ τῶν ἀδήλων τὰ μὲν *πρὸς καιρὸν ἄδηλα … τὰ δὲ φύσει ἄδηλα.*	*Sect. Int.* 5 (10,24–25; 11,8–18 Helmreich/I,77,7–9.17–78,9 K.): πάντα γὰρ δεῖσθαι τῆς ἐξ αὐτῶν γνώσεως μηδ' εἶναί τι σημεῖον *ἀδήλου φύσει* πράγματος οὐδενός … *ὁ δ' ἐπιλογισμός*, ὃν δὴ ⟨τῶν⟩ φαινομένων λόγον εἶναί φασι, χρήσιμος μὲν εἰς εὕρεσιν τῶν *προσκαίρων ἀδήλων* …, οὐ μὴν ὅ γ' *ἀναλογισμός*, φασίν, ἀλλ' ἄρχεται μὲν ἀπὸ τῶν φαινομένων, προέρχεται δ' ἐπὶ τὰ διὰ παντὸς ἄδηλα κτλ.
32, 6/101,10: καθάπερ *ἡ συμμετρία τῶν λόγῳ θεωρητῶν πόρων*	*MM* IV 4 (X,268,10 f. K.): ἐν *συμμετρίᾳ μέν τινι πόρων* τὸ ὑγιαίνειν … ἐν ἀμετρίᾳ δὲ τὸ νοσεῖν. *MM* XIII 16 (X,919,16 K.): *λόγῳ θεωρητοῖς πόροις*

It should be emphasized that these and some other parallels noted in the commentary are not of the same value as those on which the Galen hypothesis is based. Some of them are commonplace, most of them are found elsewhere, and none of them is as striking as those in sections β and γ, i.e. such as would resist the postulation of a common source or more broadly conceived intellectual background. Nonetheless, when we put these passages together and add them to the earlier considerations, the plausibility of extending the Galen hypothesis to the chapter on causes increases. Suppose that we find the arguments about β and γ convincing; and that we accept their extension to ε and η as well. Shall it not appear likely that the section on causes, with its predilection for medical theory, with a topic that occupied Galen a great deal, and with a number of views and expressions paralleled in his writings, also consists of material derived from the same author?

If we accept this conclusion, as I for my part am inclined to, we are not obliged to think that all views and distinctions presented in the chapter on causes are Galen's own. That could hardly be correct. First, these views and distinctions are not always consistent with one another; second, not all of them correspond to Galen's attested opinions.[127] Perhaps much of the material goes back to quotations and paraphrases of other sources, separated from the context in which they were quoted by the excerpted author. Thus, for example, many pages of Galen's treatise *On the Doctrines of Hippocrates and Plato* (*PHP*) are filled with quotations and paraphrases from other sources, notably Chysippus and Posidonius, and we can easily imagine an excerptor of that work who is interested in the reported views as much as in the context of their transmission.

It remains to consider the two brief sections dealing, each from a different perspective, with the sceptic notion of ἐποχή (δ and ζ). Like the rest of the material, the sections are written by a well-informed and sharp-minded writer, whose concerns are chiefly philosophical. And, like the rest of the material, they discuss topics relevant to the theory of knowledge. But there is little in these texts pointing towards Galen as a source. It is true that Galen was concerned with scepticism and it is likely that he dealt with it in *DD* in some way. So it is conceivable that section δ reflects his own polemic against the requirement of universal ἐποχή that makes demonstration impossible.[128] Section ζ, which describes ἐποχή from the perspective of a sceptic, could be a quotation from a different author, extracted by Clement from this polemic. The Galenic trace could be mildly supported by the partial concession to the usefulness of ἐποχή

127 Cf. esp. the view of the 'self-complete' cause in *Strom.* VIII (9) 25, 3 and 33, 2, which seems to be in conflict not only with Galen's views, but also with other sections of the chapter; cf. below, commentary on 98,5–9 and 101,19 f.

128 Cf. esp. Galen's spirited polemic against the sceptic Favorinus, called *On the Best Kind of Teaching* (CMG V,1,1, ed. A. Barigazzi, pp. 89–109); Galen mentions it in a list of books written after *DD* that deal, in a more compendious manner, with topics discussed in that work (*Lib. Prop.* 14, 17 and 21: 167,14–16; 168,10 f. Boudon Millot/XIX,43–44 K.). Like Clement, Galen was concerned about διαφωνία among (medical) schools, citing this concern as a reason for his interest in demonstration: "As I said in my writing *On the Demonstrative Discovery*, having become completely inundated by the mass of disagreement among doctors (ὑπὸ τοῦ πλήθους τῆς τῶν ἰατρῶν διαφωνίας), and having then set on the course of making judgement about it, I realized that first of all I must train myself in demonstrative methods. And having done this for many years since, I have subjected every doctrine to this [method] accordingly" (*MM* VII 5/X,469,14–18 K.). For the anti-sceptic motivation of *DD*, cf. my "The Purpose of Galen's Treatise *On Demonstration*," *Early Science and Medicine* 20 (2015), pp. 265–287.

at the end of the polemical passage, which does correspond to Galen's attitude.[129] But all this is conjectural: there are few similarities in points of detail between these sections and the writings of Galen and the safest course in this case at least is to suspend judgement.

Thus there are grounds for extending the Galen hypothesis, originally proposed in light of sections β and γ, to subsequent parts of the 'eighth book', or most of them, as well. The hypothesis does not say whether Clement had access to Galen's original work, or whether he used an abridged version made by somebody else. Nor does it attempt to localize the excerpted material within the source. These and other questions must be further explored, particularly against the background of the fragments and testimonies of Galen's *On Demonstration*. And of course, the hypothesis itself has to be further scrutinized. Whatever the worth of the present inquiry, its conclusions are provisional and they will probably remain so unless an unexpected stroke of luck helps us to unveil the mystery.

f *Continuity with the* Stromateis?

The focus on Clement's sources, justified by the predominantly philosophical character of the material, has kept us away from the problem of the purpose of the 'eighth book' and its place within the framework of Clement's *œuvre*. But this problem, if it can be settled at all, cannot be settled unless we ask the following question: Is there, as the title suggests, continuity between the 'eighth book' and the previous books of the *Stromateis*? As we know, many scholars (Reinkens, Zahn, and Nautin in particular) have answered this question affirmatively, citing Clement's plans for the continuation of the *Stromateis*, outlined on several occasions in that work.[130] Thus, to make a judgement about the question of continuity, we must review these plans and see whether and how they correspond to the contents of the 'eighth book'.

The first version of Clement's plan is drafted as early as *Strom.* I (1) 15, 2. In this passage, Clement makes a distinction between two parts of his literary project: a preparatory and 'cleansing' part (whose propaedeutic function is characterized by several verbs with the prefix προ-) and a part described as a progression of knowledge, whose summit is called ἐποπτική or φυσικὴ θεωρία.[131]

---

129 Cf. below, commentary on 90,4f.

130 *Strom.* I (1) 15, 2; II (1) 1, 1–2, 3; IV (1) 1, 1–3, 3; VI (1) 1, 1.4; VII (15) 89, 1; VII (18) 110, 4. An annotated list of these passages is provided by Nautin, "La fin," pp. 282–291.

131 *Strom.* I (1) 15, 2: τά τε παρὰ τῶν ἐπισήμων δογματιζόμενα αἱρέσεων παραθήσεται, καὶ τούτοις ἀντερεῖ πάνθ' ὅσα *προοικονομηθῆναι* καθήκει τῆς κατὰ τὴν ἐποπτικὴν θεωρίαν γνώσεως, ἣ προβήσεται ἡμῖν κατὰ τὸν εὐκλεῆ καὶ σεμνὸν τῆς παραδόσεως κανόνα ἀπὸ τῆς τοῦ κόσμου γενέ-

We learn that the preparatory part will include the presentation of views belonging to the most significant αἱρέσεις (a word that might refer to Christian 'heresies', but also to philosophical schools)[132] and a polemic against these views; the second part, in turn, will proceed "from the creation of the world", i.e. presumably from the *Book of Genesis* by way of its interpretation, "according to the glorious and holy rule of our tradition," as Clement puts it. It is not clear if the contents of the extant *Stromateis* can be associated with anything mentioned in this passage. It is true that much of these contents is polemical and one might think that they correspond to the 'cleansing' phase of the project. But the sequence presentation/refutation is not observed anywhere in the *Stromateis* and the announcement Clements makes at the beginning of book II does not suggest that he was planning to proceed in this way.[133] By contrast, the description of the preparatory phase reminds us of Clement's promise, made in book IV, to present and refute the views of the Greeks and the "other" (i.e. non-Christian) "barbarians" on the principles.[134] But, as we shall see below, this promise refers to a later stage of Clement's project and not to the topics covered in the *Stromateis*. Did Clement, before turning to book II, decide to postpone the plan drafted in book I? Or was the plan never meant to refer to the text that immediately followed?

In any case, at the beginning of book II, Clement draws a different picture. He starts by revealing his aim to show that the Greeks have "stolen" various doctrines from the "barbarian philosophy", i.e. from Scripture. He further specifies that he is going to speak about "virtues belonging to the truth" (such as

---

σεως προϊοῦσιν, ⟨τὰ⟩ ἀναγκαίως ἔχοντα *προδιαληφθῆναι* τῆς φυσικῆς θεωρίας *προπαρατιθεμένη* καὶ τὰ ἐμποδὼν ἱστάμενα τῇ ἀκολουθίᾳ *προαπολυομένη*, ὡς ἑτοίμους ἔχειν τὰς ἀκοὰς πρὸς τὴν παραδοχὴν τῆς γνωστικῆς παραδόσεως *προκεκαθαρμένης* τῆς γῆς ἀπό τε τῶν ἀκανθῶν καὶ τῆς πόας ἁπάσης γεωργικῶς εἰς καταφύτευσιν ἀμπελῶνος. ("Also, the views of the most significant schools will be presented, and these will be opposed by all the things we should deal with before [arriving at] knowledge according to the science of the initiated, [knowledge] progressing as we proceed, according to the glorious and holy rule of our tradition, from the creation of the world, after it [i.e. knowledge] has presented things that must be handled before [the arrival at] the science of nature and after it has removed the obstacles standing in the way of consequentiality, so that, as the ground is cleared, as by a good farmer, from thorns and every kind of weed for the plantation of a vineyard, the ears are ready to receive the tradition of knowledge"). For the notion of 'consequentiality' (ἀκολουθία) cf. Laura Rizzerio, "La nozione di ἀκολουθία come logica della verità in Clemente di Alessandria," *Rivista di Filosofia Neo-Scolastica* 79 (1987), pp. 175–195.

132 Cf. Nautin, "La fin," p. 282 n. 39.

133 See *Strom.* II (1) 1, 1–2, 3, discussed below.

134 Cf. *Strom.* IV (1) 2, 1.

faith, wisdom, knowledge, science, hope, love, repentance, self-control, and fear) and "whatever the outline in the part before us will demand", including a discussion of the symbolic genre.[135] This passage announces topics that can be discerned in the subsequent *Stromateis* up to and including book V (where the symbolic genre is discussed). When speaking of "the part before us" (ὁ προκείμενος τόπος), Clement probably indicates that these topics somehow belong to the same "place" or part of his project. More specifically, he seems to suggest that the second book marks the beginning of a section he later refers to as the "ethical part" (ἠθικὸς τόπος *vel* λόγος), a section which, as it turns out at the end of the seventh book, will eventually cover all the remaining *Stromateis*.[136] Apart from these topics, other things are promised to follow: First, a defence of the Christians regarding some issues the Greeks attack them for, presented in such a way as to convince the Jews as well. Second, a polemic against philosophers concerning their way of life and doctrines they claim to have discovered. Third, a discussion of the encyclic arts and various 'sciences' the Greeks boast of, like astrology, mathematics, and magic.[137] Nothing in the extant *Stromateis* corresponds precisely to these three topics. Whether or not Clement adhered to his promise to deal with them all, only the first is mentioned in the next version of his plan.

This version is found at the beginning of the fourth book and it is the most extensive account of Clement's project.[138] It starts by (correctly) describing the contents of the fourth and the fifth books and, like in the previous version, announces the end of the 'ethical part' after a discussion of the symbolic genre. And, like in the previous version, the next topic to follow is a debate with 'the Greeks and the Jews'. This time, Clement does not speak of a defence against their attacks; rather, he says, the debate will take the form of a brief exposition of Scripture (ἥ τε πρὸς τοὺς Ἕλληνας καὶ ἡ πρὸς τοὺς Ἰουδαίους κατ' ἐπιτομὴν τῶν γραφῶν ἔκθεσις). Subsequently, he vows that the items he had promised to deal with "in the prooemium" of the *Stromateis* will also be discussed.[139] This is presumably a reference to the very beginning of the treatise, which is missing

135 *Strom.* II (1) 1, 1–2.

136 Cf. *Strom.* IV (1) 1, 2; VI (1), 1, 1; VII (18) 110, 4. Cf. also *Strom.* II (18) 78, 1, where Clement says that for the Greeks, virtues provide a starting-point to "the whole of the ethical part" (παντὸς τοῦ ἠθικοῦ τόπου), i.e. to the ethical division of philosophy.

137 *Strom.* II (1) 2, 1–4.

138 *Strom.* IV (1) 1, 1–3, 3.

139 *Strom.* IV (1) 1, 3: ὅσα ἐν τοῖς πρὸ τούτου Στρωματεῦσι *κατὰ τὴν τοῦ προοιμίου εἰσβολὴν* ἐν ἑνὶ προθεμένοις τελειώσειν ὑπομνήματι τῷ πλήθει τῶν πραγμάτων ἀναγκαίως δουλεύσασι περιλαβεῖν οὐκ ἐξεγένετο.

from our manuscript. Clement seems to recall a plan that had already been drafted in that lost beginning and assures the reader that he still feels obliged to it. At the same time, he appears to suggest that when he is done with these topics, the original plan will be fulfilled.[140] Nevertheless, his project carries on, starting with an exposition of the Greek and other non-Christian views on the first principles and leading towards γνωστικὴ φυσιολογία, the 'gnostic science of nature', which culminates in theology.[141] This seems to echo the project outlined in *Strom.* I (1) 15, 2.

Did Clement conceive of this continuation as part of a plan for the *Stromateis*, or had he contemplated some other treatise? We can hardly be sure. At one point in *QDS*, Clement refers to a finished piece of writing of his called περὶ ἀρχῶν καὶ θεολογίας, so we may assume that he fulfilled his plans there.[142] At any rate, this interesting question has no bearing on the problem of *Stromateis* VIII. For if Clement suceeded in pursuing his plan beyond the seventh book, then the eighth book must have dealt with topics that were to follow immediately after the end of the 'ethical part', starting with a defence of Christianity against its adversaries.

This is confirmed by two other passages where Clement restates his interest to meet the objections raised against the Christian faith by the Greeks and the Jews. In the introduction to book VI, Clement expands the agenda of the 'ethical part' by stating that it will reach its conclusion in books VI and VII while at the same time proposing items not mentioned in the previous plan as topics of these books: the life of the Christian 'gnostic' and his defence against the charge of atheism.[143] Afterwards, Clement adds, "as the treatise proceeds according to the character of the *Stromateis*, other puzzles proposed to us by the Greeks and the barbarians should be solved, regarding the advent of the Lord."[144] Again, the

140 *Strom.* IV (1) 2, 1: ἐπὶ τούτοις ὕστερον πληρωθείσης ὡς ἔνι μάλιστα τῆς κατὰ τὰ προκείμενα ἡμῖν ὑποτυπώσεως, κτλ.

141 *Strom.* IV (1) 3, 1–3.

142 *QDS* 26, 8; cf. Nautin, "La fin," pp. 292f. For the question of the continuation of the *Stromateis*, see now an intriguing hypothesis of Marco Rizzi, "The End of *Stromateis* VII and Clement's Literary Project," in *The Seventh Book of the Stromateis*, ed. M. Havrda, V. Hušek, and J. Plátová (Suppl. Vig. Chr. 117; Leiden/Boston: Brill, 2012), pp. 299–311. Rizzi agrees with those who believe that Clement fulfilled his plans in the *Hypotyposeis* (consisting, according to Eusebius, of seven books), but argues that the *Hypotyposeis* was simply a different name for the books of the *Stromateis* which followed the seventh book. Rizzi suggests that the work, consisting of fifteen books of the *Stromateis*, was divided in consequence of its transference from scrolls to (two) codices.

143 *Strom.* VI (1) 1, 1.

144 *Strom.* VI (1) 1, 4: τότε ἤδη, προϊόντων τῶν ὑπομνημάτων κατὰ τὸν τῶν Στρωματέων χαρακτῆρα,

first topic on the list after the conclusion of the 'ethical part' is a defence of the Christian faith. Clement describes the objections of his opponents as "puzzles" (ἀπορίαι) and promises to "solve" them (ἐπιλύειν), mentioning "the advent of the Lord" as their subject matter. No doubt the same plan is recalled towards the end of the seventh book, at a point when the main theme of the book, the piety of the perfect Christian (the 'gnostic'), has been delivered. "The following task," Clement says, "is to make a defence against the accusations brought against us by the Greeks and the Jews."[145] In the remaining space of the seventh book, he takes up one of these accusations (the objection that disagreement among schools of Christian doctrine undermines their claim for the truth), so that, "as we proceed to the next *Stromateus*, having cleared the obstacles away, we are ready for the solutions of the puzzles (εὐτρεπεῖς ἐπὶ τὰς τῶν ἀποριῶν λύσεις)."[146] This confirms that the next book was intended to deal with these puzzles in such a way as to solve them.

Can we find any trace of this programme—or any part of the programme planned for the continuation of the *Stromateis*—in our 'eighth book'? Nautin thinks we can. He recognizes one such trace at the beginning of the first chapter, where Clement draws a contrast between "the most ancient philosophers" and their "more recent" counterparts on the subject of their attitude to disputes: Whereas the more recent philosophers "have been carried away by vain and endless ambition and reduced to useless foolery by refutations and contentious arguments", the most ancient "were not lead towards disputing and puzzling over things" (οὐδέ ... ἐπὶ τὸ ἀμφισβητεῖν καὶ ἀπορεῖν ἐφέροντο).[147] Recalling the fact that the eighth book was supposed to deal with "puzzles" (ἀπορίαι), Nautin sees a connection between these lines and the plan for the eighth book. The connection, as he sees it, is this: In the passage just mentioned, Clement addresses the same opponents whose objections he had planned to confront in the eighth book. He starts by criticising them as "sophists" and "sceptics", i.e. as people with an appetite for controversy who pretend to know nothing. It is against these opponents, Nautin argues, that the polemic against the sceptics in section ζ (15, 2–16, 1) is also directed.[148]

ἐπιλυτέον τά τε ὑπὸ Ἑλλήνων τά τε ὑπὸ βαρβάρων προσαπορούμενα ἡμῖν περὶ τῆς τοῦ κυρίου παρουσίας.

145 *Strom.* VII (15) 89, 1: ... ἀκόλουθόν ἐστι πρὸς τὰ ὑπὸ Ἑλλήνων καὶ Ἰουδαίων ἐπιφερόμενα ἡμῖν ἐγκλήματα ἀπολογήσασθαι κτλ.

146 Ibid.

147 *Strom.* VIII (1) 1, 1–2.

148 Nautin, "La fin," p. 291. Scholars favouring the continuity view often suggest that the overall agenda of the 'eighth book' is anti-sceptic, and propose this as a connecting link

Is it plausible that Clement would start the ἀπορίαι-and-λύσεις section with an attack on those who raised these ἀπορίαι, branding them as sophists and sceptics? Perhaps it is. But the point of our first chapter seems to be different. Clement does not defend the Christians against attacks of any kind, but rather points out to the Christians why and how they should search for the truth. The "more recent philosophers" (by whom Clement *could* mean the sceptics) are not invoked as critics of Christian beliefs, but rather as people who do not believe that truth can be found at all. In contrast, Clement argues, the Christians are obliged by Scripture to "seek in order to find"; in other words, they cannot give up looking for the truth. Clement even outlines a method of inquiry suited to this purpose. All this could conceivably serve as an introduction to a polemic against specific charges (like the one concerning the advent of the Lord), an introduction showing how to proceed in answering them. But the first chapter gives no indication of that: Christian inquiry is portrayed as a way towards "apprehensive knowledge", not as a way of defending the truth of Christian beliefs.

Nevertheless, as far as the first chapter goes, Nautin's suggestion of its relevance to the expected contents of the eighth book cannot be ruled out. But the task of extending the same argument to the remaining parts of the text is difficult. For there is simply no connection between these parts and Clement's declared aim. Nautin thinks that the book must have contained arguments corresponding to Clement's promise, but that some excerptor left them out. But (aside from the anti-sceptic polemic) he does not even attempt to show how the rest of the material fits into the picture. If we follow Nautin's explanation, we have to believe that all the discussion about demonstration, the method of inquiry, division and definition, the categories, and the chapters about ἐποχή were originally embedded in the context of Clement's attempt to answer the ἀπορίαι raised against the Christians and that they were later cut out of this context so carefully as to leave no trace of the original bedding. But it is hard to believe such things without a single piece of evidence. Rather than imag-

---

with Clement's polemical interests; cf. Laura Rizzerio, "Foi, Gnosis, Dialectique, Logique: notes à propos de *Stromates* VIII de Clément d'Alexandrie," in *Studia Patristica* 31, ed. E.A. Livingstone (Leuven: Peeters, 1997), pp. 522–529; Marcelo Merino Rodríguez, "El *Stromata* VIII de Clemente de Alejandría," *Scripta Theologica* 37/1 (2005), pp. 13–51. Cf. also George Karamanolis, *The Philosophy of Early Christianity* (Durham: Acumen, 2013), pp. 125–129. But it is far from clear that "the Greeks and the Jews", whom Clement had planned to address in the next *Stromateus* were sceptics. The fact that they use an argument from disagreement against the Christians says little about their own views. Celsus raises the same objection (cf. Origen, *C. Cels.* V 61–65) and he is no sceptic.

ining an excerptor who (with the exception of the first chapter) excises everything Christian, it would be safer to suppose that, in the course of writing the eighth book, Clement got carried away by philosophical methodology to such an extent that he forgot about his apologetic agenda.

This is of course possible. Clement could have started the text with the intention of providing methodological guidelines for Christian thinkers, perhaps in view of the task of solving difficult questions concerning the Christian faith. Then, in the course of the exposition, he might have become immersed in matters such as the theory of demonstration and scientific inquiry to the extent that the original intention drifted out of sight. But if this is what happened, it seems very unlikely that Clement ever declared the result to be another book of the *Stromateis*. No other instance of Clement's losing control of his writing to this degree can be found in his *œuvre*. If we believe that the contents of the 'eighth book' somehow correspond to Clement's plans for the continuation of the *Stromateis*, we must surely subscribe to some version of the 'standard view' and conclude that the text was written at most *in preparation for* this planned continuation.

g *Parallels with the* Stromateis

Before we decide on that issue, however, we must consider another phenomenon that has played an important part in the debate about the place and purpose of the 'eighth book'. It has long been recognized that its text contains several passages whose parallels are found in the regular books of the *Stromateis*. Reinkens mentions these parallels in support of the continuity view,[149] but it might seem odd to suppose that Clement would use (or plan to use) the same formulations in different sections of the same work. Do these parallels not suggest something else instead, namely that the 'eighth book' was written *before* the *Stromateis*, perhaps as a source book used by Clement for the composition of this work? Such, as we know, was the opinion of von Wedel, developed in particular by Ernst, but disputed by Nautin. However, none of these scholars has analysed these parallels in sufficient detail to give us a reliable grounds for judgement.

The most striking overlaps with the material of the 'eighth book' appear in the first, the second, and the seventh book of the *Stromateis*. Similarities have also been detected between the discussion about inquiry in the Christian 'prooemium' and a passage about the same topic in the fifth book. In what follows, we are going to inspect these parallels in turn.

---

149 Cf. above, p. 13.

Case 1: *Strom.* I (17) 82, 3–5

This passage is part of an exegesis of John 10:8 ("All those before the Lord's coming are thieves and robbers"); the verse is quoted in the context of a discussion about the origin and value of Greek philosophy.[150] Some people, Clement reports, think that philosophy was instigated by the devil or inspired by some inferior powers.[151] However, although Clement agrees that it was not bestowed on humankind directly by God, he wishes to avoid the conclusion that there is nothing good about philosophy at all. The motif of theft in the Johannine verse, which Clement assumes to be hinting at the origin of philosophy, enables him to draw a distinction between philosophy on the one hand and biblical prophecy on the other (prophets were no thiefs, but servants of God), while retaining the idea that Greek philosophy contains at least a portion of the truth: "Philosophy was not sent out by the Lord, but came, as Scripture says, as something stolen or given by a thief. Some power or angel learned a portion of the truth, but did not remain within the truth, and stole these things and taught them to human beings by way of inspiration."[152] But if God did not wish to bestow philosophy on humankind, why did he allow others to do so? Clement, of course, rejects the possibility that God, who "knows the outcome of the future from before the foundation of the world", had not known that this was going to happen. But, he explains, God did not prevent this "audacious transgression" from happening, as he knew that it could be set straight in a useful manner by providence.[153]

At this point of the argument, Clement embarks on a digression concerning the problem of whether a failure to prevent an action entails responsibility. This digression, which clearly draws on philosophical sources (von Arnim designates chunks of it as SVF II 353), also contains a couple of sentences closely paralleled in the 'eighth book'. Here is the passage in its context, with the relevant lines in italics:

> I know that many people are relentlessly pressing on us and saying that not to prevent something is to be responsible for it. They say that a person who does not keep proper guard to prevent a theft is reponsible for that theft, just as much as a person who does not put out the beginning of a dangerous fire is responsible for the fire, or a pilot who has not furled the sail is responsible for the shipwreck. Further, those responsible in

150 The exegesis extends from *Strom.* I (17) 81, 1 to 84, 7.

151 *Strom.* I (17) 80, 5.

152 *Strom.* I (17) 81, 1–5.

153 *Strom.* I (17) 81, 5.

these cases are punishable by law. Responsibility for an event attaches to anyone who had the power to prevent it.

We retort that *a cause is conceived as that which makes, acts, and performs; to the extent that something fails to prevent, it does not act*. Furthermore, a cause is a cause with respect to its activity, as the shipbuilder is a cause with respect to ship's coming into existence or the housebuilder with respect to the house's being built. But *that which does not prevent is separate from that which comes about, for this is accomplished precisely because something capable of preventing it has failed to act and prevent it. What action does one perfom when failing to prevent*?

Their argument is now passing into absurdity if they really charge that the wound is caused, not by the weapon, but by the failure of the shield to impede the passage of the weapon. That is to exempt the thief from criticism and pile the blame on the person who fails to prevent the theft. We might as well say that it was not Hector who burned the Greek ships but Achilles by his failure to prevent Hector when he could have done so. In point of fact, Achilles was perhaps partly responsible for his failure to put out the fire because of his anger (it being in his power, whether or not to succumb to anger). But the devil is free to choose and he was capable both of repenting and of stealing, and thus he is himself responsible for the theft, not the Lord who did not prevent it.[154]

A synopsis of the parallels brings out their proximity:

***Strom.* I (17) 82, 3–5**	***Strom.* VIII (9) 27, 6–28, 1**
φαμὲν δὴ πρὸς αὐτοὺς (a) *τὸ αἴτιον ἐν τῷ ποιεῖν καὶ ἐνεργεῖν καὶ δρᾶν νοεῖσθαι*, (b) *τὸ δὲ μὴ κωλῦον κατά γε τοῦτο ἀνενέργητον εἶναι.*	(b) *τὸ μὴ κωλῦον ἀνενέργητόν ἐστιν*, διὸ οὐκ ἔστιν αἴτιον τὸ μὴ κωλῦον, ἀλλὰ τὸ κωλῦον. (a) *ἐν γὰρ τῷ ἐνεργεῖν καὶ δρᾶν τι τὸ αἴτιον νοεῖται.*
ἔτι (c) *τὸ μὲν αἴτιον πρὸς τῇ ἐνεργείᾳ ἐστί*, καθάπερ ὁ μὲν ναυπηγὸς πρὸς τῷ γίγνεσθαι τὸ σκάφος, ὁ δὲ οἰκοδόμος πρὸς τῷ κτίζεσθαι τὴν οἰκίαν·	

154 *Strom.* I (17) 82, 1–83, 2; in the first and the third paragraphs I have adopted Ferguson's translation.

(d) *τὸ δὲ μὴ κωλῦον κεχώρισται τοῦ γινομένου. διὰ τοῦτο γοῦν ἐπιτελεῖται, ὅτι τὸ κωλῦσαι δυνάμενον* οὐκ ἐνεργεῖ οὐδὲ κωλύει. τί γὰρ ἐνεργεῖ ὁ μὴ κωλύων;	ἔτι (d) *τὸ μὴ κωλῦον κεχώρισται τοῦ γινομένου (διὰ τοῦτο γοῦν ἐπιτελεῖται, ὅτι τὸ δυνάμενον κωλύειν οὐ* πάρεστιν), (c) *τὸ δὲ αἴτιον πρὸς τὸ γινόμενον·* οὐκ ἂν οὖν εἴη αἴτιον τὸ μὴ κωλῦον.

How shall this proximity be explained? There seems to be no more than three options: Either *Strom.* VIII draws on *Strom.* I, or *vice versa*, or both passages derive from a common source. Now, the first option is very unlikely: The parallels concern a topic adapted to the point at issue in the first book from a philosophical source; and, since the last chapter of the 'eighth book', in which the parallel appears, discusses aspects of causation from a philosophical point of view, we may surmise that the 'eighth book' reflects the original context of the passage more faithfully than the debate about God, angels, and the devil in book I. But there are also difficulties with the opposite view, namely, that the first book draws on 'book VIII': The parallel in the first book is accompanied by details that seem to stem from the same philosophical source (the examples of the shipbuilder and the housebuilder, of Hector and Achilles), but these details are *not* found in the 'eighth book'. Shall we conclude that Clement drew from another source on both occasions, while preserving some of its features more fully in the first book? Let us see what the other cases tell us.

Case 2: *Strom.* I (20) 97,1–3

This passage, following shortly after the first one, continues to deal with Greek philosophy and its usefulness from the Christian perspective. Philosophy, Clement argues, is a search for truth, and thus contributes to the apprehension of truth. When explaining the nature of its contribution, Clement takes recourse to an analogy illustrating the notion of a joint-cause (συναίτιον), a quasi-causal factor which does not bring about an effect on its own, but only when joined with others:

> When many men draw a ship to sea, we do not say that they constitute many causes. There is one cause consisting of many factors. None of them is the cause of the ship's being drawn to sea, except with the others. In the same way, philosophy, being the search for truth, contributes to the apprehension of truth, but it is not the cause of the apprehension. It is a cause in connection with others, an auxiliary one.[155]

155 *Strom.* I (20) 97, 1.

But Clement acknowledges that a joint-cause could also be described as a cause. Moreover, he compares various factors contributing to the search for truth to multiple causes of one effect, which seem to exercise their force independently of one another:

> Further, the causes of being happy, which is one thing, are virtues, which are many. And warming up is caused by the sun, fire, bath, clothes. In the same way, truth is one and there are many factors contributing to its search, although discovery is mediated by the Son.[156]

Picking up on the point about the Son, i.e. Christ in his role as teacher, Clement further explains:

> Consider this: virtue is potentially one, but when it appears in one sort of circumstances, it happens to be called prudence, when in others, moderation, when in these, courage or justice. So by analogy, truth is one, but in geometry it is a truth of geometry, in music of music, and in the correct philosophy there is presumably a truth of the Greeks. But truth in the strict sense which cannot be tampered with is only the one in which we are educated by the Son of God.[157]

Thus the relation between Greek philosophy and Christian truth is as follows: If practised correctly, Greek philosophy does pursue the truth, but it is a partial truth, not truth in the strict sense, which is only mediated through Christ. Clement suggests that this relation is similar to the relation between a particular virtue and that which provides unity to all the multiple virtues. This unity, he further contends, is not achieved by the exercise of one particular virtue or another, but only by their joint use (σύγχρησις): even though each virtue has its own effect (τὸ καθ' ἑαυτὴν ἀποτέλεσμα), when joined together all virtues become the causes of one effect, namely happiness:

> It is in this way that when one and the same coin is given to a sea-captain we speak of money to pay for passage, when it is given to a tax-collector, we speak of a tax, when to the property owner, of a rent, when to the teacher, of a tuition, and when to the salesperson, of a deposit. Each virtue or truth, though they are called by the same name, brings about only

156 *Strom.* I (20) 97, 2.
157 *Strom.* I (20) 97, 3–4.

> the effect corresponding to it. But when they are jointly used, happy life results (surely are we not happy on account of names), if we apply the word 'happiness' to the right way of life and the word 'happy' to the person who has ordered his soul in a virtuous manner.[158]

The moral of the analogy is clear and well-known to the reader of the *Stromateis*: Greek philosophy is a legitimate but insufficient attempt to grasp the truth. It has to be set straight by the teaching of Christ, who is the Truth, thus providing access to everything that the best schools of Greek philosophy have ever aspired to.[159] However, more interesting for our purpose is the analogy itself: not only because it clearly draws on philosophical discussion about the unity of virtues (again, von Arnim marks parts of this passage as a Stoic fragment),[160] but mainly because of the following parallels with the 'eighth book':

Strom. **I (20) 97, 1–3**	***Strom.*** **VIII (9) 31, 1–3**
ὡς δὲ οἱ πολλοὶ *ἄνθρωποι* οἱ καθέλκοντες τὴν ναῦν οὐ πολλὰ αἴτια λέγοιντ' ἄν, ἀλλ' ἐκ πολλῶν αἴτιον ἕν (*οὐκ ἔστι γὰρ αἴτιος ἕκαστος τοῦ καθέλκεσθαι τὴν ναῦν, ἀλλὰ σὺν τοῖς ἄλλοις*), οὕτω καὶ ἡ φιλοσοφία πρὸς κατάληψιν τῆς ἀληθείας, ζήτησις οὖσα ἀληθείας, συλλαμβάνεται, οὐκ αἰτία οὖσα καταλήψεως, σὺν δὲ τοῖς ἄλλοις αἰτία καὶ συνεργός. *τάχα δὲ καὶ τὸ συναίτιον αἴτιον.*	ἔτι ζητεῖται εἰ πολλὰ κατὰ σύνοδον ἑνὸς αἴτια γίνεται πολλά. οὐ [L: οἱ] *γὰρ ἄνθρωποι συνελθόντες αἴτιοί εἰσι τοῦ καθέλκεσθαι τὴν ναῦν, ἀλλὰ σὺν τοῖς ἄλλοις, εἰ μή τι καὶ τὸ συναίτιον αἴτιον.*
ὡς δέ, ἑνὸς ὄντος *τοῦ εὐδαιμονεῖν, αἴτιαι τυγχάνουσιν αἱ ἀρεταὶ πλείονες ὑπάρχουσαι, καὶ ὡς τοῦ θερμαίνεσθαι* ὅ τε ἥλιος τό τε πῦρ βαλανεῖόν τε καὶ ἐσθής, οὕτω μιᾶς οὔσης τῆς ἀληθείας πολλὰ τὰ συλλαμβανόμενα πρὸς ζήτησιν αὐτῆς, ἡ δὲ εὕρεσις δι' υἱοῦ.	ἄλλοι δέ φασιν, εἰ πολλὰ αἴτια, κατ' ἰδίαν ἕκαστον ἑνὸς αἴτιον γίνεται. *τοῦ γοῦν εὐδαιμονεῖν ἑνὸς ὄντος αἴτιαι τυγχάνουσιν αἱ ἀρεταὶ πολλαὶ οὖσαι, καὶ τοῦ θερμαίνεσθαι* καὶ τοῦ ἀλγεῖν ὁμοίως πολλὰ τὰ αἴτια.

158 *Strom.* I (20) 98, 1–2.

159 Cf. e.g. *Strom.* I (5) 32, 4; VII (16) 94, 5; Alain Le Boulluec, "La rencontre de l'hellénisme et de la «philosophie barbare» selon Clément d'Alexandrie," in idem, *Alexandrie antique et chrétienne: Clément et Origène* (Paris: Institut d'Études Augustiniennes, 2006), pp. 81–93, with references.

160 Cf. SVF I, p. 86, Aristo, fr. 376.

(cont.)

εἰ γοῦν σκοποῖμεν, *μία κατὰ δύναμίν ἐστιν ἡ ἀρετή*, ταύτην δὲ συμβέβηκεν τούτοις μὲν τοῖς πράγμασιν ἐγγενομένην λέγεσθαι φρόνησιν, ἐν τούτοις δὲ σωφροσύνην, ἐν τούτοις δὲ ἀνδρείαν ἢ δικαιοσύνην.	μή τι οὖν αἱ πολλαὶ ἀρεταὶ *μία ἐστὶ κατὰ δύναμιν* [*καὶ τὰ θερμαίνοντα καὶ τὰ ἀλγύνοντα*]. καὶ τὸ πλῆθος δὲ τῶν ἀρετῶν κατὰ γένος ἓν τυγχάνον ἑνὸς αἴτιον γίνεται τοῦ εὐδαιμονεῖν.

Again, in the 'eighth book', the overlapping sentences are found in the chapter on causes. Clement reports of a controversy regarding multiple causation, specifically the question whether many factors coming together to produce one effect may be individually described as *causes* of that effect. The example of men drawing a ship to sea is used in support of the view that they cannot, unless a joint-cause is also called a 'cause'. But others, Clement says, bring up the example of virtues, which are also multiple causes of one effect (happiness), the same being true of the causes of warming up or of pain. As for the virtues, the text adduces that they are potentially one, and that the plurality of virtues, insofar as it is one by genus, is the cause of one effect.

Comparing the two passages yields similar results as in the previous case: The 'eighth book' likely preserves the original context of the discussion more faithfully than book I. However, in the first book, Clement's account of the analogy of joint-causes and virtues, which clearly draws on a philosophical source, contains several details not found in 'book eight' (examples of the causes of warmth and several aspects of the theory of virtues); these details are integrated in the account so well as to render the impression of belonging to that source, too.

Case 3: *Strom.* I (20) 99, 1–100, 1

This is a continuation of the preceding text. Further articulating his thoughts on the value of Greek philosophy, Clement freely adapts a classification of causes drafted—on more than one occasion—in the 'eighth book':

> Captiousness [of our opponents] may require us to go on making distinctions. In saying that philosophy is a *joint-cause* and *auxiliary* to the apprehension of truth, because it is a search for truth, we shall be accepting it as a kind of preparatory education for the Christian gnostic. We do not mistake a joint-cause for a cause, nor an auxiliary cause for a *synectic* cause, nor do we think that philosophy is a *prerequisite*. Almost none of us have gone through elementary education (ἐγκυκλίος παιδεία) or Greek

> philosophy; but we have been inspired by the divine philosophy of the barbarians to receive 'in power' (cf. 1 Thess 1:5), through faith, a teaching (λόγος) about God: Wisdom has educated us by its own means. Anything which operates alongside something else without being capable of acting effectively on its own is, we maintain, auxiliary and a joint-cause, since it is a cause only when joined with a cause or it receives the name of a cause only when it has accrued to something else; on its own, however, it is incapable of producing a true result. Even though in the past philosophy might have brought the Greeks to righteousness on its own accord, it was not a universal righteousness (…). Nor does it mean that by the removal of philosophy the universal teaching (λόγος) would lack anything or that truth would be destroyed. (…) Thus the teaching (διδασκαλία) according to Christ is *self-complete*, not needing anything else, being a power and wisdom of God (cf. 1 Cor 1:24), and when philosophy is added, it does not make the truth more powerful.

Apart from terminology (all the italicized terms are used and explained in the 'eighth book'),[161] the following parallels are particularly noteworthy:

Strom. **I (20) 99, 2**	***Strom.*** **VIII (9) 33, 7–9**
ὃ δὲ (a) *μεθ' ἑτέρου ποιεῖ, ἀτελὲς ὂν καθ' αὑτὸ* ἐνεργεῖν, *συνεργόν* φαμεν καὶ (b) *συναίτιον ἀπὸ τοῦ σὺν αἰτίῳ αἴτιον ὑπάρχειν* ⟨ἢ⟩ (c) *ἀπὸ τοῦ ἑτέρῳ συνελθὸν αἴτιον* γίγνεσθαι ὠνομασμένον, *καθ'* ἑαυτὸ δὲ μὴ δύνασθαι *τὸ ἀποτέλεσμα* τὸ κατ' ἀλήθειαν *παρέχειν*.	νοεῖται γὰρ (b) *σὺν ἑτέρῳ τὸ συναίτιον* οὐδ' αὐτῷ δυναμένῳ *κατ'* ἰδίαν ποιῆσαι τὸ ἀποτέλεσμα, *αἴτιον ὂν σὺν αἰτίῳ*. διαφέρει δὲ τοῦ συναιτίου (a, c) *τὸ συνεργὸν* ἐν τῷ τὸ μὲν συναίτιον ⟨*μεθ' ἑτέρου*⟩ *κατ' ἰδίαν μὴ ποιοῦντος τὸ ἀποτέλεσμα παρέχειν*, τὸ δὲ (a, c) *συνεργὸν* [ἐν τῷ] *κατ' ἰδίαν μὴ ποιεῖν*, *ἑτέρῳ δὲ προσερχόμενον* τῷ κατ' ἰδίαν ποιοῦντι συνεργεῖ⟨ν⟩ αὐτῷ πρὸς τὸ σφοδρότερον γίνεσθαι τὸ ἀποτέλεσμα.

However, in the first book, Clement is less rigorous about causal terminology than in *Strom.* VIII. For example, in the passage just quoted, he does not make a clear distinction between the auxiliary cause and the joint-cause, apparently because there is no use for it in his argument.

161 Cf. *Strom.* VIII (9) 25, 1–4; 28, 3–5; 32, 7–33, 9.

Case 4: *Strom.* II (11) 48, 1–2; 49, 2–3

In the second book of the *Stromateis*, Clement launches a discourse on Christian virtues, starting with faith (πίστις).[162] Over against the opinion that faith is determined by one's nature, he argues that it is a virtuous act of rational choice (προαιρέσεως κατόρθωμα), something that everyone is free to accept or reject.[163] Describing faith as a decision to believe God (or the "voice" of God, i.e. Scripture), Clement submits that faith is a choice of (eternal) life and, further, that it is a "criterion" which enables us to know truth.[164] The latter aspect of faith comes to the fore particularly in connection with another virtue discussed in the second book, namely knowledge.[165] Similar to his working method elsewhere in the *Stromateis* (and in keeping with the apologetic agenda announced at the beginning of book II),[166] Clement draws together biblical quotations with relevant snippets of Greek epistemology in order to show a fundamental agreement of both traditions regarding faith and knowledge. In this connection, once again, he finds a use for material paralleled in the 'eighth book':

> The knowledge of those who seem wise to themselves, whether Greek philosophers or barbarian sects, is a knowledge that "puffs up", in the words of the apostle (1 Cor 8:1). But knowledge of which one could say that it is an epistemic demonstration of things handed down according to true philosophy,[167] such knowledge is credible. We might call it *an argument that furnishes credence from things agreed to things disputed. But since there are two kinds of credence, one epistemic and one doxastic, nothing prevents us from speaking of two kinds of demonstration, one epistemic and one doxastic, since even knowledge and foreknowledge is said to be of two kinds, one whose nature has been brought to perfection and one deficient.* And there is no doubt that our demonstration is the only true one, since it is furnished out of divine Scriptures, sacred writings, and the wisdom the apostle describes as "God-taught" (1 Thess 4:9). (…) In contrast, the doxastic demonstration is human and consists of rhetorical arguments or dialectical syllogisms. For the highest form of demonstration, which

162 *Strom.* II (2) 4, 1–(6) 31, 3.

163 Cf. *Strom.* II (3) 10, 1–(4) 12, 1.

164 *Strom.* II (4) 12, 1.

165 Cf. esp. *Strom.* II (10) 46, 1–(12) 55, 6 and (17) 76, 1–3.

166 Cf. *Strom.* II (1) 1, 1 and above, pp. 51–52.

167 On 'true philosophy,' cf. below, comm. on 80,4f.

> we have referred to as epistemic, by presenting Scriptures and opening them up, *inserts credence* (πίστιν) *in the souls eager to learn*, which is no other that knowledge. For if things received are accepted as true with regard to the thing sought, on account of being divine and prophetic, it seems clear that the conclusion drawn from them consequently will be a true conclusion. And we are surely right to take knowledge as demonstration.[168]

This intriguing passage deserves a detailed commentary regarding Clement's view of demonstration as a method of teaching based on scriptural exegesis.[169] But more important for our purpose is to explore the following correspondences:

***Strom.* II (11) 48, 1–2; 49, 2–3**	***Strom.* VIII (3) 5, 1–3**
φήσαιμεν δ' ἂν αὐτὴν *λόγον εἶναι τοῖς ἀμφισβητουμένοις ἐκ τῶν ὁμολογουμένων ἐκπορίζοντα τὴν πίστιν.*	ὡσαύτως δὲ καὶ τὴν ἀπόδειξιν πάντες ἄνθρωποι ὁμολογήσαιεν ⟨ἂν⟩ *λόγον εἶναι τοῖς ἀμφισβητουμένοις ἐκ τῶν ὁμολογουμένων ἐκπορίζοντα τὴν πίστιν.*
πίστεως δ' οὔσης διττῆς, τῆς μὲν ἐπιστημονικῆς, τῆς δὲ δοξαστικῆς, οὐδὲν κωλύει *ἀπόδειξιν* ὀνομάζειν διττήν, *τὴν μὲν ἐπιστημονικήν*, τὴν δὲ δοξαστικήν, ἐπεὶ καὶ *ἡ γνῶσις καὶ ἡ πρόγνωσις διττὴ* λέγεται, ἣ μὲν ἀπηκριβωμένην ἔχουσα τὴν ἑαυτῆς φύσιν, ἣ δὲ ἐλλιπῆ. (…)	οὐ μόνον δὲ *ἀπόδειξις καὶ πίστις καὶ γνῶσις, ἀλλὰ καὶ πρόγνωσις* λέγεται διχῶς, *ἣ μὲν ἐπιστημονική* τε καὶ βεβαία, ἄλλη δὲ μόνον ἐλπιστική.
ἡ δὲ *δοξαστικὴ ἀπόδειξις* ἀνθρωπική τέ ἐστι καὶ πρὸς τῶν ῥητορικῶν γινομένη ἐπιχειρημάτων ἢ καὶ διαλεκτικῶν συλλογισμῶν. ἡ γὰρ *ἀνωτάτω ἀπόδειξις*, ἣν ᾐνιξάμεθα ἐπιστημονικήν, *πίστιν ἐντίθησι* διὰ τῆς τῶν γραφῶν παραθέσεώς τε καὶ διοίξεως *ταῖς τῶν μανθάνειν ὀρεγομένων ψυχαῖς*, ἥτις ἂν εἴη γνῶσις.	*κυριώτατα μὲν οὖν ἀπόδειξις* λέγεται ἡ τὴν ἐπιστημονικὴν *πίστιν ἐντιθεῖσα ταῖς τῶν μανθανόντων ψυχαῖς, δοξαστικὴ* δὲ ἡ ἑτέρα.

168 *Strom.* II (11) 48, 1–49, 3.

169 See my "Demonstrative Method in *Stromateis* VII: Context, Principles, and Purpose," in *The Seventh Book of the Stromateis*, ed. M. Havrda, V. Hušek, and J. Plátová (Suppl. Vig. Chr. 117; Leiden/Boston: Brill, 2012), pp. 261–275.

(cont.)

	Ibid. 6, 2
εἰ γὰρ τὰ *παραλαμβανόμενα πρὸς τὸ ζητούμενον ἀληθῆ λαμβάνεται*, ὡς ἂν θεῖα ὄντα καὶ προφητικά, δῆλόν που ὡς καὶ *τὸ συμπέρασμα τὸ ἐπιφερόμενον αὐτοῖς ἀκολούθως* ἀληθὲς ἐπενεχθήσεται.	*καὶ τὸ μὲν περαίνειν* ἐξ ὁμολογουμένων *συλλογίζεσθαί* ἐστιν, τὸ δέ γε ἐξ ἀληθῶν τι *περαίνειν* ἀποδεικνύειν ἐστίν, ὥστ' εἶναι σύνθετόν τινα τῆς ἀποδείξεως τὴν νόησιν (L: ὄνησιν) *ἔκ τε τοῦ τὰ λαμβανόμενα πρὸς τὰ ζητούμενα ἀληθῆ λαμβάνειν κἀκ τοῦ τὸ συμπέρασμα αὐτοῖς ἀκόλουθον ἐπιφέρεσθαι.*

We can see that in the second book, Clement draws from a philosophical source dealing with demonstration. A glimpse at the table above shows that the same source is represented in the 'eighth book' as well. Further, there is no doubt that the 'eighth book' provides a better picture of the original context of the paralells. Yet it is far from plain that Clement used the text of the 'eighth book' when composing the second. Both passages include a distinction between two kinds of ἀπόδειξις, πίστις, and γνῶσις, but it is not presented in the same way. On its first appearance in the 'eighth book', the distinction is referenced as something already established, the only new point being its application to πρόγνωσις. When reading this passage, we are left with the impression that Clement abbreviates his source or skips over a few sentences in which the distinction was introduced. Interestingly, the second book seems to provide precisely this missing element: after a definition of demonstration (the same as in the 'eighth book'), it draws a distinction between two kinds of demonstration, deriving it from a twofold division of πίστις and backing it up further through the same division of γνῶσις and πρόγνωσις. This order of argument cannot be easily reconstructed on the basis of the 'eighth book', but it fills in its omissions remarkably well. In addition, when speaking of the 'doxastic demonstration', the second book tells us that it consists of rhetorical arguments or dialectical syllogisms. This point is never made in the 'eighth book', but it coincides with the view, mentioned elsewhere in the text, that the deficient kind of demonstration produces "mere persuasion" (πειθὼ μόνον) in the souls of the auditors and that it draws its conclusion "from [premisses] that are merely reputable" (ἐξ ἐνδόξων μόνον).[170] Either we have to suppose that Clement's apt description of the 'doxastic demonstration' in the second book

170 *Strom.* VIII (3) 7, 7–8.

was his own invention, or we should think that he derived it from the same source as the other elements of the theory. Both options are possible, but the latter appears more likely.

Case 5: *Strom.* V (1) 11, 1–(3) 17, 2

The fifth book of the *Stromateis* mainly deals with the symbolic genre. However, at the beginning, Clement briefly returns to the topic of faith, focussing on its epistemological aspect, and argues that faith should not be "inactive and alone, but it must progress along with inquiry". In this connection, he refers to the Lord's bidding, "seek and you will find", alluding probably to Matthew 7:7.[171] This verse, as we know, is quoted *in extenso* in the first chapter of the 'eighth book' in response to the "useless foolery" of the "more recent philosophers".[172] Clement's interpretation of the verse, which covers the rest of this chapter, contains certain elements paralleled in book V. Pointing out these parallels, Ernst suggests that Clement draws from 'book VIII' in book V.[173] But, as Nautin has shown, this is surely a precipitous conclusion.[174]

One of these parallels concerns an allusion to Plato's *First Alcibiades* (109e) in *Strom.* VIII (1) 2, 3: "So one who has sought can find, but he will seek only *if he thinks that he has not known beforehand* (εἰ οἰηθείη πρότερον μὴ εἰδέναι)." This is a fitting allusion to make in connection with Matthew 7:7, as the Platonic passage describes the recognition of one's own ignorance as a condition of "seeking/inquiring" (ζητῆσαι) and, consequently, of "finding/discovering" (εὑρεῖν).[175] In *Strom.* V (3) 17, 2, Clement uses the same passage in a similar connection, this time, however, quoting it in full (even with a reference to Plato's *Alcibiades*). No doubt on both occasions Clement assumes that the *Alcibiades* passage is relevant to the topic of inquiry, but there is no indication whatsoever that he derives this assumption from the 'eighth book'. Another parallel involves the notion that 'contention' (ἔρις), 'rivalry' (φιλονικία), 'jealousy' (φθόνος), etc. are 'obstacles' (τὰ ἐμπόδων) on the path to discovery.[176] Finally, in both texts,

171 *Strom.* V (1) 11, 1: τὴν πίστιν τοίνυν οὐκ ἀργὴν καὶ μόνην, ἀλλὰ σύν ζητήσει δεῖν προβαίνειν φαμέν. (…) *ζήτει γάρ, καὶ εὑρήσεις*, λέγει.

172 *Strom.* VIII (1) 1, 2.

173 Ernst, *De Clementis*, pp. 8–11.

174 Nautin, "La fin," pp. 273–277.

175 Plato, *Alc.* 1, 109e (trans. Lamb): "Alcibiades: Do you think I could not know about what is just and unjust in any other way? Socrates: Yes, you might, supposing you discovered it. A: But do you not think I might discover it? S: Yes, quite so, if you inquired. A: And do you not think I might inquire? S: I do, if you thought you did not know."

176 Cf. *Strom.* V (1) 11, 4 and VIII (1) 1, 2; 2, 3.

Clement speaks of knowledge as a 'gift' (δόσις) given by God (παρὰ τοῦ θεοῦ, θεοδωρήτος).[177] Ernst notes that all of these motifs are found elsewhere in the *Stromateis*, too.[178] However, apart from their association with inquiry (in Clement's mind they clearly belonged to the same complex of ideas), they do not bear any trace that would point to the 'eighth book' as their source.

Case 6: *Strom.* VII (16) 95, 3–8

The last case to be mentioned is particularly interesting. It appears towards the end of the seventh book in a section presented as a preparation to the following book of the *Stromateis*. Anticipating the apologetic part of his agenda, in which "objections raised against us by the Greeks and the Jews" are to be handled, Clement decides to take up one of these issues in advance, namely the objection from disagreement between various schools (αἱρέσεις) of Christian doctrine.[179] His aim in the section is to show that the disagreement, which concerns mainly the interpretation of Scripture, can be settled by applying the correct criteria of truth.[180] In this connection, he uses the language of demonstrative theory closely reminiscent of the 'eighth book':

> For we have *the first principle of teaching*, the Lord who *guides* us from the beginning to the end of knowledge 'in many diverse ways' (Heb 1:1) through the prophets, the gospel, and the blessed apostles. And if anyone were to suppose that the first principle requires another one, it could no longer be maintained as truly the first principle. Now it is reasonable that he who is *credible by himself* will be worthy of credence in Scripture, the voice of the Lord, which works for the benefit of humankind by the agency of the Lord. We certainly use it as a *criterion for the discovery* of [the nature of] things. No object of judgement is believed before a judgement about it is made. Therefore, nothing that requires judgement is a principle. So it is reasonable that *we embrace the principle by faith as indemonstrable*, while also accepting abundant proofs about the principle from the principle itself, and let ourselves be educated by the voice of the Lord so as to know the truth. We do not pay attention to people who make simple assertions,

177 Cf. *Strom.* V (1) 12, 2 and VIII (1) 2, 1.

178 Ernst, *De Clementis*, p. 9.

179 *Strom.* VII (15) 89, 1–2.

180 For a detailed commentary on the section cf. Alain Le Boulluec, *La notion d'hérésie dans la literature grecque IIe–IIIe siècles*, II (Paris: Institut d'Études Augustiniennes, 1985), pp. 391–438. See also my "Demonstrative Method".

> *while it is equally possible to assert the contrary. If it is not sufficient to say simply what seems to be the case, but it is necessary to confirm what has been said*, we do not wait for a testimony coming from men, but we confirm the thing sought by the voice of the Lord, which is more reliable than any demonstration, or rather which is the only demonstration there is.[181]

There are points of contact with several passages of the ‘eighth book’, particularly close being those at the end of this table:

***Strom.* VII (16) 95, 3–8** ἔχομεν γὰρ *τὴν ἀρχὴν τῆς διδασκαλίας*, τὸν κύριον διά τε τῶν προφητῶν διά τε τοῦ εὐαγγελίου καὶ διὰ τῶν μακαρίων ἀποστόλων “πολυτρόπως καὶ πολυμερῶς” ἐξ ἀρχῆς εἰς τέλος *ἡγούμενον* τῆς γνώσεως. (…)	***Strom.* VIII (2) 4, 2** πᾶν οὖν τὸ προβληθὲν ὄνομα μεταλαμβάνειν χρὴ εἰς λόγον ὁμολογούμενόν τε καὶ σαφῆ τοῖς κοινωνοῦσι τῆς σκέψεως, *ἀρχὴν μὲν τῆς διδασκαλίας* ἐσόμενον, *ἐξηγησόμενον* δὲ τὴν τῶν ζητουμένων εὕρεσιν.
	***Strom.* VIII (3) 7, 2**
ὁ μὲν οὖν *ἐξ ἑαυτοῦ πιστὸς* τῇ κυριακῇ γραφῇ τε καὶ φωνῇ ἀξιόπιστος εἰκότως ἂν ⟨εἴη⟩ διὰ τοῦ κυρίου πρὸς τὴν τῶν ἀνθρώπων εὐεργεσίαν ἐνεργουμένῃ· ἀμέλει *πρὸς τὴν τῶν πραγμάτων εὕρεσιν αὐτῇ χρώμεθα κριτηρίῳ*· τὸ κρινόμενον δὲ πᾶν ἔτι ἄπιστον πρὶν κριθῆναι, ὥστ’ οὐδ’ ἀρχὴ τὸ κρίσεως δεόμενον. εἰκότως τοίνυν *πίστει περιλαβόντες ἀναπόδεικτον τὴν ἀρχήν*, ἐκ περιουσίας καὶ τὰς ἀποδείξεις παρ’ αὐτῆς τῆς ἀρχῆς περὶ τῆς ἀρχῆς λαβόντες, φωνῇ κυρίου παιδευόμεθα πρὸς τὴν ἐπίγνωσιν τῆς ἀληθείας. οὐ γὰρ ἁπλῶς ἀποφαινομένοις ἀνθρώποις προσέχομεν [L: προσέχοιμεν], οἷς καὶ *ἀνταποφαίνεσθαι ἐπ’ ἴσης ἔξεστιν. εἰ δ’ οὐκ ἀρκεῖ μόνον ἁπλῶς εἰπεῖν τὸ δόξαν, ἀλλὰ πιστώσασθαι δεῖ τὸ λεχθέν*, οὐ τὴν ἐξ ἀνθρώπων ἀναμένομεν	ὥστ’ εἴπερ ἐστὶν ἀπόδειξις, ἀνάγκη πᾶσα πρότερον εἶναί *τι πιστὸν ἐξ ἑαυτοῦ, ὃ δὴ πρῶτον καὶ ἀναπόδεικτον* λέγεται. ***Strom.* VIII (3) 8, 6** ἐν πᾶσιν οὖν τοῖς ζητουμένοις ἔστι τι προγινωσκόμενον (*ὃ πάντως ἐξ ἑαυτοῦ πιστὸν ὂν ἀναποδείκτως πιστεύεται*), ὃ χρὴ ποιεῖσθαι τῆς ζητήσεως αὐτῶν ὁρμητήριον καὶ *τῶν εὑρῆσθαι δοκούντων κριτήριον*. ***Strom.* VIII (2) 4, 1** εἰ δ’ οὐκ ἀρκεῖ τοῦτο *μόνον ἁπλῶς εἰπεῖν περὶ τοῦ ζητουμένου τὸ δόξαν* (*ἔξεστι γὰρ καὶ τὸν ἀντικαθιστάμενον ἐπ’ ἴσης ἀνταποφήνασθαι ὃ βούλεται*), *ἀλλὰ πιστώσασθαι χρὴ τὸ λεχθέν*, εἰ

181 *Strom.* VII (16) 95, 3–8.

(cont.)

μαρτυρίαν, ἀλλὰ τῇ τοῦ κυρίου φωνῇ πιστούμεθα τὸ ζητούμενον, ἣ πασῶν ἀποδείξεων ἐχεγγυωτέρα, μᾶλλον δὲ ἣ μόνη ἀπόδειξις οὖσα τυγχάνει.	μὲν εἰς ὁμοίως ἀμφισβητούμενον ἀναφέροιτο αὐτοῦ ἡ κρίσις κἀκείνου πάλιν ὁμοίως εἰς ἀμφισβητούμενον ἕτερον, εἰς ἄπειρον προβήσεται καὶ ἀναπόδεικτον ἔσται, εἰ δ' εἰς ὁμολογούμενον ἅπασιν ἡ τοῦ ⟨μὴ⟩ ὁμολογουμένου πίστις ἀναφέροιτο, ἐκεῖνο *τὴν ἀρχὴν τῆς διδασκαλίας* ποιητέον

Did Clement, when facing the objection from disagreement in the seventh book, have the 'eighth book' in front of him, picking up ideas, words, and whole phrases pertaining to the theory of demonstration? He certainly did have a text like the 'eighth book' in front of him, but perhaps not necessarily the same one. On the face of it, it seems plausible that in the above-quoted passage Clement used the text we know. Yet there are elements of philosophical origin in the final part of book VII that are not found in 'book VIII', but give us reasons to think that they might have been present in its philosophical source. I have in mind particularly the following sentence:

> *Strom.* VII (16) 93, 2: ἔστι μὲν οὖν κοινά τινα τῶν ἀνθρώπων κριτήρια καθάπερ τὰ αἰσθητήρια, τὰ δ' ἄλλα τῶν βουληθέντων καὶ ἀσκησάντων τὰ ἀληθῆ, τὰ διὰ νοῦ καὶ λογισμοῦ τεχνικὰ λόγων ἀληθῶν τε καὶ ψευδῶν.
>
> Now, there are certain criteria common to all men, such as the sensory organs, while the other, artificial criteria of true and false arguments, acquired through the intellect and reasoning, are confined to those whose have made truth their aim and practice.[182]

Although it has no parallel in the 'eighth book', this sentence can be readily compared with a passage in Galen that, in turn, seems to constitute a link with *Strom.* VIII. In this passage (a polemic against the sceptic Favorinus), Galen draws a distinction between natural and artificial criteria, i.e. between the senses and rational capacities on the one hand and various measures constructed on the basis of the natural criteria for the use of the arts on the other. He then applies this distinction to the method of demonstration,

182 Trans. Hort and Mayor, modified.

identifying, within this method, certain items whose role is analogous to that of the artificial criteria in the arts. The function of these items is exactly the same as the function of the 'artificial criteria' in Clement. Here is the Galenic passage:

Galen, *Opt. Doct.* 4, 3–4 (CMG V,1,1: 104,6–15/I,49 f. K.)	**Galen, *On the Best Kind of Teaching*[183]**
ἐπεὶ δ' ἐκ τῶν φυσικῶν κριτηρίων αἱ τέχναι πᾶσαι κατασκευάζουσιν ὄργανά τε καὶ *κριτήρια τεχνικά*, δι' ὧν τὰ μὲν αὐτοὶ συντιθέασι, τὰ δ' ὑφ' ἑτέρων συγκείμενα κρίνουσι, κἀγὼ διδάξω ⟨σ'⟩ ὅλως ὄργανά τε καὶ κριτήρια, τὰ μέν οἷς κατασκευάσεις τοὺς ἀληθεῖς λόγους, *τὰ δὲ οἷς τοὺς ὑφ' ἑτέρων γεγονότας κρινεῖς.*	Since it is from the natural criteria that all the arts construct their instruments and *artificial criteria*, through which they construct things themselves and judge the things that have been constructed by others, I too will teach instruments and criteria of arguments, [instruments] through which you will produce true arguments and [*criteria*] *through which you will judge the* [*arguments*] *produced by others.*
ἔχει γὰρ οὕτω τὸ πᾶν· εἰ μὲν ἐξ ἑαυτοῦ τι φαίνεται πρὸς αἴσθησιν ἢ νόησιν ἐναργῶς, οὐ χρῄζει τοῦτο ζητήσεως· εἰ δὲ μή τι τοιοῦτον ὑπάρχει, τῆς ἐξ ἑτέρου δὴ προσδεῖται γνώσεως. ἐγὼ μὲν ⟨οὖν⟩ ἐπαγγέλλομαί σοι διδάξειν ἔνια μὲν τοῖς κατὰ τὰς τέχνας ὀργάνοις ἀνάλογα, ἐξ ὧν εὑρήσεις τὸ ζητούμενον, *ἔνια δὲ τοῖς κριτηρίοις, ἐξ ὧν τὸ δοκοῦν εὑρῆσθαι κρινεῖς.*	It all amounts to this: if, by itself, something appears clearly to perception or thought, it need not be investigated; but if it is not like this, it requires to be known through something else. I undertake to teach you some things that are analogous to the artisan's instruments, by which you will discover the thing you are investigating, and others that are analogous to *criteria, by which you will judge what appears to have been discovered.*

The last words recall a sentence from the 'eighth book' that I have already quoted:

183 Trans. Stephen Menn (unpublished).

Opt. Doct. 4, 4 (104,13–15/I,50 K.)	*Strom.* VIII (3) 8, 6
ἐγὼ μὲν ⟨οὖν⟩ ἐπαγγέλλομαί σοι διδάξειν ἔνια μὲν τοῖς κατὰ τὰς τέχνας ὀργάνοις ἀνάλογα, ἐξ ὧν εὑρήσεις τὸ ζητούμενον, ἔνια δὲ *τοῖς κριτηρίοις, ἐξ ὧν τὸ δοκοῦν εὑρῆσθαι κρινεῖς*.	ἐν πᾶσιν οὖν τοῖς ζητουμένοις ἔστι τι προγινωσκόμενον (ὃ πάντως ἐξ ἑαυτοῦ πιστὸν ὂν ἀναποδείκτως πιστεύεται), ὃ χρὴ ποιεῖσθαι τῆς ζητήσεως αὐτῶν ὁρμητήριον καὶ *τῶν εὑρῆσθαι δοκούντων κριτήριον*.

Clement's description of the non-artificial criteria as κοινά, and his inclusion of the sensory organs among them, is also paralleled in Galen's *PHP*.[184]

All this need not mean that Clement took his description of the two kinds of criteria in the seventh book from Galen.[185] But if the Galen hypothesis is right for the 'eighth book', it is reasonable to suspect a Galenic trace here, too. And it is reasonable to believe that this trace points back to the same source from which the parallels with the 'eighth book' also spring. This source, however, cannot be the 'eighth book' itself, or the text we call by that name, as the latter does not contain the sentence about the criteria or anything similar.

To sum up, it seems abundantly clear that Clement used a text like the 'eighth book' as a source for the composition of some regular books of the *Stromateis*. Specifically, he used a text like the chapter on causes (section θ) in book I and a text like the chapter on demonstration (section β) in books II and VII. However, there are reasons to think that he did not use these chapters as they are preserved in our 'eighth book'; rather, he used a different text, stemming in all probability from the same source, but containing some other material.

184 *PHP* IX 1, 11–13 (CMG V,4,2,1: 542,8–18): "If we have no natural criterion, we shall not be able to find an artificial criterion either; but if we possess natural criteria, we could find some artificial criterion as well. Do we possess any natural criteria common to all men? For we must not call things 'natural' that are not common to all. Indeed what is natural must not only be common to all but also have a common nature.—I say that you all do have natural criteria, and in saying this I am reminding you, not teaching or demonstrating or making an assertion on my own authority. What are these criteria? Eyes in their natural state seeing what is visible; ears in their natural state hearing what is audible; the tongue sensing savours, the nostrils odours, the whole skin objects of touch; and besides thought or mind or whatever you want to call it, by which we distinguish entailment and incompatibility and other things that pertain to them …" (Trans. Phillip De Lacy, modified). With the last clause compare *Strom.* VIII (3) 7, 5 and the commentary on 83,29f.

185 The distinction is also found in Sextus (*PH* II 15; *M.* VII 31–32), who does not speak of the artificial criteria of arguments, however.

Does this conclusion shed any light on the purpose of *Stromateis* VIII and its place in Clement's *œuvre*? Our inquiry does seem to shed light on one thing. It shows, through particular examples, why Clement was interested in material such as was preserved in the 'eighth book'. In all these cases, the material is fully subordinated to Clement's apologetic purpose with little regard for its original context. This suggests the following: When approaching a philosophical source, Clement's aim was not to inform his readers of the views transmitted by that source, whether it be for the purpose of education or polemic; rather, his aim was to identify elements within that source that could be adapted to his own use. In fact, as we shall see in the next (and last) step of our inquiry, there are indications within the 'eighth book' that point in this direction as well.

h *Christian Traits in* Stromateis *VIII*

It is a delicate task to separate Christian traits from non-Christian ones in a writer like Clement, whose cultural Hellenism permeates the most hidden recesses of his thought and informs the way he articulates the very core of the Christian message.[186] Yet when we look at the parallels explored in the previous section and compare the contexts in which they are embedded, we can immediately recognize a fundamental difference: It is not a difference between the presence and absence of words and ideas of Hellenic or philosophical origin; these are omnipresent in Clement's works. Rather, it is a difference concerning *arguments*, the assumptions they imply, and the goals they pursue. And seeing that certain assumptions and goals typical of Clement's writings, particularly those that distinguish him as a Christian thinker, leave no trace for the most part of the 'eighth book', it is reasonable to ask about exceptions.

How do we recognize them? Surely the presence of biblical quotations or allusions is a safe indicator. Apart from the first chapter, which is Christian throughout, there seem to be only two other instances, both in the chapter on causes: one of them is an allusion to Genesis 1:1, made on the margin of a discussion about the Stoic notion of cause as "that which does something"; the other is a combination of Leviticus 24:20 and 24:18, again clearly a marginal gloss attached to a passage dealing with reciprocal causation.[187] Spare as they are, these two glosses are revealing, as they show what kind of questions Clement was asking when reading his philosophical source. They seem to

186 Cf. Alain Le Boulluec, "Clément d'Alexandrie et la conversion du «parler grec»," in idem, *Alexandrie antique et chrétienne: Clément et Origène* (Paris: Institut d'Études Augustiniennes, 2006), pp. 63–79.

187 *Strom.* VIII (9) 28, 5 and 30, 4. Cf. below, commentary on 98,14 f. and 99,25.

indicate that, in these instances at least, Clement read the source in view of its usability in biblical exegesis.[188] Two other passages adjacent to these quotations arguably bear the mark of the same exegetical intent.[189]

Another Christian trait can be detected in the section on demonstration and pertains to the use of the substantivized adjective ὁ γνωστικός, referring to a person with perfect knowledge, a usage not attested in Greek philosophical literature, but commonplace in Clement.[190] The remark is made on the margin of a distinction between two kinds of demonstration, the epistemic and the doxastic, the former being a demonstration in the proper sense of the word, as it is a perfect representative of its genus. In the same way, according to Clement, we arrive at the proper sense of the words 'doctor' or 'gnostic' when we think of someone whose capacity to heal is perfect or someone whose epistemic knowledge is flawless, respectively.[191] Here what seems to be Clement's addition to his source reveals his constant preoccupation with the idea of the gnostic, which is intimately linked to the notion of perfection, as far as it is attainable for a human being. This idea is developed especially in the seventh book of the *Stromateis*, where Clement also happens to associate it with the notion of a perfect doctor: "As we speak of a perfect doctor and a perfect philosopher, so, I suppose, we may speak of a perfect gnostic."[192] Since this sentence appears not far from the above-quoted passage containing parallels with (the same section of) the 'eighth book' (*Strom.* VII [16] 95, 3–8), we may suspect that its similarity with *Strom.* VIII (3) 5, 3 is not accidental. It could be due to a common source, being a similar (if flatter) adaptation of it. In any case, the sentence shows what kind of use Clement may have had in mind when adding the 'gnostic' as another example of perfection in the 'eighth book'.

There are a couple of other instances where Clement arguably added a gloss or modified something in light of his own interests.[193] All in all, these rare

188 I have developed this point in my paper "Clement's Exegetical Interests in *Stromateis* VIII," in *Clement's Biblical Exegesis. Proceedings of the Second Colloquium on Clement of Alexandria* (Olomouc, May 29–31, 2014), ed. V. Černušková, J.L. Kovacs, and J. Plátová (Suppl. Vig. Chr. 139; Leiden/Boston: Brill, 2016), pp. 162–178.

189 Cf. below, on 99,1 and 99,32–100,2.

190 *Strom.* VIII (3) 5, 3; cf. below, on 82,25 f.

191 For the philosophical background and parallels, cf. below, on 82,24 and 82,24 f.

192 *Strom.* VII (14) 88, 5: ὡς γὰρ τέλειόν φαμεν ἰατρὸν καὶ τέλειον φιλόσοφον, οὕτως, οἶμαι, καὶ τέλειον γνωστικόν. Trans. Hort and Mayor, slightly modified. Clement hastens to emphasize that, unlike the Stoics, he does not maintain that the ἀρετή of man and of God is the same.

193 Cf. esp. *Strom.* VIII (3) 7, 2–3 and commentary on 83,23 f. and 84,23 f. Cf. also perhaps 30, 2/99,19 f.

traces of Clement's presence as a Christian thinker support the impression that Clement composed the 'eighth book' chiefly for the purpose of using it in a similar way to how he used the philosophical material similar to sections β and θ in *Stromateis* I, II, and VII. Did he compose it *after* book VII, in view of the continuation of the *Stromateis*? Perhaps, and perhaps not. We simply do not know. It is possible, for example, that Clement made more than one collection of excerpts from the same source, using one for the composition of the first seven books and making another in view of their continuation. Or, perhaps, he made just one collection of excerpts, using it in the first seven books and planning to use it in their continuation as well, a collection which has been lost, but whose abridged version is our 'eighth book'. Other stories could be told, but there seems to be little at hand to substantiate any of them.

However, the picture of the 'eighth book' as a collection of source material is problematized by the first chapter. Speaking of 'Christian traits' in our text, we can hardly leave it aside. Is this not clearly a piece of Clement's own composition, as Christian and as elaborate as anything in the *Stromateis*? And should the rest of the material not be explained in its light? It is true that the chapter is elaborate and Christian, and that its position at the beginning invites us to regard it as an introduction and key to what follows. But this impinges on the problem that the first chapter is *not* an introduction to what follows, as it speaks of issues never mentioned in the following text, such as the pursuit of "true philosophy" and inquiry "concerning Scriptures", without explaining how the text is supposed to be relevant to these issues. Nor, of course, is it a key, as it does not help the reader in understanding the rest of the book at all. It does seem to show, however, why Clement himself was interested in the subsequent material: Heeding the command of Matthew 7:7, Clement addresses the problem of Christian 'inquiry' (ζήτησις) and mentions 'demonstration' (ἀπόδειξις) as its method. This word appears for the first time, rather unexpectedly, towards the end of the chapter.[194] The next sentence, which starts a discussion of the teaching on demonstration, is also the first in a series of parallels pointing to Galen's *DD*.[195] From this moment on, the concerns expressed in the first chapter are not heard of again, even in other passages where Christian traits are discernible. As we have seen, these passages are probably Clement's glosses, marking parts of the source-material he found useful for exegesis or some other purpose. Is it not likely that the function of the first chapter is similar, namely, to point out how the theme of the subsequent pages, i.e. demonstration, is relevant from

194 *Strom.* VIII (1) 2, 5 (GCS 17: 81,6).

195 See below, commentary on 81,9f.

Clement's point of view? The whole chapter, perhaps, is a gloss on the margin of Clement's source, explaining why he was attracted to a text dealing with demonstration.[196]

i *Conclusions*

Many questions have been raised and some solutions proposed in the course of this inquiry. The book-long material added to the seven books of the *Stromateis* in the Florentine codex and supplied with the title στρωματεὺς ὄγδοος remains an enigma. We have looked in some detail at the first thirty pages—the text that has been traditionally referred to as the eighth book of the *Stromateis*. This text, described by Heinse as 'liber logicus', is distinguished from the others, and from any other piece of Clement's writings, by its predominantly philosophical character. But it shares an unpolished and fragmentary form of composition with the other two texts, such as suits the 'genre' of drafts or excerpts. It has been proposed that the whole 'meta-Stromatic material' consists of excerpts from Clement's finished works. This proposal, whose most influential version was formulated by Nautin, has not been based on a detailed study of the texts, but rather on an analysis of Clement's literary plans in the *Stromateis* and on some extraneous considerations. Whatever the worth of these arguments with regard to the *Excerpta* and *Eclogae*, our investigation has shown that they do not hold up in the case of 'liber logicus'. In all probability, this text is a collection of excerpts made or acquired by Clement and held by him as a source-material for his own writing. I have argued that the excerpted source (or the main one) is Galen, most likely his lost treatise *On Demonstration*. We do not know if Clement himself made excerpts from this work, or whether he used excerpts made by somebody else. But he seems to have used the same source or excerpts therefrom in the regular books of the *Stromateis*. The reasons for his interest in this material are not difficult to discern: On the one hand they are revealed by the above-mentioned passages from books I, II, and VII, where Clement uses texts similar to some sections of 'liber logicus' for apologetic aims. On the other hand they are indicated in 'liber logicus' itself, where Clement occasionally adds a gloss to the source, pointing towards the usefulness of a particular passage for biblical exegesis or another purpose. His interest in the topic of demonstration (and perhaps in a treatise that has the word 'demonstration' in its title) is explained by the first 'chapter'; this is a comment in the form of an exegesis of Matthew 7:7, where Clement mentions demonstration as a method of inquiry leading towards knowledge, which, for him, is at the same time a

196 For more details, see commentary on 80,3 f. and 81,9 f.

method of biblical interpretation. Regrettably, Clement never explains how the teaching on demonstration set forth by his source could stand up to these expectations. But he gives us something else instead: A precious testimony of a lost treatise dealing with some fundamental issues of ancient philosophy of science.

The So-Called Eighth Stromateus (*'Liber logicus') by Clement of Alexandria: Greek Text, Translation, and Commentary*

Prefatory Note to the Greek Text and Translation

a Greek Text

As noted in the introduction, our knowledge of the text of *Strom.* VIII essentially depends on Codex Laurentianus Pluteus 5.3 (L).[1] This codex from the 11th century was the basis of the *editio princeps* of the *Stromateis*, published in 1550. Other editions of the *Stromateis* that included the 'eighth book' were subsequently prepared by Friedrich Sylburg (1592), Daniel Heinse (1616), John Potter (1715), Reinhold Klotz (1832), Wilhelm Dindorf (1869), and Otto Stählin (1909).[2] Potter's edition was reprinted in Migne's *Patrologia Graeca*, Vol. IX (1857). Stählin's edition, mildly revised by Ludwig Früchtel, was republished in 1970. In addition, all the 'meta-Stromatic material' (the 'eighth book' under the title *Isagoge dialectica*) was published in Bunsen's *Analecta Ante-Nicaena* (1854).

Stählin's edition supersedes all the previous ones in several respects: First of all, it provides as detailed and accurate information about its manuscript basis as any of them; second, while making a reasonable choice of emendations proposed by earlier scholars, it adds many others that make better sense of the text; third, it adduces new intertextual parallels.

The text printed on the following pages could be described as an adaptation of the Stählin edition for the use of the adjacent commentary. It brings no new information about the manuscript. As a matter of fact, even though I often consulted the manuscript, I did not check Stählin's reading on every line; wherever I checked his reading (that is to say, in all problematic cases), I found it very precise. Why, then, did I not simply choose to reproduce Stählin's edition? Mainly because, in the course of my work on the commentary, I did not always deem Stählin's corrections of corrupt or difficult passages satisfactory. In some cases I ventured to think that we can do justice to the text as it stands;[3] on other occasions I either adopted or proposed a different solution.[4] Also, I

---

1 Some passages are preserved in sources relying on earlier manuscripts: (a) The beginning of the first sentence (1, 1/80,3) is quoted in Photius, *Bibl.* 111; (b) 1, 3–2, 1.3 (80,11–17 and 80,24 f.) in Codex Lavra B 113, fol. 18r; (c) 2, 5 (81,3–8) in the *Sacra parallela* (ἱερά) by John of Damascus (fr. 277 Holl); (d) 16, 2–3 (89,24–90,6) in a Syriac collection of patristic testimonies (MS Add. 14,153; fol. 137a, 1st col., ll. 16–31; cf. below, pp. 210–211, n. 231); (e) 23, 2 (94,14–16) again in the *Sacra parallela* (fr. 278 Holl).

2 For the earlier editions, cf. Stählin, GCS 12, pp. LXV–LXXIII.

3 Cf. 85,2.3; 86,27; 88,4.31; 90,15; 93,4; 96,19.24; 97,9–11; 98,25; 99,29; 100,4.18; 101,21.

4 I have proposed emendations in 81,23; 83,29; 84,29; 88,4; 90,10; 90,16; 91,29; 92,9; 92,33; 93,12;

 DOI: 10.1163/9789004325289_003

believed that a slight adjustment of structure (paragraphs, punctuation) could sometimes make it easier for the reader to follow the argument.[5] And I wished to relieve the text of what I perceived as a glossator's hand.[6]

Thus the text differs from the Stählin edition on a number of occasions (for convenience, all differences from the main text in the Stählin edition are marked by an asterisk in the apparatus). At the same time, however, it bears many traces of its editorial history, including the numbers of chapters, paragraphs, and subparagraphs, as they were introduced in the editions of Potter, Klotz, and Stählin, respectively. To facilitate cross-references, it follows the page and line numbers of the Stählin edition. And it embraces many of the emendations proposed since the Renaissance. Deviations from the manuscript reading, with the exception of trifling oversights, are printed in italics.

Not all solutions mentioned in the notes found their way into the text. First of all, I cite solutions accepted by Stählin even when I disapprove of them. Second, I cite some other conjectures (including my own) which I found plausible or worth consideration, but not strong enough to justify tampering with the text. Most of the cited emendations are noted in the apparatus of Stählin's edition.[7] Others include corrections kindly suggested to me by Jonathan Barnes and Marwan Rashed, who read parts of the 'eighth book' during a seminar held in Paris in 2012/13.

93,28; 95,31 (note); and 100,5. Other solutions different from Stählin's include 80,15; 82,21; 83,4; 83,14; 83,26; 84,28; 85,17; 86,18; 87,4; 90,10; 93,5; and 94,15.

5 Cf. e.g. 82,14; 84,31; 91,6; 92,1; 93,3–6; 97,8–13; 100,7 f.

6 Cf. 84,15; 87,4; 89,20 f.; 90,16; 93,12; 93,17; 94,2; 95,8; 95,25; 100,2; 100,11. All these cases of superfluous, syntactically disruptive, or perplexing elements have the appearance of marginal notes incorporated into the text; we can hardly be sure, however, that none of them goes back to Clement himself (cf. esp. an innocuous note ὡς τῶν λημμάτων in 84,15). And *vice versa*: some notes that have been left in the text on the presumption of being Clement's own (cf. esp. 98,14 f. and 99,25) might well be due to a later glossator.

7 The corrections marked by the names of Clement's editors go back to their respective editions, mentioned above (see also the full list below, in the Bibliography section). The corrections by Theodor Heyse (made in preparation for an unfinished edition of Clement's works), Ulrich von Wilamowitz-Moellendorff, and Eduard Schwartz (proposed in their notes to a draft of Stählin's edition) are first published by Stählin (cf. GCS 12, pp. LXXXI–III). Those by Hans von Arnim appear in his study *De octavo Clementis Stromateorum libro* (Rostock: Adler, 1894) or in his *Stoicorum veterum fragmenta*. The corrections by Christiane von Wedel and Johannes Mewaldt (and some of those by Wilamowitz-Moellendorff) first appear in von Wedel's *Symbola ad Clementis Alexandrini stromatum librum VIII. interpretandum* (Weimar: R. Wagner iun., 1905). Those by Johannes Arcerius and William Lowth are first published by Heinse and Potter, respectively.

b Translation

As far as I am aware, six translations of the 'eighth book' have been produced so far: (1) Latin by Gentien Hervet, first published in 1551 and reprinted many times since, most notably in the editions of Heinse and Potter;[8] (2) English by William Wilson, first published in 1869 as part of the Ante-Nicene Christian Library Series;[9] (3) Polish by Janina Niemirska-Pliszczyńska, published in 1994;[10] (4) Spanish by Marcelo Merino Rodríguez, published in 2005;[11] (5) Italian by Annalisa Zanotti Fregonara, published in 2006 as an appendix to the second edition of Giovanni Pini's translation of the *Stromateis*;[12] and (6) Japanese by Manabu Akiyama, published in 2014 in a bulletin of the University of Tsukuba.[13]

The present rendering of the text into English follows the line of interpretation developed in the commentary. It does not aspire at being 'literal', whatever it might mean, or restricted to words whose equivalents are spelled out in the Greek. Its main goal is to express the meaning extracted from the original in reasonably standard and clear English, without adding or paraphrasing *too much*. Most of the added elements, signalled by square brackets, are taken from the immediate context. Angle brackets designate words whose equivalents are editorial additions to the manuscript. Titles of thematic sections (printed in boldface) are added for convenience.

As a non-native speaker of English, I have often sought advice from English translations of thematically and stylistically related texts, particularly the translations of Galen, Sextus Empiricus, and Porphyry made by Jonathan Barnes, Richard Bett, and Jim Hankinson. Translation of sections included in Anthony Long and David Sedley's anthology *The Hellenistic Philosophers* is largely

8 *Clementis Alexandrini omnia quae quidem extant opera, nunc primum è tenebris eruta Latinitateque donata, Gentiano Herueto Aurelio interprete* (Florence: Laur. Torrentinus, 1551), pp. 267–277.

9 *Ante-Nicene Christian Library*, XII: *The Writings of Clement of Alexandria*, II (Edinburgh: T&T Clark, 1869), pp. 490–514.

10 *Kobierce zapisków filozoficznych dotyczących prawdziwej wiedzy*, II (Warszawa: PAX/ATK, 1994), pp. 311–336.

11 Clemente de Alejandría, *Stromata VI–VIII. Vida intelectual y religiosa del cristiano* (Madrid: Ciudad Nueva, 2005).

12 Clemente di Alessandria, *Gli Stromati. Note di vera filosofia* (Milan: Paoline, 2006), pp. 855–882.

13 "Arekusandoria no Kuremensu, *'Sutromateisu'* (*'Tsuzureori'*) 8," in *Bungei gengo kenkyū. Bungei hen* [*Studies in Language and Literature. Literature*] 66 (2014), pp. 87–115.

adopted from this source. Jonathan Barnes has kindly read part of the text in an early version, supplying such comments as "not English, alas" and making useful suggestions. Sean Coughlin has been particularly helpful in reading through the text as a whole and improving it on many occasions. None of these people, of course, is liable for the outcome.

Greek Text and Translation

Στρωματεὺς ὄγδοος

GCS

p. 80,3 **1. 1** (1) Ἀλλ' οὐδὲ οἱ παλαίτατοι τῶν φιλοσόφων ἐπὶ τὸ ἀμφισβη- L 346v
τεῖν καὶ ἀπορεῖν ἐφέροντο. ἦ πού γ' ἂν ἡμεῖς, οἱ τῆς ὄντως ἀληθοῦς ἀντεχόμενοι φιλοσοφίας, οἷς ἄντικρυς ἡ γραφὴ εὑρέσεως χάριν ἐπὶ τὸ διερευνᾶσθαι τὸ ζητεῖν παρεγγυᾷ; (2) οἱ μὲν γὰρ νεώτεροι τῶν παρ' Ἕλλησι φιλοσόφων ὑπὸ φιλοτιμίας κενῆς τε καὶ ἀτελοῦς ἐλεγκτικῶς ἅμα καὶ ἐριστικῶς εἰς τὴν ἄχρηστον ἐξάγονται φλυαρίαν, ἔμπαλιν δὲ ἡ βάρβαρος φιλοσοφία τὴν πᾶσαν ἔριν ἐκβάλλουσα "ζητεῖτε" εἶπεν "καὶ εὑρήσετε, κρούετε καὶ ἀνοιγήσεται, αἰτεῖσθε καὶ δοθήσεται ὑμῖν". (3) κρούει μὲν οὖν κατὰ τὴν ζήτησιν ὁ πρὸς ἐρώτησιν καὶ ἀπόκρισιν λόγος τὴν θύραν τῆς ἀληθείας κατὰ τὸ φαινόμενον, διοιχθέντος δὲ τοῦ ἐμποδὼν κατὰ τὴν ἔρευναν ἐπιστημονικὴ ἐγγίνεται θεωρία. **2** (1) τοῖς οὕτως, οἶμαι, κρούουσιν ἀνοίγνυται τὸ ζητούμενον καὶ τοῖς οὕτως αἰτοῦσιν τὰς πεύσεις καὶ τὰς ἀποκρίσεις κατὰ τὰς γραφὰς ἐφ' ὃ βαίνουσιν ἐκ τοῦ θεοῦ γίνεται, ἡ δόσις τῆς θεοδωρήτου γνώσεως καταληπτικῶς διὰ τῆς λογικῆς ὄντως ἐκλαμπούσης ζητήσεως. (2) οὐ γὰρ εὑρεῖν μὲν οἷόν τε, μὴ ζητῆσαι δέ· οὐδὲ ζητῆσαι μέν, οὐχὶ δὲ ἐρευνήσασθαι· οὐδὲ διερευνήσασθαι μέν, οὐχὶ δὲ διαπτύξαι καὶ ἀναπετάσαι δι' ἐρωτήσεως εἰς σαφήνειαν ἄγοντα τὸ ζητούμενον, οὐδ' αὖ διὰ πάσης ἐξετάσεως χωρήσαντα μὴ οὐχὶ λοιπὸν τὸ ἔπαθλον λαβεῖν, τὴν ἐπιστήμην τοῦ ζητουμένου. (3) ἀλλ' ἔστι μὲν εὑρεῖν τὸν ζητήσαντα, ζητῆσαι δέ, εἰ οἰηθείη πρότερον μὴ εἰδέναι. πόθῳ δὴ ἐντεῦθεν ἀγόμενος πρὸς τὴν εὕρεσιν τοῦ καλοῦ εὐγνωμόνως ζητεῖ, ἀφιλονίκως, ἀφιλοδόξως || ἐρω- 347r
τώμενος καὶ ἀποκρινόμενος, πρὸς δὲ καὶ αὐτὰ ἐπισκεπτόμενος τὰ
81,1 λεγόμενα. (4) ἐχομένους γὰρ καθήκει οὐ μόνον τῶν γραφῶν τῶν θείων, ἀλλὰ καὶ τῶν ἐννοιῶν τῶν κοινῶν τὰς ζητήσεις ποιεῖσθαι εἴς τι πέρας ὠφέλιμον τῆς εὑρέσεως καταληγούσης. (5) ἐκδέχεται γὰρ ἄλλος τόπος τε καὶ ὄχλος τοὺς ταραχώδεις τῶν ἀνθρώπων καὶ τὰς ἀγοραίους εὑρησιλογίας, τὸν δὲ τῆς ἀληθείας ἐραστήν τε ἅμα καὶ γνώρι-

---

80,3 **lac. ante* ἀλλ' *stat.* Arnim **80,15** *τὰς πεύσεις καὶ τὰς ἀποκρίσεις Früchtel *sec.* Cod. Lavra B 113, fol. 18r; τὰς πεύσεις L **80,18** οὐδὲ[2] Sylburg; οὐχὶ δὲ L

 | DOI: 10.1163/9789004325289_004

The Eighth *Stromateus*

'Seek and you will find'

I. 1 (1) However, the most ancient philosophers were not lead towards disputing and puzzling over things. So how could we, who hold upon the really true philosophy and whom Scripture straightway commands to seek for the sake of finding, so that we investigate? (2) For the more recent Greek philosophers, carried away by vain and endless ambition, have been reduced to useless foolery by refutations and contentious arguments. In contrast, the barbarian philosophy rejects all contention and says: "Seek and you will find, knock and it will open, ask and it will be given to you."[1] (3) And so, in the course of seeking, the argument in the form of questions and answers knocks on the door of truth, in accordance with the apparent. And when, in the course of investigation, the obstacle is opened, epistemic contemplation comes in. **2** (1) I think that if we knock in this manner, the thing sought will open to us; and if we ask in this manner, through questions and answers concerning Scriptures, the goal we pursue will come to us from God: the gift of God-given knowledge, shining out in an illuminating way by means of a truly rational inquiry. (2) For we cannot find unless we seek and we cannot seek except by investigating. And we cannot fully investigate the thing sought unless we open and unfold it, that is, clarify it by asking questions. But once we have gone through all this examination it is impossible not to win the prize in the end, that is, not to attain knowledge of the thing sought. (3) So one who has sought can find, but he will seek only if he thinks that he has not known beforehand. Then, led by desire towards the discovery of the beautiful, he seeks prudently; he is asked and gives answers without rivalry or regard for fame, and he also examines those things that people say. (4) For if our discovery is to stop at some useful end, we should adhere in our investigations not only to divine Scriptures, but also to common notions. (5) There is another crowded place awaiting confused people and the sophistries of the market. But it is fitting for the person who is a lover of truth as

1 Matt 7:7.

μον εἰρηνικὸν εἶναι κἀν ταῖς ζητήσεσι προσῆκεν, δι' ἀποδείξεως ἐπιστημονικῆς ἀφιλαύτως καὶ φιλαλήθως εἰς γνῶσιν προσιόντα καταληπτικήν.

II. 3 (1) Τίς ἂν οὖν ἄλλη βελτίων ἢ ἐναργεστέρα μέθοδος εἰς ἀρχὴν τῆς τοιᾶσδε εἴη ἂν διδασκαλίας ἢ τὸ προταθὲν ὄνομα λόγῳ διελθεῖν οὕτω σαφῶς ὡς πάντας ἀκολουθῆσαι τοὺς ὁμοφώνους; ἆρ' οὖν τοιοῦτόν ἐστι ⟨τὸ⟩ ὄνομα τῆς ἀποδείξεως οἷόν περ τὸ βλίτυρι, φωνὴ μόνον οὐδὲν σημαίνουσα; (2) καὶ πῶς οὔθ' ὁ φιλόσοφος οὔθ' ὁ ῥήτωρ, ἀλλ' οὐδὲ ὁ δικαστὴς ὡς ἄσημον ὄνομα προφέρεται τὴν ἀπόδειξιν, οὔτε τις τῶν δικαζομένων ἀγνοεῖ τὸ σημαινόμενον ὅτι οὐχ ὑπάρχει; αὐτίκα ὡς ὑποστατὸν πορίζονται τὴν ἀπόδειξιν οἱ φιλόσοφοι, ἄλλος ἄλλως.

(3) Περὶ παντὸς τοίνυν τοῦ ζητουμένου εἴ τις ὀρθῶς διαλαμβάνοι, οὐκ ἂν ἐφ' ἑτέραν ἀρχὴν ὁμολογουμένην μᾶλλον ἀναγάγοι τὸν λόγον ἢ τὸ πᾶσι τοῖς ὁμοεθνέσι τε καὶ ὁμοφώνοις ἐκ τῆς προσηγορίας ὁμολογούμενον σημαίνεσθαι. (4) εἶτα ἐντεῦθεν ὁρμηθέντα ζητεῖν ἀνάγκη, εἰ ὑπάρχει τὸ σημαινόμενον τοῦτο π⟨ερὶ⟩ οὗ ὁ λόγος εἴτε καὶ μή· ἐφεξῆς δέ, εἴπερ || ὑπάρχειν δειχθείη, ζητητέον τούτου τὴν 347v φύσιν ἀκριβῶς, ὁποία τίς ἐστιν, καὶ μή ποτε ὑπερβαίνει⟨ν⟩ τὴν δοθεῖσαν τάξιν.

4 (1) Εἰ δ' οὐκ ἀρκεῖ τοῦτο μόνον ἁπλῶς εἰπεῖν περὶ τοῦ ζητουμένου τὸ δόξαν (ἔξεστι γὰρ καὶ τὸν ἀντικαθιστάμενον ἐπ' ἴσης ἀνταποφήνασθαι ὃ βούλεται), ἀλλὰ πιστώσασθαι χρὴ τὸ λεχθέν, εἰ μὲν εἰς ὁμοίως ἀμφισβητούμενον ἀναφέροιτο αὐτοῦ ἡ κρίσις κἀκείνου πάλιν ὁμοίως εἰς ἀμφισβητούμενον ἕτερον, εἰς ἄπειρον προβήσεται καὶ ἀναπόδεικτον ἔσται, εἰ δ' εἰς ὁμολογούμενον ἅπασιν ἡ τοῦ ⟨μὴ⟩ ὁμολογουμένου πίστις ἀναφέροιτο, ἐκεῖνο τὴν ἀρχὴν τῆς διδασκαλίας ποιητέον. (2) πᾶν οὖν τὸ προβληθὲν ὄνομα μεταλαμβάνειν χρὴ εἰς λόγον ὁμολογούμενόν τε καὶ σαφῆ τοῖς κοινωνοῦσι τῆς σκέψεως, ἀρχὴν μὲν τῆς διδασκαλίας ἐσόμενον, ἐξηγησόμενον δὲ τὴν τῶν ζητουμένων εὕρεσιν. (3) φέρε οὖν προβεβλήσθω ὁ ἥλιος τοὔνομα. φασὶν οὖν οἱ

81,9 βελτίων ἢ ἐναργεστέρα Sylburg; βελτίω καὶ ἐναργεστάτη L **81,12** ἐστι τὸ Stählin; ἐστιν L **81,16** πορίζονται] ὁρίζονται Barnes **81,21** τοῦτο περὶ οὗ Stählin; τοῦτό που L **81,23** *ὑπερβαίνειν *conieci* **81,30 f.** μὴ *add.* Potter || ὁμολογουμένου] λεχθέντος *vel* ζητουμένου Barnes **82,4** δὲ Potter; τε L

well as its disciple to be peaceful even in controversies and to proceed towards apprehensive knowledge by epistemic demonstration, not loving oneself, but the truth.

Teaching on demonstration

II. 3 (1) What other method, then, of arriving at the starting-point of such a teaching could be better or plainer than to explain the proposed name with an account so clearly that all who speak the same language will follow? Now, is the name 'demonstration' of such a kind as 'blituri', a mere sound meaning nothing? (2) But why, then, is it so that no philosopher, rhetor, or judge proposes demonstration as a meaningless name and that none of the litigants ignores the fact that the signified item exists? Philosophers, at any rate, furnish demonstration as something real, each party in a different way.

(3) Indeed, it is true of everything sought that if one wants to argue correctly about it, he will refer the argument to an agreed starting-point, which is no other than what people of the same nation and language agree to be the meaning of the noun. (4) Then, having started from there, one must inquire whether the signified item, with which the argument is concerned, exists or not. Next, if it is shown to exist, one must precisely investigate its nature, what it is like, and never transgress the given order.

4 (1) If it is not sufficient to say simply what seems to be the case about the thing sought (for an opponent can equally assert whatever he wants to the contrary), but it is necessary to confirm what has been said, then, if judgment about it is referred to something equally disputed, and judgment about the latter again to something else that is disputed as well, [the argument] will proceed to infinity and [what has been said] will be indemonstrable. Yet if credence concerning that on which there is no agreement is referred to something everyone agrees with, then the latter should be taken as the starting-point of teaching. (2) Every proposed name, then, has to be changed into an account that is agreed and clear to those who participate in the research. This will be the starting-point of teaching, one that will show the way to the discovery of the things sought. (3) For example, let us propose the name 'sun'. The Stoics say that

Στωϊκοὶ τοῦτ’ εἶναι “ἄναμμα νοερὸν ἐκ θαλαττίων ὑδάτων”. ἆρ’ οὐκ ἀσαφέστερος ὁ λόγος αὐτοῦ τοῦ ὀνόματος, ἄλλης ἀποδείξεως δεόμενος εἰ ἀληθής ἐστιν; ἄμεινον οὖν εἰπεῖν κοινῷ καὶ σαφεῖ τῷ λόγῳ ἥλιον ὀνομάζεσθαι “τὸ λαμπρότατον τῶν κατ’ οὐρανὸν ἰόντων”· πιστότερος γάρ, οἶμαι, καὶ σαφέστερος καὶ πᾶσιν ὁμοίως ὁμολογούμενος ὁ λόγος οὗτος. | **III. 5** (1) ὡσαύτως δὲ καὶ τὴν ἀπόδειξιν πάντες ἄνθρωποι ὁμολογή|σαιεν *ἂν λόγον* εἶναι τοῖς ἀμφισβητουμένοις ἐκ τῶν ὁμολογουμένων |ἐκπορίζοντα τὴν πίστιν.

(2) Οὐ μόνον δὲ ἀπόδειξις καὶ πίστις καὶ γνῶσις, ἀλλὰ καὶ πρόγνωσις λέγεται διχῶς, ἡ μὲν ἐπι‖στημονική τε καὶ βεβαία, 348r ἄλλη δὲ μόνον ἐλπιστική. (3) κυριώτατα μὲν οὖν ἀπόδειξις λέγεται ἡ τὴν ἐπιστημονικὴν πίστιν ἐντιθεῖσα ταῖς τῶν μανθανόντων ψυχαῖς, δοξαστικὴ δὲ ἡ ἑτέρα, ὡς καὶ ἄνθρωπος ὁ μὲν ὄντως ἄνθρωπος ὁ τὰς κοινὰς φρένας κεκτημένος, ὃ δὲ ἄγριος καὶ θηριώδης. οὕτω τοι καὶ ὁ κωμικὸς ἔλεγεν· |ὡς χαρίεις ἐστὶν ἄνθρωπος, ἔστ’ ἂν ἄνθρωπος ᾖ. (4) καὶ ἐπὶ βοὸς καὶ ἵππου καὶ κυνὸς ὁ αὐτὸς λόγος παρὰ τὴν ἀρετὴν τοῦ ζῴου ἢ κακίαν. εἰς γὰρ τὴν τοῦ γένους τελειότητα βλέποντες ἐπὶ τὰ κυριώτατα τῶν σημαινομένων ἐρχόμεθα. (5) αὐτίκα ἰατρὸν νοοῦμεν ᾧ μηδὲν τῆς ἰατρικῆς δυνάμεως ἐνδεῖ, γνωστικὸν δὲ ᾧ μηδὲν λείπει τῆς ἐπιστημονικῆς εἰδήσεως.

6 (1) Καὶ διαφέρει ἔνδειξις συλλογισμοῦ ᾗ τὸ μὲν ἐνδεικνύμενον ἑνός ἐστι δηλωτικὸν ἓν ὑπάρχον καὶ αὐτό, ὡς τοῦ μηκέτι εἶναι παρθένον ἔνδειξιν εἶναι τὸ κυεῖν φαμεν, τὸ δὲ συλλογισμῷ λαμβανόμενον ἓν ὑπάρχον ἕπεται πλείοσιν, ὡς τοῦ Πύθωνα προδιδόναι Βυζαντίους, εἰ οὕτως ἔτυχεν, οὐχ ἕν, ἀλλὰ πλείω λαμβάνεται τὰ δηλωτικά.

(2) Καὶ |τὸ μὲν *περαίνειν* ἐξ ὁμολογουμένων συλλογίζεσθαί ἐστιν, τὸ δέ γε ἐξ ἀληθῶν τι περαίνειν ἀποδεικνύειν ἐστίν, ὥστ’ εἶναι σύνθετόν τινα τῆς ἀποδείξεως τὴν *νόησιν* ἔκ τε τοῦ τὰ λαμβανόμενα πρὸς τὰ ζητούμενα ἀληθῆ λαμβάνειν κἀκ τοῦ τὸ συμπέρασμα αὐτοῖς ἀκόλουθον ἐπιφέρεσθαι. (3) εἰ *δ’ ἤτοι* μὴ ὑπάρχοι τὸ πρότερον ἢ μὴ ἕποιτο αὐτῷ τὸ δεύτερον, οὐκ ἀπέδειξεν μέν, συνελογίσατο δέ. (4) τὸ μὲν γὰρ οἰκεῖον ἐπενεγκεῖν συμπέ‖ρασμα τοῖς λήμμασιν συλλογίσασθαι μόνον ἐστίν, 348v τὸ δὲ καὶ τῶν λημμάτων ἕκαστον ὑπάρχειν ἀληθὲς οὐ συλλογίσασθαι μόνον, ἀλλὰ καὶ ἀποδεδειχέναι. (5) περαίνειν δ’ ἐστίν, ὡς καὶ ἐκ τοῦ

82,13 ἂν λόγον Dindorf (*cf.* Sylburg: λόγον εἶναι *vel* εὔλογον εἶναι λόγον); εὔλογον L **82,14** ἐκπορίζοντα Sylburg (*cf.* 83,32); ἐμπορίζοντα L **82,21** *χαρίεις Sylburg; χαρίης L (χαρίεν Stählin *sec.* Stobaeum, *Anth.* III 12) ‖ *ἐστὶν ἄνθρωπος, ἔστ’ ἂν ἄνθρωπος] ἐστ’ ἄνθρωπος, ἂν ἄνθρωπος Stählin (*sec.* Stobaeum) **82,30** Πύθωνα Dindorf; Πύθωνος L **83,2** περαίνειν Potter; παραινεῖν L **83,4** *νόησιν Pohlenz; ὄνησιν L **83,6** δ’ ἤτοι Sylburg; δή τοι L

it is 'intelligent ignited mass from the waters of the sea'. Is not such an account less clear than the name itself? Does it not need another proof to demonstrate if it is true? It is better to use a common and clear account and call the sun 'the brightest thing traversing the sky'. This account, I think, is more credible, clear, and agreed to by all alike. III. 5 (1) In the same vein, all human beings will agree that demonstration is an argument that furnishes credence from something agreed to something disputed.

(2) However, not only demonstration, credence, and knowledge are spoken of in two ways, but foreknowledge, too: One is epistemic and firm, and the other merely based on expectation. (3) Now demonstration in the proper sense is the one that inserts epistemic credence to the souls of students, whereas the other kind produces mere opinion. Similarly, we call 'human being' the truly human being, i.e. the one equipped with common reason, but also the one that is savage and brutish. Thus the comic writer says: "How graceful the human being, if human being it is."[2] (4) And the same is true of ox, horse or dog, with respect to the virtue or vice of the animal in question. We have to look at what is perfect within a genus to arrive at the most proper meaning of names. (5) For example, by 'doctor' we mean someone with a perfect capacity to heal, and by 'gnostic' someone with flawless epistemic knowledge.

6 (1) Further, indication differs from deduction in that the item which indicates something shows one thing while being itself one. For example, when a woman gives birth to a child we call it an indication that she is no longer a virgin. But when reached by deduction, one thing follows from more than one item. Thus, for instance, in the case of the claim that Pytho betrayed the people of Byzantium, we take into account not one but many items to show it.

(2) Further, to conclude from agreed premisses is to deduce, but in order to demonstrate we must draw a conclusion from true premisses. Thus the conception of demonstration is a composite one consisting of two things: true assumptions are assumed in view of the things sought and a conclusion following from these assumptions is drawn. (3) If either the first does not obtain, or if it is not followed by the second, one has made a deduction, but not a demonstration. (4) By drawing an appropriate conclusion from our premisses we only make a deduction; however, when each of our premisses is also true, we have not only deduced a conclusion, but also demonstrated it. (5) To conclude (*perainein*), as

2 Menander, fr. 484 (Koerte).

ὀνόματος δῆλον, τὸ ἄγειν ἐπὶ τὸ πέρας. ἔστι δὲ δήπου πέρας ἐν ἑκάστῳ λόγῳ τὸ ζητούμενον, ὃ δὴ καὶ συμπέρασμα καλεῖται. (6) οὐδεὶς δὲ ἁπλοῦς καὶ πρῶτος λόγος ὀνομάζεται συλλογισμός, κἂν ἀληθὴς ᾖ, ἀλλ' ἔστι τοὐλάχιστον ἐκ τριῶν τοιούτων *σύνθετος*, δυεῖν μὲν ὡς λημμάτων, ἑνὸς δὲ ὡς συμπεράσματος.

(7) Ἤτοι δὲ πάντα ἀποδείξεως δεῖται ἢ καί τινα ἐξ αὐτῶν ἐστι πιστά. 7 (1) ἀλλ' εἰ μὲν τὸ πρότερον, ἑκάστης ἀποδείξεως ἀπόδειξιν αἰτοῦντες εἰς ἄπειρον ἐκβησόμεθα καὶ οὕτως ἀνατραπήσεται ἡ ἀπόδειξις· εἰ δὲ τὸ δεύτερον, ταῦτα αὐτὰ τὰ ἐξ αὑτῶν πιστὰ τῶν ἀποδείξεων ἀρχαὶ γενήσονται. αὐτίκα οἱ φιλόσοφοι ἀναποδείκτους ὁμολογοῦσι τὰς τῶν ὅλων ἀρχάς. (2) ὥστ' εἴπερ ἐστὶν ἀπόδειξις, ἀνάγκη πᾶσα πρότερον εἶναί τι πιστὸν ἐξ ἑαυτοῦ, ὃ δὴ πρῶτον καὶ ἀναπόδεικτον λέγεται. ἐπὶ τὴν ἀναπόδεικτον ἄρα πίστιν ἡ πᾶσα ἀπόδειξις ἀνάγεται. (3) εἶεν δ' ἂν καὶ ἄλλαι τῶν ἀποδείξεων ἀρχαὶ μετὰ τὴν ἐκ πίστεως πηγήν, τὰ πρὸς αἴσθησίν τε καὶ νόησιν ἐναργῶς φαινόμενα. (4) τὰ μὲν γὰρ πρὸς αἴσθησιν *σύμπαντα* ἐστὶν ἁπλᾶ τε καὶ ἄλυτα, τὰ δὲ πρὸς νόησιν ἁπλᾶ τε καὶ λογικὰ καὶ πρῶτα, τὰ δὲ ἐξ αὐτῶν γεννώμενα σύνθετα μέν, οὐδὲν δ' ἧττον ἐναργῆ καὶ πιστὰ καὶ λογικώτερα τῶν πρώτων. (5) ἀκολούθου καὶ μαχομένου οὖν ⟨διαγνωστική⟩ ἐστιν ἥνπερ ἰδίαν λόγου δύναμιν πεφυκυῖαν ἅπαντες ἔχομεν φύσει. ‖
(6) ἐὰν οὖν τις εὑρεθῇ λόγος τοιοῦτος οἷος ἐκ τῶν ἤδη πιστῶν τοῖς 349r
οὔπω πιστοῖς ἐκπορίζεσθαι τὴν πίστιν δυνάμενος, αὐτὸν τοῦτον εἶναι φήσομεν οὐσίαν ἀποδείξεως.

(7) Εἴρηται δὲ ὡς καὶ τὸ τῆς πίστεως ¦καὶ τὸ τῆς ἀποδείξεως γένος διττόν, τὸ μὲν πειθὼ μόνον ταῖς ψυχαῖς ¦τῶν ἀκουόντων, τὸ δὲ ἐπιστήμην ἀπεργαζόμενον. (8) εἰ μὲν οὖν ἐκ τῶν πρὸς αἴσθησιν καὶ νόησιν ἐναργῶν ἄρξαιτό τις, κἄπειτα τὸ οἰκεῖον ἐπενέγκοι συμπέρασμα, ὄντως ἀποδείκνυσιν, εἰ δ' ἐξ ἐνδόξων μόνον, οὐ μὴν πρώτων γε, τουτέστιν οὔτε πρὸς αἴσθησιν οὔτε πρὸς νόησιν ἐναργῶν, εἰ μὲν οἰκεῖον ἐπιφέροι συμπέρασμα, συλλογιεῖται μέν, οὐ μὴν ἐπιστημονικήν γε ποιήσεται τὴν ἀπόδειξιν, εἰ δ' οὐκ οἰκεῖον, οὐδὲ συλλογιεῖται τὴν ἀρχήν.

8 (1) Διαφέρει δ' ἀναλύσεως ἀπόδειξις· ἕκαστον μὲν γὰρ τῶν ἀποδεικνυμένων διά τινων ἀποδεικνυμένων ἀποδείκνυται, προαποδεδειγμένων κἀκείνων ὑφ' ἑτέρων, ἄχρις ἂν εἰς τὰ ἐξ ἑαυτῶν πιστὰ ἀναδράμωμεν ἢ εἰς τὰ πρὸς αἴσθησίν τε καὶ νόησιν ἐναργῆ, ὅπερ

83,14 *σύνθετος Barnes; σύνθετον L **83,26** *σύμπαντα L *corr. a sec. manu*; συμβάντα L ‖ ἄλυτα] ἄλογα Barnes **83,27** ταῦτα μὲν *ante* πρῶτα *add.* Barnes **83,29** *διαγνωστική *addidi*; διακριτική *vel* διαιρετική *add.* Stählin **83,32** οὔπω Hervet; οὕτω L

the word shows, is to bring something to an end (*peras*). Now the end-point of every argument is the thing sought, and this is what we call 'conclusion'. (6) A simple and primary statement, even if true, is never called a deduction, because the latter consists of at least three such statements, two of which function as premisses and one as a conclusion.

(7) Either everything needs demonstration, or there are certain things credible by themselves. 7 (1) If the first, we will have an infinite regress by demanding demonstration of every demonstration and so demonstration will be refuted. If the second, these very things that are credible by themselves will become the principles of demonstration. For example, philosophers agree that the principles of the universe are not demonstrated. (2) Thus if demonstration exists, it is absolutely necessary that it is preceded by something credible by itself and this is called primary and undemonstrated. Every demonstration, therefore, is based on undemonstrated credence. (3) But there are supposed to be yet other principles of demonstrations beyond the source springing from credence, namely things plainly apparent to sense-perception and intellection. (4) Those [plainly apparent] to sense-perception are all simple and indissoluble, while those [plainly apparent] to intellection are simple, rational, and primary. Items generated from them are composite, but no less plain and credible, and more rational than the primary ones. (5) Indeed, it is a peculiar capacity of reason, available to all of us by nature, to recognize what is entailed and what is incompatible. (6) So if an argument is found such that it will be able to furnish credence from things already credible to those not credible yet, we will describe this very argument as the essence of demonstration.

(7) However, as we have already said, there are two kinds of credence and demonstration: one producing mere persuasion in the souls of the auditors and one producing epistemic knowledge. (8) So if you start from things plain to sense-perception and intellection, and then infer an appropriate conclusion, you will truly demonstrate. But if you start from those that are merely reputable and not primary, i.e. not plain to sense perception and intellection, then, should you draw an appropriate conclusion, you will make a deduction, but your demonstration will not be an epistemic one. If the conclusion is not appropriate, it will not even be a deduction.

8 (1) Further, demonstration differs from analysis. Each of the things that are demonstrated is demonstrated by means of certain demonstrated premisses and these have been demonstrated by means of other premisses, until we ascend to premisses credible by themselves or those plain to sense-perception

ἀνάλυσις ὀνομάζεται· ἀπόδειξις δέ ἐστιν, ὅταν ἀπὸ τῶν πρώτων ⟨τις
14 εἰς⟩ τὸ ζητούμενον ἀφικνῆται διὰ πάντων τῶν ἐν μέσῳ.
14/15 (2) Χρὴ τοίνυν ¦τὸν ἀποδεικτικὸν ἄνδρα τῆς μὲν ἀληθείας [ὡς τῶν λημμάτων]
15/16 πολλὴν ¦ποιήσασθαι πρόνοιαν, τῶν δὲ ὀνομάτων ἀφροντιστεῖν,
16/17 εἴτε ἀξιώματά ¦τις ἐθέλοι καλεῖν εἴτε προτάσεις εἴτε λήμματα, ὁμοίως δὲ καὶ τοῦ
18 τίνων ὑποκειμένων τί περαίνεται πολλὴν ὡσαύτως πεποιῆσθαι τὴν
πρόνοιαν, εἴτε δὲ περαίνοντα λόγον εἴτε περαντικὸν εἴτε συλλογιστι-
20 κὸν ἐθέλοι τις ὀνομάζειν αὐτόν, ἥκιστα ‖ φροντίζειν. (3) δύο γὰρ ταῦτα 349v
ἐν ἅπασι χρῆναί φημι τὸν ἀποδεικτικὸν φυλάττειν, τὰ μὲν λήμματα
ἀληθῆ λαμβάνειν, ἀκόλουθον δ' αὐτοῖς ἐπιφέρειν ⟨τὸ⟩ συμπέρασμα,
ὅπερ τινὲς καὶ ἐπιφορὰν καλοῦσιν, τὸ ἐπιφερόμενον τοῖς λήμμασιν.

(4) Περὶ παντὸς τοῦ ζητουμένου καθ' ἕκαστον πρόβλημα διαφόρων
25 μὲν δεῖ τῶν λημμάτων, οἰκείων δὲ τῷ προβληθέντι, καὶ ⟨τὸ⟩ προ-
βληθὲν *αὐτὸ* εἰς λόγον μεταλαμβάνειν ἀναγκαῖον, τόν τε λόγον τοῦ-
τον ὁμολογούμενον ἅπασιν εἶναι προσήκει· (5) τῶν δὲ λημμάτων μὴ
οἰκείων τῷ προβλήματι λαμβανομένων οὐκ ἐνδέχεται καλῶς οὐδὲν
αὐτῶν ἐξευρεῖν ⟨***⟩ ἀγνοουμένης ὅλου τοῦ προβλήματος, ὃ καὶ ζήτημα
30 καλεῖται, τῆς φύσεως.
30/31 (6) Ἐν πᾶσιν οὖν τοῖς ζητουμένοις ἔστι τι προ¦γινωσκόμενον (ὃ πάντως ἐξ
31/32 ἑαυτοῦ πιστὸν ὂν ἀναποδείκτως πιστεύε¦ται), ὃ χρὴ ποιεῖσθαι τῆς ζητήσεως
32/33/1 αὐτῶν ὁρμητήριον καὶ τῶν εὑ¦ρῆσθαι δοκούντων κριτήριον· ¦ **IV. 9** (1) πᾶσα
85,1 γὰρ ζήτησις ἐκ προϋπαρχούσης εὑρίσκεται γνώσεως. εἶναι
δὲ τὴν γνῶσιν τὴν προϋπάρξασαν τοῦ ζητουμένου παντὸς
ποτὲ μὲν τῆς οὐσίας ψιλῶς, ἀγνοουμένων δὲ τῶν ἔργων αὐτῆς (οἷον
λίθων, φυτῶν, ζῴων, ὧν τὰς ἐνεργείας ἀγνοοῦμεν) ἢ παθῶν ἢ δυνά-
5 μεων ἢ ἁπλῶς εἰπεῖν [ἓν] τῶν ὑπαρχόντων τοῖς οὖσιν· (2) ἐνίοτε δὲ
γιγνώσκεσθαι μέν τι τούτων τῶν δυνάμεων ἢ παθῶν ἤ τινα τούτων,
ὡς τῆς ψυχῆς τὰς ἐπιθυμίας καὶ τὰ πάθη, ἀγνοεῖσθαι δὲ καὶ ζητεῖ-
σθαι τὴν οὐσίαν· (3) ἐν πολλοῖς δέ, τῆς νοήσεως αὐτῆς τῆς ἡμετέρας

84,13 f. τις εἰς *add.* Stählin **84,15** *ὡς τῶν λημμάτων *delevi*; ὡς *del.* Stählin **84,22** τὸ *add.* Wilamowitz **84,23** τοῖς λήμμασιν *del.* Barnes **84,24** καὶ *post* ζητουμένου *add.* Barnes **84,25** τὸ *add.* Bunsen **84,26** αὐτὸ Bunsen; αὐτῶι L **84,28** **lac. post* καλῶς *ind.* Schwartz **84,29** *αὐτῶν *conieci*; αὐτῶι L ‖ **lac. post* ἐξευρεῖν *indicavi* ‖ ἀγνοουμένης Wilamowitz; ἀγνοουμένου L **84,31** ὃ Stählin; τὸ L **85,2** *δυνατὸν *post* δὲ *add.* Schwartz **85,3** *δὲ *del.* Schwartz **85,4** φυτῶν Sylburg; πάντων L **85,5** ἓν τῶν ὑπαρχόντων] ἓν *del.* Stählin; τῶν ἐνυπαρχόντων Potter

and intellection; and this is called 'analysis'. But when we reach the thing sought through all the intermediates, it is demonstration.

(2) Thus a man practiced in demonstration must be very careful about truth, but he need not be concerned about names, whether one prefers to speak of axioms, propositions, or premisses. Likewise, he must be very careful about what conclusion is drawn from what suppositions, but not be at all concerned whether one prefers to call it a 'concluding argument', or a 'concludent', or a 'syllogistic' one. (3) These, I claim, are two things that a man practiced in demonstration must always observe: he must assume true premisses and infer a conclusion that follows from them, i.e. he must reach what some call an 'inference', that which is inferred from the premisses.

Method of discovery

8 (4) For everything sought in the case of each problem we need premisses that are different, but appropriate to what is proposed. What is proposed must itself be changed into an account and the account should be agreed to by all. (5) If the assumed premisses are not appropriate to the problem, it is impossible to succeed in finding anything on their basis[3] (...) if the nature of the problem, also called the question, is unknown as a whole.

(6) In everything we seek, then, there is something known in advance (which, insofar as it is credible by itself, is necessarily believed without demonstration), and this should be taken as a base of inquiry about the thing sought and as the criterion of everything that seems to have been found. IV. 9 (1) For everything sought is found on the basis of pre-existent knowledge. And sometimes pre-existent knowledge regarding everything sought concerns solely the substance, while we are ignorant of its works (as in the case of stones, plants, or animals, whose activities we do not know), affections, or powers, or generally speaking attributes of beings. (2) At other times, some of these powers, affections, or other such things, like desires and affections of the soul, are known, but we are ignorant about, and seek for, the substance. (3) In many cases, while this understanding of ours presupposes all these things, we seek which of these

3 Text uncertain.

ὑποτιθεμένης ἑαυτῇ ταῦτα πάντα, τὴν ζήτησιν εἶναι, τίνι ‖ τῶν οὐσιῶν 350ʳ ἂν οὕτω μὲν ὑπάρχῃ· (4) ἀμφοτέρων γάρ, τῆς τε οὐσίας τῆς τε ἐνεργείας, τὰς ἐπινοίας ἐν τῇ διανοίᾳ λαβόντες οὕτως ἐπὶ τὴν ζήτησιν ἐρχόμεθα. (5) ἔστιν δὲ ὧν καὶ τὰς ἐνεργείας εἰδότες ἅμα ταῖς οὐσίαις ἀγνοοῦμεν τὰ παθήματα.

(6) Ἔστιν οὖν ἡ μέθοδος τῆς εὑρέσεως τοιαύτη· ἀρκτέον γὰρ ἐξ αὐτοῦ τοῦ γνωρίζειν τὰ προβλήματα. (7) πολλάκις γοῦν ἐξαπατᾷ τὸ τῆς λέξεως σχῆμα καὶ συγχεῖ καὶ ταράττει τὴν διάνοιαν, ὥστε μὴ ῥᾳδίως εὑρίσκειν ἐκ ποίας ἐστὶ διαφορᾶς, οἷον εἰ ⟨ζῷον ἢ⟩ μὴ ζῷον τὸ κυούμενον· (8) ἔχοντες γὰρ καὶ ζῴου τι νόημα καὶ κυουμένου ζητοῦμεν εἰ τῷ κυουμένῳ ζῴῳ εἶναι ὑπάρχει, τοῦτο δ' ἐστίν, εἰ τῇ κυουμένῃ οὐσίᾳ τό τε κινεῖσθαι δύνασθαι καὶ ἔτι τὸ αἰσθάνεσθαι ὑπάρχει. (9) ὥστε ἐνεργειῶν ἐστι καὶ παθῶν ἡ ζήτησις ἐπὶ προγινωσκομένης οὐσίας.

10 (1) Εὐ|θέως οὖν τὸν προβάλλοντα ἀντερωτητέον τί ποτε καλεῖ ζῷον· μάλιστα γὰρ τοῦτο ποιητέον, ἐπειδὰν εἰς διαφόρους χρήσεις ἠγμένον ἴδωμεν τοὔνομα· καὶ διερευνητέον εἴτε ἀμφισβητούμενόν ἐστι τὸ σημαινόμενον ἐκ τῆς προσηγορίας εἴθ' ὁμολογούμενον ἅπασιν. (2) εἰ γὰρ ζῷον εἴποι καλεῖν ὅτιπερ ἂν αὐξάνῃ καὶ τρέφηται, πάλιν αὖ προσανερωτήσομεν εἰ καὶ τὰ φυτὰ *νομίζει* ζῷα, κἄπειτα φάντος μὲν οὕτως ἐπιδεικνύειν ἤδη χρὴ [τι] τὸ κυούμενον ⟨αὐξανόμενόν⟩ τε καὶ τρεφόμενον. (3) Πλάτων γὰρ καὶ τὰ φυτὰ ζῷα καλεῖ "τοῦ τρίτου τῆς
86,1 ψυχῆς εἴδους", τοῦ ἐπιθυμητικοῦ, μόνου μετέχοντα, Ἀριστοτέλης δὲ τῆς *φυτικῆς* τε καὶ θρεπτικῆς ψυχῆς μετέχειν οἴεται ‖ τὰ φυτά, ζῷα 350ᵛ δ' ἤδη προσαγορεύειν οὐκ ἀξιοῖ· τὸ γὰρ δὴ τῆς ἑτέρας ψυχῆς τῆς αἰσθητικῆς μετέχον τοῦτο μόνον ἀξιοῖ καλεῖσθαι ζῷον. (4) οὐ μὴν οἵ γε Στωϊκοὶ τὴν φυτικὴν δύναμιν ἤδη ψυχὴν ὀνομάζουσιν. (5) ἀποφήσαντος δὲ τοῦ προβαλόντος εἶναι ζῷα καὶ τὰ φυτά, δείξομεν ἑαυτῷ μαχόμενα λέγειν. τῷ γὰρ τρέφεσθαί τε καὶ αὐξάνεσθαι τὸ ζῷον ὁρισάμενος, ἀλλὰ καὶ τὸ φυτὸν οὐκ εἶναι ζῷον ἀποφηνάμενος, οὐδὲν ἄλλο ἔοικεν λέγειν ἢ ὅτι τὸ τρεφόμενόν τε καὶ αὐξανόμενον καὶ ζῷόν ἐστι καὶ οὐ ζῷον. (6) τί οὖν βούλεται μαθεῖν, εἰπάτω, ἆρά γε εἰ αὔξεται καὶ τρέφεται τὸ κατὰ γαστρὸς ἢ *εἰ* αἰσθήσεώς τινος ἢ καὶ τῆς καθ' ὁρμὴν κινήσεως αὐτῷ μέτεστιν. (7) κατὰ μὲν γὰρ Πλάτωνα τὸ φυτὸν ἔμψυχόν τε καὶ ζῷον, κατὰ δὲ Ἀριστοτέλη ζῷον μὲν οὔπω,

---

85,10 οὕτω μὲν ὑπάρχῃ] οὕτως ἐνυπάρχῃ Bunsen **85,17** *ζῷον ἢ *add.* Schwartz **85,27** νομίζει Sylburg; νομίζοι L **85,28** τι *del.* Wilamowitz; ὅτι *coni.* Schwartz ‖ αὐξανόμενον *add.* Stählin **85,29** μὲν *post* Πλάτων *add.* Barnes **86,2** φυτικῆς Potter; φυσικῆς L **86,6** ἑαυτῷ] ἑαυτοῖς Barnes **86,9** φυτὸν *post* αὐξανόμενον *add.* Rashed **86,11** εἰ Sylburg; εἰς L ‖ ἢ² *del.* Barnes (*cf.* 87,32)

attributes belong to which substances: (4) Having conceived in our mind the notions of both a substance and an activity, this is how we embark on inquiry. (5) There are also cases in which we know activities and substances, but do not know affections.

(6) Now the method of discovery is like this. We must start by recognizing the problems. (7) For the form of expression is often deceptive and confounds and confuses our mind, and it is, therefore, not easy to find the kind of thing it expresses, as with the problem whether the embryo is an animal or not. (8) Here we have a certain concept of an animal and of an embryo and we inquire whether it belongs to the embryo to be an animal, i.e. whether the capacity to move and also to perceive belongs to the substance conceived in the womb. (9) So the inquiry concerns activities and affections of a substance known in advance.

10 (1) Thus we should immediately respond to the person who has proposed this question by a counter question what it is he calls 'animal'. This is particularly important when we see that the name has been applied in different uses. Next we must investigate whether the meaning expressed by the noun is disputed, or whether everyone agrees with it. (2) If he answers that he gives the name 'animal' to everything that grows and is nourished, we should again respond by asking whether he considers plants to be animals, too. If he affirms this, then it is already time to show that the embryo grows and is nourished. (3) As a matter of fact, Plato calls even plants 'animals', such as partake only of "the third kind of the soul", the appetitive. Aristotle, on the other hand, thinks that plants partake of the vegetative and nutritive soul, but he does not deem it right to call them 'animals' yet. For he only deems it right to call 'animals' those beings which partake of a different soul, viz. the sentient one. (4) As for the Stoics, they do not give the name 'soul' even to the vegetative capacity yet. (5) But if the person who has proposed the question denies that plants are animals, too, we will show him that he contradicts himself. For if he defines the animal as that which grows and is nourished, but asserts that it is not the case that plants are also animals, he appears to be saying nothing less than this: that something which is nourished and grows is an animal and is not an animal at the same time. (6) So let him tell us what he wants to learn: whether the thing in the womb grows and is nourished, or rather whether it partakes of some sense-perception or movement by impulse? (7) For according to Plato, plants have souls and they are animals; whereas according to Aristotle, they are not

λείπει γὰρ αὐτῷ τὸ αἰσθητικόν, ἔμψυχον δὲ ἤδη· ἔστι γοῦν αὐτῷ τὸ ζῷον οὐσία ἔμψυχος αἰσθητική· (8) κατὰ δὲ τοὺς Στωϊκοὺς οὔτε ἔμψυχον οὔτε ζῷόν ἐστι τὸ φυτόν· ἔμψυχος γὰρ οὐσία τὸ ζῷον. **11** (1) εἰ τοίνυν ἔμψυχον τὸ ζῷον, ἡ ψυχὴ δὲ φύσις αἰσθητική, δῆλον ὡς αἰσθητικὸν ἤδη τὸ ζῷον.

(2) Εἰ μὲν οὖν ἀντερωτηθεὶς ὁ ζητήσας εἰ ζῷον τὸ κατὰ γαστρὸς *εἴποι καλεῖν* ζῷον [ἐπὶ] τὸ τρεφόμενον καὶ αὖξον, ἔχει τὴν ἀπόκρισιν· εἰ δὲ φάσκοι τοῦτ' εἶναι ὃ ζητεῖ, πότερον αἰσθητικὸν ἤδη τὸ κυούμενον ἢ καὶ καθ' ὁρμήν τινά ἐστι κινητικὸν ἑαυτοῦ, ἤδη *σαφὴς* γίνεται ἡ διερεύνησις τοῦ πράγματος, οὐ μενούσης ἔτι τῆς περὶ τοὔνομα ἀπάτης· (3) μὴ ἀποκρινόμενος δὲ πρὸς τὸ ἀντερωτηθὲν μηδὲ βουλόμενος τί ποτέ ἐστιν εἰπεῖν ὃ νοῶν ἢ κατὰ τίνος ἐπιφέρων πρά‖γματος τὸ ζῷον ὄνομα τὴν πρότασιν ἐποιήσατο, κελεύων δ' 351r
ἡμᾶς αὐτοὺς διελέσθαι, ἐριστικὸς ὑπάρχων γνωρισθείη. (4) εἰ δὴ δυοῖν τρόποιν ὑπαρχόντοιν, ἑτέρου μὲν τοῦ κατ' ἐρώτησίν τε καὶ ἀπόκρισιν, ἑτέρου δὲ τοῦ κατὰ διέξοδον, ἤρνηται τὸ ἕτερον, ἐπακουσάτω πάντα τὰ εἰς τὸ πρόβλημα διεξιόντων ἡμῶν· εἶτ' ἐπειδὰν τελεώσωμεν, ἐξέσται αὐτῷ τότε περὶ ἑνὸς ἑκάστου ἐν μέρει διαλαμβάνειν. **12** (1) εἰ δὲ διακόπτειν ἐπιχειροίη τὴν ἐξέτασιν πυνθανόμενος, δῆλός ἐστιν *οὐδὲ* ἀκούειν βουλόμενος.

87,1 (2) Ἀλλ' εἰ μὲν ἀποκρίνεσθαι ἕλοιτο, πάντων πρῶτον ἐρωτητέον αὐτὸν ἐφ' ὅ τι φέρει πρᾶγμα τὸ ζῷον ὄνομα, κἀπειδὰν τοῦτο ἀποκρίνηται, πάλιν ἐρωτητέον ὅ τι ποτὲ αὐτῷ σημαίνει τὸ κυούμενον ἢ τὸ κατὰ γαστρός, εἰ [τὰ διαπεπλασμένα ἤδη καὶ τὰ ζῷα] καὶ τὸ σπέρμα αὐτὸ ⟨τὸ⟩ καταβεβλημένον τὸ κατὰ γαστρὸς αὐτῷ σημαίνειν βούλεται ἢ μόνα τὰ διηρθρωμένα τε καὶ ἤδη διαπεπλασμένα, τὰ ἔμβρυα καλούμενα. (3) ἀποκριναμένου δὲ καὶ πρὸς τοῦτο περαίνειν ἤδη τὸ προκείμενον ἐφεξῆς καὶ διδάσκειν χρή.

(4) Εἰ δὲ ἡμᾶς λέγειν βούλοιτο αὐτὸς μὴ ἀποκρινάμενος, ἀκουσάτω· ἐπεὶ σὺ μὴ βούλει λέγειν καθ' ὅτου σημαινομένου λέγεις ὃ προὔβαλες (οὕτω γὰρ ἂν οὐ περὶ σημαινομένων ἐγὼ ἐποιούμην τὸν λόγον, ἀλλ' ὑπὲρ αὐτῶν ἤδη τῶν πραγμάτων ἐσκοπούμην), γίνωσκε τοιοῦτόν τι ποιήσας οἷον εἰ καὶ προὔβαλες εἰ ζῷον ὁ κύων. (5) εἰκότως γὰρ ἂν

86,18 *ζῷον Barnes; ἔμψυχον L **86,20** εἴποι καλεῖν Stählin (*cf.* 85,26); ἔτι καλεῖ τὸ L ‖ ἐπὶ *del.* Stählin **86,22** σαφὴς Heyse; σαφῶς L **86,25** ὃ Potter; ὁ L ‖ νοῶν] νοεῖ Barnes **86,27** *δὴ] δὲ Wilamowitz; *lectionem ms. def.* Barnes **86,33** οὐδὲ Wilamowitz; οὐδὲν L **87,4** εἰ Bunsen; ἢ L ‖ *τὰ διαπεπλασμένα ἤδη καὶ τὰ ζῷα *del.* Barnes ‖ τὰ διαπεπλασμένα ἤδη καὶ τὰ ζῷα καὶ *del.* Pohlenz ‖ μὴ *post* τὰ *add.* Bernays ‖ καὶ τὰ ζῷα *del.* Stählin **87,5** αὐτὸ] αὖ τὸ Potter ‖ [αὐ]τὸ Barnes ‖ τὸ *post* αὐτὸ *add.* Bunsen

animals yet, since they lack a capacity to perceive, but they already have souls. For, in his view, the animal is an ensouled substance having perception. (8) But according to the Stoics, plants are neither ensouled nor animals, for the animal is an ensouled substance. **11** (1) Thus if the animal has a soul, and if the soul is a nature having perception, plainly the animal already has perception.

(2) So if someone who has raised the inquiry whether the embryo is an animal responds to the counter question by saying that he attaches the name 'animal' to that which is nourished and grows, then he has got the answer. But if he says he is seeking something else, namely whether the embryo already has perception, or whether it moves itself by some impulse, then the name is no longer misleading and it becomes clear how the matter should be investigated. (3) If he fails to respond to the counter question, not wishing to explain what he had in mind or what thing he was referring to by the name 'animal' when proposing the question, and tells us to decide it ourselves, then he has shown himself to be an eristic person. (4) So if there are two modes [of argument], one by question and answer, the other by exposition, and [our opponent] has declined one of them, then let him listen to everything we have to say about the problem while we present our exposition. Then, when we have finished, he will have an opportunity to argue about each single point in turn. **12** (1) If he tries to interrupt the exposition by asking questions, it is clear that he does not even want to listen.

(2) However, if he chooses to respond, first of all we must ask him what thing he is referring to by the name 'animal'. And when he answers this, we must ask again what he takes to be the meaning of 'embryo' or 'the thing in the womb'. Is 'the thing in the womb' supposed to mean even the seed deposited [in the womb], or only beings that have been differentiated and formed already, the so-called fetuses? (3) As soon as he responds to this, we should already bring the proposed matter to a conclusion and then teach.

(4) But if he fails to respond and wants us to speak, let him listen: Since you do not want to say in what sense you mean what you have proposed—for otherwise I would not be making an argument about meanings, but would already be looking into the things themselves—you should realize that what you have done is similar to proposing the question whether the dog is an animal. (5) Of course I will say: 'Which dog?' For I will go through them one after the other

εἴποιμι· ποίου κυνός; ἐγὼ γὰρ καὶ περὶ τοῦ χερσαίου καὶ τοῦ θαλαττίου καὶ τοῦ κατ' οὐρανὸν ἄστρου, ἀλλὰ καὶ Διογένους ‖ καὶ τῶν 351v
ἄλλων ἐφεξῆς δίειμι κυνῶν. οὐ γὰρ ἂν μαντευσαίμην πότερον ὑπὲρ πάντων ἐρωτᾷς ἢ τινός. (6) ὅπερ οὖν ἐξ ὑστέρου ποιήσεις, τοῦτ' ἤδη μαθεῖν περὶ ποίου ζητεῖς, σαφῶς εἰπέ. (7) εἰ δὲ περὶ ὀνόματα στρέφῃ, τὸ κυούμενον αὐτὸ δὴ τοῦτο τοὔνομα παντὶ δῆλον ὅτι μήτε ζῷόν ἐστι μήτε φυτόν. ἀλλ' ὄνομά τε καὶ φωνὴ καὶ σῶμα καὶ ὂν καὶ τὶ καὶ πάντα μᾶλλον ἢ ζῷον. καὶ εἴπερ τοῦτο προὔβαλες, ἔχεις τὴν ˡἀπόκρισιν.

13 (1) οὐ μὴν οὐδὲ τὸ σημαινόμενον ἐκ τοῦ "κυούμενον" ὀνόματός ἐστι ζῷον, ἀλλ' ἐκεῖνο μὲν ἀσώματόν τέ ἐστι καὶ *λεκτὸν* καὶ πρᾶγμα καὶ νόημα καὶ πάντα μᾶλλον ἢ ζῷον. (2) ἄλλη δέ τις ⟨ἂν⟩ εἴη τοῦ ζῴου φύσις· ἐναργῶς γὰρ ἐδείκνυτο τοῦ πράγματος αὐτοῦ τοῦ ζητουμένου, λέγω δὲ *τοῦ ἐμβρύου*, τῆς φύσεως ὁποία τίς ἐστιν. ἕτερον ὑπάρχον πρόβλημα τὸ περὶ τῶν σημαινομένων ἐκ τοῦ "ζῷον" ὀνόματος. (3) λέγω τοίνυν, εἰ τοῦτο λέγεις ζῷον τὸ δυνάμενον αἰσθέσθαι τε καὶ κινηθῆναι καθ' ὁρμήν, ζῷόν ἐστιν οὐχ ἁπλῶς τὸ κινούμενον καθ' ὁρμὴν καὶ αἰσθανόμενον· (4) δύναται γὰρ καὶ κοιμᾶσθαι ἢ μὴ παρόντων τῶν αἰσθητῶν μὴ αἰσθάνεσθαι, τὸ δὲ *δύνασθαι* ἤτοι ὁρμᾶν ἢ καὶ αἰσθάνεσθαι πεφυκὸς ζῴου γνώρισμα. (5) τοιοῦτον γάρ τι σημαίνεται ἐκ τούτων, πρῶτον μέν, εἰ τὸ κυούμενον *ἤδη* αἰσθάνεσθαι ἢ *καθ' ὁρμὴν* κινεῖσθαι δύναται, ὅπερ πρόκειται σκοπεῖσθαι, ἕτερον δέ, εἰ τὸ κυούμενον αἰσθ⟨ήσ⟩εσθαι ποτὲ ἢ κινηθήσεσθαι δύναται καθ' ὁρμήν, καθ' ὃ σημαινόμενον οὐδεὶς ζητεῖ *ἐναργὲς* ὄν. (6) ἐζήτητο δὲ πότερον ζῷόν ἐστιν ἤδη τὸ ἔμβρυον ἢ φυτὸν ἔτι, κἄπειτα μετελήφθη τοῦ ζῴου
τοὔνομα εἰς λόγον, ‖ ἵν' ᾖ σαφές. (7) αἰσθήσει δὴ καὶ κινήσει τῇ καθ' 352r
ὁρμὴν εὑρόντες αὐτὸ διαφέρον τοῦ μὴ ζῴου, πάλιν τοῦτο διωρισάμεθα τῶν παρακειμένων αὐτῷ πραγμάτων ἕτερον μὲν εἶναι φάμενοι τὸ δυνάμει τοιοῦτον, ὃ μήπω μέν ἐστιν αἰσθανόμενόν τε καὶ κινούμενον, ἔσται δέ ποτε τοιοῦτον, ἕτερον δὲ τὸ κατ' ἐνέργειαν ὑπάρχον ἤδη τοιοῦτον, τούτου δὲ τὸ μὲν ἤδη ἐνεργοῦν, τὸ δὲ ἐνεργεῖν μὲν δυνάμενον, ἡσυχάζον δὲ ἢ κοιμώμενον. (8) τοῦτο δέ ἐστι τὸ ζητούμενον. οὐ γὰρ ἐκ τοῦ τρέφεσθαι τὸ ἔμβρυον ζῷον εἶναι λεκτέον, ὃ τῆς οὐ-

---

87,24 λεκτὸν Stählin; λεκτέον L 87,25 ἂν *post* τις *add.* Dindorf 87,27 τοῦ ἐμβρύου Heinse; τῶι ἐμβρύωι L 87,32 *δύνασθαι Stählin *app. crit.*; δυνάμενον L ‖ ἤτοι ὁρμᾶν] ἤδη κινεῖσθαι καθ' ὁρμὴν Barnes ‖ ἢ *del.* Barnes (*cf.* 86,11) 88,2 τούτων] τούτου Barnes ‖ ἤδη αἰσθάνεσθαι ἢ καθ' ὁρμὴν Stählin; μὴ αἰσθάνεσθαι ἢ καθ' ὁρμὴν Potter; καθ' ὁρμὴν μὴ αἰσθάνεσθαι ἢ L 88,4 *αἰσθήσεσθαι *conieci*; αἰσθέσθαι L ‖ *κινηθήσεσθαι δύναται] κινηθῆναι δυνήσεται Stählin 88,5 ἐναργὲς Stählin; ἐναργῶς L

and speak about the terrestrial dog, the dogfish, the dog-star in the sky, even about Diogenes, and all the other 'dogs'. I can hardly divine whether you are asking about all or some one of them. (6) So do now what you would have to do later and tell me clearly which one is the dog you are looking to learn about. (7) But if you are concerned with names, it is obvious to everyone that the word 'embryo' itself is not an animal neither a plant, but a name, a sound, a body, a being, a 'something', anything but an animal. So if this is the question you have proposed, then you have got the answer. **13** (1) Neither, of course, is the meaning of the word 'embryo' an animal, since this is incorporeal, a 'sayable', a 'thing', a concept, anything but an animal. (2) The nature of an animal is surely something else. For, as regards the nature of the very thing we are seeking (I mean the fetus), we have clearly shown what it is like. But there is a different problem concerning the meaning of the name 'animal'. (3) Thus if you say an animal is that which can perceive and move by impulse, I say an animal is not that which moves by impulse and perceives without qualification. (4) It may also be asleep or it may not be perceiving because no perceptible objects are present; but the mark by which we recognize something as an animal is that it is capable by nature either of being moved by impulse or of perceiving. (5) For the meaning conveyed by these words is, roughly, this: first, whether the embryo can already perceive or move by impulse (which is the problem we must explore); and another one, whether the embryo can perceive or move by impulse in the future, a meaning according to which nobody inquires, as the matter is obvious.

(6) An inquiry was raised whether the embryo is already an animal, or still a plant. The name 'animal' was then changed into an account in order to make it clear. (7) We have discovered that it differs from non-animals in perception and movement by impulse. We again distinguished it from similar things by pointing out the difference between, on the one hand, something that is such potentially, i.e. when something does not yet perceive and move, but will at some point; and, on the other hand, something that is already such in act. The latter is either something active at this moment, or something capable of being active, but at rest or sleeping. (8) And this is what we are seeking. It is wrong to say that the embryo is an animal because it is nourished: those who say this

σίας ἐστὶν ἀποχωρούντων τοῦ ζητουμένου, τοῖς δ' ἄλλως συμβεβηκόσιν προσεχόντων τὸν νοῦν.

14 (1) Κοινὴ δ' ἐπὶ πᾶσι τοῖς εὑρίσκεσθαι λεγομένοις τέτακται ἡ ἀπόδειξις, ἥτις ἐστὶ λόγος ἐξ ἑτέρων ἕτερόν τι πιστούμενος. (2) ἐξ ὧν δὲ χρὴ πιστοῦσθαι τὸ ζητούμενον, ὁμολογεῖσθαί τε καὶ γινώσκεσθαι δεῖ τῷ μανθάνοντι. (3) ἀρχὴ δὲ τούτων ἁπάντων ἐστὶ τὸ πρὸς αἴσθησίν τε καὶ νόησιν ἐναργές. ἡ μὲν οὖν πρώτη ἀπόδειξις ἐκ τούτων [ἁπάντων] σύγκειται, ἡ δ' ἐκ τῶν ἤδη φθασάντων ἀποδεδεῖχθαι διὰ τῆσδε πάλιν ἕτερόν τι περαίνουσα πιστὴ μὲν οὐδὲν ἧττόν ἐστι τῆς προτέρας, οὐ μὴν καὶ πρώτη ὀνομάζεσθαι δύναται, διότι μηδ' ἐκ πρώτων περαίνεται προτάσεων.

25/26 (4) Τὸ μὲν οὖν πρῶτον εἶδος τῆς τῶν ˈζητουμένων διαφορᾶς,
26/27 τριῶν ὄντων, ἐδείχθη, λέγω δὲ τὸ (τῆς οὐσίας ˈγινωσκομένης)
27 ἀγνοεῖσθαί τι τῶν ἔργων ἢ παθῶν αὐτῆς, δευτέρα δ'
ἣν διαφορὰ προβλημάτων ἐφ' οὗ τὰ μὲν ἔργα καὶ πάθη γινώ‖σκομεν 352ᵛ
ἅπαντες, ἀγνοοῦμεν δὲ τὴν οὐσίαν, οἷον ἐν τίνι τοῦ σώματος μορίῳ τὸ ἡγεμονικόν ἐστι τῆς ψυχῆς. **V. 15** (1) ἡ δὲ αὐτὴ ἐπιχείρησις τῆς ἀποδείξεως κἀπὶ τούτου τοῦ προβλήματος. φασὶν οὖν τινες μὴ ἐγχωρεῖν πλείους ἀρχὰς ἑνὸς εἶναι ζῴου. ὁμογενεῖς μὲν οὖν ἀρχὰς οὐκ ἐγχωρεῖ πλείους ὑπάρχειν ἑνὸς ζῴου, διαφερούσας δὲ τοῖς γένεσιν οὐδὲν ἄτοπον.

89,1 [Πρὸς τοὺς Πυρρωνίους]

(2) *Εἴ φησιν* ἡ ἐποχὴ βέβαιον εἶναι μηδέν, δῆλον ὅτι ἀφ' ἑαυτῆς ἀρξαμένη πρῶτον ἀκυρώσει ἑαυτήν. (3) ἢ τοίνυν δίδωσιν ἀληθές τι εἶναι καὶ οὐ περὶ πάντων *ἐφεκτέον*, ἢ ἐνίσταται μηδὲν εἶναι ἀληθὲς λέγουσα, καὶ δῆλον ὅτι οὐδ' αὐτὴ πρότερον ἀληθεύσει. (4) ἤτοι γὰρ αὐτὴ ἀληθεύει ἢ οὐκ ἀληθεύει. ἀλλ' εἰ μὲν ἀληθεύει, δίδωσιν ἄκουσά τι εἶναι ἀληθές, εἰ δὲ μὴ ἀληθεύει, ἀληθῆ ἀπολείπει ἅπερ ἀνελεῖν ἐβούλετο. (5) ἐν ᾧ γὰρ ψευδὴς δείκνυται ἡ ἀναιροῦσα ἐποχή, ἐν τούτῳ τὰ ἀναιρούμενα ἀληθῆ δείκνυται, ὡς ὁ ὄνειρος ὁ λέγων ψευδεῖς εἶναι πάντας τοὺς ὀνείρους. (6) *ἑαυτῆς* γὰρ ἀναιρετικὴ οὖσα τῶν ἄλλων γίνε-

---

88,22 ἁπάντων *del.* Stählin (*cf.* 88,20) **88,31** *τούτου τοῦ] τοῦ τρίτου Stählin; *lectionem ms. def.* Barnes **89,1** **delevi*; Πυρρωνείους Dindorf **89,2** εἴ φησιν Stählin; εἰ, ᾗ φησιν Arnim; εἴγε φησὶν Wedel; εἴη φασὶν L **89,4** ἐφεκτέον Sylburg; ἀφεκτέον L **89,10** ἑαυτῆς Sylburg; αὐτοῦ L

are missing the essence of what is sought, while turning attention to things incidental to it in another way.

14 (1) Demonstration which applies in common to everything said to have been discovered is an argument confirming one thing on the basis of others. (2) Those items on whose basis the thing sought is supposed to be confirmed must be agreed to and known by the student. (3) The principle of all of them is that which is plain to sense-perception and intellection. The primary demonstration is composed of these items; the one composed of the items previously demonstrated by the primary demonstration and producing yet another conclusion is no less credible, but it cannot be called the primary demonstration, because it does not draw its conclusion from the primary propositions.

(4) Well, then, of the kinds distinguished among things we seek (there being three such kinds), the first one has been exhibited: I mean, when the substance is known, but some of its works or affections are not. The second kind of problems was the one concerning things whose works and affections we all know, but do not know the substance, like 'In which part of the body is the ruling faculty of the soul?'. **V. 15** (1) The reasoning of demonstration is the same in the case of this problem, too. Some people say that it is impossible for one animal to have more than one beginning. It is surely impossible for one animal to have more than one beginning of the same genus, but if the beginnings are of different genera, there is nothing absurd about it.

Suspension of judgement I

(2) If suspension says that nothing is firm, clearly if it starts with itself it will invalidate itself first. (3) Either it grants that something is true and we should not suspend judgement about everything, or it objects by saying that nothing is true, and clearly it will not speak the truth in the first place. (4) For it either speaks the truth or does not speak the truth. And if it speaks the truth, it inadvertently grants that something is true. But if it does not speak the truth, then it admits that the very matters it wanted to abolish are true. (5) For insofar as it is shown that the abolishing suspension is false, it is also shown that the abolished matters are true, like [in the case of] a dream which says that all dreams are false. (6) Thus if suspension abolishes itself, it validates other

ται *κυρωτική*. καὶ ὅλως εἰ ἔστιν ἀληθής, ἀφ' ἑαυτῆς ποιήσεται τὴν ἀρχήν, οὐκ ἄλλου τινὸς οὖσα ἐποχή, ἀλλ' ἑαυτῆς πρῶτον.

(7) Ἔπειτα εἰ *καταλαμβάνει τις* ὅτι ἄνθρωπός ἐστιν ἢ ὅτι ἐπέχει, δῆλός ἐστι μὴ *ἐπέχων*. (8) πῶς δ' ἂν καὶ τὴν ἀρχὴν εἰς τὴν ἀμφισβήτησιν ἀφίκετο περὶ πάντων ἐπέχων; πῶς δ' ἂν καὶ ἀπεκρίνετο πρὸς τὸ ἐρωτηθέν; (9) περὶ γὰρ αὐτοῦ τούτου δῆλός ἐστιν οὐκ ἐπέχων, ναὶ μὴν καὶ ἀποφαίνεται ὅτι ἐπέχει, ⟨δι'⟩ ὃ καὶ εἰ δεῖ πειθόμενον αὐτοῖς περὶ πάντων ἐπέχειν, περὶ αὐτῆς πρότερον ‖ τῆς ἐποχῆς ἐφέξομεν, 353r
εἴτε πειστέον αὐτῇ εἴτε καὶ μή.

16 (1) Ἔτι εἰ τοῦτο αὐτό ἐστι τὸ ἀληθὲς τὸ μὴ εἰδέναι τὸ ἀληθές, [οὐδὲ τὴν ἀρχὴν] ἀληθές τι παρ' ἐκείνου δίδοται. εἰ δὲ καὶ τοῦτο ἀμφισβητήσιμον ἐρεῖ τὸ ἀγνοεῖν τἀληθές, ἐν αὐτῷ τούτῳ δίδωσιν εἶναι τὸ ἀληθὲς γνώριμον, ἐν ᾧ τὴν περὶ αὐτοῦ ἐποχὴν φαίνεται μὴ βεβαιῶν.

(2) Αἵρεσίς ἐστι πρόσκλισις δογμάτων ἤ, ὥς τινες, πρόσκλισις δό-
90,1 γμασι πολλοῖς ἀκολουθίαν πρὸς ἄλληλα καὶ τὰ φαινόμενα *περιέχουσι* πρὸς τὸ εὖ ζῆν συντείνουσα. καὶ τὸ μὲν δόγμα ἐστὶ κατάληψίς τις λογική, κατάληψις δὲ ἕξις καὶ συγκατάθεσις τῆς διανοίας.

(3) Οὐ μόνον οἱ *ἐφεκτικοί*, ἀλλὰ καὶ πᾶς δογματικὸς ἔν τισιν ἐπέχειν εἴωθεν ἤτοι παρὰ γνώμης ἀσθένειαν ἢ παρὰ πραγμάτων ἀσάφειαν ἢ παρὰ τὴν τῶν λόγων ἰσοσθένειαν.

Αἱ τῶν ζητήσεων ἔφοδοι καὶ ἀρχαὶ περὶ ταῦτα καὶ ἐν ¦τούτοις εἰσίν

VI. 17 (1) Προτακτέον δὴ καὶ τῶν ὅρων καὶ τῶν ἀποδείξεων καὶ διαιρέσεων ⟨τὸ⟩ ποσαχῶς λέγεται τὸ ζητούμενον, τά τε ὁμώνυμα *χωριστέον* καὶ τὰ συνώνυμα εὐκρινῶς τακτέον κατὰ *τὰς* σημασίας. (2) ἔπειτα ζητητέον εἰ τῶν πρὸς ἕτερα θεωρουμένων ἐστὶν τὸ προκείμενον ἢ καθ' αὐτὸ λαμβάνεται, ἐπὶ τούτοις, εἰ ἔστι, τί ἐστι, τί αὐτῷ συμβέβηκεν, ἢ καὶ οὕτως, εἰ ἔστι, τί ἐστι, διότι ἐστίν. (3) πρὸς δὲ τὴν τούτων θεω-

---

89,11 κυρωτική Höschel; κυριωτική L **89,13** καταλαμβάνει τις Arnim; καταλαμβάνεται L **89,14** ἐπέχων Arnim; ἐπέχειν L **89,17** δι' *add.* Heyse **89,19** πειστέον Sylburg; πιστέον L **89,20 f.** *οὐδὲ τὴν ἀρχὴν *delevi* **90,1** περιέχουσι Potter; περιέχουσα L **90,4** ἐφεκτικοί Sylburg (cf. MS Add. 14,153, fol. 137a: *hwpwqṭīqw*); ἐφεκτοί L **90,10** *τὸ *ante* ποσαχῶς *addidi* ‖ *χωριστέον Sylburg; χειριστέον L, *agn.* Hervet (*tractanda sunt homonyma*), Stählin **90,11** τὰς Heyse; τῆς L

matters. And generally, if it is true, it will start from itself and from the start it will be a suspension about nothing but itself.

(7) Further, if someone apprehends that he is a human being or that he suspends judgement, he is clearly not suspending judgement [about it]. (8) But how could he arrive at controversy at all, when he suspends judgement about everything? How could he even answer the question? (9) He is plainly not suspending judgement about this very [fact]; indeed, he even asserts that he is suspending judgement. Therefore, even if one who believes them ought to suspend judgement about everything, we will start by suspending judgement about suspension itself, [doubting] whether we should believe it or not.

16 (1) Also, if this very [fact] is the truth, viz. that he does not know the truth, then he has granted that something is true. But if he says that this is disputable, too, namely that he does not know the truth, then he grants that truth is known insofar as he appears not to affirm that he is suspending judgement about it.

(2) School is an inclination to beliefs, or, as some put it, an inclination to many beliefs containing congruence with one another and with appearances, aiming towards good life. And belief is a sort of rational apprehension, while apprehension is a state of assent of the mind.

(3) Not only the 'suspenders', but every 'dogmatist' too is used to suspending judgement in some matters, either from weakness of thought, the obscurity of things, or the equipollence of arguments.

PROCEDURES AND PRINCIPLES OF INQUIRIES INVOLVE AND CONSIST IN THESE MATTERS

Division and definition

VI. 17 (1) So before definitions, demonstrations, and divisions, we should first [inquire] in how many ways the thing sought is said, separate the homonyms and discriminate the synonyms correctly according to their meanings. (2) Then, we should inquire if that which is proposed belongs to items studied in relation to others, or if it is taken by itself. After this: if it is, what it is, and what belongs to it; or, alternatively, if it is, what it is, and why it is. (3) With respect

ρίαν ἡ τῶν καθ’ ἕκαστα γνῶσις καὶ ἡ τῶν καθόλου συνυποβάλλει
καὶ τὰ πρότερα καὶ *τὰς εἰδικὰς* διαφοράς [αἱ διαιρέσεις αὐτῶν]. (4) ἡ μὲν
οὖν ἐπαγωγὴ φέρει τὸ ‖ καθόλου καὶ τὸν ὁρισμόν, αἱ δὲ διαιρέσεις τὰ 353v
εἴδη καὶ τὸ τί ἦν εἶναι καὶ τὸ ἄτομον, ἡ δὲ θεωρία τοῦ ποσαχῶς
τὴν λῆψιν τοῦ οἰκείου, αἱ δὲ διαπορήσεις τὰς κατ’ αὐτὰς διαφορὰς
καὶ τὰς ἀποδείξεις καὶ ἄλλως τὴν θεωρίαν συναύξουσι τά τε παρε-
πόμενα αὐτῇ. ἐκ πάντων δὲ ἐπιστήμη τε παρίσταται καὶ ἀλήθεια.

(5) Πάλιν τὸ κεφαλαιωθὲν ἐκ τῆς διαιρέσεως ὅρος γίνεται· λαμβάνεται γὰρ καὶ πρότερος τῆς διαιρέσεως καὶ ὕστερος ὁ διορισμός· πρότερος μὲν ὅταν συγχωρηθῇ ἢ προτεθῇ, ὕστερος δὲ ὅταν ἀποδειχθῇ. (6) καὶ κατὰ τὴν αἴσθησιν δὲ ἐκ τοῦ καθ’ ἕκαστα κεφαλαιοῦται τὸ καθόλου. ἀρχὴ γὰρ τῆς ἐπαγωγῆς ἡ αἴσθησις, πέρας δὲ τὸ καθόλου. (7) ἡ μὲν οὖν ἐπαγωγὴ οὐ τὸ τί ἐστι δείκνυσιν, ἀλλ’ ὅτι ἐστὶν ἢ οὐκ ἔστιν, ἡ διαίρεσις δὲ ὅ τι ἐστὶ παρίστησιν (8) ὅ τε διορισμὸς ὁμοίως τῇ διαιρέσει τὴν οὐσίαν καὶ τὸ τί ἐστι διδάσκει, οὐχὶ δὲ τὸ εἰ ἔστιν, ἥ τε ἀπόδειξις τὰ τρία, τό τε εἰ ἔστιν καὶ τὸ τί ἐστιν καὶ τὸ διὰ τί ἐστιν, σαφηνίζει. εἰσὶ δὲ ἔνιοι καὶ τῆς αἰτίας ἐμπεριεκτικοὶ ὅροι. | **18** (1) ἐπεὶ δὲ ἐπίστασθαι ἔστιν ὅταν ἴδωμεν τὴν αἰτίαν, αἰτίαι δὲ |τέσσαρες, ἡ ὕλη, τὸ κινοῦν, τὸ εἶδος, τὸ τέλος, τετραχῶς ἔσται ὁ |διορισμός.

(2) Ληπτέον οὖν πρῶτον τὸ γένος, ἐν ᾧ ἐστι τῶν ἐπάνω τὰ ἐγγύτατα, μετὰ τοῦτο τὴν προσεχῆ διαφοράν. ἡ δὲ συνέχεια τῶν διαφορῶν τεμνομένη καὶ ἀποδιαιρουμένη τὸ τί ἦν εἶναι ἐκπληροῖ.
(3) οὐκ ἀνάγκη δὲ πάσας λέγειν ἑκάστου τὰς διαφοράς, ἀλλὰ τὰς εἰδοποι-
οὔσας. (4) ἡ γεωμετρικὴ ἀνάλυσις καὶ σύνθεσις τῇ διαλεκτικῇ διαιρέσει ‖
καὶ ὁρισμῷ ἔοικεν, καὶ ἀπὸ μὲν τῆς διαιρέσεως ἐπὶ τὰ ἁπλούστερα 354r
καὶ ἀρχικώτερα ἀνατρέχομεν. (5) τὸ γοῦν γένος τοῦ ζητουμένου πρά-
γματος διαιροῦμεν εἰς τὰ ἐνυπάρχοντα αὐτῷ εἴδη, οἷον ἐπὶ τοῦ ἀν-
θρώπου τὸ ζῷον γένος ὂν εἰς τὰ ἐμφαινόμενα εἴδη διαιροῦμεν, τὸ θνητὸν καὶ ἀθάνατον, καὶ οὕτως ἀεὶ τὰ σύνθετα δοκοῦντα εἶναι γένη εἰς τὰ ἁπλούστερα εἴδη τέμνοντες ἐπὶ τὸ [μὴ] ζητούμενον καὶ μηκέτι τομὴν ἐπιδεχόμενον παραγινόμεθα. (6) τὸ γὰρ ζῷον εἰς τὸ θνητὸν καὶ ἀθάνατον διελόντες, εἶτα μέντοι τὸ θνητὸν εἰς τὸ χερσαῖον καὶ ἔνυδρον, καὶ πάλιν τὸ χερσαῖον εἰς τὸ πτηνὸν καὶ πεζόν, καὶ οὕτως τὸ προσεχὲς τῷ ζητουμένῳ εἶδος, ὃ καὶ περιλαμβάνει τὸ ζητούμενον, διαιροῦντες ἀφιξόμεθα τέμνοντες ἐπὶ τὸ ἁπλούστατον εἶδος,

---

90,15 *συνυποβάλλει] συνεπιβάλλει Lowth; συμβάλλεται Stählin **90,16** *τὰς εἰδικὰς *conieci* (*cf.* 92,9); τὰ ἐπὶ τὰς L; αἱ ἐπὶ τὰς Stählin (1909); αἱ κατὰ τὰς Stählin ('Nachträge und Berichtigungen' in Clemens Alexandrinus, IV: Register, GCS 39, 1936, p. LXIX); ‖ *αἱ διαιρέσεις αὐτῶν *delevi*; αἱ *del.* Stählin **91,16** μὴ *del.* Potter

to the study of these [questions], knowledge of the particulars and of the universals provides both what is prior and the specific differences.[4] (4) Induction, then, conveys the universal and a definition; divisions [convey] the species and 'what it is to be', i.e. the unsplittable [species]; the study of 'in how many ways' [conveys] the grasp of what is appropriate; and going through the difficulties [conveys] corresponding differences as well as demonstrations and, generally, it helps to add to the study and its consequences. From all these things science and truth come about.

(5) Again, the summing up of a division becomes a definition. For definition is taken both prior to division and after it; prior to it, when accepted or proposed; after it, when demonstrated. (6) And it is on the basis of perception that the universal is summed up from the particular. For perception is the starting-point of induction, the universal its end. (7) Induction, then, does not show what something is, but that something is or is not the case. Division, on the other hand, establishes what something is. (8) Definition, like division, teaches the essence and what something is, but not if it is. And demonstration makes these three things clear: if it is, what it is, and why it is. But some definitions also include the cause. 18 (1) And, since to know scientifically is to know the cause, and there are four causes, the matter, the mover, the form, and the goal, definition will be [said] in four ways.

(2) So let us first take the genus which contains the nearest of the [species] above [the defined thing], and then a proximate difference. The continuity of differences, split and divided apart, adds up to what it is [for the thing in question] to be. (3) We need not, however, recount all the differences pertaining to each thing, but only the species-forming ones. (4) Geometrical analysis and synthesis is similar to dialectical division and definition: by division we also ascend to what is more simple and primitive. (5) For we divide the genus of the thing sought into its inherent species. For example, in the case of 'man', we divide 'animal', which is the genus, into the species mirrored in it, 'mortal' and 'immortal', and in this manner, by splitting the apparently composite genera into more simple species, we arrive at the thing sought, i.e. at that which does not allow for more splitting. (6) For when we divide 'animal' into 'mortal' and 'immortal', and then 'mortal' into 'terrestrial' and 'aquatic', and again 'terrestrial' into 'winged' and 'pedestrian', and when we divide, in this manner, the species that is proximate to and also includes the thing sought, we will, by splitting the

4 Text uncertain.

ὅπερ ἄλλο μὲν οὐδέν, μόνον δὲ τὸ ζητούμενον περιλαμβάνει. (7) πάλιν γὰρ τὸ πεζὸν εἰς λογικὸν καὶ ἄλογον διαιρούμεθα κἄπειτα τὰ προσεχῆ τῶν ἐκ τῆς διαιρέσεως λαμβανομένων εἰδῶν τῷ ἀνθρώπῳ ἐκλεγόμενοι καὶ ἐπισυνθέντες εἰς ἕνα λόγον τὸν ὅρον τοῦ ἀνθρώπου ἀποδίδομεν, ὅς ἐστι ζῷον θνητὸν χερσαῖον πεζὸν λογικόν.

19 (1) Ὅθεν ὕλης μὲν τάξιν ἐπέχει ἡ διαίρεσις τῷ ὅρῳ, τὴν ἁπλότητα τοῦ ὀνόματος ἀναζητοῦσα, τεχνίτου δὲ καὶ δημιουργοῦ ὁ ὅρος ἐπισυντιθεὶς καὶ κατασκευάζων ⟨τὸν λόγον⟩ καὶ τὴν γνῶσιν τοῦ ὄντος παριστάς. (2) *οὔτ' αὐτῶν* τῶν πραγμάτων οὔτε τῶν ἰδεῶν οἱ ὅροι, ἀλλὰ γὰρ *ὧν* πραγμάτων [ὧν] ἔχομεν καθολικὰς διανοίας, τούτων τῶν δια‖νοιῶν τοὺς 354ᵛ
ἑρμηνευτικοὺς λόγους ⟨ὅρους⟩ εἶναί φαμεν· τούτων γὰρ τῶν διανοιῶν καὶ αἱ διαιρέσεις γίνονται.

(3) Τῶν δὲ διαιρέσεων ἣ μέν τις εἰς εἴδη |διαιρεῖ τὸ διαιρούμενον ὡς γένος, ἣ δέ τις εἰς μέρη ὡς ὅλον, ἣ δὲ |εἰς τὰ συμβεβηκότα. (4) ἡ μὲν οὖν τοῦ ὅλου εἰς τὰ μέρη διαίρεσις ὡς |ἐπὶ τὸ πλεῖστον κατὰ μέγεθος ἐπινοεῖται, ἡ δὲ εἰς τὰ συμβεβηκότα οὐδέποτε ὅλη δύναται διαληφθῆναι, εἴ γε καὶ οὐσίαν ἑκάστῳ δεῖ πάντως τῶν ὄντων ὑπάρχειν. (5) ὅθεν ἀδόκιμοι ἄμφω αὗται αἱ διαιρέσεις, μόνη δὲ εὐδοκιμεῖ ἡ τοῦ γένους εἰς εἴδη τομή, ὑφ' ἧς χαρακτηρίζεται ἥ τε ταὐτότης ἡ κατὰ γένος ἥ τε ἑτερότης ἡ κατὰ τὰς *εἰδικὰς* διαφοράς. (6) τὸ εἶδος ἀεὶ ἔν τινι μέρει θεωρεῖται, οὐ μὴν ἀνάπαλιν, εἴ τι μέρος ἐστί τινος, τοῦτο καὶ εἶδος γενήσεται. ἡ γὰρ χεὶρ μέρος μέν ἐστι τοῦ ἀνθρώπου, εἶδος δὲ οὐκ ἔστιν. (7) καὶ τὸ μὲν γένος ἐν τοῖς εἴδεσιν ἐνυπάρχει, τὸ γὰρ ζῷον καὶ ἐν τῷ ἀνθρώπῳ καὶ ἐν τῷ βοΐ, τὸ δὲ ὅλον ἐν τοῖς μέρεσιν οὐκ ἐνυπάρχει, οὐ γὰρ ὁ ἄνθρωπος ἐν τοῖς ποσὶν ὑπάρχει. (8) διὸ κυριώτερον τὸ εἶδος τοῦ μέρους, καὶ ὅσα τοῦ γένους κατηγορεῖται, ταῦτα πάντα καὶ τοῦ εἴδους κατηγορηθήσεται.

20 (1) Ἄριστον μὲν οὖν εἰς δύο *διαιρεῖν* εἴδη τὸ γένος, εἰ δὲ μή, εἰς τρία. τὰ τοίνυν εἴδη γενικώτερον μὲν [διαιρούμενα χαρακτηρίζεται] τῷ τε ταὐτῷ καὶ θατέρῳ, ἔπειτα δὲ διὰ τῶν γενικῶς σημαινομένων διαιρούμενα χαρακτηρίζεται. (2) ἕκαστον γὰρ τῶν εἰδῶν ἤτοι οὐσία ἐστὶν (ὥσπερ ὅταν λέγωμεν· τῶν ὄντων τὰ μὲν σώματά ἐστι, τὰ δὲ ἀσώματα) ἢ ποσὸν ἢ ποιὸν ἢ πρός τι ἢ ποῦ ἢ πότε ⟨ἢ κεῖσθαι ἢ ἔχειν⟩ ἢ ποιεῖν ἢ πάσχειν.

91,29 *τὸν λόγον *addidi* **91,30** οὔτ' αὐτῶν Stählin (*cf.* 93,2); οὐ τούτων L ‖ ὧν Wilamowitz; τῶν L **91,31** ὧν *del.* Wilamowitz **91,32** ὅρους *add.* Pohlenz (*cf.* 92,29) **92,9** *εἰδικὰς *correxi*; ἰδικὰς L **92,17** διαιρεῖν Arcerius; θεωρεῖν L **92,18** *διαιρούμενα χαρακτηρίζεται *del.* Stählin *app. crit.* **92,22** ἢ κεῖσθαι ἢ ἔχειν *add.* Stählin (*cf.* 94,26)

species, arrive at the simplest species, which includes nothing but the thing sought. (7) For we divide again 'pedestrian' into 'rational' and 'non-rational' and then, selecting those species taken from division that are proximate to 'man' and putting them together into one account, we produce a definition of 'man', which is: 'mortal, terrestrial, pedestrian, rational animal'.

19 (1) Thus with respect to definition, division has the role of matter, seeking for the simplicity of the [proposed] name, whereas definition [has the role] of an artisan or a craftsman, since it puts together and constructs ⟨an account⟩ and supplies the knowledge of a being. (2) Definitions are [definitions] neither of things themselves, nor of [their] forms, but regarding those things of which we have universal thoughts, we say that ⟨definitions⟩ are accounts expressing these thoughts. For divisions too are divisions of these thoughts.

(3) One sort of division divides the divided item as a genus into species, another one as a whole into parts, another one into attributes. (4) Now the division of the whole into parts is most usually conceived in terms of magnitude, whereas [the object of] the division into attributes can never be discerned as a whole, if, as is the case, every being must necessarily also have an essence. (5) Thus neither of these two divisions is acceptable, the only respectable one being the splitting of a genus into species, which characterizes the [species in terms of] sameness according to genus and diversity according to specific differentiae. (6) A species is always observed in some part, not, of course, the other way round, viz., that if something is a part of something, it is also a species. Thus hand is a part of a man, but it is not a species. (7) And genus is inherent in the species (animal is in the man and in the ox), but the whole is not inherent in the parts (man is not in the feet). (8) Therefore, species are more important than parts, and everything predicated of a genus will be predicated of the species, as well.

20 (1) It is best to divide a genus into two species; otherwise, into three. Now, on a more general level, species are characterized by the same and the diverse, and then by being divided through generic meanings. (2) For every species is either a substance (as when we say, 'some beings are bodies, some are incorporeal') or a quantity or a quality or relative or where or when ⟨or in a position or having⟩ or an activity or being affected.

(3) Παντὸς οὑτινοσοῦν ἐπιστήμων ‖ τίς ἐστι, τούτου 355r
καὶ τὸν ὅρον ἀποδώσει, ὡς ὅ γε μὴ δυνάμενος λόγῳ περιλαβεῖν καὶ ὁρίσασθαι ὁτιοῦν, τούτου ἐπιστήμων οὐκ ἄν ποτε εἴη. (4) ἐκ δὲ τῆς τοῦ ὅρου ἀγνοίας καὶ τὰς πολλὰς ἀμφισβητήσεις γίνεσθαι καὶ τὰς ἀπάτας συμβαίνει. (5) εἰ γὰρ ὁ τὸ πρᾶγμα εἰδὼς ἔχει κατὰ διάνοιαν τὴν εἴδησιν αὐτοῦ, δύναται δὲ καὶ λόγῳ σαφηνίσαι ὃ διανοεῖται, ἡ δὲ τῆς διανοίας ἑρμηνεία ὅρος ἐστίν, ἀναγκαῖόν ἐστι τὸν εἰδότα τὸ πρᾶγμα δύνασθαι αὐτοῦ καὶ τὸν ὅρον ἀποδιδόναι.

21 (1) Προσλαμβάνεται δὲ ὅροις καὶ ἡ διαφορὰ σημείου τάξιν ἐπέχουσα τῷ ὅρῳ. τῷ γοῦν τοῦ ἀνθρώπου ὅρῳ τὸ γελαστικὸν προστεθὲν ποιεῖ τὸ ὅλον ζῷον [λογικὸν] θνητὸν χερσαῖον πεζὸν ⟨λογικὸν⟩ γελαστικόν. (2) σημεῖα γάρ ἐστι τὰ κατὰ διαφορὰν προσαγόμενα τῷ ὅρῳ οἰκείων πραγμάτων, τὴν δὲ φύσιν αὐτῶν τῶν πραγμάτων οὐκ ἐνδείκνυται. αὐτίκα ἰδίου ἀπόδοσίν φασιν εἶναι τὴν διαφοράν· (3) καὶ καθ' ὃ τῶν ἄλλων ἁπάντων διαφέρει τὸ ἔχον τὴν διαφοράν, ὃ μόνῳ ὑπάρχει καὶ ἀντικατηγορεῖται τοῦ πράγματος ⟨***⟩ ἐν τοῖς ὅροις ἀνάγκη τὸ πρῶτον γένος ὡς ἀρχηγὸν καὶ ὑποστατὸν [δεῖ] παραλαμβάνειν. (4) ἐν μὲν οὖν τοῖς μακροτέροις ὅροις τὸ πλῆθος τῶν εἰδῶν τῶν ἐξευρισκομένων διὰ τῶν δέκα κατηγοριῶν, ἐν δὲ τοῖς ἐλαχίστοις τὰ προηγούμενα τῶν προσεχῶν εἰδῶν ληφθέντα τὴν οὐσίαν καὶ φύσιν διασημαίνει τοῦ πράγματος, ὁ δὲ ἐλάχιστος ἐκ τριῶν συνέστηκεν, τοῦ γένους καὶ δυεῖν τῶν ἀναγκαιοτάτων εἰδῶν. (5) γίνεται δὲ τοῦτο διὰ συντομίαν. φαμὲν οὖν, ‖
ἄνθρωπός ἐστι τὸ ζῷον ⟨θνητὸν λογικόν⟩ [γελαστικόν]· τό τε ἐξαιρέτως 355v
συμβεβη|κὸς τῷ ὁριζομένῳ προσπαραληπτέον, ἢ τὴν ἰδίαν ἀρετὴν αὐτοῦ ἢ τὸ ἴδιον ἔργον αὐτοῦ καὶ τοιούτων τινῶν ἄλλων. (6) ἐξηγητικὸς οὖν ὁ ὅρος ὢν τῆς τοῦ πράγματος οὐσίας περιλαβεῖν μὲν ἀκριβῶς τὴν φύσιν τοῦ πράγματος ἀδυνατεῖ, διὰ δὲ τῶν κυριωτάτων εἰδῶν τὴν δήλωσιν τῆς οὐσίας ποιεῖται καὶ σχεδὸν ἐν ποιότητι [ὁ ὅρος] τὴν οὐσίαν ἔχει.

VII. 22 (1) Τὰ ποιητικὰ τῆς ἐποχῆς αἴτια δύο ἐστὶν τὰ ἀνωτάτω, ἓν μὲν τὸ πολύτροπον καὶ ἄστατον τῆς ἀνθρωπίνης γνώμης, ὅπερ γεννητικὸν εἶναι πέφυκεν τῆς διαφωνίας ἤτοι τῆς ἀλλήλων πρὸς ἀλλήλους ἢ τῆς ἑαυτῶν πρὸς ἑαυτούς, δεύτερον δὲ ἡ ἐν τοῖς οὖσι δια-

---

92,33 *λογικόν *post* πεζόν *transposui* (*cf.* 91,26) **93,4** *ὃ] ἢ Stählin **93,5** **post* πράγματος *lac. indicavi* ‖ ἀνάγκη] ἀνάγκῃ Sylburg **93,6** δεῖ *del.* Stählin **93,12** *θνητὸν λογικόν *addidi*; λογικόν *add.* Stählin ‖ *γελαστικόν *delevi* **93,17** *ὁ ὅρος *delevi*

(3) If anyone has scientific knowledge of anything, he will also give its definition, just as, if someone is unable to grasp something with an account and define it, he will never have scientific knowledge of it. (4) Ignorance of a definition also gives rise to many controversies and brings about much deception. (5) For if one who knows something has the knowledge of it in his thought, and he is also able to explain what he thinks with an account, and if the expression of the thought is a definition, then, necessarily, one who knows something will also be able to give its definition.

21 (1) Added to definitions is also a difference which has the role of a sign with respect to definition. For example, the addition of the predicate 'laughing' to the definition of 'man' makes the whole 'mortal, terrestrial, pedestrian, rational, laughing animal'. (2) For predicates added to a definition on the basis of a difference are signs of appropriate things, but do not indicate the nature of these things. Thus it is said that difference is an account of a property. (3) And it is in this respect that the thing having a difference differs from all the other things, namely, in respect of that which belongs only to the thing and is counterpredicated of it, ⟨whereas?⟩ in definitions we must use the first genus as the fundamental and substantial entity. (4) Now, in longer definitions, the nature and essence of the thing is signallized by a number of species discovered through the ten categories, whereas in the shortest ones the main proximate species are taken to that effect. The shortest definition consists of three items, genus and the two most necessary species. This happens for the sake of brevity. (5) So we say that 'man' is a '⟨mortal, rational⟩ animal'. And, in addition, we should employ something that exclusively belongs to the defined thing, its proper virtue or its proper activity and other items of this sort. (6) So a definition expressing the essence of the thing is unable to grasp the nature of the thing precisely, but it provides an explanation of the essence through the most important species and almost gets hold of the essence in a quality.

Suspension of judgement II

VII. 22 (1) The most general causes producing suspension of judgement are two: One is the fickleness and instability of the human mind; it naturally generates disagreement, either that of one with another, or that of one with oneself.

φωνία, ἣ καὶ εἰκότως ἐμποιητικὴ καθέστηκε τῆς ἐποχῆς. (2) μήτε γὰρ πάσαις ταῖς φαντασίαις πιστεύειν δυνηθέντες διὰ τὴν μάχην μήτε πάσαις ἀπιστεῖν διὰ τὸ καὶ τὴν λέγουσαν πάσας ἀπίστους ὑπάρχειν ἐξ ἁπασῶν οὖσαν συμπεριγράφεσθαι πάσαις μήτε τισὶ μὲν πιστεύειν, τισὶ δὲ ἀπιστεῖν διὰ τὴν ἰσότητα, κατήχθημεν εἰς ἐποχήν. (3) τούτων δὲ αὐτῶν τῶν ἀρχικωτάτων ⟨αἰτίων⟩ τῆς ἐποχῆς τὸ μὲν ἀβέβαιον τῆς διανοίας γεννητικόν ἐστι διαφωνίας, ἡ δὲ διαφωνία προσεχὲς αἴτιον τῆς ἐποχῆς, ὅθεν πλήρης μὲν ὁ βίος δικαστηρίων τε καὶ βουλευτηρίων καὶ ἐκκλησιῶν καὶ καθόλου τῆς περὶ τὰ λεγόμενα ἀγαθὰ καὶ κακὰ αἱρέσεως ⟨καὶ φυγῆς⟩, ἅπερ ἠπορημένης ἐστὶ διανοίας καὶ πρὸς τὴν
94,1 τῶν ἀντικειμένων πραγμάτων *ἰσοσθένειαν* μετοκλαζούσης τεκμή||ρια. 356r
(4) πλήρεις δ' αἱ θῆκαι τῶν βιβλίων [καὶ αἱ συντάξεις καὶ αἱ πραγματεῖαι] τῶν διαφωνούντων ἐν τοῖς δόγμασι καὶ πεπεικότων ἑαυτοὺς τὴν ἐν τοῖς οὖσιν ἀλήθειαν γινώσκειν.

VIII. 23 (1) Τρία ἐστὶ περὶ τὴν φωνήν· τά τε ὀνόματα σύμβολα ὄντα τῶν νοημάτων κατὰ τὸ προηγούμενον, κατ' ἐπακολούθημα δὲ καὶ τῶν ὑποκειμένων, δεύτερον δὲ τὰ νοήματα ὁμοιώματα καὶ ἐκτυπώματα τῶν ὑποκειμένων ὄντα (ὅθεν ἅπασι καὶ τὰ νοήματα τὰ αὐτά ἐστι διὰ τὸ τὴν αὐτὴν ἀπὸ τῶν ὑποκειμένων ἅπασιν ἐγγίνεσθαι τύπωσιν, οὐκέτι δὲ καὶ τὰ ὀνόματα διὰ τὰς διαλέκτους τὰς διαφόρους)· τρίτον δὲ τὰ ὑποκείμενα πράγματα, ἀφ' ὧν ἡμῖν τὰ νοήματα ἐντυποῦνται. (2) τὰ μὲν οὖν ὀνόματα τῇ γραμματικῇ εἰς τὰ καθολικὰ στοιχεῖα κδ' *ἀνάγεται*· ὡρισμένα γὰρ χρὴ εἶναι τὰ στοιχεῖα· τῶν γὰρ καθ' ἕκαστα ἀπείρων ὄντων μὴ εἶναι ἐπιστήμην, ἴδιον δὲ ἐπιστήμης καθολικοῖς ἐπερείδεσθαι θεωρήμασι καὶ *ὁρισμοῖς*. ὅθεν τὰ καθ' ἕκαστα εἰς τὰ καθόλου ἀνάγεται. (3) ἡ δὲ τῶν φιλοσόφων πραγματεία περί τε τὰ νοήματα καὶ τὰ ὑποκείμενα καταγίνεται. *ἐπεὶ* δὲ τούτων τὰ καθ' ἕκαστα ἄπειρα, στοιχεῖά τινα καὶ τούτων εὑρέθη, ὑφ' ἃ πᾶν τὸ ζητούμενον ὑπάγεται. (4) καὶ εἰ μὲν φαίνοιτο ὑφ' ἕν τι *ὑποδεδυκὸς* τῶν στοιχείων ἢ καὶ πλείω, ἀποφανούμεθα αὐτὸ εἶναι, εἰ δὲ πάντα διαφύγοι, μηδαμῇ εἶναι. (5) τῶν λεγομένων τὰ μὲν ⟨μετὰ συμπλοκῆς

93,28 *αἰτίων *addidi* **93,32** καὶ φυγῆς *add.* Stählin **94,1** ἰσοσθένειαν Heyse, Wedel (*cf.* 90,6); εἰς ἀσθένειαν L **94,2** *καὶ αἱ συντάξεις καὶ αἱ πραγματεῖαι *delevi* **94,13** ἀνάγεται Stählin; ἀνάγει L **94,15** *ὁρισμοῖς *Sacr. Par.* 278; ὡρισμένοις L **94,17** ἐπεὶ Hervet; ἐπὶ L **94,19** ὑποδεδυκὸς Sylburg; ὑποδεδοικὼς L **94,21 f.** μετὰ … τὰ δὲ *add.* Potter *sec.* Arist. *Cat.* 2, 1a17 f.

Second is the disagreement in things, which also, with good reason, comes down to inducing suspension of judgement. (2) We can neither believe all presentations, because of their conflict; nor disbelieve them all, because the one saying that all [presentations] are untrustworthy, is bracketed along with the 'all', being one of them; nor can we believe some and disbelieve others, because of their equality; thus we are brought to suspension of judgement. (3) Out of these most principal ⟨causes⟩ of suspension, the infirmity of the mind is the one that generates disagreement, and disagreement is the proximate cause of the suspension of judgement. As a consequence, life is full of courts, councils, assemblies, and, generally speaking, choice ⟨and avoidance⟩ regarding so-called good and evil things. These are tokens [showing] that the mind is puzzled and that it wavers vis-à-vis the equipollence of opposite things. (4) And libraries are full of books that disagree with one another in their beliefs, each one having convinced itself that it knows the truth in things.

Categories

VIII. 23 (1) Speech involves three items: names, which are primarily symbols of concepts, but consequently also of the underlying things; second, concepts, which are copies and imprints of the underlying things (that is why concepts are the same for all men, because the same imprints of the underlying things arise in all men, but names and different languages are not); third, the underlying things, from which concepts are imprinted in us. (2) Now grammar reduces names to twenty-four universal elements; for elements must be limited; for there is no knowledge of particulars, as they are infinite, and it is a property of knowledge to be based on universal theorems and definitions. That is why particulars are reduced to universals. (3) But philosophers are occupied with concepts and things; and since the particular instances of these are infinite, certain elements have been discovered for them, too, and everything sought is brought under them. (4) And if it turns out that [the thing sought] has fallen under one or more of these elements, we will assert that it exists; but if it misses them all, [we will assert] that it does not exist in any way. (5) Of things that are

λέγεται, ὡς τὸ “ἄνθρωπος τρέχει”, τὰ δὲ⟩ ἄνευ συμπλοκῆς λέγεται, ὡς τὸ “ἄνθρωπος” ‖ καὶ ὡς τὸ “τρέχει” καὶ ὅσα λόγον οὐκ ἀποτελεῖ 356ᵛ *οὐδὲ* τἀληθὲς ἢ τὸ ψεῦδος ἔχει. (6) τῶν δὲ ⟨μὴ⟩ μετὰ συμπλοκῆς λεγομένων τὰ μὲν οὐσίαν σημαίνει, τὰ δὲ ποιόν, τὰ δὲ ποσόν, τὰ δὲ πρός τι, τὰ δὲ ποῦ, τὰ δὲ ποτέ, τὰ δὲ κεῖσθαι, τὰ δὲ ἔχειν, τὰ δὲ ποιεῖν, τὰ δὲ πάσχειν, ἃ δὴ καὶ στοιχεῖα τῶν ὄντων φαμὲν τῶν ἐν ὕλῃ καὶ μετὰ τὰς ἀρχάς. ἔστι γὰρ λόγῳ θεωρητὰ ταῦτα, τὰ δὲ ἄυλα νῷ μόνῳ ληπτά ἐστι κατὰ τὴν πρώτην ἐπιβολήν.

24 (1) Τῶν δὲ ὑπὸ τὰς δέκα κατηγορίας ὑποτασσομένων τὰ μὲν καθ’ αὑτὰ λέγεται, ὡς αἱ ἐννέα κατηγορίαι, τὰ δὲ πρός τι. (2) καὶ πάλιν τῶν ὑπὸ ταύτας τὰς δέκα κατηγορίας τὰ μέν ἐστι συνώνυμα, ὡς βοῦς καὶ ἄνθρωπος, καθὸ ζῷον· ἔστι γὰρ συνώνυμα ὧν τό τε ὄνομα ἀμφοῖν κοινόν, τὸ ζῷον, καὶ ὁ λόγος ὁ αὐτός, [τουτέστιν ὁ ὅρος,] τουτέστιν οὐσία ἔμψυχος· (3) ἑτερώνυμα δὲ ὅσα περὶ τὸ αὐτὸ ὑποκείμενον ἐν διαφόροις ἐστὶν ὀνόμασιν, οἷον ἀνάβασις καὶ κατάβασις· ὁδὸς γὰρ ἡ αὐτή, ἤτοι εἰς τὸ ἄνω ἢ εἰς τὸ κάτω· (4) τὸ δ’ ἄλλο εἶδος τῶν ἑτερωνύμων, οἷον ἵππος καὶ μέλας, καὶ ὄνομα καὶ λόγον ἕτερον ἀλλήλων ἔχοντα μηδὲ τοῦ ὑποκειμένου κοινωνοῦντα, ἕτερα *δὴ* λεκτέον, οὐχ ἑτερώνυμα. (5) πολυώνυμα δὲ τὰ τὸν μὲν λόγον τὸν αὐτὸν ἔχοντα, ὄνομα δὲ διάφορον ὡς ἄορ, ξίφος, φάσγανον. (6) παρώνυμα δ’ ἐστὶ τὰ παρά τι ἕτερον ὠνομασμένα, ὡς ἀπὸ τῆς ἀνδρείας ὁ ἀνδρεῖος. (7) τὰ δὲ ὁμώνυμα τῷ αὐτῷ ὀνόματι χρώμενα, λόγον δὲ οὐ τὸν αὐτὸν ἔχοντα, οἷον ἄνθρωπος τό τε ζῷον καὶ γεγραμμένος. (8) τῶν δὲ ὁμωνύμων ‖ τὰ μὲν ἀπὸ τύχης ὁμωνυμεῖ, ὡς Αἴας ὁ Λοκρὸς καὶ ὁ Σαλα- 357ʳ μίνιος, τὰ δὲ ἀπὸ διανοίας, καὶ τούτων τὰ μὲν καθ’ ὁμοιότητα, ὡς ἄνθρωπος τό τε ζῷον καὶ ὁ γεγραμμένος, τὰ δὲ κατὰ ἀναλογίαν, ὡς “πόδες Ἴδης” καὶ οἱ ἡμέτεροι πόδες διὰ τὸ κατωτέρω εἶναι, τὰ δὲ κατ’ ἐνέργειαν, ὡς ποὺς πλοίου, δι’ οὗ τὸ πλοῖον πλεῖ, καὶ πούς ὁ ἡμέτερος, δι’ οὗ κινούμεθα. (9) λέγεται ὁμώνυμα ἀπὸ τοῦ αὐτοῦ καὶ πρὸς τὸ αὐτό, ὡς [ἀπὸ τοῦ ἰατροῦ] τὸ βιβλίον καὶ τὸ σμιλίον ἰατρικὰ ἀπό τε τοῦ χρωμένου ἰατροῦ καὶ πρὸς τὸν αὐτὸν λόγον τὸν ἰατρικόν.

IX. 25 (1) Τῶν αἰτίων τὰ μὲν προκαταρκτικά, τὰ δὲ συνεκτικά, τὰ δὲ συνεργά, τὰ δὲ ὧν οὐκ ἄνευ. (2) προκαταρκτικὰ μὲν τὰ πρώτως ἀφορ-

---

94,24 οὐδὲ Stählin; ὃ δὴ L ‖ μὴ *add.* Potter **95,8** *τουτέστιν ὁ ὅρος *delevi* **95,13** δὴ Heyse; δὲ L **95,25** *ἀπὸ τοῦ ἰατροῦ *del.* Potter; ἀπὸ *del.* Stählin

said, some involve combination, like 'man runs', while others are said without combination, like 'man' and 'runs' and all expressions that do not make up a statement, being neither true nor false. (6) Of things said without combination, each signifies either substance or quality or quantity or a relative or where or when or being-in-a-position or having or doing or being affected. These, we say, are also the elements of things in matter and after the principles. For they can be understood by reasoning, whereas the immaterial entities can only be grasped by the intellect immediately.

24 (1) Of items subordinated to the ten categories, some are said by themselves, like the nine categories, some in relation to something. (2) Again, of items under these ten categories, some are synonyms, like ox and man, each as 'animal'. For synonyms are items that have a name common to both, 'animal', and the same account, that is, 'an ensouled being'. (3) Heteronyms are items expressed by different names while referring to the same underlying thing, like 'ascent' and 'descent'. For the way is the same, leading either up or down. (4) The other kind of heteronyms, like 'horse' and 'black', which have different names as well as accounts and share no underlying thing, should not be called 'heteronyms', but 'diverse' (*hetera*). (5) Polyonyms are items that have the same account, but different names, like 'hanger', 'sword' and 'sabre'. (6) Paronyms are items whose name is derived from something else, like 'brave' from 'bravery'. (7) Homonyms bear the same name, but do not have the same account, e.g. 'man' is both the animal and the one on the picture. (8) Of homonyms, some are homonymous from coincidence, like 'Ajax' from Locris and 'Ajax' from Salamis, others from thought. And of the latter group, some [are homonymous] according to similarity, like 'man' being both the animal and the one on the picture; some according to analogy, like the 'feet' of a mountain and our 'feet', because they are lower; some according to activity, like the 'foot' (i.e. rudder) of a ship, by which the ship is kept on course, and our 'foot', by which we move. (9) Homonyms are called after the same thing and in view of the same thing, like a book and a scalpel are 'medical' after the medical man using them and in view of the same account of the 'medical'.

Causes

IX. 25 (1) Of causes, some are procatarctic, some synectic, some auxiliary, some prerequisite. (2) Procatarctic are those that provide the first stimulus towards

μὴν παρεχόμενα εἰς τὸ γίγνεσθαί τι, καθάπερ τὸ κάλλος τοῖς ἀκολάστοις τοῦ ἔρωτος· ὀφθὲν γὰρ αὐτοῖς τὴν ἐρωτικὴν διάθεσιν ἐμποιεῖ μόνον, οὐ μὴν κατηναγκασμένως. (3) συνεκτικὰ δὲ ἅπερ *συνωνύμως* καὶ
96,1 αὐτοτελῆ καλεῖται, ἐπειδήπερ αὐτάρκως δι' *αὐτῶν ποιητικά* ἐστι τοῦ ἀποτελέσματος. (4) ἑξῆς δὲ πάντα τὰ αἴτια ἐπὶ τοῦ μανθάνοντος δεικτέον. ὁ μὲν πατὴρ αἴτιόν ἐστι προκαταρκτικὸν τῆς μαθήσεως, ὁ διδάσκαλος δὲ συνεκτικόν, ἡ δὲ τοῦ μανθάνοντος φύσις συνεργὸν αἴτιον, ὁ δὲ χρόνος τῶν ὧν οὐκ ἄνευ λόγον ἐπέχει.

(5) Αἴτιον δὲ κυρίως λέγεται τὸ παρεκτικόν τινος ἐνεργητικῶς, ἐπεὶ καὶ τὸν σίδηρον τμητικόν φαμεν εἶναι οὐ μόνον ἐν τῷ τέμνειν, ἀλλὰ καὶ ἐν τῷ μὴ τέμνειν· οὕτως οὖν καὶ τὸ παρεκτικὸν ἄμφω σημαίνει καὶ τὸ ἤδη ἐνεργοῦν καὶ τὸ μη||δέπω μέν, δυνάμει δὲ κεχρημένον τοῦ 357v
ἐνεργῆσαι.

26 (1) Οἳ μὲν οὖν σωμάτων, οἳ δ' ἀσωμάτων φασὶν εἶναι τὰ αἴτια· οἳ δὲ τὸ μὲν σῶμα κυρίως αἴτιόν φασι, τὸ δὲ ἀσώματον καταχρηστικῶς καὶ οἷον αἰτιωδῶς· ἄλλοι δ' ἔμπαλιν ἀναστρέφουσι, τὰ μὲν ἀσώματα κυρίως αἴτια λέγοντες, καταχρηστικῶς δὲ τὰ σώματα, οἷον τὴν τομὴν ἐνέργειαν οὖσαν ἀσώματον εἶναι καὶ αἰτίαν εἶναι τοῦ τέμνειν, ἐνεργείας οὔσης καὶ ἀσωμάτου, καὶ τοῦ τέμνεσθαι ὁμοίως τῇ τε μαχαίρᾳ καὶ τῷ τεμνομένῳ σώμασιν οὖσιν.

(2) Τὸ "*τινῶν* ἐστιν αἴτιον" λέγεται τριχῶς, τὸ μὲν ὅ ἐστιν αἴτιον, οἷον ὁ ἀνδριαντοποιός, τὸ δὲ οὗ ἐστιν αἴτιον, τοῦ γίνεσθαι τὸν ἀνδριάντα, *τὸ* δὲ ᾧ ἐστιν αἴτιον, ὥσπερ *τῇ ὕλῃ*· τῷ χαλκῷ γὰρ αἴτιός ἐστι τοῦ γίνεσθαι τὸν ἀνδριάντα. (3) τὸ γίνεσθαι οὖν καὶ τὸ τέμνεσθαι, τὰ οὗ ἐστιν *αἴτιον*, ἐνέργειαι οὖσαι ἀσώματοί εἰσιν.

(4) Εἰς ὃν λόγον κατηγορημάτων ἤ, ὥς τινες, λεκτῶν (λεκτὰ γὰρ τὰ κατηγορήματα καλοῦσιν Κλεάνθης καὶ Ἀρχέδημος) τὰ αἴτια· ἤ, ὅπερ καὶ μᾶλλον, τὰ μὲν κατηγορημάτων αἴτια λεχθήσεται, οἷον *τοῦ "τέμνεται"*, οὗ πτῶσις τὸ τέμνεσθαι, τὰ δ' ἀξιωμάτων, ὡς τοῦ "ναῦς
97,1 γίνεται", οὗ πάλιν [ἡ] πτῶσίς ἐστι τὸ ναῦν γίνεσθαι· Ἀριστοτέλης δὲ *προσηγοριῶν*, οἷον τῶν τοιούτων, οἰκίας, νεώς, καύσεως, τομῆς. ἡ πτῶσις δὲ ἀσώματος εἶναι ὁμολογεῖται· (5) διὸ καὶ τὸ σόφισμα ἐκεῖνο οὕτως λύεται· "ὃ λέγεις, διέρχεταί σου διὰ τοῦ στόματος", ὅπερ ἀλη-

---

95,31 οὐ μὴν κατηναγκασμένως *post* εἰς τὸ γίγνεσθαί τι (95,29) *transposuerim* ‖ συνωνύμως Sylburg; συνώνυμα L **96,1** αὐτῶν ποιητικά Arnim; αὐτοῦ ποιητικόν L **96,18** τινῶν Wedel; τίνων L **96,19** *οἷον *post* αἴτιον *add.* Mewaldt **96,20** τὸ Bunsen; ὁ L ‖ τῇ ὕλῃ Wilamowitz; ἡ ὕλη L **96,22** αἴτιον Wedel; αἴτια L **96,24** *αἴτια *ante* τὰ αἴτια *add.* Stählin **96,26** τοῦ[1] Wilamowitz; τὸ L ‖ τέμνεται Hervet; τέμνεσθαι L **97,1** γίνεται Hervet; γίνεσθαι L ‖ ἡ *del.* Arnim, Wilamowitz **97,2** προσηγοριῶν Stählin; προσηγορίαν L

something's coming about, as beauty [is a cause] of love to intemperate people: when seen, it creates in them an erotic condition only, not necessarily.[5] (3) Cohesive causes are also synonymously called 'self-complete', because they produce their effect sufficiently by themselves. (4) Next, we must exhibit all of these causes with the example of a student. The father is the procatarctic cause of learning, the teacher the synectic cause, the student's nature is an auxiliary cause, and time has the status of prerequisites.

(5) 'Cause' is properly said of that which actually brings something about; for in the case of a knife, as well, we say that it is a cutting instrument not only when it is cutting, but also when it is not. In this way, too, 'that which brings something about' means two things: that which already does so, and that which does not do so yet, but has the capacity of doing so in the future.

26 (1) Some say that causes are causes of bodies, others of incorporeal items. Some say that body is a cause in the proper sense, whereas that which is incorporeal is a cause in a loose sense and, as it were, quasi-causally. Others turn it around, saying that incorporeal items are causes in the proper sense, whereas bodies in a loose sense. For example, they say, a cut, being an activity, is incorporeal, and it is the cause of cutting, itself an activity and incorporeal, to a knife, i.e. a body, and, likewise, [it is the cause] to the cut object, itself a body, of being cut.

(2) By saying that something is the cause of certain things, we refer to three items: that which is the cause, e.g. the sculptor; that of which it is a cause, e.g. becoming a sculpture; and that to which it is a cause, e.g. the matter. For it is to bronze that the sculptor is the cause of becoming a sculpture. (3) Hence becoming and being cut, that of which something is a cause, are activities, and therefore incorporeal.

(4) This is the reasoning according to which causes are causes of predicates or, as some have it, of 'sayables' (for Cleanthes and Archedemus call predicates 'sayables'). Or we shall say, and this is more fitting, that some causes are causes of predicates, e.g. 'is cut', whose case is 'being cut', and others of propositions, e.g. 'a ship is built', whose case, in turn, is 'a ship's being built'. Aristotle thinks that causes are causes of nouns, as, for example, the following: house, ship, burning, cut. But everyone agrees that a case is incorporeal. (5) This is how the well-known sophism will be solved, which says: "Whatever you say, goes through your mouth, which is true; but you say 'house', therefore house goes

5 "Not necessarily" should perhaps be moved after "something's coming about".

θές, “οἰκίαν δὲ λέγεις, οἰκία ἄρα διὰ τοῦ στόματός σου διέρχεται”, ὅπερ ψεῦδος· *οὐδὲ* γὰρ τὴν οἰκίαν λέγομεν ‖ σῶμα οὖσαν, ἀλλὰ τὴν 358r
πτῶσιν ἀσώματον οὖσαν, ἧς οἰκία τυγχάνει. ǀ **27** (1) καὶ τὸν οἰκοδόμον οἰκοδομεῖν λέγομεν κατὰ τὴν ἐπὶ τὸ γενησόǀμενον ἀναφοράν· οὕτως φαμὲν χλαμύδα ὑφαίνεσθαι· τὸ γὰρ ποιǀοῦν ἐνεργείας δηλωτικὸν ὑπάρχει, (2) οὐκ ἔστι δὲ ἑτέρου μὲν τὸ ποιοῦν, ǀἑτέρου δὲ τὸ αἴτιον, ἀλλὰ ταὐτοῦ, τῆς χλαμύδος καὶ τῆς οἰκίας· καθ'ǀ ὃ γὰρ αἴτιος τοῦ γίνεσθαι, κατὰ τοῦτο καὶ ποιητικός ἐστι τοῦ γίνεǀσθαι.

(3) Τὸ [δὲ] αὐτὸ ἄρα αἴτιον καὶ ποιητικόν [καὶ δι' ὅ]. καὶ εἰ μέν τί ἐστιν αἴτιον καὶ ποιητικόν, τοῦτο πάντως ἐστὶ καὶ δι' ὅ, εἰ δέ τί ἐστι δι' ὅ, οὐ πάντως τοῦτο καὶ αἴτιον. πολλὰ γοῦν ἐφ' ἓν ἀποτέλεσμα συντρέχει, δι' ἃ γίνεται τὸ τέλος, ἀλλ' οὐκ ἔστι πάντα αἴτια. (4) οὐ γὰρ ἂν ἐτεκνοκτόνησεν Μήδεια, εἰ μὴ ὠργίσθη, οὐδ' ἂν ὠργίσθη, εἰ μὴ ἐζήλωσεν, οὐδὲ τοῦτο, εἰ μὴ ἠράσθη, οὐδὲ τοῦτο, εἰ μὴ Ἰάσων ἔπλευσεν εἰς Κόλχους, οὐδὲ τοῦτο, εἰ μὴ Ἀργὼ κατεσκευάσθη, οὐδὲ τοῦτο, εἰ μὴ τὰ ξύλα ἐκ τοῦ Πηλίου ἐτμήθη. (5) ἐν τούτοις γὰρ ἅπασιν τοῦ “δι' ὅ” τυγχάνοντος οὐ πάντα τῆς τεκνοκτονίας αἴτια τυγχάνει, μόνη δὲ ἡ Μήδεια.

(6) [Διὸ] τὸ μὴ κωλῦον ἀνενέργητόν ἐστιν, διὸ οὐκ ἔστιν αἴτιον τὸ μὴ κωλῦον, ἀλλὰ τὸ κωλῦον. ἐν γὰρ τῷ ἐνεργεῖν καὶ δρᾶν τι τὸ αἴτιον νοεῖται. **28** (1) ἔτι τὸ μὴ κωλῦον κεχώρισται τοῦ γινομένου (διὰ τοῦτο γοῦν ἐπιτελεῖται, ὅτι τὸ δυνάμενον κωλύειν οὐ πάρεστιν),
98,1 τὸ δὲ αἴτιον πρὸς τὸ γινόμενον· οὐκ ἂν οὖν εἴη αἴτιον τὸ μὴ κωλῦον.

(2) Τετραχῶς τὸ αἴτιον λέγεται τὸ ποιοῦν ὡς ὁ ἀνδριαντοποιός, καὶ ἡ ὕλη ὡς ὁ χαλκός, καὶ τὸ εἶδος ὡς ὁ χα‖ρακτήρ, καὶ τὸ τέλος 358v
ὡς ἡ τιμὴ τοῦ γυμνασιάρχου. (3) τῶν ὧν οὐκ ἄνευ λόγον ὁ χαλκὸς ἐπέχει πρὸς τὸ γενέσθαι τὸν ἀνδριάντα καὶ ὁμοίως ἐστὶν αἴτιον. πᾶν γὰρ οὗ χωρὶς οὐκ ἐνδεχόμενον γενέσθαι τὸ ἀποτέλεσμα, κατὰ ἀνάγκην ἐστὶν αἴτιον, αἴτιον δὲ οὐχ ἁπλῶς· οὐ γάρ ἐστι συνεκτικὸν τὸ οὗ μὴ ἄνευ, συνεργὸν δέ. (4) πᾶν τὸ ἐνεργοῦν παρέχει τὸ ἀποτέλεσμα μετ' ἐπιτηδειότητος τοῦ πάσχοντος· διατίθησι μὲν γὰρ τὸ αἴτιον, πάσχει δὲ ἕκαστον εἰς ὃ πέφυκέν τι, παρεκτικῆς τῆς ἐπιτηδειότητος οὔσης καὶ τῶν ὧν οὐκ ἄνευ λόγον ἐπεχούσης. (5) ἄπρακτον οὖν τὸ αἴτιον ἄνευ τῆς ἐπιτηδειότητος, οὐκ ἔστιν δὲ αἰτία, ἀλλὰ συνεργός, ἐπεὶ πᾶν αἴτιον ἐν τῷ δρᾶν νοεῖται. (αὐτὴν δὲ οὐκ ἂν ποιοίη ἡ γῆ,

---

97,6 οὐδὲ Dindorf; οὔτε L 97,9 *lac. post* ὑφαίνεσθαι *ind.* Wedel 97,11 *τῆς χλαμύδος καὶ τῆς οἰκίας *del.* Stählin 97,13 δὲ *del.* Mewaldt ‖ καὶ δι' ὅ *del.* Schwartz 97,23 διὸ[1] *del.* Wilamowitz

through your mouth, which is false." For we do not say house as a body, but a case, something incorporeal, whose object is the house. **27** (1) Also, when saying that a housebuilder builds a house, we refer to something that will come into existence in the future. In the same way, we say that a mantle is being weaved. For 'agent' indicates an activity, (2) but there is no difference between what the agent does and what the cause causes, since it is the same item: the mantle and the house. Insofar as someone is the cause for something to come into existence, it is by his activity that it comes into existence.

(3) So that which causes something is the same as that which does it. And if something causes and does P, then, necessarily, it is also that on account of which P. But if something is that on account of which P, it is not necessarily also the cause of P. For there are many factors contributing to one effect, on account of which the end result is what it is, but not all of them are its causes. (4) Thus Medea would not have killed her children, had she not been furious; but she would not have been furious, had she not been jealous; but that would not have happened, had she not been in love; nor would that have happened, had Iason not travelled to Colchis; nor that, had Argo not been built; not that, had trees from Pelion not been cut. It is on account of all these factors that Medea killed her children, but not all of them have caused the killing, only Medea herself.

(6) That which does not prevent does not act. Therefore, that which does not prevent is not a cause, but that which prevents is. For a cause is conceived as active and doing something. **28** (1) Moreover, that which does not prevent is separate from what is coming about (for the latter is accomplished precisely on account of the fact that something capable of preventing it is not present). But a cause is [a cause] in relation to what is coming about. Thus that which does not prevent is not a cause.

(2) Cause is said in four ways: with regard to an agent, e.g. a sculptor, a matter, e.g. bronze, a form, e.g. features, and a goal, e.g. to honour the gymnasiarch. (3) Bronze has the status of prerequisites with regard to the coming to be of a statue and it is a cause accordingly: everything without which an effect cannot come about is necessarily a cause, but it is not a cause without qualification. For what is prerequisite is not synectic, but auxiliary. (4) Everything active brings about an effect together with a suitability of that which is acted upon. For a cause fashions its object, but every object undergoes that to which it is naturally fit; this is what suitability provides, having, therefore, the status of prerequisites. (5) So without suitability, a cause can do nothing. Nevertheless, it is not a cause, but auxiliary, because every cause is conceived as doing something.

ὥστε οὐδὲ αἰτία ἂν εἴη ἑαυτῆς.) (6) καταγέλαστον δὲ τὸ λέγειν μὴ τὸ πῦρ αἴτιον τῆς καύσεως, ἀλλὰ τὰ ξύλα, μηδὲ τὴν μάχαιραν τῆς τομῆς, ἀλλὰ τὴν σάρκα, μηδὲ τοῦ καταπαλαισθῆναι τὸν ἀθλητὴν τὴν ἰσχὺν τοῦ ἀντιπάλου, τὴν ἑαυτοῦ δὲ ἀσθένειαν.

(7) Τὸ συνεκτικὸν αἴτιον οὐ δεῖται χρόνου· τὸ γὰρ καυτήριον ἅμα τῷ κατὰ τῆς σαρκὸς ἐπερεισθῆναι τὴν ἀλγηδόνα παρέχει. τῶν προκαταρκτικῶν τὰ μὲν χρόνου δεῖται, ἄχρις ἂν γένηται τὸ ἀποτέλεσμα, τὰ δὲ οὐ δεῖται, ὡς ἡ πτῶσις τοῦ κατάγματος. μή τι οὖν οὐ κατὰ στέρησιν χρόνου ἄχρονα λέγεται ταῦτα, ἀλλὰ κατὰ μείωσιν, ὡς καὶ τὸ ἐξαίφνης, μηδὲ αὐτὸ χωρὶς χρόνου γενόμενον.

29 (1) Πᾶν αἴτιον ὡς αἴτιον διανοίᾳ ληπτὸν τυγχάνει, *ἐπεὶ* τινὸς καὶ πρός *τινι* νοεῖται, τινὸς μέν, τοῦ ἀποτελέσματος, καθάπερ ἡ μάχαιρα τοῦ τέμνειν, πρός τινι δέ, [καθάπερ] τῷ ἐπιτηδείως ‖ ἔχοντι, 359r
καθάπερ τὸ πῦρ τῷ ξύλῳ· τὸν ἀδάμαντα γὰρ οὐ καύσει. (2) τὸ αἴτιον τῶν "πρός τι"· κατὰ γὰρ τὴν πρὸς ἕτερον νοεῖται σχέσιν, ὥστε δυεῖν ἐπιβάλλομεν, ἵνα τὸ αἴτιον ὡς αἴτιον νοήσωμεν.

(3) Ὁ αὐτὸς καὶ περὶ τοῦ δημιουργοῦ καὶ ποιητοῦ λόγος καὶ πατρός. οὐκ ἔστιν αὐτό τι ἑαυτοῦ αἴτιον οὐδὲ ἑαυτοῦ τις πατήρ, ἐπεὶ τὸ πρῶτον γενήσεται δεύτερον· τό γε μὴν αἴτιον ἐνεργεῖ καὶ διατίθησι, τὸ ὑπὸ τοῦ αἰτίου γενόμενον πάσχει καὶ διατίθεται. (4) οὐ δύναται δὲ τὸ αὐτὸ πρὸς ἑαυτῷ λαμβανόμενον ἐνεργεῖν ἅμα καὶ διατίθεσθαι, οὐδὲ υἱὸς εἶναι καὶ πατήρ. (5) καὶ ἄλλως τὸ αἴτιον τοῦ ὑπ' αὐτοῦ γινομένου προχρονεῖ κατὰ τὴν οὐσίαν, ὡς ἡ μάχαιρα τῆς τομῆς. οὐ δύναται δὲ τὸ αὐτὸ κατὰ τὸ αὐτὸ καὶ προχρονεῖν τῇ ὕλῃ, καθὸ αἴτιόν ἐστιν, ἅμα καὶ ὑστερεῖν καὶ ὑστεροχρονεῖν, καθὸ τῆς αἰτίας ἐστὶν ἔργον. (6) διαφέρει τε τὸ εἶναι τοῦ *γίνεσθαι*· οὕτως καὶ αἴτιον μὲν τοῦ γινομένου, πατὴρ δὲ υἱοῦ. οὐκ ἐνδέχεται γὰρ τὸ αὐτὸ κατὰ τὸ αὐτὸ καὶ εἶναι ἅμα καὶ γίνεσθαι. οὐθὲν οὖν ἐστιν ἑαυτοῦ αἴτιον.

30 (1) Ἀλλήλων οὐκ ἔστι τὰ αἴτια, ἀλλήλοις δὲ αἴτια. ἡ γὰρ σπληνικὴ διάθεσις προϋποκειμένη οὐ πυρετοῦ αἴτιος, ἀλλὰ τοῦ γίνεσθαι τὸν πυρετόν· καὶ ὁ πυρετὸς προϋποκείμενος οὐ σπληνός, ἀλλὰ τοῦ αὔξεσθαι τὴν διάθεσιν. (2) οὕτως καὶ αἱ ἀρεταὶ ἀλλήλαις αἴτιαι *τοῦ* μὴ χωρίζεσθαι διὰ τὴν ἀντακολουθίαν, καὶ οἱ ἐπὶ τῆς ψαλίδος λίθοι ἀλλήλοις εἰσὶν αἴτιοι τοῦ μένειν κατηγορήματος, ἀλλήλων δὲ οὐκ εἰσὶν αἴτιοι, καὶ ὁ διδάσκαλος δὲ καὶ ὁ μανθάνων ἀλλήλοις εἰσὶν αἴτιοι τοῦ προκόπτειν κατηγορήματος. (3) λέγεται δὲ ἀλλήλοις ‖ αἴτια 359v

---

98,25 *διπλῆ *ante* διανοίᾳ *add.* Stählin ‖ ἐπεὶ Wilamowitz; ἐπὶ L **98,26** τινι Wedel; τι L ‖ μέν, τοῦ Heyse; μέντοι L **98,27** καθάπερ *del.* Wedel **99,10** γίνεσθαι Wilamowitz; γενέσθαι L **99,16** τοῦ Sylburg; τῶι L

(Earth cannot make itself, and so it cannot be a cause of itself.) (6) It would be ridiculous to say that the cause of burning is not a fire, but wood, or that the cause of a cut is not a knife, but flesh, or that a wrestler was thrown to the ground not by the strength of his rival, but by his own weakness.

(7) Synectic causes do not need time. For a cautery produces pain immediately as it is pressed onto flesh. Of the procatarctic causes, some need time before the effect comes about, whereas some do not, as when a fall causes a fracture. Of course, when saying that they do not need time, we do not mean it in the sense of a negation of time, but of its diminution, like with a sudden event, which also does not come about without time.

29 (1) Every cause is grasped by the mind as a cause, since it is conceived as a cause of something and in relation to something, namely as a cause of an effect, for example a knife of the act of cutting, and in relation to that which is suitable, for example fire in relation to wood (it will not burn a diamond). (2) Causes belong to the category of relation, for they are conceived in relation to something else. Thus in order to think of a cause as a cause, we focus on two items.

(3) The same argument can be made about an artisan and a creator and a father. Nothing can be its own cause and no one can be his own father, for otherwise that which is first would come second. For a cause indeed acts and fashions its object, whereas that which comes about by that cause is affected and fashioned by it. (4) It is impossible for anything to act and be fashioned at the same time, or to be a son and a father, in relation to itself. (5) And generally, a cause is, in essence, chronologically prior to that which comes about by it, as a knife is prior to a cut. But it is impossible for something to be both chronologically prior to matter, as a cause, and, at the same time and in the same respect, posterior and later, as an effect of a cause. (6) And there is a difference between being and becoming. In the same way, a cause differs from that which comes about by it and a father differs from a son. For it is impossible for the same thing to be and to become at the same time and in the same respect. Thus nothing is its own cause.

30 (1) Causes are not causes of each other, but there are causes to each other. For the pre-existing condition of the spleen is the cause, not of fever, but of the fever's coming about; and the pre-existing fever is the cause, not of the spleen, but of its condition's being intensified. (2) In the same way, virtues are the causes to each other of not being separated, owing to their inter-entailment, and the stones in the vault are the causes to each other of the predicate 'remaining', but they are not the causes of each other. And the teacher and the pupil are the causes to each other of the predicate 'making progress'. (3) Things are said to be the causes to each other sometimes of the

ποτὲ μὲν τῶν αὐτῶν, ὡς ὁ ἔμπορος καὶ ὁ κάπηλος ἀλλήλοις εἰσὶν αἴτιοι τοῦ κερδαίνειν, ποτὲ δὲ ἄλλου καὶ ἄλλου, καθάπερ ἡ μάχαιρα καὶ ἡ σάρξ· ἣ μὲν γὰρ τῇ σαρκὶ τοῦ τέμνεσθαι, ἡ σὰρξ δὲ τῇ μαχαίρᾳ τοῦ τέμνειν. |(4) (τὸ "ὀφθαλμὸν ἀντὶ ὀφθαλμοῦ" καὶ "ψυχὴν ἀντὶ ψυχῆς".) ὁ μὲν γὰρ πλήξας τινὰ θανασίμως αἴτιός ἐστιν αὐτῷ τοῦ θανάτου ἢ *τοῦ* γίνεσθαι τὸν θάνατον, ἀντιπληγεὶς δὲ ὑπ' αὐτοῦ θανασίμως ἔσχεν αὐτὸν ἀνταίτιον, οὐ καθὸ ἐγένετο αὐτῷ αἴτιος, καθ' ἕτερον δέ. (5) αἴτιος μὲν γὰρ αὐτῷ θανάτου γίνεται, οὐχ ὁ θάνατος δὲ τούτῳ πάλιν τὴν θανασίμην ἀντεπέθηκεν πληγήν, ἀλλ' αὐτὸς ὁ τρωθείς, ὥστ' ἄλλου μὲν αὐτὸς γέγονεν αἴτιος, ἄλλον δὲ ἔσχεν αἴτιον. καὶ ὁ ἀδικήσας *ἄλλῳ* μὲν αἴτιος καθίσταται τοῦ ἀδικηθέντος, ὁ δὲ ἀντιτιμωρεῖσθαι κελεύων νόμος οὐκ ἀδικήματος, ἀλλὰ τῷ μὲν ἐκδικίας, τῷ δὲ παιδείας. ὥστε οὐκ ἀλλήλων τὰ αἴτια [ὡς αἴτια], ἀ⟨λλήλοις⟩ δέ ἐστιν αἴτια.

31 (1) Ἔτι ζητεῖται εἰ πολλὰ κατὰ σύνοδον ἑνὸς αἴτια γίνεται |πολλά. *οὐ* γὰρ ἄνθρωποι συνελθόντες αἴτιοί εἰσι τοῦ καθέλκε|σθαι τὴν ναῦν, ἀλλὰ σὺν τοῖς |ἄλλοις, εἰ μή τι καὶ τὸ συναίτιον αἴτιον. ἄλλοι δέ φασιν· "εἰ πολλὰ αἴτια, κατ' ἰδίαν ἕκαστον ἑνὸς αἴτιον γίνεται· (2) τοῦ γοῦν εὐδαιμονεῖν ἑνὸς ὄντος αἴτιαι τυγχάνουσιν αἱ ἀρεταὶ πολλαὶ οὖσαι, καὶ τοῦ θερμαίνεσθαι καὶ τοῦ ἀλγεῖν ὁμοίως πολλὰ τὰ αἴτια. (3) μή τι οὖν αἱ πολλαὶ ἀρεταὶ μία ἐστὶ κατὰ δύναμιν· [καὶ τὰ θερ‖μαίνοντα καὶ τὰ ἀλγοῦντα] 360r καὶ τὸ πλῆθος δὲ τῶν ἀρετῶν κατὰ γένος ἓν τυγχάνον ἑνὸς αἴτιον γίνεται τοῦ εὐδαιμονεῖν." (4) τῷ ὄντι δὲ προκαταρκτικὰ μὲν αἴτια ἑνὸς γίνεται πλείονα κατὰ γένος καὶ κατ' εἶδος, καὶ κατὰ γένος μὲν τοῦ νοσεῖν ὁπωσοῦν, οἷον ψῦξις, *ἔγκαυσις*, κόπος, ἀπεψία, μέθη, κατ' εἶδος δὲ τοῦ πυρετοῦ. (5) τὰ δὲ συνεκτικὰ αἴτια κατὰ γένος μόνον, οὐκέτι δὲ καὶ κατ' εἶδος. τοῦ γὰρ εὐωδιάζεσθαι κατὰ γένος ἑνὸς ὄντος πολλὰ τὰ αἴτια κατ' εἶδος, οἷον λιβανωτός, ῥόδον, κρόκος, στύραξ, σμύρνα, μύρον· τὸ γὰρ ῥόδον οὐκ ἂν οὕτως εὐῶδες εἴη ὡς ἡ σμύρνα.

32 (1) Τὸ δὲ αὐτὸ τῶν ἐναντίων αἴτιον γίνεται, ποτὲ μὲν παρὰ τὸ μέγεθος τοῦ αἰτίου καὶ τὴν δύναμιν, ποτὲ δὲ παρὰ τὴν ἐπιτηδειότητα τοῦ πάσχοντος. (2) παρὰ μὲν τὴν ποιὰν δύναμιν· ἡ αὐτὴ χορδὴ παρὰ

99,22 προκόπτειν κατηγορήματος ... ἀλλήλοις εἰσὶν αἴτιοι τοῦ (99,20–22) *post* αἴτιοι τοῦ *in ms. repetitur* **99,27** τοῦ Bunsen; τὸ L **99,29** *γίνεται] ἐγένετο Wedel **99,32** ἄλλῳ Stählin; ἄλλου L **100,2** ὡς αἴτια *delevi* **100,2 f.** ἀλλήλοις Bywater; ἃ L **100,4** *αἴτια *post* γίνεται *add.* Stählin **100,5** *οὐ *scripsi*; οἱ L ‖ *οἱ *post* ἄνθρωποι *add.* Stählin **100,6** **lac. post* ναῦν *ind.* Arnim **100,11** *καὶ τὰ θερμαίνοντα καὶ τὰ ἀλγοῦντα *delevi*; ἀλγοῦντα] ἀλγύνοντα *corr.* Arnim; ὁμοίως ἓν *post* καὶ[1] *add.* Heyse **100,15** ἔγκαυσις Wilamowitz; ἔκλυσις L **100,18** *κατ' εἶδος *del.* Wedel

same effects, as the wholesaler and the retailer are the causes to each other of making a profit; but sometimes of different effects, as in the case of the knife and the flesh; for the knife is the cause to the flesh of being cut, while the flesh is the cause to the knife of cutting. (4) (The 'eye for an eye' and 'life for a life'). For when one person fatally strikes another, he is the cause of the other one's death, or the death's coming about; but when he is hit back by the other person, he has the cause of his death in the other person; [this cause is] not that what he was the cause of [to the other person], but [it is] something else. (5) For he is the cause of death to the other person, but it was not death which hit him back with a fatal blow, but the wounded person himself. Thus he has been the cause of something else than what has been the cause [of death] to him. And when someone commits an injury, he becomes the cause to another of the injury the other person suffers, but the law that orders retaliation is not a cause of an injury, but of satisfaction to one person, and of education to the other. So causes are not causes of each other, but there are causes to each other.

31 (1) It is also inquired whether many factors coming together are many causes of one effect. For men coming together are ⟨not⟩ the causes of the ship's having been drawn to the sea, except with the others, unless indeed the joint-cause is also a cause. Others say: 'If many [factors are] causes, then each of them is separately a cause of one effect. (2) For example, of being happy, which is one thing, the causes are virtues, which are many. Similarly, there are many causes of warming up or feeling pain. (3) Not to mention the fact that many virtues are potentially one; and since the multitude of virtues is one according to genus, it becomes the cause of one effect, namely, being happy.' (4) But in fact, in the case of procatarctic causes, there can be many [causes] of one effect according to genus and according to species. [Many of one] according to genus are the causes of becoming ill in any way whatsoever, for example cold, heat, fatigue, indigestion, drunkenness; [many of one] according to species are the causes of fever. (5) As far as synectic causes are concerned, there are only [many of one effect] according to genus, but not according to species. Thus smelling good is one effect according to genus, but it has many causes according to species, for example frankincense, rose, saffron, storax, myrrh, or sweet oil. For rose certainly does not smell the same as myrrh.

32 (1) The same item becomes the cause of opposite effects, sometimes due to the magnitude and power of the cause, sometimes due to the suitability of that which is acted upon. (2) Due to the power's being such-and-such: the same

τὴν ἐπίτασιν ἢ τὴν ἄνεσιν ὀξὺν ἢ βαρὺν ἀποδίδωσι τὸν φθόγγον.
(3) παρὰ δὲ τὴν ἐπιτηδειότητα τῶν πασχόντων· τὸ μέλι γλυκάζει μὲν
101,1 τοὺς ὑγιαίνοντας, πικράζει δὲ τοὺς πυρέσσοντας, καὶ εἷς καὶ ὁ αὐτὸς
οἶνος τοὺς μὲν εἰς ὀργήν, τοὺς δὲ εἰς διάχυσιν ἄγει, καὶ ὁ αὐτὸς ἥλιος
τήκει μὲν τὸν κηρόν, ξηραίνει δὲ τὸν πηλόν.

(4) Τῶν οὖν αἰτίων τὰ μέν ἐστι πρόδηλα, τὰ [δὲ] ἐπιλογισμῷ λαμ-
βανόμενα, τὰ δὲ ἄδηλα, τὰ [δὲ] ἀναλογισμῷ. (5) καὶ τῶν ἀδήλων τὰ μὲν
πρὸς καιρὸν ἄδηλα, ⟨τὰ⟩ κατὰ μέν τινα καιρὸν ἀποκεκρυμμένα, κατὰ
δέ τινα πάλιν ἔκδηλα βλεπόμενα, τὰ δὲ φύσει ἄδηλα, τὰ κατὰ μηδένα
καιρὸν πρόδηλα γενέσθαι ‖ δυνάμενα. (6) καὶ τῶν μὲν φύσει τὰ μὲν κατα- 360v
ληπτά, ἅπερ οὐκ ἄδηλά τινες ἐκάλουν διὰ σημείων ἀναλογιστικῶς
λαμβανόμενα, καθάπερ ἡ συμμετρία τῶν *λόγῳ* θεωρητῶν πόρων, τὰ
δὲ ἀκατάληπτα, τὰ κατὰ μηδένα τρόπον ὑπὸ κατάληψιν πεσεῖν δυνά-
μενα, ἃ δὴ καὶ ἄδηλα ἐν τῷ καθάπαξ *λέγεται*.

(7) Καὶ τὰ μὲν προκαταρκτικά, τὰ δὲ συνεκτικά, τὰ δὲ συναίτια, τὰ
δὲ συνεργά. καὶ τὰ μὲν τοῦ κατὰ φύσιν, τὰ δὲ τοῦ παρὰ φύσιν, τὰ
δὲ νόσου, καὶ κατὰ τὸ *ὑποβεβηκὸς* τὰ μὲν παθῶν, τὰ δὲ τοῦ μεγέ-
θους αὐτῶν, τὰ δὲ χρόνων καὶ καιρῶν.

33 (1) Τῶν μὲν οὖν προκαταρκτικῶν αἱρομένων μένει τὸ ἀποτέλεσμα,
συνεκτικὸν δέ ἐστιν αἴτιον, οὗ παρόντος μένει τὸ ἀποτέλεσμα καὶ
αἱρομένου αἴρεται. (2) τὸ δὲ συνεκτικὸν συνωνύμως καὶ αὐτοτελὲς κα-
λοῦσιν, ἐπειδὴ αὐτάρκως δι' αὐτοῦ ποιητικόν ἐστι τοῦ ἀποτελέσματος.
(3) εἰ δὲ τὸ αἴτιον αὐτοτελοῦς ἐνεργείας ἐστὶ δηλωτικόν, τὸ
συνεργὸν ὑπηρεσίαν σημαίνει καὶ τὴν σὺν ἑτέρῳ λειτουργίαν. (4) εἰ μὲν οὖν
μηδὲν παρέχεται, οὐδὲ συνεργὸν λεχθήσεται, εἰ δὲ παρέχεται, τούτου
πάντως γίνεται αἴτιον οὗ καὶ παρέχεται, τουτέστιν τοῦ δι' αὐτοῦ
γινομένου. (5) ἔστιν οὖν συνεργὸν οὗ παρόντος ἐγίνετο τὸ ἀποτέλεσμα,
προδήλῳ μὲν οὖν παρόντος ⟨προδήλου⟩, ἀδήλῳ δὲ ἀδήλου. (6) καὶ τὸ
συναίτιον δὲ ἐκ τοῦ γένους ἐστὶ τῶν αἰτίων, καθάπερ ὁ συστρατιώτης
102,1 στρατιώτης καὶ ὁ συνέφηβος ἔφηβος. (7) τὸ μὲν οὖν συνεργὸν αἴτιον
τῷ συνεκτικῷ πρὸς τὴν ἐπίτασιν βοηθεῖ τοῦ ὑπ' αὐτοῦ γινομένου,
τὸ δὲ συναίτιον οὐκ ἐπὶ τῆς αὐτῆς ἐστιν ἐννοίας· δύναται γὰρ συν-
αίτιον ὑπάρχειν, κἂν μὴ ‖ συνεκτικὸν αἴτιον ᾖ τι. (8) νοεῖται γὰρ σὺν 361r
ἑτέρῳ τὸ συναίτιον οὐδ' *αὐτῷ* δυναμένῳ κατ' ἰδίαν ποιῆσαι τὸ ἀπο-
τέλεσμα, αἴτιον ὂν σὺν αἰτίῳ. (9) διαφέρει δὲ τοῦ συναιτίου τὸ συνεργὸν

101,4 δὲ *del.* Stählin **101,5** δὲ[2] *del.* Stählin **101,6** τὰ *add.* Wilamowitz **101,10** λόγῳ Potter; λόγων L **101,12** λέγεται Potter; λέγεσθαι L **101,15** ὑποβεβηκὸς Sylburg; ὑπεβηκὸς L **101,21** *αὐτοτελὲς *ante* αἴτιον *add.* Stählin, Schwartz **101,26** προδήλου *add.* Potter **102,5** αὐτῷ Hervet; αὖ τῶι L

string makes a high-pitched or a low-pitched sound, depending on the degree of tension. (3) Due to the suitability of things acted upon: Honey is sweet for the healthy, but bitter for the feverish; one and the same wine makes some people angry, others merry; the same sun melts the wax and dries the mud.

(4) Of causes, some are clear, viz. those grasped by epilogism, whereas some are unclear, viz. those [grasped] by analogism. (5) And of the unclear ones, some are unclear for the moment, viz. those hidden on one occasion, but clearly observed again on another, whereas some are unclear by nature, viz. those incapable of becoming clear on any occasion. (6) And of those [unclear] by nature, some are apprehensible, these being sometimes referred to as 'not unclear', as they are grasped analogically through signs, for example the symmetry of pores discernible by reason, whereas some are inapprehensible, viz. those incapable of becoming an object of apprehension in any way, these being also referred to as unclear once and for all.

(7) And some are procatarctic, some synectic, some joint-causes, some auxiliary. And some are causes of effects that are natural, some of those that are unnatural, some of disease, and, on a lower level, some of affections, some of their magnitude, some of temporal intervals.

33 (1) When procatarctic causes are removed, the effect remains, whereas a synectic cause is one during whose presence the effect remains and on whose removal the effect is removed. (2) The synectic cause is synonymously called the 'self-complete' cause, since it produces its effect sufficiently by itself. (3) If 'cause' is indicative of a self-complete activity, the 'auxiliary' signifies assistance and service alongside something else. (4) Hence if no result is produced it will not even be called auxiliary, but if one is produced it does become a cause of what is actually being produced, that is, of what is coming about through it. (5) So an auxiliary cause is one that was present while an effect was coming about. When present to something clear, it is clear, when present to something unclear, it is unclear. (6) The joint-cause is another member of the genus of causes (in the way that a fellow soldier is a soldier and a fellow trainee a trainee). (7) Whereas the auxiliary cause aids the synectic cause, so as to intensify what comes about through the latter, the joint-cause does not correspond to the same notion, since a joint-cause can exist even if there is no synectic cause. (8) For the joint-cause is conceived jointly with another which is itself likewise incapable of producing the effect by its own, since it is jointly that they are causes. (9) The difference between the joint-cause and the auxiliary cause lies in the fact that

ἐν τῷ τὸ μὲν συναίτιον ⟨μεθ' ἑτέρου⟩ κατ' ἰδίαν μὴ ποιοῦντος τὸ
ἀποτέλεσμα παρέχειν, τὸ δὲ συνεργὸν [ἐν τῷ] κατ' ἰδίαν μὴ ποιεῖν,
ἑτέρῳ δὲ προσερχόμενον τῷ κατ' ἰδίαν ποιοῦντι συνεργεῖ⟨ν⟩ αὐτῷ
10 πρὸς τὸ *σφοδρότερον* γίνεσθαι τὸ ἀποτέλεσμα. μάλιστα δὲ τὸ ἐκ
11/12 προκαταρκτικοῦ συνεργὸν γεγονέναι τὴν τοῦ |αἰτίου διατείνειν δύναμιν
12 παρίστησιν.

Αἱ τῶν ζητήσεων ἔφοδοι καὶ ἀρχαὶ περὶ ταῦτα καὶ ἐν τούτοις

102,7 μεθ' ἑτέρου *add.* Stählin (*cf.* Arnim: μεθ' ἑτέρου αἰτίου) **102,8** ἐν τῷ *del.* Stählin **102,9** συνεργεῖν *corr.* Stählin **102,10** σφοδρότερον Arnim; σφοδρότατον L **102,11** *πρὸς τὸ συνεργὸν *post* γεγονέναι *add.* Stählin **102,13** *cf.* 90,7 f.

the joint-cause produces the effect along with another cause which does not produce it by its own, whereas the auxiliary cause, in producing the effect not by its own but by accruing to another, is acting as auxiliary to the very cause which is producing the effect by its own, so that the effect becomes stronger. It is above all an auxiliary cause's having grown out of the procatarctic one that shows that it is intensifying the force of the cause.

Commentary

(I) 1, 1–2, 5: 'Seek and You Will Find'

This section is about inquiry (ζήτησις). It is an argument in favour of the view that inquiry leads to discovery (εὕρεσις) and an outline of a method by which this is done. The argument is based on the authority of Scripture which says "seek and you will find" (Matt 7:7). The method consists of two things: (a) questions and answers concerning Scripture (1, 3–2, 1) and (b) an investigation of "things said" (τὰ λεγόμενα), in the course of which one adheres not only to Scripture, but also to common notions (2, 3–4). The discussion is grafted in the Greek philosophical polemic against scepticism (vouching for the authority of the "most ancient philosophers"), but its overall character is unmistakably Christian. It also abounds with doctrinal and stylistic parallels with Clement's other writings, especially books V and VII of the *Stromateis*.

1, 1–3

GCS **80,3 f.** **ἀλλ' οὐδὲ οἱ παλαίτατοι ... ἐπὶ τὸ ἀμφισβητεῖν καὶ ἀπορεῖν ἐφέροντο.** The text starts abruptly, as if in the middle of an argument, invoking the authority of "the most ancient philosophers". The argument is probably directed against scepticism of some sort, as the point at issue is the old philosophers' disinclination to "dispute" and "puzzle over" things (the implication presumably being: everything). Cf. below, 4, 1/81,27–30, on disputes without end; 15, 2–16, 1, on the view that "nothing is certain". The sceptics' inclination to ἀπορία is reflected in their description as οἱ ἀπορητικοί.[1] For the opposition between the 'old' and the 'new' cf. below, on 80,6–8.

Following the arguments of von Arnim,[2] Stählin thinks that the beginning of our text has been lost. It is true that the beginning is abrupt;[3] it is also true,

---

1 Cf. DL IX 69; Sextus *M.* VIII 76.78.80.99; X 67; *PH* I 7; Galen, *Nat. Fac.* II 9 (193,14 Helmreich/II,127 K.); *Diff. Puls.* IV 2 (VIII,710,18 and 711,1 K.), etc.

2 *De octavo*, p. 9.

3 Zahn, *Untersuchungen*, pp. 115 f., challenges this view (already proposed by Reinkens) and suggests that the beginning of book VIII is no more abrupt than that of book III (Οἱ μὲν οὖν ἀμφὶ τὸν Οὐαλεντῖνον κτλ.). But this is not convincing, as the beginning of book III takes on the discussion started at the end of the previous book. Zahn also refers to Raphael Kühner, *Ausführliche Grammatik der griechischen Sprache*, II, § 526 (1870: II, p. 806; 1904: II/2, p. 263), who registers two instances in Xenophon where a whole book starts with ἀλλά: *On the Spartan Constitution* ('Αλλ' ἐγὼ ἐννοήσας ποτέ) and *Symposium* ('Αλλ' ἐμοὶ δοκεῖ). Cf. also

 | DOI: 10.1163/9789004325289_005

as von Arnim points out, that the beginning of chapter two presupposes, on the part of the reader, knowledge of something that has not been revealed in our text, namely that the topic of the ensuing discussion is ἀπόδειξις (cf. below, on 81,9f.). But if much of the book consists of excerpts occasionally glossed by the Christian writer, we need not assume that it was supposed to be coherent. The abrupt beginning could be due to the fact that this sentence, too, is an excerpt further commented upon by Clement (cf. also below, on 80,6–8). We have no way of establishing if, in the original form of Clement's text, it had been preceded by other excerpts (or anything else) or not. In any event, if anything is missing, it must have been lost before the 9th century, as the initial words (ἀλλ' οὐδὲ οἱ παλαίτατοι τῶν φιλοσόφων) are attested by Photius (*Bibl.* 89b35) as the beginning of the eighth book of the *Stromateis*.

80,4f. ἦ πού γ' ἂν ἡμεῖς οἱ τῆς ὄντως ἀληθοῦς ἀντεχόμενοι φιλοσοφίας. Clement often refers to Christians as ἡμεῖς, as opposed to Greeks, Jews, other barbarians, or the 'heretics'.[4] 'True philosophy' (also called 'divine philosophy', the Lord's teaching, etc.) is a term that, in Clement's view, captures the distinction between Greek philosophical schools and the pursuit of knowledge and perfection that characterize Christians.[5] Distinguishing features of this philosophy include the beliefs that (a) it is of divine origin; (b) it is mediated by Scripture; (c) it is mediated by the teaching of Christ (as transmitted by the church).[6] Cf. also below, on 80,8–10. For the expression ἀντέχομαι with the genitive, cf. Prov.

J.D. Denniston, *The Greek Particles* (Oxford: Clarendon Press, 1953), pp. 20f. There are parallels with this in Latin prose; cf. Apuleius, *Metamorphoses* (*At ego*) and *Golden Ass* (*Sed ego* at the beginning of the *Cupid et Psyche* narrative); cf. Luca Graverini, "A *lepidus susurrus*. Apuleius and the Fascination of Poetry," in R.R. Nauta, (ed.), *Desultoria Scientia. Genre in Apuleius' Metamorphoses and Related Texts* (Leuven/Paris/Dudley, MA: Peeters, 2006), pp. 1–18, here 1f. However, neither the form of our sentence nor the genre of our text is similar to these instances.

4 Cf. e.g. *Protr.* 6, 4; 27, 2–3; 59, 2–3; 82, 7; 116, 3; *Paed.* I (5) 19, 4; I (6) 37, 2; II (2) 32, 1; II (4) 42, 3; *Strom.* I (20) 98, 4; II (2) 5, 1; II (4) 12, 1; II (6) 28, 4; III (12) 82, 2; V (1) 1, 5; V (4) 19, 1; V (13) 88, 2; V (14) 89, 3; V (14) 127, 4; VI (5) 41, 7; VII (2) 6, 2; VII (16) 96, 1. Cf. Alain Le Boulluec, "L'identité chrétienne en autodéfinition chez Clément d'Alexandrie," in N. Belayche and S. Mimouni (eds.), *Les identités religieuses dans l'Antiquité grecque et romaine, Collection de la Revue des Études Juives* (Leuven: Peeters, 2009), pp. 437–458.

5 Cf. *Strom.* I (2) 21, 2; I (5) 32, 4; I (18) 90, 1; I (19) 94, 1; I (20) 99, 1; II (11) 48, 1; IV (3) 9, 5; V (9) 56, 3; V (11) 67, 2; V (13) 87, 1; VI (7) 58, 2; VI (7) 59, 3; VI (8) 67, 1; VI (11) 89, 3; VI (15) 117, 1; VI (16) 147, 3; VI (17) 153, 1; VII (2) 11, 3; VII (16) 98, 2. The expression 'true philosophy' goes back to Plato; cf. *Resp.* VII, 521c7–8, quoted in *Strom.* V (14) 133, 5.

6 Clement shares the first belief with such thinkers as Numenius of Apamea (fr. 14 Des Places)

3:18: "[Wisdom] is a tree of life to them that lay hold upon her (τοῖς ἀντεχομένοις αὐτῆς)."[7] For the (interrogative) phrase ἦ πού γ' ἂν ἡμεῖς (*scil.* ἐπὶ τὸ ἀμφισβητεῖν κτλ. φεροίμεθα), cf. e.g. Clement, *Protr.* 62, 3.[8]

80,5 f. ἡ γραφὴ εὑρέσεως χάριν ἐπὶ τὸ διερευνᾶσθαι τὸ ζητεῖν παρεγγυᾷ. Here it is explained why Christians cannot rest content with sceptical doubt about the possibility of finding the truth: Scripture commands them to seek *for the sake of finding* (the reference is to the verse quoted below, 1, 2/80,9 f.). ἐπὶ τὸ διερευνᾶσθαι (*scil.* ἡμᾶς) explains the purpose of this command and thus also fixes the meaning of ζήτησις in the scriptural passage.[9] Cf. below, 2, 2/80,18.

80,6–8 οἱ μὲν γὰρ νεώτεροι τῶν παρ' Ἕλλησι φιλοσόφων ... εἰς τὴν ἄχρηστον ἐξάγονται φλυαρίαν. This sentence introduces a contrast between the activity of "the more recent philosophers" (οἱ μέν ... ἐξάγονται) and the bidding of the 'barbarian philosophy' (ἔμπαλιν δέ ... εἶπεν in 80,8 f.). This contrast explicates the one already implied in the preceding lines, namely the contrast between that which 'the most ancient philosophers' did *not* do (οὐκ ἐφέροντο in 80,4) and the command given by Scripture (παρεγγυᾷ in 80,6).

As already mentioned (on 80,3 f.), the opponents in this discussion are likely sceptic philosophers. Clement seems to build upon traditional arguments against scepticism, developed by the Greeks. At least since the time of Antiochus of Ascalon, these arguments had included an appeal to the authority of the 'ancients', whose line of thinking had allegedly been abandoned by innovators like Arcesilaus.[10] In our passage, 'the more recent philosophers' probably include the Pyrrhonian sceptics as well; cf. below, 15, 2–16, 1, and Aristocles, F 4,29 (Chiesara), on Aenesidemus.[11] The charge of 'ambition' (φιλοτιμία) or

---

and the second with, for example, Philo of Alexandria (cf. e.g. *Post. Cain.* 102). The third belief is a matter of dispute with the 'heretics' (cf. e.g. *Strom.* VI [15] 123, 3).

7 The verse is quoted in *Strom.* V (11) 72, 4.

8 ἦ πού γ' ἂν ἔτι τὴν Πραξιτέλους Δήμητρα καὶ Κόρην καὶ τὸν Ἴακχον τὸν μυστικὸν θεοὺς ὑπολάβοιμεν ἢ τὰς Λυσίππου τέχνας ἢ τὰς χεῖρας τὰς Ἀπελλικάς, αἳ δὴ τῆς θεοδοξίας τὸ σχῆμα τῇ ὕλῃ περιτεθείκασιν;

9 For ἐπὶ τό in the sense of 'for the purpose of', cf. *Strom.* VI (8) 65, 6: ... ἄνθρωπος ... γεωργεῖ καὶ γεωμετρεῖ καὶ φιλοσοφεῖ, ὧν τὸ μὲν ἐπὶ τὸ ζῆν, τὸ δὲ ἐπὶ τὸ εὖ ζῆν, τὸ δὲ ἐπὶ τὸ μελετᾶν τὰ ἀποδεικτικὰ γεγένηται.

10 Cf. Cicero, *Ac.* I 13.43.46; II 15. The authority of 'the ancients' (in this case, mainly the Presocratics, Socrates, and Plato) was invoked by both sides of the dispute; cf. Cicero, *Ac.* II 14; 72–74.

11 "One Aenesidemus began to revive this nonsense yesterday or the day before (ἐχθὲς καὶ πρώην) ..."

'rivalry' (φιλονεικία) was put forward by both sides of the dispute, who both regarded the views of their opponents as a digression from tradition; cf. Cicero, *Ac.* I 44; Numenius, fr. 25,10f.62f. (Des Places); Plutarch, *Cicero* 4, 1–2; Galen, *PHP* III 4, 31 (CMG V,4,1,2: 198,27 f./V,319 K.).[12]

There is a parallel to Clement's distinction of the two groups in Galen. In *MM* I 4 (X,36–37 K.), Galen draws a contrast between "the ancient philosophers" (οἱ παλαιοὶ φιλόσοφοι) on the one hand and "some combative philosophers" (ἔνιοι τῶν φιλοσόφων ἐρίζοντες) on the other, with respect to their attitude to the principles of demonstration. Whereas the former acknowledge two kinds of principles, namely those apparent to the senses and those apparent to reason (see below, on 7, 3/83,24 f.), the latter dispute (ἀμφισβητοῦσιν) the rational principles. Here the expression 'ancient philosophers' probably includes or refers to Aristotle and his school, while the second, 'combative' group (cf. above, 1, 2/80,7 f.: ἐλεγκτικῶς ἅμα καὶ ἐριστικῶς) includes the sceptics.[13] Galen also likes to speak of the "useless foolery" (ἄχρηστος φλυαρία) of his opponents, just as Clement does here (εἰς τὴν ἄχρηστον ... φλυαρίαν) and nowhere else.[14]

80,8–10 ἔμπαλιν δὲ ἡ βάρβαρος φιλοσοφία κτλ. The grounds for Christian optimism in the face of sceptical doubt are now made explicit. The quoted verse is Matt 7:7; cf. *Strom.* I (11) 51, 4, where the context of the quotation is similar to our passage.[15] The rest of the chapter (1, 3/80,11–2, 5/81,8) may be regarded as a commentary on this verse.[16]

---

12 Cf. George Boys-Stones, *Post-Hellenistic Philosophy* (Oxford: Oxford University Press, 2001), pp. 137–140.

13 Cf. R.J. Hankinson, *Galen: On the Therapeutic Method* (Oxford: Oxford University Press, 1991), pp. 126 and 128.

14 Cf. Galen, *Adv. Typ. Scr.* 1 (VII,476,13 K.); *Hipp. Epid. III*, II 4 (XVIIa,611,5 K.). The only other instance found in the Greek thesaurus is in Cyril of Alexandria. Closely related is Galen's expression ματαία φλυαρία (*Diff. Feb.* I 3/VII,281,2 f. K.; *Adv. Typ. Scr.* 3/VII,488,11 K.; *Diff. Puls.* IV 1/VIII,699,4 K.; *Dig. Puls.* IV 4/VIII,951,2 K.; *SMT* I 37/XI,449,3 K.), found nowhere else in the thesaurus. See also *Strom.* I (11) 51, 4 (κενὴ φλυαρία).

15 *Strom.* I (11) 51, 4: "ζητεῖτε" γὰρ "καὶ εὑρήσετε" λέγει, ἀλλὰ τὴν ζήτησιν εἰς εὕρεσιν περαιοῖ, τὴν κενὴν ἐξελάσας φλυαρίαν κτλ. For the importance of Matt 7:7 in early Christian thought, cf. Antonio Orbe, *Parábolas evangélicas en San Ireneo*, I (Madrid: La Editorial Católica, 1972), pp. 34–74; for Clement, cf. Alain Le Boulluec, *La notion d'hérésie dans la littérature grecque IIe–IIIe siècles*, II (Paris: Institut d'Études Augustiniennes, 1985), pp. 385–389.

16 Cf. Pierre Nautin, "La fin des *Stromates* et les *Hypotyposes* de Clément d'Alexandrie," *Vigiliae Christianae* 30 (1976), pp. 268–302, here 268 and 291.

The phrase 'barbarian philosophy' is used in the same sense as 'true philosophy' above, 80,4f.[17] While often distinguishing Christians from both Greeks and barbarians (here especially Jews), at other times Clement proudly includes Christians among the barbarians.[18]

80,11f. ὁ πρὸς ἐρώτησιν καὶ ἀπόκρισιν λόγος. "Argument in the form of questions and answers"[19] is here shown as an instrument by which one may comply with the bidding of Matt 7:7 and "knock on the door of truth" in the course of inquiry (κατὰ τὴν ζήτησιν). Clement probably alludes to a passage in Plato's *Seventh Letter*, where "questions and answers without envy" are described as instruments of "friendly refutations", in the course of which "names, definitions, sights and perceptions" are "rubbed against one another" until "wisdom and understanding shines out".[20] The same passage seems to be reflected below, 2, 1/80,16f. and 2, 3/80,24f. In *Strom.* VI (15) 116, 3, Clement indicates that ὁ κατ' ἐρώτησιν καὶ ἀπόκρισιν λόγος is used by the Christian gnostic in his dual role as educator and interpreter of the hidden meaning of Scripture.[21] Cf. also *Strom.* I (9) 45, 1–4. It seems likely that a similar context is presupposed here; cf. below, on 80,14f. For the philosophical background to 'questions and answers' cf. below, on 86,28f.

---

17 Cf. *Strom.* I (20) 99, 1; II (2) 5, 1; V (9) 56, 2–3; VI (8) 67, 1–2.

18 Cf. esp. *Strom.* VI (17) 151, 2: ἐπεὶ τοίνυν δύο εἰσὶν ἰδέαι τῆς ἀληθείας, τά τε ὀνόματα καὶ τὰ πράγματα, οἳ μὲν τὰ ὀνόματα λέγουσιν [...], οἱ παρ' Ἕλλησι φιλόσοφοι, τὰ πράγματα δὲ παρ' ἡμῖν ἐστι τοῖς βαρβάροις. Cf. also *Strom.* V (2) 15, 3; V (3) 16, 3; V (11) 70, 7. Cf. Alain Le Boulluec, "La rencontre de l'hellénisme et de la «philosophie barbare» selon Clément d'Alexandrie," in idem, *Alexandrie antique et chrétienne: Clément et Origène* (Paris: Institut d'Études Augustiniennes, 2006), pp. 81–93, here esp. 82–85.

19 For the meaning of πρός with the accusative, cf. DL II 106, on the Megarian school: ὕστερον δὲ διαλεκτικοί [*scil.* προσηγορεύοντο], οὓς οὕτως ὠνόμασε πρῶτος Διονύσιος ὁ Χαλκηδόνιος διὰ τὸ *πρὸς ἐρώτησιν καὶ ἀπόκρισιν τοὺς λόγους διατίθεσθαι*.

20 *Ep.* VII 344b: μόγις δὲ τριβόμενα πρὸς ἄλληλα αὐτῶν ἕκαστα, ὀνόματα καὶ λόγοι ὄψεις τε καὶ αἰσθήσεις, ἐν εὐμενέσιν ἐλέγχοις ἐλεγχόμενα καὶ ἄνευ φθόνων ἐρωτήσεσιν καὶ ἀποκρίσεσιν χρωμένων, ἐξέλαμψε φρόνησις περὶ ἕκαστον καὶ νοῦς, συντείνων ὅτι μάλιστ' εἰς δύναμιν ἀνθρωπίνην. In *Strom.* V (11) 77, 1, Clement refers to the *Seventh Letter* as ἡ μεγάλη ἐπιστολή. Other allusions to this text include *Strom.* V (10) 66, 3; V (12) 78, 1; VI (17) 150, 1. Cf. also *Strom.* I (11) 54, 1, on "charitable refutations" (οἱ μετ' ἀγάπης ἔλεγχοι). An interesting parallel with our passage is found in Origen's polemic against Celsus; while responding to his opponent's quotation of *Ep.* VII 344b, Origen refers, among other passages, to Matt 7:7, in order to show that "divine Logos exhorts us to dialectics" (*C. Cels.* VI 7).

21 For the context, cf. *Strom.* VI (15) 115, 1–117, 1. The exegetical role of the gnostic is discussed ibid. 115, 5–116, 2.

80,12 κατὰ τὸ φαινόμενον. In Greek epistemological debates, the word τὸ φαινόμενον usually designates a datum of experience which precedes and guides judgment. The word is used in this sense by the Pyrrhonian sceptics, who even concede in calling τὸ φαινόμενον their criterion of conduct; cf. Sextus, *PH* I 22. Cf. also Plutarch, *Adv. Col.* 1118a–b. In our passage, the phrase κατὰ τὸ φαινόμενον seems to qualify the way in which the dialectical argument "knocks on the door of truth". The idea, presumably, is that there are certain clearly apparent data that have the role of starting-points (and, perhaps, criteria) of inquiry; cf. below, 2, 4/81,1f., where two candidates for such φαινόμενα are mentioned: "divine Scriptures" and "common notions". For Scripture, cf. below, on 80,14f. For the role of φαινόμενα as starting-points of inquiry, cf. also below, 7, 3/83,24f.

80,12f. διοιχθέντος δὲ τοῦ ἐμποδὼν. The metaphor of opening continues the idea of knocking on "the door of truth" through questions and answers (80,11). Insofar as the door is closed, it is an obstacle preventing us from seeing the truth, an obstacle that should be removed "in the course of investigation" (κατὰ τὴν ἔρευναν). Clement does not tell us what obstacle he means, but elsewhere he speaks of τὰ ἐμποδὼν as *internal* obstacles of understanding, such as the irrational desires and false opinions generated by them; cf. *Protr.* 114, 1; *Strom.* VII (13) 82, 4–5.[22] According to one passage, obstacles preventing the soul from finding the truth include such passions as rivalry (φιλονικία), envy (φθόνος), and sophistry (ἔρις);[23] cf. above, 1, 1/80,7f., as opposed to below, 2, 3/80,24f. Elsewhere, Clement points out how difficult it is "to understand and explain what is covertly said by the Spirit" (τὰ ἐπικεκρυμμένως πρὸς τοῦ πνεύματος εἰρημένα νοήσειν τε καὶ διασαφήσειν), i.e. in Scripture, saying that the only person fit for this task is one who has subdued "brutish passions" within himself.[24] A similar complex of ideas seems to be behind our passage as well. Cf. also below, on 80,13f., on the opening of the thing sought.

80,13 ἐπιστημονικὴ ἐγγίνεται θεωρία. In the course of inquiry, as the door of truth is being opened, something called ἐπιστημονικὴ θεωρία "comes in" (as it

---

22 Cf. Plato, *Phaedo* 65a9–67b2; Ps.Plato, *Sisyphus* 389b6–c1; Plutarch, *Plat. Quaest.* 1000a1–2.

23 *Strom.* V (1) 11, 4: ἀλλὰ τὸ μὲν διορατικὸν τῆς ψυχῆς ἀποτείνειν πρὸς τὴν εὕρεσιν χρὴ καὶ τὰ ἐμποδὼν διακαθαίρειν φιλονικίαν τε αὖ καὶ φθόνον καὶ τὴν ἔριν αὐτὴν τὴν κάκιστα ἐξ ἀνθρώπων ὀλουμένην ἀπορρῖψαι τέλεον.

24 Cf. *Strom.* VI (15) 115, 2–116, 2, here 115, 6 and 115, 2. For passions as an obstacle to a correct understanding of Scripture, cf. also *Strom.* VII (16) 94, 4: πάντων δὲ ἀνθρώπων τὴν αὐτὴν κρίσιν ἐχόντων οἱ μὲν ἀκολουθοῦντες τῷ αἱροῦντι λόγῳ ποιοῦνται τὰς πίστεις, οἱ δὲ ἡδοναῖς σφᾶς αὐτοὺς ἐκδεδωκότες βιάζονται πρὸς τὰς ἐπιθυμίας τὴν γραφήν.

were, from behind the door). This (somewhat pleonastic) expression clearly refers to the goal of inquiry, i.e. to that which the followers of the 'true philosophy' are supposed to find (cf. above 1, 2/80,9f.; below, 2, 2/80,21f.). Elsewhere Clement speaks of ἐπιστημονικὴ θεωρία as one goal of the gnostic life (the contemplative part of the "twofold goal", as opposed to πρᾶξις) and something that the Christian gnostic "trains himself in" (ἐγγυμναζόμενος); this training includes better understanding of the divine will, as communicated through Scripture.[25] For the goal of inquiry, see also below, on 80,16f.

2, 1–5

80,13f. τοῖς οὕτως … κρούουσιν ἀνοίγνυται τὸ ζητούμενον. If the followers of the 'true philosophy' seek in the way just described (i.e. by means of questions and answers in accordance with the apparent), then 'the thing sought' will open to them. Here, for the first time, the object of inquiry is mentioned. Given that the goal of inquiry is θεωρία (80,11–13), we may assume that it corresponds to the understanding of this object; cf. below, 2, 2/80,21f.: τὴν ἐπιστήμην τοῦ ζητουμένου.[26] For the metaphor of 'opening', cf. *Strom.* II (11) 49, 3, where Clement applies it to scriptural exegesis: διὰ τῆς τῶν γραφῶν παραθέσεώς τε καὶ διοίξεως. Cf. also below, on 80,19 and 19f.

80,14f. τοῖς οὕτως αἰτοῦσιν τὰς πεύσεις καὶ τὰς ἀποκρίσεις κατὰ τὰς γραφάς. In L, the words καὶ τὰς ἀποκρίσεις are missing, but they are happily restored by Ludwig Früchtel on the basis of Codex Lavra B 113, a rare witness of an independent textual tradition.[27]

Having commented on 'knocking' in Matt 7:7, Clement turns to the third bidding: "… ask and it will be given to you." He interprets this by pointing to the description of inquiry already given (οὕτως referring to 80,11–13), which seems to suggest that 'asking' and 'knocking' are the same activities. The words τὰς πεύσεις καὶ τὰς ἀποκρίσεις cannot be external accusatives with αἰτοῦσιν, unless we think of questions and answers as something to ask *for*.[28] Presumably, they are internal accusatives with instrumental force, reminding the reader of the contents of οὕτως ("through questions and answers").

---

25 Cf. *Strom.* VII (16) 102, 2; VII (11) 60, 2–61, 1. The expression ἐπιστημονικὴ θεωρία is already found in Philo, *Praem.* 51.

26 For τὸ ζητούμενον, cf. also below, p. 149 and n. 78.

27 Cf. Früchtel's introduction to *Clemens Alexandrinus, Bd. II: Stromata Buch I–VI* (GCS 52, 1960, 3nd edn), pp. VIII–IX. Früchtel notes the reading of Cod. Lavra in the apparatus to the revised GCS edition (GCS 17, 1970), marking it as "richtig".

28 Following L, we *could* construct τὰς πεύσεις as an external accusative, if we took πεῦσις

There are at least two ways of explaining the phrase κατὰ τὰς γραφάς. It could refer to the biblical verse commented in this passage (thus supporting the claim that "the goal we pursue will come to us from God"), or it could qualify the "questions and answers" as being concerned with scriptural passages. I find the latter interpretation more likely, partly for syntactic reasons (it is easier to link κατὰ τὰς γραφάς with the preceding than with the following words) and partly because it explains why a few lines later it is already taken for granted that the participants of the inquiry "adhere to divine Scriptures"; cf. below, on 81,1 f. (with the quotation of *Strom.* VII [16] 96, 4, where the phrase κατὰ τὰς γραφάς seems to be used in a relevant sense). If this is correct, it indicates that when speaking of inquiry in this section, Clement has in mind the sort of inquiry whose aim is to elucidate some of the difficulties that the followers of 'true philosophy' encounter in Scripture.[29] This also helps us to refine the concept of τὸ ζητούμενον, which (again, in this section) seems to refer to the meaning of difficult scriptural passages; cf. below, on 80,19 f.

80,15 ἐφ' ὃ βαίνουσιν. For the meaning of (τὸ) ἐφ' ὅ, cf. LSJ, s.v. ἐπί C.III.1 ("the object *for* which one goes"). Cf. below, 2, 2/80,21: τὸ ἔπαθλον λαβεῖν.

80,16 f. ἡ δόσις τῆς θεοδωρήτου γνώσεως καταληπτικῶς ... ἐκλαμπούσης κτλ. The goal of inquiry is knowledge (γνῶσις), of which we learn that (a) it is a gift from God and (b) it "shines out" in a way that conveys apprehension (καταληπτικῶς ... ἐκλαμπούσης, *scil.* τῆς γνώσεως).[30] This illuminating activity is further quali-

---

in the sense of 'information', cited by LSJ, s.v. However, unlike the common pair πεῦσις—ἀπόκρισις, this usage seems to be unattested in relevant (philosophical/theological) sources.

29 It is hardly surprising that Clement should speak of ζήτησις in this sense; cf. Le Boulluec, *La notion d'hérésie*, p. 387: "[l]a règle « chercher pour trouver » s'applique tout particulièrement à l'exégèse des textes sacrés." Among other references, Le Boulluec quotes *Strom.* VI (15) 126, 1, where the obscurity of scriptural meaning is partly explained as an incentive for Christians to "become inquirers" and "be awake to discovery": διὰ πολλὰς τοίνυν αἰτίας ἐπικρύπτονται τὸν νοῦν αἱ γραφαί, πρῶτον μὲν ἵνα *ζητητικοὶ* ὑπάρχωμεν καὶ προσαγρυπνῶμεν ἀεὶ τῇ τῶν σωτηρίων λόγων *εὑρέσει* κτλ. As mentioned above (on 80,11 f.), Clement speaks of "the argument in the form of questions and answers" in connection with exegesis; cf. *Strom.* VI (15) 116, 3 and 115,5–116,2; similarly in *Strom.* I (9) 45, 1–4. It is worth noting in this connection that Origen (in the passage mentioned above, on 80,11 f.) explains Matt 7:7 as Jesus' command to "seek for the meaning of Scriptures and to ask God about them and to knock for the truths locked inside them" (ζητεῖν τὴν διάνοιαν τῶν γραφῶν καὶ αἰτεῖν περὶ αὐτῶν ἀπὸ τοῦ θεοῦ καὶ κρούειν αὐτῶν τὰ κεκλεισμένα); *C. Cels.* VI 7; trans. Chadwick.

30 It seems more natural to link ἐκλαμπούσης with ζητήσεως, which immediately follows it.

fied as something coming about "through inquiry that is truly rational" (διὰ τῆς λογικῆς ὄντως ζητήσεως).

For knowledge as a divine gift, cf. *Strom.* V (1) 12, 2: "For those who seek in a true inquiry while praising the Lord will be filled by a gift given by God, i.e. knowledge, and their soul will live."[31] The idea of the goal of inquiry as an illuminating insight echoes the above-mentioned passage from the *Seventh Letter* (Cf. *Ep.* VII 344b: ἐξέλαμψε φρόνησις περὶ ἕκαστον καὶ νοῦς); cf. above, on 80,11 f. The word καταληπτικῶς suggests that this insight has the character of 'apprehension' (κατάληψις). This is not a surprising word to use in connection with inquiry; cf. e.g. Cicero, *Ac.* II 26 (trans. Brittain): "Thus ⟨reason⟩ contains both the starting point for investigation and its result, i.e. apprehension."[32] However, for Clement, it normally has a special connotation, congruent with his idea of 'true philosophy' as one based on Scripture and the tradition of the church (cf. above, on 80,4 f.). So he speaks of "the apprehension of the truth" (κατάληψις τῆς ἀληθείας) as something to which Greek philosophy contributes, at best, as an auxiliary cause;[33] he describes apprehension as something given by Christ/Logos and mediated by Him in the flesh, by the Prophets, and by the apostolic tradition.[34] Furthermore, rather uniquely, Clement regards κατάληψις as a culmination of a mental act called 'anticipation' (πρόληψις), i.e. an act by which the mind of a believer anticipates the meaning of the words he believes and seeks to understand.[35] Finally, on one occasion, Clement speaks of κατάληψις in connection with scriptural exegesis in the course of which "that which is hidden under many covers may, examined and understood, come to light and shine forth".[36] This, I believe, is also the context in which the word καταληπτικῶς is used in our passage.

But this would create the problem of the syntactic function of καταληπτικῶς, in addition to the fact that the idea of an inquiry's "shining out" seems odd. It is reasonable to think that καταληπτικῶς qualifies ἐκλαμπούσης, but then ἐκλαμπούσης must be connected with γνώσεως. For Clement's use of hyperbaton, cf. Henricus Steneker, *ΠΕΙΘΟΥΣ ΔΗΜΙΟΥΡΓΙΑ. Observation sur la fonction du style dans le Protreptique de Clément d'Alexandrie* (Nijmegen/Utrecht: Dekker & Van de Weckt, 1967), pp. 57–61.

31 οἱ γὰρ ζητοῦντες κατὰ τὴν ζήτησιν τὴν ἀληθῆ αἰνοῦντες κύριον ἐμπλησθήσονται τῆς δόσεως τῆς παρὰ τοῦ θεοῦ, τουτέστι τῆς γνώσεως, καὶ ζήσεται ἡ ψυχὴ αὐτῶν. Cf. also *Strom.* V (11) 71, 5.

32 *Ac.* II 26: *sic et initium quaerendi et exitus percipiundi et comprendendi tenetur. Compre[he]nsio* is Cicero's equivalent for κατάληψις; cf. *Ac.* II 31 and 145.

33 *Strom.* I (20), 97, 1 and 99, 1.

34 Cf. *Strom.* V (1) 7, 8; VI (7) 54, 1; VI (7) 61, 1–3; VI (8) 68, 1–3; VI (15) 121, 3–123, 4, here esp. 121, 4–122, 1 and 123, 1; VI (18) 162, 4.

35 Cf. *Strom.* II (2) 8, 4 and (4) 17, 1–3.

36 Cf. *Ecl. Proph.* 32, 1–3: δεῖ τοίνυν τὰς γραφὰς ἀκριβῶς διερευνωμένους …, ἵνα ἡμῖν ἀκριβῶς

The qualification of inquiry as "truly rational" again seems to be linked with the notion of 'true philosophy' in that it differentiates the inquiry performed by Christians from that performed by Greek philosophers.[37] The word 'rational' (λογικός) hints at the belief that 'true philosophy' is based on the teaching of Logos, i.e. Christ.[38]

80,17 f. οὐ γὰρ εὑρεῖν μὲν οἷόν τε, μὴ ζητῆσαι δέ. Cf. *Strom.* V (1) 11, 1: "We say that faith should not be idle and isolated, but it should progress along with inquiry [...] As Scripture says: 'Seek and you will find.' "[39]

80,18 οὐδὲ ζητῆσαι μέν, οὐχὶ δὲ ἐρευνήσασθαι. It is unclear if the words ζητέω and ἐρευνάω (both of which may be translated as 'to search') designate different kinds or phases of inquiry. In any case, Clement seems to suggest that it is impossible to heed the bidding of Matt 7:7, unless one seeks by way of investigating things. Cf. above, 80,5 f. By using the word ἐρευνάω, Clement could be alluding to 1 Cor 2:9: "The Spirit investigates everything (πάντα ἐρευνᾷ), even the depth of God."[40]

80,19 διαπτύξαι καὶ ἀναπετάσαι. Lit. 'unfold and spread out', perhaps like a scroll. The word διαπτύσσω is often used in connection with interpretation; cf. Philo, *Vit. Cont.* 78, on symbols (τὰ ... σύμβολα [*scil.* ἡ λογικὴ ψυχή] διαπτύξασα καὶ διακαλύψασα); Nicomachus, *Harm. Ench.* 8 (250,4 f. von Jan), on Plato, *Tim.* 36a (τὴν Πλατωνικὴν ... διαπτύξαι λέξιν); Clement, *Strom.* VI (15) 131, 3 (τὴν

ἐξεταζόμενα διαπτύσσηται μὲν τὰ ὀνόματα πολυσήμως εἰρημένα, *τὸ δ' ἐγκεκρυμμένον ἐν πολλοῖς τοῖς σκέπουσι ψηλαφώμενον καὶ καταμανθανόμενον ἐκφαίνηται καὶ ἀναλάμψῃ*. ὡς γὰρ καὶ ὁ μόλιβδος τοῖς τρίβουσιν ἐξανθεῖ λευκὸν ἐκ μέλανος, τὸ ψιμύθιον, οὕτως καὶ ἡ γνῶσις, φέγγος καὶ λαμπρότητα καταχέουσα τῶν πραγμάτων, ἡ τῷ ὄντι θεία σοφία εἴη ἄν, τὸ φῶς τὸ εἰλικρινές, τὸ φωτίζον τοὺς καθαροὺς τῶν ἀνθρώπων ὡς κόρην ὀφθαλμοῦ εἰς ὄψιν καὶ *κατάληψιν τῆς ἀληθείας βεβαίαν*. Cf. Carlo Nardi, "Socratismo evangelico nell' *Ottavo Stromateus* (cap. I) di Clemente Alessandrino," *Annali del Dipartimento di Filosofia* 5 (1989), pp. 23–36, here 29 f.

37 Cf. e.g. *Strom.* I (20) 91, 1–4, here esp. 97, 2: μιᾶς οὔσης τῆς ἀληθείας πολλὰ τὰ συλλαμβανόμενα πρὸς ζήτησιν αὐτῆς, ἡ δὲ εὕρεσις δι' υἱοῦ. On 'true inquiry', cf. *Strom.* V (1) 12, 2, quoted above, p. 136.

38 Cf. e.g. *Strom.* II (2) 9, 4; II (4) 12, 1. For the idea of exegesis being based on Logos, cf. e.g. *Strom.* I (9) 45, 4–5.

39 τὴν πίστιν τοίνυν οὐκ ἀργὴν καὶ μόνην, ἀλλὰ σὺν ζητήσει δεῖν προβαίνειν φαμέν. οὐ γὰρ τοῦτο λέγω μηδ' ὅλως ζητεῖν· "ζήτει γάρ, καὶ εὑρήσεις" λέγει (Matt 7:7).

40 Cf. *Strom.* V (4) 25, 4–5; VI (18) 166, 3. One also thinks of John 5:39 (ἐραυνᾶτε τὰς γραφάς), but this verse is never quoted by Clement.

διάπτυξιν … τὴν γνωστικὴν τῶν γραφῶν) and *Ecl. Proph.* 32, 2 (ἵνα … διαπτύσσηται … τὰ ὀνόματα πολυσήμως εἰρημένα), on Scripture.

80,19 f. δι' ἐρωτήσεως εἰς σαφήνειαν ἄγοντα τὸ ζητούμενον. In Clement, the word σαφήνεια ('clarification') invariably refers to biblical interpretation; cf. *Strom.* I (9) 45, 1; I (20) 99, 4; VI (7) 59, 3; VI (16) 133, 1; VII (18) 109, 6. Cf. also VI (15) 115, 5, on the capacity of the gnostic to 'clarify' (διασαφήσειν) things uttered by the Spirit in a cryptic manner. On 'questions', cf. above, on 80,11 f. On τὸ ζητούμενον, cf. above, on 80,13 f. and 14 f.

80,22 ἀλλ' ἔστι μὲν εὑρεῖν τὸν ζητήσαντα. For the idea that inquiry should lead to discovery, cf. *Strom.* VII (15) 92, 1: "It is possible to find the truth for those who wish to do so" (τοῖς μὲν γὰρ βουλομένοις ἔξεσται καὶ τὸ εὑρεῖν τὴν ἀλήθειαν).[41] Cf. also below, on 81,2 f. A similar thought is expressed by Cicero in the context of an anti-sceptic polemic: "Investigation is the impulse for apprehension, ending in discovery" (*quaestio autem est adpetitio cognitionis, quaestionisque finis inventio*).[42]

80,22 f. εἰ οἰηθείη πρότερον μὴ εἰδέναι. Clement alludes to Plato, *Alcib. I*, 109e1–7, quoted in a similar connection in *Strom.* V (3) 17, 2.

80,23 f. πόθῳ δὴ ἐντεῦθεν ἀγόμενος πρὸς τὴν εὕρεσιν τοῦ καλοῦ. For desire as a motivating force of inquiry, cf. *Strom.* VII (11) 60, 2: "Desire arises with the progression of faith along with inquiry with which it is mixed" (πόθος … κατὰ προκοπὴν πίστεως ἅμα ζητήσει κραθεὶς συνίσταται); V (1) 8, 3: "… until we reach the object of our desire" (ἄχρις ἂν περιτύχωμεν τῷ ποθουμένῳ). For 'the beautiful' as the goal of inquiry, cf. Plato, *Symp.* 209e5–212a7; Alcinous, *Did.* 10, 6 (H 164,27–165,34).[43]

80,24 f. ἀφιλονίκως, ἀφιλοδόξως ἐρωτώμενος καὶ ἀποκρινόμενος. It is important for Clement that inquiry is conducted without rivalry or ambition, in contrast to what is claimed of the sceptics; cf. above, 1, 2/80,6–8. Cf. *Strom.* V (1) 8,

41 For other parallels, cf. Le Boulluec, *La notion d'hérésie*, pp. 386 f.

42 Cicero, *Ac.* II 26 (trans. Brittain); cf. Witt, *Albinus*, p. 32.

43 Alcinous' description of the *via eminentiae*, while closely following Diotima's speech in the *Symposium*, also includes a motif of light, probably inspired by the *Seventh Letter*. Cf. *Did.* 10, 6 (H 165,30–33): εἶτα ἐπὶ τὸ πολὺ πέλαγος τοῦ καλοῦ, μεθ' ὃ αὐτὸ τὸ ἀγαθὸν νοεῖ καὶ τὸ ἐραστὸν καὶ ἐφετὸν ὥσπερ φῶς φανὲν καὶ οἷον ἐκλάμψαν τῇ οὕτως ἀνιούσῃ ψυχῇ (cf. the commentary of John Dillon, *Alcinous*, p. 110). Cf. above, on 80,16 f.

3, commenting on Jer 6:16: "Ask, he says, and learn without rivalry or strife (ἀφιλονίκως καὶ ἀδηρίτως) from those who know." Cf. also *Strom.* V (1) 11, 4, on φιλονικία, φθόνος and ἔρις as obstacles preventing the soul from finding the truth (cf. above, on 80,12 f.). For a similar concern, cf. Plato, *Ep.* VII, 344b5–6, on elenctic dialogue "without envy" (ἄνευ φθόνων).

Interestingly, Clement says that the person who conducts the inquiry "is asked and gives answers". This indicates that the inquiry takes the form of a dialogue in which one party asks questions and the other gives answers. This conforms to the rules of dialectic as performed in the Platonic dialogues and elaborated in Aristotle's *Topics* (esp. book VIII). If Clement has this model in mind, then, presumably, the person who "is asked and gives answers" assumes the role of student. But Clement could also be thinking of a different situation in which the student asks questions and the teacher gives answers; cf. *Strom.* V (1) 8, 3 (quoted above in this note): "Ask ... and learn ... from those who know." Cf. also *Strom.* VI (10) 82, 3, where Clement suggests that the task of the interpreter of difficult scriptural passages is "to answer correctly" (τὸ ὀρθῶς ἀποκρίνεσθαι).[44] Even according to this model, the teacher continues to learn new things about the subject matter of inquiry; cf. *Strom.* I (1) 12, 3: "One who teaches learns more and as he speaks, he often listens along with those who listen to him."[45] Cf. below, 30, 2/99,19 f.

80,25–81,1 αὐτὰ ἐπισκεπτόμενος τὰ λεγόμενα. In addition to giving answers to questions pertinent to the matter at hand, the inquirer also examines "the things said". Unfortunately, Clement does not tell us what he means by this expression, but αὐτά suggests that it is something either previously mentioned or well known. If it is something previously mentioned, then, presumably, τὰ λεγόμενα must be the answers given by the inquirer himself, and so the idea would be that he subjects these answers to further examination. But it is not clear why Clement should have made this point and nothing supports the view that he did. If τὰ λεγόμενα is something well known, we may hope to find a reference to it elsewhere in Clement's writings. Now, Clement speaks of 'things said' in two contexts that could be relevant to our passage. In *Strom.* I (6) 35, 2, he says that without learning it is impossible to understand τὰ ἐν τῇ πίστει λεγόμενα. This expression probably refers to the content of Christian

44 This interpretation is also supported by the fact that the one who gives answers is supposed to do so ἀφιλοδόξως, without regard for fame. This seems to be more appropriate to a teacher than to a student.

45 διδάσκων τις μανθάνει πλεῖον καὶ λέγων συνακροᾶται πολλάκις τοῖς ἐπακούουσιν αὐτοῦ.

faith, as articulated in the confession formula.[46] If, as argued above, inquiry in this chapter concerns difficult scriptural passages, it would make sense to say that, apart from explaining these passages, the Christian philosopher is also expected to examine the beliefs of the Christian community in order to understand them.

But a more likely interpretation is suggested by another use of τὰ λεγόμενα, namely in passages where it refers to things *other* people say on matters pertaining to the Christian faith. In the first book of the *Stromateis*, after quoting 1 Thess 5:21 ("test everything; hold fast what is good"), Clement says that 'spiritual people' (i.e. those who have come to understand what they believe) "judge all things, whether they seem to be said in accordance with truth (κατὰ ἀλήθειαν λεγόμενα), or whether they really adhere to truth (ὄντως ἔχεται τῆς ἀληθείας)".[47] These people, he explains, are capable of "charitably refuting" the erroneous views of their opponents, inflated by their pretended knowledge, these unworthy opponents being likely the Christian 'heretics'.[48] In the sixth book, while defending Greek philosophy against the claim that it comes from the devil, Clement appeals to his fellow Christians not to condemn things people say because of who says them, but to "examine the things said" in order to find out "if they adhere to truth" (ἀλλὰ τὰ λεγόμενα σκοπητέον, εἰ τῆς ἀληθείας ἔχεται).[49] Finally, in *Strom.* VII (15) 91, 7, in the context of a polemic against the sceptic argument from disagreement, Clement says that "we have many starting-points given to us by nature for the examination of things said" (πολλὰς ἐκ φύσεως ἀφορμὰς πρὸς τὸ ἐξετάζειν τὰ λεγόμενα); here Clement's point is that Christians need not be disturbed by the fact that "different schools say different things about truth" (ἄλλα ἄλλων περὶ ἀληθείας λεγόντων), again thinking mainly of Christian

46 For traces of the baptismal confession formula in Clement, cf. esp. *Strom.* V (1) 1, 1; see the commentary of Alain Le Boulluec, *Clément d'Alexandrie: Les Stromates. Stromate V, Tome II (commentaire)*, (SC 279; Paris: Éditions du Cerf, 1981), p. 11, and my "Some Observations on Clement of Alexandria, *Stromata*, Book Five," *Vigiliae Christianae* 64/1 (2010), p. 4 n. 9. Cf. also *Strom.* VII (15) 90, 1–2, on the ὁμολογία of the believers; cf. Le Boulluec, *La notion d'hérésie*, pp. 367–370. This could also be the background to the following description of πίστις: εἰ τοίνυν ἡ πίστις οὐδὲν ἄλλο ἢ πρόληψίς ἐστι διανοίας περὶ τὰ λεγόμενα καὶ τοῦτο ὑπακοή τε εἴρηται σύνεσίς τε καὶ πειθώ, οὐ μὴ μαθήσεταί τις ἄνευ πίστεως, ἐπεὶ μηδὲ ἄνευ προλήψεως (*Strom.* II [4] 17, 3).

47 *Strom.* I (11) 53, 3. For the 'spiritual people' (οἱ πνευματικοί) who "judge all things", cf. 1 Cor 2:15.

48 Cf. *Strom.* I (11) 54, 1–4.

49 *Strom.* VI (8) 66, 5. Clement applies the same principle to the 'Phrygian' (i.e. Montanist) prophesies: ὅπερ καὶ ἐπὶ τῶν προφητεύειν νῦν δὴ λεγομένων παρατηρητέον (ibid.). Cf. *Strom.* IV (13) 93, 1; VII (17) 108, 2.

'heresies'.[50] It is tempting to think that our text refers to a similar kind of examination, i.e. to an examination of views pertaining to the thing sought (the meaning of an obscure biblical passage, for example), but formulated by people of other persuasions.

81,1f. ἐχομένους … οὐ μόνον τῶν γραφῶν τῶν θείων, ἀλλὰ καὶ τῶν ἐννοιῶν τῶν κοινῶν. Here Clement makes it clear that the followers of 'true philosophy' conduct their inquiry while "adhering to Scriptures" (for ἔχομαι with the genitive, cf. the expression τῆς ἀληθείας ἔχεται in *Strom.* I [11] 53, 3 and VI [8] 66, 5, quoted in the previous note). This adherence seems to be constituted by their willingness to inquire about scriptural passages on the basis of Scripture itself. Cf. above, on 80,14f.; *Strom.* VII (16) 96, 4: "But the truth is discovered not by altering the meaning of words (…) but by considering what is perfectly fitting and appropriate to the Lord and the almighty God, and by confirming everything demonstrated concerning Scriptures from similar passages in Scriptures themselves (κἀν τῷ βεβαιοῦν ἕκαστον τῶν ἀποδεικνυμένων κατὰ τὰς γραφὰς ἐξ αὐτῶν πάλιν τῶν ὁμοίων γραφῶν)."[51] Cf. ibid. VII (16) 96, 1: "So we, perfectly demonstrating from Scriptures about Scriptures themselves (ἀπ' αὐτῶν περὶ αὐτῶν τῶν γραφῶν τελείως ἀποδεικνύντες), also derive our conviction from faith in a demonstrative manner." At the same time, Clement indicates that this view of inquiry was already at the background to the preceding discussion, as it is only now that he introduces a new point, namely that we should also adhere to the common notions. It is likely that this new point is related to the examination of τὰ λεγόμενα, mentioned in the previous sentence: 81,1f. (ἐχομένους γὰρ κτλ.) is explanatory of πρὸς δὲ καὶ κτλ. in 80,25. The relation is presumably such that this examination involves the adherence to the common notions, similar to the adherence to Scripture in that it provides the examination with guidelines.

What are the 'common notions'? In the post-Hellenistic philosophy, the Euclidean context apart, the term is mainly associated with the Stoics and their theory of the naturally formed 'preconceptions'.[52] In scientific literature, as

50 Cf. *Strom.* VII (15) 91, 5: … ἄλλα ἄλλων περὶ ἀληθείας λεγόντων οὐκ ἀποστατέον, ἐπιμελέστερον δὲ θηρατέον τὴν ἀκριβεστάτην περὶ αὐτῆς γνῶσιν. For the context, see my "Demonstrative Method in *Stromateis* VII: Context, Principles, and Purpose," in M. Havrda, V. Hušek, and J. Plátová (eds.), *The Seventh Book of the Stromateis* (Suppl. Vig. Chr. 117; Leiden/Boston: Brill, 2012), pp. 261–275.

51 Trans. Hort and Mayor, modified.

52 The intricate connection between preconceptions and common notions is thoroughly investigated by Henry Dyson, *Prolepsis and Ennoia in the Early Stoa* (Berlin/New York: de

exemplified by Galen, it is often used in the sense of "generally held views about things, in accordance with which science must proceed".[53] This meaning falls in with a view that Clement regards the common notions as guidelines in the examination of τὰ λεγόμενα. It is likely that Clement (like Galen) derives these notions from the nature of the human capacity to think, thus regarding them as the natural criteria of truth.[54] We have seen (in the previous note) that in *Strom.* VII (15) 91, 7, Clement speaks of 'starting-points' (ἀφορμαί) for the examination of τὰ λεγόμενα, given by nature. Here the vocabulary is of Stoic origin[55] and refers to the idea that rational beings, by virtue of being rational, are equipped with certain preconceptions that enable them to "differentiate between that which is inconsistent, inappropriate, unnatural, and false, and that which is true, consistent, appropriate, and according to nature", as Clement puts it. It is these 'starting points', he adds, that we should use when searching for the truth.[56] It is likely that the expression has a similar meaning in our passage.

81,2 f. εἴς τι πέρας ὠφέλιμον τῆς εὑρέσεως καταληγούσης. One reason why it is expedient to adhere to the common notions is that they help us to make the discovery "stop at some useful end". For the end of inquiry, cf. *Strom.* VI (15) 121, 4: "Inquiry is an impulse towards apprehension (…) whereas discovery is the end and cessation of inquiry that has reached apprehension (ἡ εὕρεσις δὲ πέρας καὶ ἀνάπαυσις ζητήσεως ἐν καταλήψει γενομένης) and this [discovery] is called knowledge."[57] As far as the usefulness of inquiry is concerned, it marks a contrast with the "useless foolery" of the sceptics, mentioned above, 1, 2/80,8.

Gruyter, 2009), *passim*, esp. 53–71. For the origin of the term 'common notions', cf. ibid., pp. 5–8.

53 R.J. Hankinson, *Galen: On Antecedent Causes* (Cambridge: Cambridge University Press, 1998), p. 244; cf. also Galen, *MM* I 5 (X,40,12–16 K.) and the commentary of Hankinson, *Therapeutic Method*, p. 131.

54 For Clement's view of the natural criteria, see *Strom.* VII (16) 93, 2 and my "Demonstrative Method," p. 273.

55 Cf. e.g. *Strom.* II (21) 129, 4: πρὸς τούτοις ἔτι Παναίτιος τὸ ζῆν *κατὰ τὰς δεδομένας ἡμῖν ἐκ φύσεως ἀφορμὰς* τέλος ἀπεφήνατο. See also SVF II 988; III 214, 228, 264; SVF I 566 (Cleanthes); Epictetus, *Diss.* IV 1, 51; Marcus Aurelius, *Ad se ipsum* IX 1, 5, etc.

56 *Strom.* VII (15) 91, 7–8: ἔχοντες οὖν πολλὰς ἐκ φύσεως ἀφορμὰς πρὸς τὸ ἐξετάζειν τὰ λεγόμενα καὶ τῆς ἀληθείας τὴν ἀκολουθίαν ἐξευρίσκειν ὀφείλομεν. διὸ καὶ εἰκότως κρινόμεθα, οἷς δέον πείθεσθαι μὴ συγκατατιθέμενοι, μὴ διαστέλλοντες τὸ μαχόμενον καὶ ἀπρεπὲς καὶ παρὰ φύσιν καὶ ψεῦδος ἀπὸ [τε] τἀληθοῦς καὶ τοῦ ἀκολούθου καὶ τοῦ πρέποντος καὶ τοῦ κατὰ φύσιν, αἷς ἀφορμαῖς καταχρηστέον εἰς ἐπίγνωσιν τῆς ὄντως οὔσης ἀληθείας.

57 Cf. Le Boulluec, *La notion d'hérésie*, pp. 386 f., who points out that, for Clement, the end of inquiry amounts to salvation.

In Clement, the word ὠφέλιμος often refers to things that help someone to reach the goal of life according to the Christian faith, i.e. salvation or true knowledge.[58] In our passage, too, the 'useful end' seems to amount to such a presentation of scriptural meaning as benefits the addressees on their way to salvation or knowledge, unlike the "confused people" mentioned dissuasively in 2, 5/81,4, who are misled by the "sophistries of the market".

How do the common notions contribute to the usefulness of inquiry? We have seen that they were introduced in connection with the examination of τὰ λεγόμενα, in addition to the inquiry about Scripture. This seems to indicate that when it comes to the understanding of scriptural meaning, common notions are not indispensable, but they do play an important role in the context of teaching, enabling the teacher to present his case in a more convincing manner than his adversaries.

81,4 τόπος τε καὶ ὄχλος. A hendiadys for 'crowded place'.[59] Clement probably hints at the idea of post-mortem punishment; cf. *Strom.* V (14) 90, 4–6; VII (12) 78, 3. For τόπος (and εἰρηνικὸν εἶναι in 81,6) cf. Ps 36:35–37 (LXX), quoted in *Strom.* IV (6) 32, 2.

81,4f. τοὺς ταραχώδεις … καὶ τὰς ἀγοραίους εὑρησιλογίας. Two groups of people are liable to be punished: those confused by the sophists and the sophists themselves. For the confusing effect of sophistries, cf. below, 9, 7/85,15f. When speaking of sophists, Clement sometimes fails to distinguish between the (unspecified) Greek adversaries and people he calls βάρβαροι σοφισταί (*Strom.* I [8] 40, 2), i.e. the 'heretics'.[60] At other points, he clearly has Christian heretics in mind.[61] In our passage, he could still be referring to the particular group mentioned above, 1, 2/80,6–8, i.e. the sceptics, but it is also possible that he speaks more generally.

81,5f. τὸν δὲ τῆς ἀληθείας ἐραστήν τε ἅμα καὶ γνώριμον. The expression ἀληθείας ἐραστής goes back to Plato, *Resp.* VI, 501d2.[62] Clement uses it in *Strom.* VII (16)

---

58 Cf. *Protr.* 85, 3 (1Tim 4:8); 87, 2 (2Tim 3:16); *Paed.* I (8) 64, 1; II (8) 69, 2–3; III (12) 96, 1 (1Tim 4:8); *Strom.* VI (1) 2, 2; VI (17) 158, 3.

59 For another example of hendiadys in Clement, cf. *Strom.* V (4) 24, 2 (ὄνειροί τε καὶ σύμβολα); cf. my "Some Observations," p. 13.

60 Cf. esp. *Strom.* I (3) 22–24; (8) 39–41; (10) 47.

61 Cf. Le Boulluec, *La notion d'hérésie*, pp. 280–288. Cf. also *Strom.* VII (16) 92, 7.

62 The expression is used by Philo of Alexandria, *Spec. Leg.* I 59 and 309; *Virt.* 182. It is also very common in Galen; for our passage, cf. esp. *At. Bil.* 7, 1 (CMG V,4,1,1: 84,30f./V,130f.

94, 5, where it refers to the Christian gnostic, as opposed to the heretics who have "given themselves up to pleasures and violate Scripture so as to suit their desires" (94, 4). The idea that the 'lover of truth' is also its 'disciple' derives from the view of a Christian philosopher as a disciple of Christ, who takes the criteria of truth from "Truth itself" (cf. *Strom.* VII [16] 94, 5), i.e. from Christ, who is regarded as the truth (cf. John 14:6; *Strom.* I [5] 32, 4).[63]

81,6 εἰρηνικὸν εἶναι κἀν ταῖς ζητήσεσι. For being 'peaceful', cf. Ps 36:37, mentioned above, on 81,4. For ζητήσεις, cf. 1 Tim 6:4.

81,6f. δι' ἀποδείξεως ἐπιστημονικῆς. Now, for the first time, we are told that the Christian inquirer proceeds by way of 'epistemic demonstration'. In what follows, we will learn more about the meaning of this expression in Greek philosophy; cf. esp. below, 5, 1–3; 7, 6–8. In our passage, it seems to be used in order to draw a contrast between the method of inquiry outlined in the preceding lines (1, 3–2, 1; 2, 3–4) and other methods that, in Clement's view, lead to results that are less convincing. For Clement, one way of producing convincing results, especially when it comes to the investigation of τὰ λεγόμενα, is to adhere to the 'common notions' (cf. above, 2, 3–4/80,25–81,2; for the convincingness of this inquiry, cf. above, on 81,2f.). Another important element is the adherence to "divine Scriptures": while asking questions concerning Scripture, the inquirer derives the answers from Scripture itself (cf. above, on 80,14f. and 81,1f.). Clement seems to think that if these conditions are met, inquiry proceeds κατὰ τὸ φαινόμενον (cf. above, on 80,12), thus making its results immune against two groups of people mentioned in a relevant connection in the seventh book of the *Stromateis*: (a) those who believe to have found the truth "without having any true demonstration" (οὐκ ἔχοντες ἀπόδειξιν οὐδεμίαν ἀληθῆ) and (b) "those aware of not knowing anything, who nevertheless darken the

---

K.): Ἱπποκράτης μὲν οὖν φαίνεται καλός τε καὶ ἀγαθός τις ἀνὴρ γεγονέναι, *μὴ φιλοτιμίας ἢ φιλοδοξίας, ἀλλ' ἀληθείας ἐραστής*. Cf. also *Ord. Lib. Prop.* 1, 12 (91,7 Boudon-Millot/XIX,53 K.), in connection with *DD*.

63 For the Christian gnostic as a disciple of Christ, cf. also *Protr.* 115, 4 (trans. Butterworth, slightly modified): "Let the light then shine in the hidden part of man, in his heart; and let the rays of knowledge (τῆς γνώσεως αἱ ἀκτῖνες) rise, revealing and illuminating the hidden man within, the student of the light, the disciple of Christ (τὸν μαθητὴν τοῦ φωτός, τὸν Χριστοῦ γνώριμον) and joint-heir with Him." Cf. also *Strom.* V (4) 25, 5 (on 1 Cor 2:9f.): πνευματικὸν γὰρ καὶ γνωστικὸν οἶδεν [*scil.* ὁ ἀπόστολος] τὸν τοῦ ἁγίου πνεύματος μαθητὴν τοῦ ἐκ θεοῦ χορηγουμένου, ὅ ἐστι νοῦς Χριστοῦ. Ibid. VII (16) 93, 5, on the gnostic: μαθητὴς θεοῦ ὁμοῦ καὶ φίλος καὶ συγγενής, etc.

truth with plausible sophisms".[64] Against both groups (the 'dogmatist' and the 'sceptic' varieties of error), Clement defends the existence of demonstration, inviting his audience to "undertake the task of investigating and learn demonstratively, directly from Scriptures, that heresies went astray and that the most exact knowledge and really the best school of thought is found only in the Truth [i.e. Christ] and the ancient church".[65] On ἐπιστημονικὴ ἀπόδειξις, cf. also *Strom.* II (11) 49, 3, quoted below, on 82,16–18.

81,7 ἀφιλαύτως καὶ φιλαλήθως. For the contrast between 'love for oneself' and 'love for the truth', cf. Plutarch, *Plat. Quaest.* 1000a1–c6.[66] Clement applies this opposition in his polemic against Greek philosophical schools as well as against Christian heresies.[67]

81,7f. εἰς γνῶσιν ... καταληπτικήν. Cf. above, on 80,16f.

(II) 3, 1–(III) 8, 3: Teaching on Demonstration

3, 1–2

81,9f. μέθοδος εἰς ἀρχὴν τῆς τοιᾶσδε ... διδασκαλίας ... τὸ προταθὲν ὄνομα λόγῳ διελθεῖν. This sentence prepares us for a discussion about a method by which we may arrive at a starting-point of a teaching. By speaking of this teaching as τοιάδε διδασκαλία, it indicates that the reader has just heard about it. But it is

---

64 *Strom.* VII (15) 92, 4–5: τῶν δὲ ἀπὸ τῆς ἀληθείας ἐκτρεπομένων οἳ μὲν σφᾶς αὐτοὺς μόνους, οἳ δὲ καὶ τοὺς πέλας ἐξαπατᾶν ἐπιχειροῦσιν. οἳ μὲν οὖν, δοξόσοφοι καλούμενοι, [οἱ] τὴν ἀλήθειαν εὑρηκέναι νομίζοντες, οὐκ ἔχοντες ἀπόδειξιν οὐδεμίαν ἀληθῆ, ἑαυτοὺς οὗτοι ἀπατῶσιν ἀναπεπαῦσθαι νομίζοντες [...] οἳ δέ, τοὺς προσιόντας ἐξαπατῶντες, πανοῦργοι σφόδρα, οἳ καὶ παρακολουθοῦντες αὑτοῖς ὅτι μηδὲν ἐπίστανται, πιθανοῖς ὅμως ἐπιχειρήμασι σκοτίζουσι τὴν ἀλήθειαν. Surprisingly, the second ('sceptic') group seems to include Christian heretics, too; cf. ibid. 92, 7, where people called οἱ σοφισταί, who are clearly the heretics, probably still correspond to the members of this group. As for the first opponents (δοξόσοφοι καλούμενοι), cf. ibid. 98, 1, where δοξοσοφία is ascribed to the heretics. Cf. Le Boulluec, *La notion d'hérésie*, pp. 388f.

65 *Strom.* VII (15) 92, 3: ἀποδείξεως δ' οὔσης ἀνάγκη συγκαταβαίνειν εἰς τὰς ζητήσεις καὶ δι' αὐτῶν τῶν γραφῶν ἐκμανθάνειν ἀποδεικτικῶς, ὅπως μὲν ἀπεσφάλησαν αἱ αἱρέσεις, ὅπως δὲ ἐν μόνῃ τῇ ἀληθείᾳ καὶ τῇ ἀρχαίᾳ ἐκκλησίᾳ ἥ τε ἀκριβεστάτη γνῶσις καὶ ἡ τῷ ὄντι ἀρίστη αἵρεσις. For a more detailed discussion of Clement's perspective on proof, see my "Demonstrative Method".

66 Cf. Jan Opsomer, *In Search of the Truth: Academic Tendencies in Middle Platonism* (Brussels: Koninklijke Academie voor Wetenschappen, 1998), pp. 150–161.

67 Cf. Le Boulluec, *La notion d'hérésie*, pp. 374–378.

not clear what the expression denotes. In the preceding chapter, Clement dealt with an inquiry leading to knowledge, describing its method. Does the word διδασκαλία refer to this process of inquiry, suggesting that it is also a process of teaching? Or does it rather refer to Clement's description of this process, as given in the preceding lines (a teaching *about* inquiry), or perhaps to the biblical command on which this description is based? Or is the teaching an outcome of the inquiry, corresponding to 'apprehensive knowledge'? The last option could be supported by a passage in *Strom.* VII (16) 95, 3, where Clement calls Christ ἀρχὴ τῆς διδασκαλίας.[68] In that context, διδασκαλία is probably another name for the 'true philosophy' of the church, the word being used in the same sense as ἡ κατὰ Χριστὸν ... ἡμῶν διδασκαλία in *Strom.* IV (18) 113, 5, and similar phrases elsewhere (cf. above, on 80,4 f.). But this cannot be the meaning of διδασκαλία in our passage, as the following lines make clear: the method of arriving at its starting-point is to explain "the proposed name" in a certain way (81,10 f.). Taken in connection with the next sentence, this statement probably does not concern the proposed names in general, but a particular name proposed for this investigation, namely 'demonstration' (81,12: τὸ ὄνομα τῆς ἀποδείξεως).[69] But if the starting-point of the teaching is the explanation of the word 'demonstration', it is reasonable to suppose that the teaching itself is concerned with this particular subject. Thus, to all appearances, τοιάδε διδασκαλία is a teaching about demonstration.[70]

This conclusion enables us to construct the connection between this and the previous section after all. As we have seen, while describing the method of inquiry, Clement remarked that the inquiry proceeds by means of 'epistemic demonstration' (above, 81,6 f.). We can hardly say that by making this remark Clement introduced a new theme. But it is likely that he made it in view of the theme of the discussion starting in chapter 2. Since, in marked contrast to the first chapter, this discussion is almost entirely devoid of anything typically Clementine or Christian (exceptions will be noted in the comments below), we may assume that by making this remark, Clement wanted to indicate how the theme is relevant from the Christian point of view. If this is correct, it appears that the first chapter is not intended as an introduction to the ensuing debate, but rather as a note Clement made on the margin of this debate, a debate which did not originate with him and did not fully correspond to his interests, but

68 The parallel with *Strom.* VII (16) 95, 3 is noted by Ernst, *De Clementis*, p. 14.

69 This is correctly noted by von Arnim, *De octavo*, p. 9.

70 This conclusion, I believe, is confirmed by what follows: in 3, 3 and 4, 1, ἡ ἀρχὴ τῆς διδασκαλίας is identified as τὸ ὁμολογούμενον and in 5, 1, ἀπόδειξις is defined in terms of what "all people agree" to be the meaning of this word, i.e. in terms of τὸ ὁμολογούμενον.

one in which he assumed the role of a reader, an excerptor, and an occasional glossator. It is possible that, in some earlier version of this text, Clement himself introduced the main topic of the discussion.[71] But it is equally possible that the topic (demonstration) was introduced in his source only, whereas Clement himself skipped this introduction (perhaps using a sentence or two from it as a starting-point for his comments; cf. above, on 80,3 f. and 6–8).

There are close and consistent parallels to this and the following passages in Galen. For ἀρχὴ τῆς διδασκαλίας, cf. *MM* I 5 (X,40,2 f. K.), where Galen uses the same expression in the introduction to the teaching on the therapeutic method. Galen's subsequent description of the method of finding the starting-point of this teaching (*MM* I 5/X,40,2–41,1 K.) is very similar to our text with respect to subject, vocabulary, and syntax; see above, pp. 35–39. Τὸ προταθὲν ὄνομα is a *dis legomenon*, the other instance being Galen, *Diff. Puls.* II 3 (VIII,569,10 f. K.). For λόγῳ διελθεῖν, cf. Galen, *MM* I 5 (X,40,9 f. K.): χρὴ δήπου καὶ νῦν ὅ τί ποτέ ἐστι νόσημα τῷ λόγῳ διελθεῖν.[72]

81,11 οὕτω σαφῶς ὡς πάντας ἀκολουθῆσαι τοὺς ὁμοφώνους. Cf. Galen, *Diff. Puls.* IV 2 (VIII,704,11 f. K.): τοῦ μὲν τὴν ἔννοιαν ἑρμηνεύοντος [*scil.* ὅρου] ὁμολογεῖσθαί … πᾶσι τοῖς ὁμοφώνοις. Cf. below, 3, 3/81,17–20.

81,11–13 ἆρ' οὖν … τὸ ὄνομα τῆς ἀποδείξεως … φωνὴ μόνον οὐδὲν σημαίνουσα; This question implies a view, on the part of the speaker's (real or imagined) opponents, that there is no demonstration. This was the view of some pre-Sextan sceptics (cf. e.g. DL IX 90–91, on Agrippa), as well as of Empiricist doctors.[73] According to Sextus, the argument from meaning had already been used by some Epicureans: "Either you understand what demonstration is, or you do not. And if you understand it and have a conception of it (εἰ μὲν νοεῖτε καὶ ἔχετε ἔννοιαν αὐτῆς), there is demonstration" (*M.* VIII 337, trans. Bett). The

71 Cf. above, on 80,3 f.

72 In *MM* I 5, as well as in our passage, λόγῳ διελθεῖν does not mean to "go over by way of a concept" (as translated by Johnston and Horsley in *Galen: Method of Medicine*, LCL 2011, Vol. I, p. 65), but rather "to explain by an account"; λόγος being equivalent to (a notional and/or essential) definition; cf. below, on 81,17–24 and 82,2 f.

73 Cf. Galen, *Sect. Int.* 5 (10,26–11,2 Helmreich/I,77 K., trans. Frede): "They even go on and criticise the postulates of logic and definition, claiming that there is no such thing as proof, anyway (οὐδὲ τὴν ἀρχὴν ἀπόδειξιν εἶναί φασιν)." Cf. Sextus, *M.* VIII 327. For the history of the debate, cf. Jonathan Barnes, "Proof Destroyed," in M. Schofield, M. Burnyeat, and J. Barnes (eds.), *Doubt and Dogmatism* (Oxford: Clarendon Press, 1980), pp. 161–181, here esp. 162–164.

opponents implied in our passage could be the same as those addressed below, 15, 2–16, 1, who claim that "nothing is firm" (cf. below, on 89,2).

81,12 οἷόν περ τὸ βλίτυρι. 'Bliturі' is a stock example of a word without meaning; see, e.g. Artemidorus, *Onir.* IV 2, 47; DL VII 57; Sextus, *M.* VIII 133. It is used in this sense by Galen in *MM* II 7 (X,144,9–11 K., trans. Hankinson): "Do the words 'animal' and 'illness' seem to you to signify nothing, but have a sense similar to 'blituri' and 'scindapsus'?" However, elsewhere Galen notes that the word is not entirely devoid of meaning, as it signifies some kind of sound produced by striking a stringed instrument; *Diff. Puls.* III 4 (VIII,662,4–8 K.): … τὸ βλίτυρι κροῦμά τι δηλοῖ.

81,13f. καὶ πῶς οὔθ' ὁ φιλόσοφος … ὁ ῥήτωρ … ὁ δικαστὴς … ὡς ἄσημον ὄνομα προφέρεται τὴν ἀπόδειξιν. To establish an agreed starting-point of the teaching, one has to determine whether the proposed word has a meaning and then to explore what this meaning is (for the latter step, cf. below, 3, 3/81,17–20). The claim that the word 'demonstration' does have a meaning is supported by the fact that it is meaningfully used. No philosopher, rhetor, or judge would present something as a proof unless they believed that the word had a meaning.[74] This everyday use is presumably also the point of departure for the notional definition proposed below (5, 1/82,12 14). Cf. Galen, *MM* I 5 (X,42,3f. K.): "We shall derive our interpretation of names from ordinary Greek usage."

81,15 οὔτε τις τῶν δικαζομένων ἀγνοεῖ τὸ σημαινόμενον ὅτι οὐχ ὑπάρχει. The phrase ἀγνοεῖ ὅτι οὐχ ὑπάρχει ('ignores that it exists') seems to be a case of superfluous negation, a phenomenon often observed after words expressing denial, contradiction, doubt, etc.[75] Litigants (οἱ δικαζόμενοι probably include both plaintiffs and defendants) testify not only to the fact that the word 'demonstration' signifies something, but also to the fact that the signified item exists. Cf. below, 3, 4/81,21: εἰ ὑπάρχει τὸ σημαινόμενον τοῦτο. In their case, of course,

---

74 For the construction καὶ πῶς οὐ + ind. pres., cf. e.g. Galen, *Temp.* I 3 (7,24f. Helmreich/I,520 K.): καὶ πῶς οὐχὶ πέντε λέγετε τὰς πάσας εἶναι κράσεις, ἀλλὰ τέτταρας, εἴπερ τῆς ἀρίστης μέμνησθε;

75 Cf. Kühner, *Ausführliche Grammatik*, II/2 (1904) § 514.2 and 3b, pp. 207 and 209f. Cf. also H.W. Smyth, *Greek Grammar* (Cambridge, MA: Harvard University Press, 1956) § 2743. There is another example of superfluous negation with ἀγνοεῖν (different from the ἀγνοῶ ὅτι οὐ type) in Galen, *MM* V 1 (X,306,6f. K.): ἀλλ' οὐχὶ κοινὸν μὲν τοῦτο καὶ πρὸς τοὺς ἰδιώτας ὑπάρχον, *οὐδεὶς οὖν ἀγνοεῖ τῶν ῥηθέντων οὐδέν* ("But this is not something common and shared by the laymen, for *none knows any of the things mentioned*").

everything depends on whether something is a proof or not. But if proof does not exist, nothing is a proof.

81,16 f. ὡς ὑποστατὸν πορίζονται τὴν ἀπόδειξιν …, ἄλλος ἄλλως. The philosophers' practice of furnishing demonstration (each school according to its own understanding of what demonstration is) shows that, like litigants, they take demonstration to be something real. For the phrase 'furnish demonstration', cf. *Strom.* VII (16) 103, 4: οἱ παρὸν τὰς οἰκείας ταῖς θείαις γραφαῖς ἐξ αὐτῶν τῶν γραφῶν πορίζεσθαι ἀποδείξεις.[76] It is used in the same sense as ἐκπορίζεσθαι τὴν πίστιν below, 5, 1/82,14; 7, 6/83,32. For the meaning of ὑποστατός, cf. Sextus, *M.* X 60.[77]

3, 3–5, 1

81,17–24 Having pointed out that the teaching on demonstration should start with the agreed account of the name (and having broached the question of whether the name has a meaning and whether the item signified by the name exists), the text makes an excursus on the correct way of dealing with "everything sought" (περὶ παντὸς τοῦ ζητουμένου).[78] This general line of reasoning continues until 4, 3/82,12 and will be picked up again below, 8, 4/84,24. In our passage, three stages of inquiry are distinguished, which correspond, respectively, to (a) the agreed meaning of the proposed name (cf. above, 3, 1/81,10 f.; for προσηγορία, see the next note); (b) the existence of the object signified by that name (cf. above, 3, 2/81,15–17); and (c) the essence of that object. This scheme goes back to Aristotle who makes a distinction between two kinds of account, one that conveys a thing's essence (λόγος τοῦ τί ἐστι) and one conveying the meaning of the thing's name (λόγος τοῦ τί σημαίνει τὸ ὄνομα); Aristotle says that

---

76 Cf. also Simplicius, *In De caelo* II 8 (CAG VII: 460,29 f.): πῶς δὲ τοῦτο παρῆκεν ὁ Ἀριστοτέλης ζητεῖν ὁ ἐκ τῶν ἐναργῶς φαινομένων ἀποδείξεις πανταχοῦ ποριζόμενος; Cf. also idem, *In Phys.* IV 8 (CAG IX: 668,12).

77 "But those who do not agree with there being any partless things […] will say that such motion [i.e. the motion of partless things] goes ahead at the level of conception (μέχρις ἐπινοίας), but that it has to be examined in the case of subsisting bodies (ἐπὶ ὑποστατῶν … σωμάτων)" (trans. Bett).

78 Throughout our text, τὸ ζητούμενον either refers to the object about which we wish to find something or to that what we wish to find about that object. For the first use, cf. e.g. 17, 1/90,10; 18, 5–7/91,12–26; 23, 3/94,18 f., for the second cf. e.g. 4, 2/82,4 f.; 13, 8/88,13. Sometimes the word refers to a proposition set forth for demonstration (cf. 14, 2/88,18 f.) or to a demonstrated conclusion (6, 5/83,11 f.; 8, 1/84,13 f.). In our passage, τὸ ζητούμενον seems to be the object of inquiry, the word προσηγορία (81,19 f.) referring to the name by which the object is called. Cf. also below, 4, 1/81,25 f.; 8, 4.6/84,24.30; 9, 1–5/85,1–13, where τὸ ζητούμενον seems to be used in the same sense.

we can only explore the thing's essence when we know that it exists (ὅτι ἔστι).[79] Referring to this passage in Aristotle (and his own commentary on it), Galen distinguishes between two kinds of definition: "one that expresses clearly the notion that those who call a thing by a name have of this thing" (ἐξηγούμενον σαφῶς τὴν τοῦ πράγματος ἔννοιαν, ἣν ἔχουσιν οἱ ὀνομάζοντες αὐτό) and "one that teaches of essence" (τὸ τὴν οὐσίαν διδάσκον).[80] Galen's interpretation of meaning (τί σημαίνει τὸ ὄνομα) as the speaker's ἔννοια of the object in question has no precise parallel in our text. The word τὸ σημαινόμενον, as used above, 3, 2/81,15, and here (81,21), seems to refer to the signified object. However, below, 13, 1/87,23–25, τὸ σημαινόμενον is explained as νόημα. See also below, 9, 8/85,18 and especially 23, 1/94,5 f.[81] Presumably, the question raised at the first stage of inquiry is how the object signified by the name is commonly understood. In a phrase that closely resembles the vocabulary of our passage, Galen notes that the notional definition "is agreed upon by all who speak the same language" (ὁμολογεῖσθαί … πᾶσι τοῖς ὁμοφώνοις); *Diff. Puls.* IV 2 (VIII,704,5–706,3 K.). The same distinction between the two kinds of definition is also made in *MM* I 5 (X,40,12–16 K.), a passage whose affinity to our text has already been noted. For the question whether a thing called by a certain name exists or not, cf. *MM* II 7 (X,144,9–13 K., trans. Hankinson): "Do the words 'animal' and 'illness' seem to you to signify nothing (…) or is it the case that they signify, *but there is no object underlying the words* (ἢ σημαίνουσι μὲν, οὐδὲν δ' ὑπόκειται πρᾶγμα ταῖς φωναῖς), as is the case with 'Scylla', and 'centaur'?" For the meaning of τὸ ὁμολογουμένον, see also Galen, *MM* I 4 (X,32,2–6 K.) and the commentary of Hankinson, *Therapeutic Method*, pp. 115 f. (*ad* I 4, 3). See further below, on 82,12–14.

81,19 f. ἐκ τῆς προσηγορίας. In our text, the words ὄνομα ('name') and προσηγορία ('noun') are used interchangeably, in the (narrower) sense of the latter, i.e. as "that which signifies a common substance, e.g. 'man', 'horse'".[82] While normally speaking of ὄνομα in connection with the notional definition (see the previous note), Galen sometimes uses the word προσηγορία, too: "In the *Sophist* and the *Statesman*, [Plato] teaches us how a noun may be replaced and its meaning

79 *An. Post.* II 10, 93b30–33.

80 *Diff. Puls.* IV 2 (VIII,704,7–10 K.).

81 For νόημα, cf. Galen, *MM* I 6 (X,46,16–18 K., trans. Hankinson slightly modified): "It is possible to produce for one thing many definitions that differ in words only, and not in concepts (ἐν ταῖς φωναῖς μόνον, οὐκ ἐν τοῖς νοήμασι διαφέροντας)."

82 Cf. Dionysius Thrax, *Ars gram.* I 11 (23,2 f. Uhlig): ἡ … προσηγορία ὡς εἶδος τῷ ὀνόματι ὑποβέβληται. I 12 (34,1 f.): προσηγορικὸν δέ ἐστι τὸ τὴν κοινὴν οὐσίαν σημαῖνον, οἷον ἄνθρωπος ἵππος.

expressed (πῶς ἄν τις ἀντὶ τῆς προσηγορίας ἑρμηνεύοι ... τὸ σημαινόμενον αὐτῆς) by a clear and brief account (λόγῳ), which later philosophers called ὁρισμός and ὅρος par excellence."[83]

81,23 f. μή ποτε ὑπερβαίνειν τὴν δοθεῖσαν τάξιν. The manuscript reading μή ποτε ὑπερβαίνει κτλ. is difficult: If there is no lacuna in the text, the subject of ὑπερβαίνει should be the same as in the preceding ὁποία τίς ἐστιν, i.e. the φύσις of the investigated matter. The syntactic function of the two phrases should also be the same, which means that μή ποτε ὑπερβαίνει should work as the explanatory appositive. Finally, the combination of μή with the indicative suggests a doubtful assertion. I find it difficult to meet these three requirements in the form of an intelligible statement. As an alternative solution, I propose adding the letter ν to the verb and to interpret the phrase μή ποτε ὑπερβαίνειν as an equivalent of δεῖ μή + infinitive, where δεῖ is substituted by the verbal adjective in -τέον in 81,22, or where the infinitive, in the mind of the writer, still depends on ἀνάγκη in 81,21.[84]

The phrase δοθεῖσα τάξις refers to the order of steps prescribed in this passage.[85] Cf. Galen, *Opt. Corp. Const.* 1 (IV,738,18–739,3 K.): "... if [someone] starts from the agreed notion (ἀπό τῆς ὁμολογουμένης ἐννοίας ἄρξηται) and, proceeding to the discovery of essence, conducts the inquiry in a certain order and method (ἐν τάξει τινὶ καὶ μεθόδῳ ποιῆται τὴν ζήτησιν), I will praise such a person." Cf. also *MM* I 4 (X,32,2–9 K.): "Those who have made reason and order (λόγον ... καὶ τάξιν) their guides to discovery (...) must begin from something agreed by everybody (ἀπὸ πρώτου τινὸς ... ὁμολογουμένου) and proceed from there to the discovery of the rest."[86]

---

83 *PHP* IX 5, 12 (CMG V,1,4,2: 566,15–17/V,753 K.); trans. De Lacy, modified. Cf. also *Diff. Puls.* IV 3 (VIII,720,7–9 K.): καλοῦσι δ' οὕτως [*scil.* ὑποτυπώσεις and ὑπογραφαί] αὐτοὶ τοὺς λόγους, ὅσοι διὰ βραχέων ἑρμηνεύουσι τὴν ἔννοιαν τοῦ πράγματος, οὗ τὴν προσηγορίαν φθεγγόμεθα.

84 Cf. also Smyth, *Greek Grammar* § 2013b, on the infinitive used for a third-person imperative. It is possible for such an infinitive to be combined with an adjective in -τέον; cf. Athenaeus quoted by Oribasius, *Libri incerti* 39, 11–12 (CMG VI,2,2: 140,2–7): διὸ γυμνασίοις *χρηστέον πᾶσι ... πειρᾶσθαι δὲ τὰς ὁρμὰς καταστέλλειν καὶ μὴ ταῖς προθυμίαις ὑπερθέσθαι τὰς* ἑαυτῶν δυνάμεις ("For this reason, one should make use of all exercises ... Try to restrain the impulses and not to let the other capacities be exceeded by these desires"); trans. by Sean Coughlin, who kindly drew my attention to this passage.

85 A.Z. Fregonara translates 'categoria assegnata', τάξις being effectively synonymous to 'definition'. But I doubt whether τάξις could mean that.

86 There is an interesting parallel to these passages in Simplicius, *In Phys.* IV 10 (CAG IX: 696,10–13): "Furthemore, even in the case of things agreed to exist, if their names are not entirely clear, the question of what the name means precedes the question if the

81,25–82,2 εἰ δ' οὐκ ἀρκεῖ … ποιητέον. This convoluted sentence has the following structure: εἰ … πιστώσασθαι χρὴ τὸ λεχθέν (81,25–27) is a protasis expressing a condition whose alternative consequences (81,29 f.: [τὸ λεχθέν] … ἀναπόδεικτον ἔσται and 82,1 f.: [τὸ ὁμολογούμενον] τὴν ἀρχὴν … ποιητέον) are further conditioned by their respective subordinate clauses (81,27–30: εἰ μὲν εἰς … ἀμφισβητούμενον ἀναφέροιτο … ἡ κρίσις κτλ. and 81,30–82,1: εἰ δ' εἰς ὁμολογούμενον … ἡ … πίστις ἀναφέροιτο).[87]

81,25–27 εἰ δ' οὐκ ἀρκεῖ … ἀλλὰ πιστώσασθαι χρὴ τὸ λεχθέν. The discussion remains at the general level introduced in 3, 3/81,17. The protasis makes a distinction between two ways of speaking about the thing sought: we may either simply express our opinion about it (ἁπλῶς εἰπεῖν … τὸ δόξαν), or we may additionally do something to confirm (lit. 'make credible') what we have said (πιστώσασθαι τὸ λεχθέν). The reason why the first option is not sufficient (οὐκ ἀρκεῖ) is given in the parentheses: "For an opponent can equally (ἐπ' ἴσης) assert whatever he wants to the contrary." Thus the insufficiency of simply saying P is revealed by the fact that anyone can say Q, where it is impossible that both Q and P. The words ἐπ' ἴσης explain why this is so: if P and Q are simply asserted, they are asserted 'equally', the equality presumably consisting in the fact that neither is more credible than the other (cf. below, on 81,30–82,1).

The distinction between simple and confirmed assertions reminds us of the 'dogmatist' opponents of Sextus, defending the existence of signs: "[F]or the person who says that there is no sign maintains this either by mere assertion or by demonstration (ἤτοι φάσει μόνον … ἢ ἀποδείξει). And if he maintains it by assertion, he will face the opposite assertion (φάσει μὲν ἀξιῶν φάσιν ἕξει, τὴν ἀντιτιθεμένην); but if he demonstrates that what he says is true, then (…) he will agree that there is some sign" (*M.* VIII 281–282; trans. Bett). For examples of statements based on unconfirmed opinions, cf. Galen, *Diff. Puls.* IV 2 (VIII,708,9–16 K.), who censors some "more recent doctors" for producing (what they regard as) an essential definition of pulse before they propose a definition according to a notion (κατὰ τὴν ἔννοιαν), which should serve them as a crite-

---

thing exists. This is the order of problems (αὐτὴν τὴν τάξιν τῶν προβλημάτων) we have adopted in our treatise *On Demonstration*." Cf. *In Phys.* IV 7 (CAG IX: 653,6–19). Later in his commentary, Simplicius refers to Galen's own Ἀποδεικτική (*In Phys.* IV 11/CAG IX: 708,27 f.) and it is possible that his description of the proper order of research in *In Phys.* IV 10 is (directly or indirectly) influenced by this source (perhaps through Alexander of Aphrodisias, quoted in CAG IX: 696,3–8).

87 I have considered taking the clause ἀλλὰ πιστώσασθαι κτλ. as an apodosis (with εἰ δ' οὐκ ἀρκεῖ κτλ. as a protasis), but this would create an asyndeton in 81,27 (εἰ μὲν).

rion for the essential definition (κριτήριον τοῦ κατὰ τὴν οὐσίαν); as a result of this error, "they assert what irrationally seems to them to be the case" (ἀποφαίνονται τὸ δόξαν ἀλόγως ἑαυτοῖς).[88] Cf. idem, *Tum. Pr. Nat* 2 (VII,707,9 f. K.), on the causes of an inflamed tumour: "For the majority of them [i.e. doctors] do not conduct inquiry according to a method (οὐδὲ ... ὁδῷ τινι ποιοῦνται τὴν ζήτησιν), but they simply assert what seems to them to be the case (ἀλλὰ τὸ δόξαν αὐτοῖς ἁπλῶς ἀποφαίνονται)."[89] For the methods of 'confirmation' (πιστώσασθαι ... τὸ λεχθέν), cf. e.g. Galen, *MM* I 3 (X,28,18–29,3 K., trans. Hankinson): "Logical methods have the power to discover what is sought, while there are two criteria of confirmation of things that have been properly discovered for all men (τοῦ πιστώσασθαι τὰ καλῶς εὑρημένα ... ἅπασιν ἀνθρώποις κριτήρια), namely reason and experience." Cf. below, 7, 3/83,25.

There is an echo of this sentence or its source in *Strom.* VII (16) 95, 7–8, in the context of a polemic against the heretics: οὐ γὰρ ἁπλῶς ἀποφαινομένοις ἀνθρώποις προσέχομεν, *οἷς καὶ ἀνταποφαίνεσθαι ἐπ' ἴσης ἔξεστιν. εἰ δ' οὐκ ἀρκεῖ μόνον ἁπλῶς εἰπεῖν τὸ δόξαν, ἀλλὰ πιστώσασθαι* δεῖ *τὸ λεχθέν* κτλ. Cf. above, pp. 69–70.

81,27–30 εἰ μὲν εἰς ... ἀμφισβητούμενον ἀναφέροιτο ... ἡ κρίσις ... ἀναπόδεικτον ἔσται. In order to confirm τὸ λεχθέν, we must refer the judgement about it (ἡ κρίσις αὐτοῦ, *scil.* τοῦ λεχθέντος) to something, in view of which τὸ λεχθέν will be accepted as true.[90] If this item is also a matter of dispute, the judgement about it must be further referred to something else, and so on. If there were no end to this process, τὸ λεχθέν would never be proved. Our passage oddly frames the matter as if the subject of the infinite regress (εἰς ἄπειρον προβήσεται) were the same as the disputed item (ἀναπόδεικτον ἔσται). Perhaps the intended subject of the first verb is the argument as a whole (cf. 81,18 f.: ἀναγάγοι τὸν λόγον).

For infinite regress, cf. Aristotle, *An. Post.* II 3 and Jonathan Barnes' note in idem, *Aristotle: Posterior Analytics*, 2nd edition (Oxford: Clarendon Press, 1993,

88 For 'criterion' in this sense, cf. below, 4, 2/82,4 f.; 8, 6/84,32 f.

89 For the phrase 'to assert what seems to be the case', cf. also Alexander, *In Top.* I 15, 106b29 ff. (CAG II/2: 104,20–22): "For he who judges according to one's opinion and asserts what seems to him to be just (τὸ δόξαν αὐτῷ δίκαιον εἶναι ἀποφηνάμενος) is said to have judged justly, even if it is not really just."

90 Alternatively, the referent of αὐτοῦ could be τὸ ζητούμενον (ἡ κρίσις αὐτοῦ picking up on περὶ τοῦ ζητουμένου τὸ δόξαν in 81,25 f.); but I find this solution less likely, partly because τὸ λεχθέν is nearer and partly because of a similar construction in 81,30–82,1 (εἰ δ' εἰς ὁμολογούμενον ἅπασιν ἡ τοῦ μὴ ὁμολογουμένου πίστις ἀναφέροιτο), where πίστις is in an analogous position as κρίσις in our passage and its object is equivalent to τὸ λεχθέν.

2nd edn), pp. 103 f. (on 72b5). For our passage, cf. esp. Sextus, *M.* VIII 347; *PH* I 166; Aspasius, *In Eth. Nic.* III 5 (CAG XIX/1: 74,21 f.); Galen, *MM* II 7 (X,151,15 f. K.) in a passage noted below, on 83,26 f. Cf. also below, 6, 7–7, 1/83,16–19. For ἀναπόδεικτον, cf. below, on 83,20.

81,30–82,1 εἰ δ' εἰς ὁμολογούμενον ἅπασιν ἡ τοῦ ⟨μὴ⟩ ὁμολογουμένου πίστις ἀναφέροιτο. The manuscript reading ἡ τοῦ ὁμολογουμένου πίστις does not seem to make sense, but Potter's addition of μή solves the problem. Confronted with the opposite assertion (81,26 f.), τὸ λεχθέν turns out to be disputed (cf. 81,28: ὁμοίως ἀμφισβητούμενον), which amounts to saying that it is "not agreed upon". We have seen that in order to confirm 'what has been said', we must refer it to an item in view of which it will be confirmed. This will provide τὸ λεχθέν with something it lacked before, namely a πίστις. This word is notoriously difficult to translate. Generally speaking, πίστις is an item on account of which we believe X (πιστεύω) or consider X trustworthy (πιστός). This item may be a state of mind or something that brings about this state of mind by warranting or showing that X is credible. Our text seems to employ this word in both ways, without making a clear distinction between them; so it describes πίστις as an item that generates persuasion or knowledge (7, 7/83,33–84,2), while on another occasion saying that proof inserts πίστις of a certain kind into the soul (5, 3/82,16 f.).[91] πίστις also plays a crucial role in the definitions of demonstration given below (5, 1/82,12–14; 7, 1/83,31–33), where it is described as something furnished from that which is agreed/credible to that which is disputed/not credible yet. The process of furnishing πίστις to something disputed seems to be the same as the one posited in our passage, in the course of which πίστις of a disputed statement is "referred" to something agreed. This formulation reflects a common Greek usage. In order to give credibility to our statement, we refer to a trusted person who can confirm it: this may be described by saying that we "refer the πίστις" of our statement (i.e. that which confirms it) to someone or something credible.[92] This also seems to be the meaning of the phrase in our passage. For the point of reference as τὸ ὁμολογούμενον, cf. above, 3, 3/81,18–20.

---

91 One English equivalent that fits both of these contexts is 'credence', and this is how I usually render πίστις in this text.

92 Cf. e.g. Plutarch, *Cato Minor* 25, 1: ἐπράχθη δὲ τοῦτον τὸν τρόπον, ὡς ἱστορεῖ Θρασέας, εἰς Μουνάτιον, ἄνδρα Κάτωνος ἑταῖρον καὶ συμβιωτήν, ἀναφέρων τὴν πίστιν. *Schol. Hom. Od.* I 188 (I,37,1 Dindorf): εἰς ἀξιόχρεων δὲ μάρτυρα τὴν πίστιν ἀναφέρει.

82,1 τὴν ἀρχὴν τῆς διδασκαλίας. Unlike above, 2, 3/81,10, where διδασκαλία seems to be specifically the teaching on demonstration, here (and below, 82,4) the word is applied more generally to a teaching on any subject matter (cf. 81,17: περὶ παντὸς τοῦ ζητουμένου, 82,2: πᾶν τὸ προβληθὲν ὄνομα).

82,2f. πᾶν οὖν τὸ προβληθὲν ὄνομα μεταλαμβάνειν χρὴ εἰς λόγον ὁμολογούμενόν τε καὶ σαφῆ τοῖς κοινωνοῦσι τῆς σκέψεως. A similar instruction, with regard to the problem of demonstration, is given above, 3, 1/81,9–11, where it is presented as a way towards the starting-point of the doctrine; the starting-point is reached by defining the proposed name "so clearly (σαφῶς) that all who speak the same language will follow". Later, 3, 3/81,17–20, we learn, with regard to "everything sought", that the argument about it should be referred to an agreed starting point, namely the meaning of the proposed name; there, too, the starting-point is plainly an agreed definition of that meaning. Our passage describes the same procedure again, the only difference being that it limits the agreement to "those who participate in the research". This is probably not a compromise on the view of the starting-point as something agreed by "all who speak the same language". The text seems to presuppose that the participants of research sufficiently represent this common understanding.

Galen uses a similar phrase in a passage referring to *DD* (*MM* I 5/X,39,9f. K.): *ὡς ἐπὶ πάντων τῶν ζητουμένων εἰς λόγον χρὴ μεταλαμβάνεσθαι τοὔνομα*, δι' ἐκείνων (*scil.* τῶν περὶ τῆς ἀποδείξεως ὑπομνημάτων) ἀποδέδεικται.[93]

82,4f. ἐξηγησόμενον δὲ τὴν τῶν ζητουμένων εὕρεσιν. The definition of meaning provides a guide-line for the discovery of things sought.[94] Cf. Galen, *Diff. Puls.* IV 2 (VIII,708,13f. K., quoted above, on 81,25–27), on the definition of meaning as a criterion. Cf. below, 8, 6/84,32f.[95]

---

93 For the vocabulary, cf. Aristotle, *Top.* V 2, 130a39 (ἄν τις μεταλαμβάνῃ τοὺς λόγους ἀντὶ τῶν ὀνομάτων); VI 4, 142b2f.; VI 9, 147b14; VI 11, 149a24f.

94 For the meaning of τὸ ζητούμενον in this passage, cf. above, p. 149 n. 78 (the second use).

95 For the idea of guidance on the way to discovery, cf. Galen, *UP* II 7 (I,85,23–25 Helmreich/III,117 K.), trans. May, modified: "For (...) we will keep in mind just one thing—something which (...) will be a shining light leading us whither we must go and guiding us surely to the discovery of things we seek (ὥσπερ τι φῶς λαμπρόν, ἄγον τε ἡμᾶς οἷ χρὴ καὶ ποδηγοῦν ἑτοίμως ἐπὶ τὴν τῶν ζητουμένων εὕρεσιν)." Here the guideline is the principle that the activity and construction of bodily parts should be understood by "becoming a first-hand careful observer of what appears in dissections" (τῶν ἐν ταῖς διαιρέσεσι φαινομένων αὐτόπτην γενόμενον ἐπιμελῶς). Cf. below, on 83,24f.

82,5 f. ἄναμμα νοερὸν ἐκ θαλαττίων ὑδάτων. This description of the sun is often quoted in doxographic sources where it is variously ascribed to Zeno (SVF I 121), Cleanthes (SVF I 501 = 'Aetius' II 20, 4/*Dox. Gr.* 349), Chrysippus (SVF II 652), Posidonius (DL VII 145), the Stoics in general (SVF II 655–656, Ps.Galen, *Hist. Phil.* 62/*Dox. Gr.* 626,2), or to other writers ('Aetius' II 20, 16/*Dox. Gr.* 351). See also *Anon. Lond.* 30,19 f. (69 Manetti).[96]

82,9 f. τὸ λαμπρότατον τῶν κατ' οὐρανὸν ἰόντων. In Plato, *Tht.* 208d, Socrates proposes a similar description of the sun (τὸ λαμπρότατόν τῶν κατὰ τὸν οὐρανὸν ἰόντων περὶ γῆν) as an example of an account that states "some mark (σημεῖον) by which the thing one is asked about differs from all other things". This, according to Socrates, is what the 'the majority of people' (οἱ πολλοί) would say an account (λόγος) should be able to do. Possibly echoing this passage, Aristotle cites another variant of the same description (ἄστρον φερόμενον ὑπὲρ γῆς τὸ λαμπρότατον) as an account in terms of a 'property' (τὸ ἴδιον), whose presence is only manifest to the senses (*Top.* V 3, 131b25–30). In our text, of course, this is not meant to be a definition of essence, but it is supposed to meet the criteria of the correct (i.e. clear and agreed) definition of meaning. For the role of τὰ ἴδια in this regard, cf. Galen, *Diff. Puls.* IV 2 (VIII,705,14–17 K.): "For [Aristotle] says that [the essential definition] expresses what it is for each defined thing to be what it is (τί μὲν γάρ ἐστιν ἑκάστῳ τῶν ὁριζομένων τὸ εἶναι τοῦτον ἑρμηνεύειν φησί), whereas the notional one renders accidents that uniquely belong to it (τὰ συμβεβηκότα δὲ ἰδίως αὐτῷ συνδιέρχεσθαι τὸν ἐννοηματικόν)."

82,12–14 τὴν ἀπόδειξιν ... λόγον εἶναι τοῖς ἀμφισβητουμένοις ἐκ τῶν ὁμολογουμένων ἐκπορίζοντα τὴν πίστιν. After a methodological detour (3, 3–4, 3), the discussion returns to the problem of demonstration (cf. above, 3, 1–2). The definition given here does not aspire at grasping the essence of demonstration, but rather at establishing an agreed starting-point of the teaching on this subject: We prove a disputed claim by providing it with πίστις, a reason to believe it, coming from that on which there is agreement. All elements of this definition have already been mentioned; cf. above, 4, 1 and note on 81,30–82,1. Clement quotes the same definition in *Strom.* II (11) 48, 1.[97] See also below, 7, 6/83,31–33. For

---

96 For the doxographic tradition, see Jaap Mansfeld and David Runia, *Aëtiana: The Method and Intellectual Context of a Doxographer*, II/1 (PhA 114; Leiden/Boston: Brill, 2009), pp. 514–533, esp. 522 f.

97 φήσαιμεν δ' ἂν αὐτὴν (*scil.* τὴν ἐπιστημονικὴν ἀπόδειξιν) λόγον εἶναι τοῖς ἀμφισβητουμένοις ἐκ τῶν ὁμολογουμένων ἐκπορίζοντα τὴν πίστιν.

parallels, cf. Cicero, *Ac.* II 26 (trans. Brittain): "[Proof] is defined as an argument from apprehended premisses leading to something that wasn't apprehended at all."[98] DL VII 45: "Proof is an argument which concludes to what is less well apprehended by way of things better apprehended."[99] Sextus, *M.* VIII 314: "Proof is an argument that by means of agreed premisses uncovers by way of conclusive reasoning a consequence that is unclear."[100] Cf. also *M.* VIII 422–423. All these passages likely go back to Stoic sources.[101] For the concept of 'agreed' premisses, see already Aristotle, *An. Pr.* II 14, 62b31 f.: ἡ δὲ δεικτικὴ [*scil.* ἀπόδειξις] ἄρχεται ἐξ ὁμολογουμένων θέσεων, cf. *An. Post.* II 7, 92a35–37. Interestingly, Alexander of Aphrodisias employs this concept in a definition of *deduction* that closely resembles the preliminary definition of demonstration in our passage; cf. *In An. Pr.* I 4, 25b32 (CAG II/1: 44,14 f.): "Deduction is a proof of something disputed by means of agreed items (ὁ συλλογισμὸς δεῖξίς τινός ἐστιν ἀμφισβητουμένου διά τινων ὁμολογουμένων)." Cf. below, on 83,1–3. For the expression ἐκπορίζεσθαι τὴν πίστιν, cf. Aristotle, *Rhet.* I 2, 1356a1; Galen, *Dig. Puls.* IV 3 (VIII,954,10 K.).

5, 2–5

82,14–16 **οὐ μόνον δὲ ἀπόδειξις καὶ πίστις καὶ γνῶσις, ἀλλὰ καὶ πρόγνωσις λέγεται διχῶς, ἡ μὲν ἐπιστημονική τε καὶ βεβαία, ἄλλη δὲ μόνον ἐλπιστική.** This formulation implies that a distinction between two kinds of ἀπόδειξις, πίστις, and γνῶσις has already been made, which, however, is not the case in our text. Clement possibly skips over a few sentences in his source or abbreviates them with a special focus on 'foreknowledge' (to which alone the specific distinction ἡ μὲν ... ἄλλη δὲ ... pertains). Remarkably, in *Strom.* II (11) 48, 2, in a passage immediately following the definition of demonstration quoted above (on 82,12–14), Clement introduces the distinction implied here (as well as in the subsequent discussion: see below, 5, 3/82,16–18; 7, 7/83,33–84,2); cf. above, pp. 64–67.

For two kinds of πίστις and ἀπόδειξις, see the next note. For two kinds of πρόγνωσις (and γνῶσις), cf. Galen, *Hipp. Prog.* I 4 (CMG V,9,2: 203,24–204,5.17 f. /XVIIIb,12–14 K.) and *CAM* 17, 3–13 (CMG V,1,3: 112–116/I,289–293 K.), speaking

98 *Argumenti conclusio, quae est Graece ἀπόδειξις, ita definitur: 'ratio quae ex rebus perceptis ad id quod non percipiebatur adducit'.*

99 τὴν δ' ἀπόδειξιν λόγον διὰ τῶν μᾶλλον καταλαμβανομένων τὸ ἧττον καταλαμβανόμενον περαίνοντα.

100 ἀπόδειξίς ἐστι λόγος δι' ὁμολογουμένων λημμάτων κατὰ συναγωγὴν ἐπιφορὰν ἐκκαλύπτων ἄδηλον.

101 Cf. Jacques Brunschwig, "Proof Defined," in M. Schofield, M. Burnyeat, and J. Barnes (eds.), *Doubt and Dogmatism* (Clarendon Press: Oxford, 1980), pp. 125–160, here 125.

of διττὴ πρόγνωσις, the 'sure' one (βεβαία) and the one that is not sure, the latter being an expectation (ἐλπίς) regarding matters that happen in a certain way "for the most part" only. Galen applies the same distinction to knowledge (γνῶσις).[102] For the vocabulary, cf. also Galen, *Hipp. Off. Med.* I 1 (XVIIIb,636,13–16 K.), referring to the sure prediction as πρόγνωσις ἐπιστημονική.[103] The description of the deficient kind of πρόγνωσις as ἐλπιστική may be inspired by Aristotle, *Mem.* 1, 449b12, who describes, perhaps half-mockingly, divination as ἐπιστήμη τις ἐλπιστική.[104]

82,16–18 κυριώτατα μὲν οὖν ἀπόδειξις λέγεται ἡ τὴν ἐπιστημονικὴν πίστιν ἐντιθεῖσα ταῖς τῶν μανθανόντων ψυχαῖς, δοξαστικὴ δὲ ἡ ἑτέρα. Again, the reader is supposed to be familiar with the distinction between two kinds of proof, derived from the distinction between two kinds of credence (see *Strom.* II 48, 2, quoted in the previous note). Both kinds of proof meet the conditions outlined above, 5, 1/82,12–14, insofar as they furnish credence from something agreed to something disputed. But if the credence is 'epistemic', i.e. such as gives rise to knowledge, the proof itself will be 'epistemic'; whereas if it is 'doxastic', yielding merely a δόξα, the proof will be 'doxastic' (cf. below, 7, 7/83,3–84,2, where the doxastic proof is described as that which creates mere persuasion). With this distinction in mind, the author tells us that the epistemic demonstration is called demonstration in the most proper sense (κυριώτατα). At this point of the discussion, this is left unexplained. What additonal conditions an argument must meet to qualify as demonstration in this sense will come to light later, the crucial point being the distinction between agreed and true premisses (6, 2–4/83,1–10) and a related distinction between premisses that are 'reputable' (ἔνδοξα) and those that are 'evident' (7, 8/84,2–84,8). See further, on 83,1–3.

By saying that the epistemic πίστις is inserted into the souls of students, the text brings out the didactic background of the exposition, not only in the sense that it teaches on demonstration, but also in the sense that it regards demonstration as a tool of teaching. Cf. below, 7, 7/84,1 f.; 14, 2/88,20. The formu-

---

102 Cf. e.g. *Hipp. Prog.* I 4 (CMG V,9,2: 203,24–27.31–33/ XVIIIb,12–13 K.): ἡ γάρ τοι γνῶσις ἡ κατὰ τὸ τῆς προγνώσεως ὄνομα περιεχομένη λέγεται μὲν ὡς τὰ πολλὰ κατὰ τὸ βέβαιον, ἔστι δ' ὅτε μήν, εἰ καὶ λείπει τι πρὸς τὸ βεβαιότατον αὐτῇ, καὶ τόθ' ὡσαύτως ὀνομάζεται … *οὕτως οὖν καὶ ἡ πρόγνωσίς ἐστιν διττή, βεβαία μέν*, ὅτι μετὰ τὸν χειμῶνα γενήσεται τὸ ἔαρ, εἶθ' ἑξῆς θέρος, εἶτα φθινόπωρον, *οὐ βεβαία δέ*, κτλ. Ibid. 204,17 f./14 K.: εἴθισται δὲ καὶ *τὴν ὡς τὸ πολὺ περὶ τῶν μελλόντων ἔσεσθαι ἐλπίδα* καὶ τὴν ἀσφαλῆ καλεῖσθαι πρόγνωσιν.

103 ὁμολογήσαντες δὲ *βεβαιοτέραν τὴν πρόγνωσιν* ἕξομεν, οὐ μὴν ὁμοίως γε *ἐπιστημονικήν*, ὡς εἰ καὶ γιγνώσκοντες αὐτοὶ τύχωμεν τὸν κάμνοντα καὶ μᾶλλόν γέ τοι συμφύσαντες αὐτῷ.

104 Cf. my "Galenus Christianus?" pp. 21 f.

lation also accentuates the subjective aspect of πίστις (cf. above, on 81,30–82,1), a crucial element in Clement's reception of these tenets. Cf. *Strom.* II (11) 49, 3: "For the most superior demonstration, which we have called 'epistemic', inserts, by way of citing and explaining Scripture, credence into the souls yearning to learn, that is hardly other than knowledge."[105] Cf. below, 7, 7/84,2, where, in a similar phrase, ἐπιστημονικὴ πίστις is replaced by ἐπιστήμη.

The expression ἐπιστημονικὴ πίστις is rare in the extant literature, but it is found in Galen, *Ut. Resp.* 3 (IV,492,10 f. K.): "Of this, there is neither epistemic credence nor necessary demonstration, such as we always try to pursue when dealing with other matters, but it is not devoid of the plausible."[106] For the vocabulary of our passage, cf. also Galen, *Hipp. Epid. III*, I 4 (CMG V,10,2,1: 23,4–6/XVIIa,517 K.): "… one who would speak of all things by asserting what seems to him without demonstration would not insert any scientific teaching into the souls of the students."[107]

82,18–21 ὡς καὶ ἄνθρωπος ὁ μὲν ὄντως ἄνθρωπος … ὃ δὲ ἄγριος καὶ θηριώδης. οὕτω τοι καὶ ὁ κωμικὸς κτλ. The quoted verse is a (metrically unsound) version of Menander, fr. 484 (Koerte).[108] There is a close parallel to this passage in Galen, *Di. Dec.* I 9, presumably quoting a different verse of the same play: "Similar to this is what the comic says: 'And rare a thing is a human being that is a human being'. For we do not think that anyone who is savage, brutish, mindless or stupid corresponds to the nature of the human being, but we apply this name to those whose nature is correct."[109] The expression κοιναὶ φρένες is probably of Stoic origin; cf. Plutarch, *Comm. Not.* 1077e4.

---

105 ἡ γὰρ ἀνωτάτω ἀπόδειξις, ἣν ᾐνιξάμεθα ἐπιστημονικήν, πίστιν ἐντίθησι διὰ τῆς τῶν γραφῶν παραθέσεώς τε καὶ διοίξεως ταῖς τῶν μανθάνειν ὀρεγομένων ψυχαῖς, ἥτις ἂν εἴη γνῶσις. For the meaning of παράθεσις cf. Lampe, *Patristic Greek Lexicon*, s.v., 4 ("citation, adducing of parallels"). For δίοιξις, cf. above, 1, 3/80,12 f.; 2, 1/80,14.

106 ταῦτ' ἐπιστημονικὴν μὲν οὐκ ἔχει τὴν πίστιν οὐδ' ἀναγκαίαν τὴν ἀπόδειξιν, οἵαν ἐν τοῖς ἄλλοις λόγοις ἀεὶ μεταχειρίζεσθαι σπεύδομεν, οὐ μὴν ἐστέρηται τοῦ πιθανοῦ.

107 εἰ μὲν ἐπὶ πάντων τις ἀναποδείκτως ἀποφαίνοιτο τὰ δόξαντα, διδασκαλίαν οὐδεμίαν ἐπιστημονικὴν ἐνθήσει ταῖς ψυχαῖς τῶν μαθητῶν κτλ.

108 The fragment is preserved in several versions, the one closest to the original being presumably Stobaeus, *Anth.* III 12 (III,195,2 Wachsmuth, Hense): ὡ̃ χαρίεν ἐστ' ἄνθρωπος, ἂν ἄνθρωπος ᾖ. In some respects, Clement's version is closer to the one preserved in Menander's *Sententiae* (852 Pernigotti = 562 Meineke): ὡς χαρίεις ἐστὶν ἄνθρωπος, ὅταν ὑιὸς (*sic!*) ᾖ.

109 *Di. Dec.* I 9 (IX,815,2–6 K.): τοιοῦτον δ' ἐστὶ καὶ τὸ παρὰ τῷ κωμικῷ· 'καὶ σπάνιόν ἐστ' ἄνθρωπος ὅτ' ἄνθρωπος'. τὸν γὰρ *ἄγριον* καὶ τὸν *θηριώδη* καὶ τὸν ἀνόητον καὶ τὸν σκαιὸν οὐ κατὰ τὴν ἀνθρώπου φύσιν ὑπάρχειν νομίζοντες ἐπὶ τοὺς κατορθοῦντας ἐν τῇ φύσει τὸ ὄνομα φέρομεν. Jonathan Barnes pointed out this parallel to me.

82,22 f. καὶ ἐπὶ βοὸς καὶ ἵππου καὶ κυνὸς ὁ αὐτὸς λόγος παρὰ τὴν ἀρετὴν κτλ. Cf. Galen, *Temp.* I 6 (24,10–15 Helmreich/I,547 K.): "The excellence of a fig-tree, for example, consists in its bringing to fruition the most and the best figs; in exactly the same way, that of a vine consists in its producing the most and the best grapes; that of a horse in running very fast, and that of a dog in extreme spiritedness in hunting and guarding, combined with very great docility towards the members of its own household." Ibid. I 9 (36,12–16/I,566 K.): "And indeed it is in this way, too, that sculptors, painters and makers of statues, and makers of images in general, achieve the greatest beauty in their painting or sculpting of each species—the most well-formed human being, for example, or horse, or ox, or lion—by investigating the middle within that particular genus."[110]

82,24 εἰς γὰρ τὴν τοῦ γένους τελειότητα βλέποντες ἐπὶ τὰ κυριώτατα τῶν σημαινομένων ἐρχόμεθα. Looking at the most perfect representative of things of the same genus, we arrive at the most proper meaning of their name.[111] This amounts to looking at that which most perfectly corresponds to their essence.[112] Cf. Galen, *Hipp. Aph.* III 2 (XVIIb,565 K.): "Hippocrates used the word 'nature' in the most proper and primary meaning (κατὰ τὸ κυριώτατόν τε καὶ πρῶτον σημαινόμενον), which corresponds to the very essence of 'nature' in the highest degree (καθ' αὐτὴν μάλιστα τὴν οὐσίαν ἐστὶ τῆς φύσεως)." As Galen notes, however, the insistence on using words in their most proper meaning is not always in place; cf. *PHP* VI 1, 20 (CMG V,4,1,2: 364,26–30/V,511 K.), on the use of ἐνέργεια (trans. De Lacy): "… some persons through lack of training in semantics claim that walking is not an activity of the legs but an activity performed by means of the legs. Their statement is indeed true, for they go back to prior and more proper meaning of the term (ἐπὶ … τὸ πρότερόν τε καὶ κυριώτερον ἔρχονται σημαινόμενον); but because they think that one should speak only in this way and do not allow the other expression also, they are in error."

82,24 f. αὐτίκα ἰατρὸν κτλ. The example goes back to Aristotle, *Met.* Δ 16, 1021b14–17: "[We call perfect] that which in respect of excellence and goodness cannot be excelled in its kind, e.g. a doctor is perfect and a flute-player is per-

110 Both passages translated by Peter Singer.

111 That is why, as Galen notes, it sometimes happens that "some outstandingly eminent items usurp the name of the whole genus (ἐξόχως καθ' ὑπεροχὴν ἔνια τὴν τοῦ γένους ὅλου προσηγορίαν σφετερίζεται)" (*Di. Dec.* I 9/IX,814,15 f. K.).

112 For genus and essence, cf. below, on 93,5 f.

fect, when they lack nothing in respect of their proper kind of excellence (ὅταν κατὰ τὸ εἶδος τῆς οἰκείας ἀρετῆς μηθὲν ἐλλείπωσιν)."[113]

82,25f. γνωστικὸν δὲ ᾧ μηδὲν λείπει τῆς ἐπιστημονικῆς εἰδήσεως. This is likely Clement's addition to (or modification of) his source, reflecting his idea of the 'perfect gnostic'.[114] Cf. *Strom.* VII (14) 88, 5, quoted above, p. 74 n. 192

6, 1–6

82,27f. καὶ διαφέρει ἔνδειξις συλλογισμοῦ ... ἓν ὑπάρχον καὶ αὐτό. The theme of this and the following sentences (6, 1–6) is deduction; in the first step, indication is distinguished from deduction (6, 1); in the second, the difference between valid deduction and demonstration is explained (6, 2–4), as well as some other aspects of deduction relevant to the theory of proof (6, 5–6). Thus there is a slight thematic shift in this section from demonstration to a more general level, after which (in 6, 7/83,16f.) the discussion returns to the main theme (proof) again. In Clement's source, the shift was presumably introduced in a passage where demonstration was characterized as a kind of deduction. The contrast with indication shows that deduction must have at least two premisses (cf. below, 6, 6/83,14f.); this is probably the reason why indication is mentioned here and described from this particular perspective. A more detailed explanation is provided by Galen: "'Indication' is what they call the discovery of what is sought on the basis of the nature of the object, in consequence of plain phenomena." In Galen's view, this kind of inference is different from demonstration, which reaches its conclusion "by means of true premisses".[115] This does not mean, of course, that ἔνδειξις cannot be spelled out in the form of a syllogism with true premisses; but it suggests that, unlike proof, ἔνδειξις need not be spelled out in this manner in order to be valid. Thus it may take the form of an argument with one premiss, whose conclusion is validated by the nature of the object to which the premiss pertains.[116] Cf. Galen,

113 Trans. Ross, slightly modified. A candidate suiting this definition could be Asclepius, whom Galen regards as one of the founders of medicine and as his own model. Cf. e.g. Galen, *Protr.* 1, 2; 9, 2 (CMG V,1,1: 115,13–16; 132,18–21/I,2.22 K.); *MM* III 7 (X,207,1–5 K.).

114 The phrase ἐπιστημονικὴ εἴδησις is unusual. Apart from this passage, it is only found in Neoplatonist writers: once in Hermias of Alexandria (5th century A.D.) and several times in Simplicius (6th cent. A.D.).

115 *Inst. Log.* 11, 1 (24,14–17 Kalbfleisch): ἔνδειξιν μὲν γὰρ καλοῦσι τὴν ἐκ τῆς τοῦ πράγματος φύσεως εὕρεσιν τοῦ ζητουμένου κατ' ἀκολουθίαν τῶν ἐναργῶς φαινομένων, ἀπόδειξιν δὲ λόγον ⟨δι'⟩ ἀληθῶν λημμάτων περαίνοντα.

116 Commenting on an earlier draft of this commentary, Jonathan Barnes explains the dif-

Dig. Puls. IV 1 (VIII,923,12–14 K.), where the distinction between indication and demonstration seems to be connected with the number of signs they start with: ὥστ' οὐδ' αἰσθήσει διαγνωστὸς ὁ τόνος, ἀλλ', εἴπερ ἄρα, διὰ σημείου τινὸς, ἢ σημείων, ἤτοι κατ' ἔνδειξιν, ἢ κατ' ἀπόδειξιν πιστός.[117]

82,28 f. ὡς τοῦ μηκέτι εἶναι παρθένον ἔνδειξιν εἶναι τὸ κυεῖν φαμεν. There is a similar example of the indicative sign in Ps.Galen, *Hist. Phil.* 9 (*Dox. Gr.* 605,15–18): "Indicative sign (ἐνδεικτικὸν σημεῖον) is one that was not previously observed along with the thing signified, but when it appears we come to know the signified thing, like when a woman is giving milk we immediately recognize that she has given birth to a child." This example goes back to Aristotle, who cites the fact that a woman gives milk as a necessary sign (τεκμήριον or 'token') that she has given birth to a child (*Rhet.* I 2, 1357b15 f., see also *An. Pr.* II 27, 70a6–16). Ps.Galen distinguishes indicative signs from recollective signs (τὰ ὑπομνηστικά), explaining the latter as signs of things with which they were previously observed (e.g. smoke as a sign of fire). The same distinction is known to Sextus, who attributes it to his opponents, described on this occasion as "dogmatist philosophers and rational doctors"; cf. *M.* VIII 151–156; *PH* II 100–101. See also below, on 101,4–12. The origin of this sign theory is probably medical.[118]

---

ference between indication and proof as follows: "… in an indication the inference is validated by the nature of the objects which the premisses describe, whereas in a syllogism the validity depends not on the content of the premisses but on their form, i.e. it's essentially the distinction between what the mediaevals called material and formal consequence."

117 For the history of the term ἔνδειξις and its use in Galenic medicine, see Michael Frede, "The Method of the So-Called Methodical School of Medicine," in idem, *Essays in Ancient Philosophy* (Minneapolis: University of Minnesota Press, 1987), pp. 261–278, here 263–266; Jonathan Barnes, "Galen on Logic and Therapy," in F. Kudlien and R.J. Durling (eds.), *Galen's Method of Healing* (Leiden: Brill, 1991), pp. 50–102, here 98–100; Fridolf Kudlien, "'Endeixis' as a Scientific Term: (A) Galen's Usage of the Word in Medicine and Logic," ibid., pp. 103–111; Hankinson, *Therapeutic Method*, pp. 202–206; Philip van der Eijk, "Therapeutics," in R.J. Hankinson (ed.), *The Cambridge Companion to Galen* (Cambridge: Cambridge University Press, 2008), pp. 283–303, here 292 f.; R.J. Hankinson, "Galen on the Limitations of Knowledge," in C. Gill, T. Whitmarch, and J. Wilkins (eds.), *Galen and the World of Knowledge* (Cambridge: Cambridge University Press, 2009), pp. 206–242, here 231–233.

118 Cf. David Sedley, "On Signs," in J. Barnes, J. Brunschwig, M. Burnyeat, and M. Schofield (eds.), *Science and Speculation* (Cambridge: Cambridge University Press, 1982), pp. 239–272, here 241 f. and n. 8; James Allen, *Inference from Signs* (Oxford: Oxford University Press, 2001), pp. 87–146, esp. 106–122.

The particular example of indication mentioned in our text is also found in Cicero, *De inv.* 72: *si peperit, virgo non est.* However, unlike our text, Cicero describes the inference as an incomplete syllogism. It may seem odd to find this example in an author who shares the belief that Jesus was born from a virgin.[119] The fact that Clement leaves the example without further comment underlines the impression that the territory of his own argument has been abandoned.

82,30 ὡς τοῦ Πύθωνα προδιδόναι Βυζαντίους. Pytho of Byzantium, a pupil of Isocrates, was a diplomat in the service of Philip II of Macedon. For the accusation that he betrayed his hometown in the course of Philip's (unsuccessful) campaign against Byzantium (340/339 B.C.), see esp. Philostratus, *Vitae Soph.* I 20 (II,27,16–26 Kayser).[120]

83,1 εἰ οὕτως ἔτυχεν. This phrase, introducing a supposed event or an example, is occasionally used by Aristotle[121] and Aristotelian commentators and it is very common in Galen.[122]

83,1–3 καὶ τὸ μὲν περαίνειν ἐξ ὁμολογουμένων συλλογίζεσθαί ἐστιν κτλ. Within the broad definition of demonstration proposed above, 5, 1/82,12–14, our text further clarifies the distinction between 'epistemic' and 'doxastic' demonstration (5, 3/82,16–18) with respect to their premisses. Both kinds of demonstration are deductions (valid arguments), which also means here that their premisses are agreed upon. Nevertheless, for a deduction to qualify as demonstration in the proper sense, its premisses, apart from being agreed, must also be true.[123]

---

119 Cf. e.g. *Strom.* III (17) 102, 1; VI (15) 127, 1; VII (16) 93, 7.

120 Cf. also *Codex Vaticanus Graecus* 96 (12th cent.), fol. 98ᵛ, ed. Brad L. Cook, "The Essential Philip of Macedon: A Byzantine Epitome of His Life," *Greek, Roman and Byzantine Studies* 45/2 (2005), pp. 189–211, here 194,7–9: Ἐπεχείρησε δὲ [*scil.* ὁ Φίλιππος] καὶ Βυζαντίοις πολεμεῖν· Πύθωνος προδιδόντος καὶ στρατηγοῦντος Λέοντος.

121 Cf. *Cat.* 7, 8b12; *EN* III 5, 1114a15; IX 1, 1164a4.

122 Cf. e.g. *Ars. Med.* 19, 4 (332,8 Boudon/I,353 K.); 28, 9 (362,12 f./I,382 K.); 28, 15 (363,17/I,383 K.); 34, 13 (378,11/I,398 K.); *San. Tu.* II 4 (CMG V,4,2: 50,7/VI,110 K.); II 11 (68,13 f./154 K.); IV 4 (113,11 f./256 K.); *MM* I 4 (X,33,3 K.); II 7 (X,133,5 K.); III 10 (X,223,5 K.); VIII 1 (X,532,13 f. K.); XII 1 (X,813,9 K.); *PHP* VI 1, 13 (CMG V,4,1,2: 362,25/V,509 K.).

123 Cf. Alexander, *In An. Pr.* I 27 (CAG II/1: 292,32–34): οὐ γὰρ ταὐτὸν ἀποδεῖξαί τε καὶ συλλογίσασθαί τι, διότι ἔδοξέ τινι ἢ συνεχώρησε· συγχωρεῖται γὰρ πολλὰ καὶ ψευδῆ, δι' ὧν συλλογισμὸς μὲν ἂν γένοιτο, ἀπόδειξις δὲ οὔ. Of course, on the Aristotelian view, it is not sufficient for the premisses of a demonstrative syllogism to be true, either; cf. Alexander, *In Top.* I 1, 100a27 f. (CAG II/2: 16,1–3): οὐ γὰρ πᾶς ὁ ἐξ ἀληθῶν συλλογισμὸς καὶ κυρίως ἀπόδειξις, ἀλλ' εἰ πρὸς τῷ ἀληθῆ εἶναι τὰ δι' ὧν ὁ τοιοῦτος δείκνυται συλλογισμὸς καὶ πρῶτα εἴη.

It is likely that the agreed premisses that do not qualify as 'epistemic' correspond to the Aristotelian ἔνδοξα, and thus 'doxastic demonstration' amounts to a dialectical deduction according to the Aristotelian scheme; cf. *Top.* I 1, 100a27–30. See also below, on 84,4–7.[124]

83,3 f. σύνθετόν τινα … τὴν νόησιν. The manuscript speaks of a "composite benefit" (σύνθετόν … τὴν ὄνησιν), but this makes little sense; Pohlenz's simple correction is supported by parallels in Galen.[125]

83,6 f. εἰ δ' ἤτοι μὴ ὑπάρχοι τὸ πρότερον ἢ μὴ ἕποιτο αὐτῷ τὸ δεύτερον κτλ. This sentence implies that an argument with true premisses, whose conclusion does not follow from them, still qualifies as deduction (συνελογίσατο δέ). But this is obviously false. Perhaps the excerptor (or a copyist) has left something out.[126] See below, 7, 8/84,7 f.: "If the conclusion is not appropriate, it will not even be a deduction."

83,7 f. τὸ μὲν γὰρ οἰκεῖον ἐπενεγκεῖν συμπέρασμα. Appropriate conclusion is one that "follows from [its] premisses" (6, 2/83,5 f.; cf. Aristotle, *Top.* I 1, 100a26:

124 Alternatively, δοξαστικὴ ἀπόδειξις could also include deductions based on 'apparently reputable premisses' (τὰ φαινόμενα ἔνδοξα), i.e. those classified as 'eristic' by Aristotle (*Top.* I 1, 100b23–26). For this interpretation see *Strom.* II (10) 49, 2–3, a passage clearly influenced by our text or its source to some extent (see above, on 82,14–16): *ἡ δὲ δοξαστικὴ ἀπόδειξις ἀνθρωπική τέ ἐστι καὶ πρὸς τῶν ῥητορικῶν γινομένη ἐπιχειρημάτων ἢ καὶ διαλεκτικῶν συλλογισμῶν.* ἡ γὰρ ἀνωτάτω ἀπόδειξις, ἣν ᾐνιξάμεθα ἐπιστημονικήν, πίστιν ἐντίθησι διὰ τῆς τῶν γραφῶν παραθέσεώς τε καὶ διοίξεως ταῖς τῶν μανθάνειν ὀρεγομένων ψυχαῖς, ἥτις ἂν εἴη γνῶσις.

125 Cf. *Diff. Puls.* I 27 (VIII,555,14–17 K.): *σύνθετοι* γὰρ αἵ τε *νοήσεις* καὶ αἱ γενέσεις αὐτῶν, τοῦ τε κυματώδους καὶ τοῦ σκωληκίζονος καὶ τοῦ κλονώδους καὶ τοῦ σπασμώδους [σφυγμοῦ]. *Syn. Puls.* 21 (IX,494,8 f. K.): καὶ ὅτι μάλιστα *σύνθετός* ἐστιν ἡ τῆς σφοδρότητος *νόησις* ἐξ ἰσχύος τε καὶ τόνου. *MMG* II 1 (XI,77,4–8 K.): ἅπαντα γὰρ ταῦτα φλεγμοναὶ μέν τινές εἰσιν … διαφόρου δ' αὐτῶν ἕκαστον ἔτυχε προσηγορίας, τῶν πρώτως θεμένων αὐτὰς *σύνθετον νόησιν* … δι' ἑνὸς ἑρμηνεῦσαι βουληθέντων ὀνόματος. In the predicate we would rather expect the neuter (σύνθετόν τι); but see Galen, *Diff. Puls.* II 4 (VIII,578,3–5 K.), re. Archigenes' classification of the qualities of pulses: εἰ μὲν οὖν *σύνθετόν τινα* ἐξ ἁπλῶν εἰργάζετο ποιοτήτων *αὐτήν* [*scil.* τὴν σκληρότητα], τάχ' ἄν τις ἀπελογήσατο ὑπὲρ αὐτοῦ κτλ.—Pohlenz's emendation is quoted by Ernst, *De Clementis*, p. 27 n. 1; Pohlenz is not aware of the Galenic parallels mentioned above, but he refers to *Strom.* VIII (4) 9, 3.8 and Sextus, *PH* III 26.48 as passages where the words νόησις, νόημα, or νοεῖσθαι are used in a relevant sense (he might also have added *Strom.* VIII 29, 2).

126 Francesco Ademollo (personal communication) proposes to supplement the text, *exempli gratia*, in the following manner: εἰ δ' ἤτοι μὴ ὑπάρχοι τὸ πρότερον ἢ μὴ ἕποιτο αὐτῷ τὸ δεύτερον, οὐκ ἀπέδειξεν ⟨. εἰ δὲ μὴ ὑπάρχοι τὸ πρότερον, ἀλλὰ τὸ δεύτερον, οὐκ ἀπέδειξεν⟩ μέν, συνελογίσατο δέ.

ἐξ ἀνάγκης συμβαίνει. *An. Pr.* I 1, 24b18–20); for the vocabulary, cf. Galen, *Hipp. Elem.* I 4, 12 (CMG V,1,2: 90,5 f./I,445 K.): οὐκ ἐπιφέροντες οἰκεῖον συμπέρασμα. *Ven. Sect. Er. Rom.* 7 (XI,230,1 f. K.): προτάσεις ... τινες ἀληθεῖς ἅμα τοῖς οἰκείοις συμπεράσμασι. Alexander, *In Top.* I 1, 100a25–27 (CAG II/2: 14,2–4): ἀλλ' οὐδὲ ἐν οἷς μὴ τὸ οἰκεῖον συμπέρασμα ἐπιφέρεται, ἄλλο δέ τι ἑπόμενον ἐκείνῳ ἢ συμβεβηκός, οὐδ' οὗτοι συλλογισμοί.

83,11 f. ἔστι δὲ δήπου πέρας ἐν ἑκάστῳ λόγῳ τὸ ζητούμενον. The goal of the argument is to prove an assertion about τὸ ζητούμενον (cf. above, 4, 1/81,27: τὸ λεχθέν). When proved, the assertion is the conclusion of the argument.[127] Since the assertion articulates something sought about τὸ ζητούμενον, our text describes it as τὸ ζητούμενον, too, both in its role as a proposition set forth for demonstration (cf. below, 14, 2/88,19) and in its role as a demonstrated conclusion.[128] Cf. also below, 8, 1/84,13 f.

83,13 ἁπλοῦς καὶ πρῶτος λόγος. Cf. Aristotle, *Int.* 5, 17a20–26, on 'simple' vs. 'compounded' assertions (ἁπλῆ/συγκειμένη ἀπόφανσις). Cf. below, on 83,26 f.

83,14 τοὐλάχιστον ἐκ τριῶν τοιούτων σύνθετος. L reads σύνθετον, but, as noticed by Barnes, this is probably a scribal error. For the requirement according to which a syllogism must consist of at least two premisses, cf. Aristotle, *An. Pr.* I 14, 34a16–19; 23, 40b30–37; II 2, 53b16–23; *An. Post.* I 3, 73a7–11; Alexander, *In An. Pr.* I 1, 24b18 (CAG II/1: 17,10 f.); I 23, 40b30 (CAG II/1: 257,8–13); *In Top.* I 1 (CAG II/2: 8,14–9,19).[129] Interestingly, our text does not mention the possibility of a deduction including compounded assertions, as is the case with hypothetical syllogisms. But this need not be taken to mean that syllogisms can be composed of simple propositions *only*.

6, 7–7, 8

83,16 f. ἤτοι δὲ πάντα ἀποδείξεως δεῖται ἢ καί τινα ἐξ αὑτῶν ἐστι πιστά. Cf. Aristotle, *Top.* I 100b18–21: "Such items are true and first that have credence not from other items, but from themselves (τὰ μὴ δι' ἑτέρων ἀλλὰ δι' αὑτῶν ἔχοντα

127 Cf. Alexander, *In An. Pr.* I 4, 25b32 (CAG II/1: 44,19–23, trans. Barnes et al.): "When a proposition is put forward for a proof as not being known, it is called a problem. (...) ... if [a proposition] has been proved, it is a conclusion (for anything which has been proved is a conclusion from the items by which it was proved)."

128 Cf. above, on 81,17–24 and n. 78

129 Cf. Barnes et al. in Alexander of Aphrodisias, *On Aristotle's Prior Analytics* 1,1–7 (Ithaka, NY: Cornell University Press, 1991), p. 64 n. 72.

τὴν πίστιν)." Cf. also Galen, *MM* I 4 (X,33,16–18 K.): "… until we arrive at the first [premisses], which have credence not from others, nor from demonstration, but from themselves (ἐξ ἑαυτῶν ἔχουσι τὴν πίστιν)." The contrast between items that "need demonstration" and those "credible by themselves" implies that the latter, if they exist, do not need demonstration; cf. below, on 83,20 f.

83,18 εἰς ἄπειρον ἐκβησόμεθα. For infinite regress, cf. references above, on 81,27–30.

83,20 f. αὐτίκα οἱ φιλόσοφοι ἀναποδείκτους ὁμολογοῦσι τὰς τῶν ὅλων ἀρχάς. Whatever principles philosophers postulate as the principles of the universe, they agree on them being ἀναπόδεικτοι.[130] Unlike above, 81,29 f., where the word ἀναπόδεικτος refers to something that can never be proved, here its meaning is determined by the contrast with items that "need demonstration" (cf. above, 83,16 f.).[131] Thus, according to our passage, philosophers agree that the principles of the universe need not be demonstrated.[132] This illustrates the point of

130 Explicit statements to this effect are not as common as one might expect. At any rate, the point would have been readily conceded by the Peripatetics; cf. e.g. Theophrastus, fr. 159 (Fortenbaugh) in Proclus, *In Tim.* III 2 (II,120,10–18 Diehl): "For if we look for the causes of the things that are primary and set our minds on the genesis of things that are self-subsistent, we shall fail to see that we shall be going on to infinity and have no end of our inquiries. For just as the person who thinks that all things are capable of demonstration actually does away with demonstration itself, so too the person who seeks after causes in this manner entirely overturns all the existent things and the order in which they proceed from some definite principle" (εἰ γὰρ καὶ τῶν πρώτων τὰ αἴτια ζητήσομεν καὶ τῶν αὐθυποστάτων γενέσεις ἐπινοήσομεν, εἰς ἄπειρον προϊόντες λησόμεθα καὶ τέλος οὐδὲν ἕξον τῆς θεωρίας· ὥσπερ γὰρ ὁ πάντα ἀποδεικτὰ νενομικὼς αὐτὴν μάλιστα τὴν ἀπόδειξιν ἀναιρεῖ, τοῦτον τὸν τρόπον καὶ ὁ πάντων αἰτίας ἐπιζητῶν ἄρδην ἀνατρέπει τὰ ὄντα πάντα καὶ τὴν τάξιν αὐτῶν τὴν ἀπό τινος ὡρισμένης ἀρχῆς προϊοῦσαν); trans. Dirk Baltzly, *Proclus: Commentary on Plato's Timaeus*, vol. IV (Cambridge: Cambridge University Press, 2009), p. 78. However, insofar as natural philosophy depends on certain preconceptions that need not be demonstrated, since they are plain by themselves, the Epicureans and the Stoics would have agreed, too; cf. LS 17 and 40, on the criteria of truth; cf. also Sextus, *PH* III 1–167 and *M.* IX–X, *passim*, contrasting various views on the first principles, some of which are allegedly based on such preconceptions. For the Platonists, cf. e.g. Alcinous, *Did.* 5, 5 (H 157,12–14.21–27), on universally agreed propositions about the first principle, reached by way of analysis.

131 Barnes plausibly proposes translating ἀναπόδεικτος in this sense as 'unproved' (rather than 'uprovable', 'indemonstrable' and the like); cf. his *Truth, etc.: Six Lectures on Ancient Logic* (Oxford: Clarendon Press, 2007), pp. 369–373.

132 This seems to be implied by the description of these principles as ἀναπόδεικτοι. Refering to the Stoic primary argument-forms, Sextus articulates this implication as one of the

there being principles of demonstration of the same description. There is a parallel to this passage in *Strom.* II (4) 13, 4–14, 1, where the principle of the universe (ἡ τῶν ὅλων ἀρχή) is described as an undemonstrated principle of knowledge. Cf. also *Strom.* VII (16) 95, 3–6, where the 'undemonstrated principle' is identified as 'the Lord'.

83,22 f. ὃ δὴ πρῶτον καὶ ἀναπόδεικτον λέγεται. Cf. Aristotle, *An. Post.* I 2, 71b27 and Barnes, *Posterior Analytics*, pp. 94 f. For Galen, see Hankinson, *Therapeutic Method*, pp. 117–119 (*ad MM* I 4, 6).

83,23 f. ἐπὶ τὴν ἀναπόδεικτον ἄρα πίστιν ἡ πᾶσα ἀπόδειξις ἀνάγεται. After stating that demonstration is preceded by something πιστὸν ἐξ ἑαυτοῦ, the text now draws a further inference (marked by ἄρα), according to which every demonstration is referred to an undemonstrated πίστις. There is something odd about this inference. Earlier, the text spoke about πίστις as an item that, in the course of providing a proof of Q, is 'furnished' from P to Q, where P is something agreed (5, 1/82,13 f.: ἐκ τῶν ὁμολογουμενων … ἐκπορίζοντα τὴν πίστιν), or as an item belonging to Q in virtue of being 'referred to' P, where P is something agreed (4, 1/81,30–82,1: εἰς ὁμολογούμενον … ἡ … πίστις ἀναφέροιτο). For the meaning of πίστις in these passages, cf. above, on 81,30–82,1. It is reasonable to suppose that when speaking of the principles of demonstration in 7, 1/83,19 f., our text refers to items that have the same function as P above, i.e. to items from which πίστις is furnished to something else. The point of describing these items as undemonstrated is that nothing needs to furnish πίστις to them, as they have πίστις (are πιστά) by themselves. But it is odd to apply this description to πίστις, as there is really no question of demonstrating (i.e. furnishing πίστις to) πίστις. Unless, of course, πίστις is used in a different sense in our passage. The only relevant sense I can think of is considering πίστις neither as a warranty nor as a cognitive state, but rather as a particular belief, a proposition held to be true. The text seems to argue the following: Since every demonstration refers to an undemonstrated P, credible by itself, it also refers to an undemonstrated belief that P. This is a correct inference, but (a) it uses the word πίστις in a different sense than in the previous passages and (b) it is not clear of what use it is in the context of the present argument. On the other hand, it serves remarkably well the interests

---

meanings of ἀναπόδεικτος; cf. *M.* VIII 223 (trans. Bett): "[A]rguments are spoken of as ἀναπόδεικτοι in two ways, covering both arguments that have not been demonstrated and those that have no need of demonstration (οἵ τε μὴ ἀποδεδειγμένοι καὶ οἱ μὴ χρείαν ἔχοντες ἀποδείξεως), given that, in their case it is immediately clear that they reach a conclusion." Cf. Barnes, *Truth, etc.*, pp. 369–371.

of Clement, whose aim in the *Stromateis* is to persuade the critics of Christian faith that πίστις can play the role of a starting-point of demonstration.[133] Thus I suspect that the inference is drawn by Clement, not by his philosophical source.

83,24f. εἶεν δ' ἂν καὶ ἄλλαι τῶν ἀποδείξεων ἀρχαί … τὰ πρὸς αἴσθησίν τε καὶ νόησιν ἐναργῶς φαινόμενα. Here, in addition to πίστις, the text postulates other principles of demonstration, namely items plainly apparent to sense-perception and intellection. If we ascribe the preceding sentence to Clement himself, this one is best explained as an attempt to bridge between the inference about πίστις he makes there and the description of ἀρχαί given here, which likely goes back to his source (see parallels below). The optative form of the sentence seems to suggest that Clement acknowledges his dependence on a source.[134]

Epistemological presuppositions of this passage are more fully articulated in *Strom.* II (4) 13, 2: "There are four things in which there is truth: sense-perception, intelligence, knowledge, opinion. Intelligence is primary by nature, whereas sense-perception is primary for us and in relation to us. The essence of knowledge then consists of sense-perception and intelligence, evidence being common to sense-perception and intelligence (κοινὸν δὲ νοῦ τε καὶ αἰσθήσεως τὸ ἐναργές)." The last sentence seems to reflect the view of Theophrastus; cf. Sextus, *M.* VII 217 (trans. Bett): "But Aristotle and Theophrastus and in general the Peripatetics also allow the criterion to be twofold (…): sense-perception for perceptible things, intelligence for intelligible things, and common to both, as Theophrastus said, what is plain (αἴσθησιν μὲν τῶν αἰσθητῶν, νόησιν δὲ τῶν νοητῶν, κοινὸν δὲ ἀμφοτέρων, ὡς ἔλεγεν ὁ Θεόφραστος, τὸ ἐναργές)."[135] But the specific wording of our text is more closely reminiscent of Galen; cf. esp. *MM* I 5 (X,39,7–10 K.): "… the starting-points of all demonstrations are those items

---

133 See my "Demonstrative Method," pp. 266–270.

134 Clement typically uses the optative of εἶναι with ἂν when proposing an explanation of, or drawing an inference from, another text, or to mark content taken from elsewhere. Cf. *Strom.* I (5) 29, 3; II (22) 133, 4; III (4) 28, 1; III (10) 68, 5 and 70, 1; III (18) 106, 2; V (6) 35, 4; V (6) 37, 3; V (11) 73, 2. For the last mentioned case, cf. *Strom.* I (21) 134, 4.

135 Theophrastus, frs. 301 A and B (Fortenbaugh). Cf. the commentary of Pamela Huby, *Theophrastus of Eresus: Sources for his Life, Writings, Thought and Influence. Commentary, IV: Psychology* (Leiden: Brill, 1999), pp. 93–100. For the history of the concept of evidence, see Woldemar Görler, "Les «évidences» dans la philosophie hellenistique," in C. Lévy and L. Pernot, *Dire l'évidence* (*philosophie et rhétorique antiques*) (Paris/Montréal: L'Harmattan, 1997), 131–143; Katerina Ierodiakonou, "The notion of enargeia in Hellenistic philosophy," in K. Ierodiakonou and B. Morison (eds.), *Episteme, etc.: Essays in Honour of Jonathan Barnes* (Oxford: Oxford University Press, 2011), pp. 60–73.

which are plainly apparent to sense-perception and intellection" (ἀρχαὶ πάσης ἀποδείξεώς εἰσι τὰ πρὸς αἴσθησίν τε καὶ νόησιν ἐναργῶς φαινόμενα).

83,26f. τὰ μὲν γὰρ πρὸς αἴσθησιν σύμπαντα ἐστὶν ἁπλᾶ τε καὶ ἄλυτα, τὰ δὲ πρὸς νόησιν ἁπλᾶ τε καὶ λογικὰ καὶ πρῶτα. Two kinds of plainly apparent items are postulated: the perceptible and the intelligible.[136] The former are described as 'simple' and 'indissoluble', the latter as 'simple', 'rational', and 'primary'. The passage is difficult, but a helpful parallel is provided by Galen, *MM* I 4: "The ancient philosophers say that there are two kinds of apparent things, one of which (...) is discerned by sense-perception, e.g. pale and dark, hard and soft, hot and cold, and the like; the other being the undemonstrated kind of things that immediately present themselves to intellection."[137] For Galen, the 'phenomena' of sense-perception are simple perceptible qualities; this could explain their two attributes (ἁπλᾶ τε καὶ ἄλυτα) in our passage.[138] What about the objects of intellection? Galen's examples are axioms.[139] Cf. also idem, *Inst.*

---

136 After τὰ ... πρὸς αἴσθησιν, I prefer to read σύμπαντα, a manuscript (second-hand) correction of the original συμβάντα, as the latter reading, to my mind, sits ill within the sentence. I doubt that συμβαίνειν πρός τινα can mean 'occur to' (cf. Fregonara: "... le cose che si presentano alla percezione sensibile") or anything relevant; and I am puzzled by the force of the aorist.

137 *MM* I 4 (X,36,10–15 K.): οἱ δ' αὖ παλαιοὶ φιλόσοφοι διττὸν γένος εἶναί φασι τῶν φαινομένων, ἓν μέν, ὅπερ καὶ τοῖς ἐμπειρικοῖς ὁμολογεῖται, τῶν αἰσθήσει τινὶ διαγινωσκομένων, οἷον λευκοῦ καὶ μέλανος καὶ σκληροῦ καὶ μαλακοῦ καὶ θερμοῦ καὶ ψυχροῦ καὶ τῶν ὁμοίων, ἕτερον δὲ τῶν ὑποπιπτόντων νοήσει κατὰ πρώτην ἐπιβολὴν ἀναπόδεικτον. For the phrase κατὰ πρώτην ἐπιβολήν, cf. below, on 95,3.

138 It is tempting to read ἄλογα for ἄλυτα, as Barnes has suggested (see apparatus *ad loc.*), and render the pair of attributes symmetrical to the description of the intelligible phenomena (ἁπλᾶ τε καὶ λογικά). Another possible equivalent of ἄλυτα is 'irrefutable'; this could indicate that Clement conceives of the perceptible phenomena as plainly true simple propositions. Cf. Sextus, *M.* VIII 316, citing some stock examples of a simple proposition ('It is day', 'This is a human being') as examples of things believed to be plainly the case. Cf. also *PH* II 97; Clement, *Strom.* V (1) 5, 3. For simple propositions (ἁπλοῦν ἀξίωμα), cf. DL VII 68; Sextus, *M.* VIII 93; Galen, *Inst. Log.* 5, 1–2. But the word ἄλυτος, when used in the sense of 'irrefutable', normally refers to arguments, not simple propositions. Cf. Aristotle, *An. Pr.* II 27, 70a29f.; *Rhet.* I 2, 1357b17; II 25, 1403a14. Cf. also Clement, *Strom.* V (1) 5, 3, on ἄλυτον ἐνθύμημα.

139 Cf. *MM* I 4 (X,36,15–37,6): "... such as, for example, that two quantities equal to a given quantity are equal to each other, and that equals added to equals yield equals, and that when equals are subtracted from equals, the remainders are equal. And they say that 'nothing occurs causelessly' is of this type, and similarly 'everything comes to be from something existent', and that 'nothing comes to be from the absolutely non-existent'. Equally, that

Log. 1, 5: "When there is a sentence credible by itself to intellection (τῇ νοήσει τίς ... ἐξ αὑτοῦ πιστὸς λόγος), they have called it an axiom, e.g. 'things equal to the same thing are equal to one another'." For the description of these objects as 'rational' and 'primary', cf. also Galen, *MM* I 4 (X,37,6f. K.): "... these sorts of principles, which we also call rational (ἃς δὴ καὶ λογικὰς ὀνομάζομεν) ..." However, the word 'simple' in our lemma seems to suggest that the intelligible phenomena are conceived as something more elementary than axioms or propositions of any sort. A possible clue is provided by another passage in Galen, *MM* II 7 (X,151,14–152,1): "And this is what the ancient philosophers meant by 'giving the definition of the word'. They do not proceed to infinity, but only ascend as far as the simple things (ἄχρι τῶν ἁπλῶν ἀνέρχονται). So when you have dissolved (διαλύσῃς) 'man' into 'animal', 'rational', and 'mortal', you will further dissolve 'animal' into 'sentient substance'; but you cannot dissolve 'substance' any further, and equally not 'sentience'. For each of these is already simple and primary (ἁπλοῦν ... καὶ πρῶτον)." Here Galen is not concerned with the intelligible phenomena, but with the simple constituents of which a concept associated with a given name ('man', 'fever' etc.) is composed (see further the next note). But of course, these simple constituents are also concepts and thus they may be fairly described as 'simple and primary' objects of intellection.

83,27–29 τὰ δὲ ἐξ αὐτῶν γεννώμενα σύνθετα μέν, οὐδὲν δ' ἧττον ἐναργῆ καὶ πιστὰ καὶ λογικώτερα τῶν πρώτων. According to Galen, "whenever you have the conception of each of the simple things (τῶν ... ἁπλῶν ἑκάστου τὴν ἔννοιαν ἔχῃς) according to its name, you can then make one composite concept out of two simple ones (ἐκ δυοῖν ἁπλοῖν ἔννοιαν αὖθις ἑτέραν ἐργάσασθαι σύνθετον) and then apply a name to it. Furthermore, you can conjoin another simple thing to this complex name, making a single concept out of three simple ones, and assign a name to this as well."[140] Galen's examples of composite concepts are 'man' and 'fever', the former being composed of the concepts of 'animal', 'mortal', and 'rational', and the latter of 'heat' and 'damage to an activity'.[141] These examples show that the relation between the composite concepts and their constituents is one of entailment: 'man' is followed by 'animal', 'rational', and 'mortal', just as 'fever' is followed by 'heat' and 'damage to an activity'. This could explain why,

'nothing is annihilated into the absolutely non-existent' and that 'it is necessary that everything must be either affirmed or denied', and many other propositions of this sort which they discuss in the logical works, and which I too have recorded in my *On Demonstration* with all the clarity of which I am capable" (trans. Hankinson).

140 *MM* II 7 (X,150,3–9 K.); trans. Hankinson.

141 *MM* II 7 (X,150,11–151,15 K.).

in our passage, the composites of the primary intelligible items are described as "more rational" than their constituents. The simple concepts are 'rational', as they can only be grasped by reason; but the composite ones are 'more rational', as they can be defined (by exercising the peculiar capacity of reason mentioned in the next sentence). Insofar as they are recognized as *composite* concepts, i.e. insofar as they are defined, they are just as plain and credible as the primary ones.

83,29 f. ἀκολούθου καὶ μαχομένου οὖν ⟨διαγνωστική⟩ ἐστιν ἥνπερ ἰδίαν λόγου δύναμιν πεφυκυῖαν ἅπαντες ἔχομεν φύσει. Reasoning enables us to produce plain and credible propositions, such as definitions, and arguments, such as demonstrations (see the next note). This description of the reasoning faculty is traditional, probably of Stoic origin.[142] In the manuscript, the sentence is incomplete, as the word characterising the relation of the reasoning faculty to its object ("entailment and incompatibility") is missing. In view of similar formulations in Galen, I propose that, e.g. διαγνωστική could have been left out (more likely than διακριτική or διαιρετική proposed by Stählin).[143]

83,31–33 ἐὰν οὖν τις εὑρεθῇ λόγος τοιοῦτος ... αὐτὸν τοῦτον εἶναι φήσομεν οὐσίαν ἀποδείξεως. Unlike above, 83,29 f., where the word λόγος refers to the reasoning faculty, here it is rather an argument produced by that faculty. Insofar as it meets the definition of ἀπόδειξις above, 5, 1/82,12–14, this argument may be described as the essence of demonstration. The word 'essence' is presumably used to underscore the understanding that the text no longer speaks merely of the agreed meaning of the word ἀπόδειξις (as it does in 5, 1), but rather of

---

142 Cf. Allen, *Inference*, p. 95 and n. 11, citing SVF II 135 (from Galen); Cicero, *Leg.* I 45; *Ac.* II 22; Sextus, *M.* VIII 276.

143 Cf. Galen, *PHP* III 5, 13 (CMG V,1,4,2: 202,34–36): τὸ μὲν γὰρ λογίζεσθαί τε καὶ νοεῖν τὰ λεγόμενα καὶ τὸ *μαχόμενον ἢ ἀκόλουθον ἐπίστασθαι* τῆς λογιστικῆς δυνάμεως ἔργον ("for reasoning and comprehending what is said and understanding incompability and entailment are the work of the rational faculty"). Ibid. IX 1, 13 (CMG V,1,4,2: 542,16–18): τὴν γνώμην ἢ ἔννοιαν ἢ ὅ τί ποτ' ἂν ἐθέλῃ τις ὀνομάζειν, ᾧ *διαγιγνώσκομεν ἀκόλουθόν τε καὶ μαχόμενον* καὶ ἄλλα ἃ καταπέπτωκε τούτοις ("... thought or mind or whatever you wish to call, by which we recognize entailment and incompatibility and other things that pertain to them"). Cf. also *PHP* IX 9, 20; *Hipp. Off. Med.* I 3 (XVIIIb,649,16–650,5 K.); *Cons.* 4, 25 (CMG Suppl. III: 26,9 f.); *Subf. Emp.* 12 (87,5–7 Deichgräber). For the periphrastic construction διαγνωστική ἐστι re. δύναμις, cf. *UP* VIII 6 (I,463 Helmreich/III,640,1–3 K.): "For these faculties are such that one recognizes odours, another flavours, another sounds, another colours" (αἱ μὲν γὰρ δυνάμεις, ἡ μὲν ὀδμῶν, ἡ δὲ χυμῶν, ἡ δὲ φωνῶν, ἡ δὲ χρωμάτων ἐστὶ διαγνωστική); *Loc. Aff.* IV 7 (VIII,260,5 f. K.): ἡ δὲ αἰσθητικὴ δύναμις, διαγνωστικὴ τῶν κατὰ τὸ μόριον οὖσα παθῶν κτλ.

the signified object itself (cf. above, 3, 4/81,22–24). Exploration of this essence includes distinctions not included in the agreed definition, e.g. the distinctions between two kinds of πίστις and ἀπόδειξις, and their explanations in terms of different kinds of premisses (cf. below, 7, 7–8/83,33–84,8). The passage is echoed in *Strom.* VII (16) 98, 3: ἡ παρ' ἡμῖν δὲ τῆς ἀληθείας ἐπίγνωσις ἐκ τῶν ἤδη πιστῶν τοῖς οὔπω πιστοῖς ἐκπορίζεται τὴν πίστιν, ἥτις οὐσία ὡς εἰπεῖν ἀποδείξεως καθίσταται.

83,33–84,2 εἴρηται δὲ ὡς καὶ τὸ τῆς πίστεως καὶ τὸ τῆς ἀποδείξεως γένος διττὸν κτλ. In our text, this distinction has not really been explained yet, but it is presupposed above, 5, 2–3. The passage referred to here has probably not been preserved. Cf. above, on 82,14–18.

84,2–8 Cf. above, 6, 2–4, and notes on 83,1–7 f.

84,4–7 εἰ δ' ἐξ ἐνδόξων … συλλογιεῖται μέν, οὐ μὴν ἐπιστημονικήν γε ποιήσεται τὴν ἀπόδειξιν. Deductions from the ἔνδοξα are demonstrations of sorts, different from epistemic demonstrations. Cf. above, on 82,16–18. In contrast, above, 6, 2–4, the word ἀπόδειξις seems to be used in a more restricted sense.[144] For the ἔνδοξα, cf. above, on 83,1–3.

8, 1–3

84,9–13 ἕκαστον μὲν γὰρ τῶν ἀποδεικνυμένων διά τινων ἀποδεικνυμένων ἀποδείκνυται … ἄχρις ἂν εἰς τὰ ἐξ ἑαυτῶν πιστὰ ἀναδράμωμεν … ὅπερ ἀνάλυσις ὀνομάζεται. This explanation of analysis is reminiscent of Galen, *MM* I 4: "Each of these [premisses] needs to be proved on the basis of further premisses, which themselves are based on others still, until we arrive at the primary ones, which derive their credence neither from others, nor from demonstration, but from themselves."[145] See also below, on 91,10–12.

---

144 This practice is similar to Aristotle, who sometimes reserves the name ἀπόδειξις for epistemic deduction only (cf. *Top.* I 1, 100a27–30; *An. Post.* I 2, 71b23 f.), but on other occasions uses the word in a looser sense; cf. e.g. *Rhet.* I 1, 1355a4–7: ἡ δὲ πίστις ἀπόδειξίς τις … ἔστι δ' ἀπόδειξις ῥητορικὴ ἐνθύμημα. For various uses of the word ἀπόδειξις in Aristotle, cf. G.E.R. Lloyd, "The Theories and Practices of Demonstration," in idem, *Aristotelian Explorations* (Cambridge: Cambridge University Press, 1996), pp. 7–37.

145 X,33,14–18 K.: ἑκατέρα δὲ πάλιν τούτων ἑτέρων τινῶν εἰς ἀπόδειξιν ἐδεῖτο προτάσεων, εἶτ' ἐκεῖναι πάλιν ἑτέρων, ἄχρι περ ἂν ἐπὶ τὰς πρώτας ἀνέλθωμεν, αἱ δὲ οὐκ ἐξ ἄλλων, οὐδὲ δι' ἀποδείξεως, ἀλλ' ἐξ ἑαυτῶν ἔχουσι τὴν πίστιν. Trans. Hankinson, slightly modified. The parallel is noted by Teun Tieleman, *Galen and Chrysippus On the Soul: Argument and Refutation in* De

84,13 f. ἀπόδειξις δέ ἐστιν, ὅταν ἀπὸ τῶν πρώτων ⟨τις εἰς⟩ τὸ ζητούμενον ἀφικνῆται διὰ πάντων τῶν ἐν μέσῳ. Having reached the first premisses, we may then arrive at 'the thing sought' by showing how the proposition we wish to demonstrate follows from these first premisses by means of all the intermediate ones discovered through analysis. This explains how demonstration "furnishes credence" to something disputed (cf. above, 4, 1; 7, 6). For the 'thing sought', cf. above, on 83,11 f.

84,14 f. χρὴ ... τὸν ἀποδεικτικὸν ἄνδρα. For the vocabulary, cf. Galen, *Sem.* II 6, 27 (*CMG* V 3,1: 204,8 f./IV,649 K.): τὸν δ' ἀποδεικτικὸν ἄνδρα ... προσήκει.[146]

84,15 f. τῆς μὲν ἀληθείας [ὡς τῶν λημμάτων] πολλὴν ποιήσασθαι πρόνοιαν, τῶν δὲ ὀνομάτων ἀφροντιστεῖν. As preserved in the manuscript, the Greek text is not sound. Stählin deletes ὡς before τῶν λημμάτων, but I think it is more likely that the whole phrase ὡς τῶν λημμάτων is originally a scribal gloss, correctly explaining the vague opposition τῆς μὲν ἀληθείας ... τῶν δὲ ὀνομάτων in Clement's text.

84,16 f. εἴτε ἀξιώματά τις ἐθέλοι καλεῖν εἴτε προτάσεις εἴτε λήμματα. Purported laxness over terminology is a distinctive feature of Galen's style.[147] See also below, 12, 7/87,18. For the terminology of premisses, cf. Galen, *SMT* I 13 (XI,401, 18–402,1 K.): εἰς τὴν ἀπόδειξιν ἀρχῶν, ἢ *λημμάτων, ἢ ἀξιωμάτων, ἢ προτάσεων,* ἢ ὅπως ἂν *ἐθελήσῃς ὀνομάζειν. PHP* II 3, 12 (CMG V,4,1,2: 112,4 f./V,222 K.): περὶ τῶν *σοφιστικῶν εἴτε προτάσεων ἐθέλοις ὀνομάζειν εἴτε ἀξιωμάτων* εἴτε λόγων, οὐδὲν γὰρ εἰς τὰ παρόντα διαφέρε. For ἀξιώματα in the (Stoic) sense of propositions, see also Galen, *Inst. Log.* 3, 3 (8,18 Kalbfleisch).[148]

---

placitis, *Books II–III* (PhA 68; Leiden/New York/Cologne: Brill, 1996), 33 f. For Galen's concept of analysis, cf. Hankinson, "Galen on the Limitations of Knowledge," pp. 222–225.

146 Commenting on this passage in Galen, Phillip De Lacy writes: "LSJ cite no other passage where ἀποδεικτικός is used of a person, and I have found none" (CMG V,3,1, p. 251). But in fact, this use of ἀποδεικτικός is pretty common; cf. Galen, *Opt. Med.* 4, 1 (292,13 Boudon-Millot/I,62 K.); *PHP* II 3, 17 (CMG V,4,1,2: 112,30/V,224 K.); III 2, 9 (176,29/V,295 K.); IX 9, 18 (602,6/V,796 K.); *Ord. Lib. Prop.* 1, 12 and 2, 3 (91,1 f. and 92,4 Boudon-Millot/XIX,52.54 K.); Alexander, *In Top.* VIII 1, 155b7 (CAG II/2: 520,34 f., 521,3–4.7); *In Met.* Γ 3, 1005b5 (CAG I: 268,3).

147 Cf. Benjamin Morison, "Language," in Hankinson (ed.), *The Cambridge Companion to Galen*, p. 148.

148 For a contrast to our passage, cf. Alexander, *In An. Pr.* I 4, 25b32 (CAG II/1: 44,16–23), who points out that each of these terms (πρότασις, λῆμμα, ἀξίωμα) has a distinct meaning.

84,19 f. εἴτε δὲ περαίνοντα λόγον εἴτε περαντικὸν εἴτε συλλογιστικὸν ἐθέλοι τις ὀνομάζειν αὐτόν, ἥκιστα φροντίζειν. In Stoic logic, 'syllogistic' arguments (συλλογιστικοί) are described as a subclass of 'concludent' (i.e. valid) ones (περαντικοί), and distinguished from arguments 'concludent in the specific sense' (περαντικοὶ εἰδικῶς); cf. DL VII 77–78.[149] In *Inst. Log.* 19, 6, Galen refers to an earlier writing of his in which he refuted Chysippus' classification of 'concludent' arguments (here περαντικοί in the 'specific' sense), presumably by showing that their difference from syllogistic arguments is a matter of words, not meaning.[150] Elsewhere (*Adv. Lyc.* 4, 5/CMG V,10,3: 15,9/XVIIIa,219 K.), Galen uses the word περαντικοί in the general sense,[151] as it is used in our text.

84,20–22 δύο γὰρ ταῦτα ἐν ἅπασι χρῆναί φημι τὸν ἀποδεικτικὸν φυλάττειν, τὰ μὲν λήμματα ἀληθῆ λαμβάνειν, ἀκόλουθον δ' αὐτοῖς ἐπιφέρειν ⟨τὸ⟩ συμπέρασμα. Cf. above, 6, 2–4; 7, 8. Underlining the didactic framework of his work, the author repeats the main outcomes of the preceding discussion in the form of methodological advice. The summary marks the end of a section concerned with the definition of demonstration.

84,23 ὅπερ τινὲς καὶ ἐπιφορὰν καλοῦσιν. Aristotelian terminology is set against that of the Stoics; cf. Ammonius, *In An. Pr.* (CAG IV/6: 68,13 f.): ὃ δ' ἡμεῖς λέγομεν συμπέρασμα, ἐκεῖνοι ἐπιφορὰν καλοῦσι.

(III) 8, 4–(V) 15, 1: Method of Discovery

In this section, Clement plainly excerpts the same source as in the previous one. Indeed, continuity of topic and vocabulary is so close that editors disagree as to where exactly the borderline between the two sections should be located. But the matter is quite clear: 8, 3 is a summary of the preceding discussion, while 8, 4 starts a new theme (introduced by an asyndetic sentence). In the previous section, demonstration was shown to be a deductive argument whose premisses are plainly true. Now we are taught how to proceed in order to find epistem-

---

149 Cf. Susanne Bobzien in K. Algra, J. Barnes, J. Mansfeld, and M. Schofield (eds.), *The Cambridge History of Hellenistic Philosophy* (Cambridge: Cambridge University Press, 1999), pp. 127, 151 f.

150 Cf. Jonathan Barnes, "Galen on the Utility of Logic," in J. Kollesch and D. Nickel (eds.), *Galen und das hellenistische Erbe* (Stuttgart: Steiner, 1993), pp. 33–52, esp. 41–52; Morison, "Logic," in Hankinson (ed.), *The Cambridge Companion to Galen*, pp. 92 f.

151 Cf. Barnes, "Utility of Logic," p. 41 n. 47.

ically credible solutions to specific problems. Problems are formulated in the form of questions, two examples of which are found in our text: 'Is the embryo an animal or not?' (9, 7–13, 8) and 'In which part of the body is the ruling faculty of the soul?' (14, 4). In order to tackle these problems, one must find appropriate premisses, starting with aspects of the problem known in advance. These previously known aspects may concern a substance and/or some of its attributes, such as works/activities, affections, or powers. Our text distinguishes four kinds of problems from this point of view (9, 1–5; cf. 14, 4–15,1). The embryo question is then discussed as an example of the first kind: We know the substance and ask whether such attributes as the capacity to perceive and move by impulse hold of it or not (9, 8–9). The question is introduced as a paradigmatically "confusing" one, which, nevertheless, may be settled if approached correctly. Most of the subsequent discussion serves the purpose of showing, by means of this example, how to "recognize the problems" in order to discover demonstrative premisses appropriate to the sought conclusion. The discussion is relatively detailed, but there are some missing points, notably an explanation (implied in 13, 2, but never clearly stated) that the word 'embryo' is used in the restricted sense of a 'fetus'. Also, even though the text shows the way towards discovering the appropriate premisses, it never reaches a stage in which the premisses are in fact laid down. After a brief summary, the section abruptly ends at the beginning of a discussion of another problem (14, 1–15, 1).

8, 4–5

84,24f. **περὶ παντὸς τοῦ ζητουμένου καθ' ἕκαστον πρόβλημα.** The reader has already been instructed how to proceed if he wants to argue correctly about "everything sought" (3, 1/81,17; cf. 4, 1/81,25f.). Now these instructions are applied to the inquiry of προβλήματα. In Aristotle's *Topics*, πρόβλημα is described as a subject matter of deductive reasoning (*Top*. I 4, 101b16), a research question in the form of 'is P the case or not?' (I 4, 101b32f.; I 11, 104b5–17). While respecting this form in the example discussed below (cf. on 85,17b), our text applies the word πρόβλημα to other kinds of questions as well (cf. esp. 14, 4/29f.); at the same time, it focusses on questions that (supposedly at least) can be answered by means of demonstrative premisses (cf. below, 14, 1–15, 1). Although the word τὸ ζητούμενον is sometimes used as an equivalent of τὸ πρόβλημα (cf. esp. below, 14, 4/88,25f. and 28), here it seems to refer to any item involved in the problem that needs to be investigated prior to the solution of the problem itself.[152]

---

152 Cf. Alexander, *In An. Pr.* I 29, 45a23f. (CAG II/1: 317,25–28), summing up *An. Pr.* I 27–28 (trans. Mueller): "He has shown how in the case of each problem we can be well supplied

84,25 οἰκείων δὲ τῷ προβληθέντι. We have seen that the premisses of the demonstrative argument must be evidently true; now they are further qualified as "appropriate" to the problem. Presumably, for premisses to be appropriate to the problem, it means that they must be appropriate to the sought conclusion; this, in turn, means that they must be "primary in the kind with which the proof is concerned" (Aristotle, *An. Post.* I 6, 74b25 f.).[153] For the task of discovering appropriate premisses "in order to get to grips with problems" (πρὸς τὸ ἔχειν τὰ προβλήματα), cf. also *An. Post.* II 14–17; *An. Pr.* I 27–30.[154] In our text, the method of discovering these premisses will be outlined in the following pages. Galen tells us that he had discussed this issue extensively in *DD* (*PHP* II 2, 3/CMG V 4,1,2: 102,25–104,2; *PHP* II 3, 1–3/CMG V 4,1,2: 108,21–31); however, it is essential to any treatment of the demonstrative method.

84,25 f. ⟨τὸ⟩ προβληθὲν αὐτὸ εἰς λόγον μεταλαμβάνειν. τὸ προβληθέν picks up on 84,25 (οἰκείων τῷ προβληθέντι), where it is an equivalent of τὸ πρόβλημα (cf. 84,28: οἰκείων τῷ προβλήματι). Strictly speaking, however, the item "changed into a definition" is the proposed *name* (cf. above, 4,2/82,2: τὸ προβληθὲν ὄνομα), or rather names by which the terms of the problem are called. Cf. below, on 85,18–20. Cf. Alexander of Aphrodisias, *In An. Pr.* I 5, 26b37 (CAG II/1, 75,24–26; trans. Barnes et al.): "For everyone who syllogizes first defines for himself what he wants to prove (ὁρίσας παρ' αὑτῷ πρῶτον, ὃ βούλεται δεῖξαι) and so obtains premisses appropriate to it."

84,27–30 τῶν δὲ λημμάτων μὴ οἰκείων … λαμβανομένων … οὐκ ἐνδέχεται καλῶς οὐδὲν … ἐξευρεῖν ἀγνοουμένης ὅλου τοῦ προβλήματος … τῆς φύσεως. As it stands in the manuscript, this text is probably not what Clement wrote. The sentence (if it is one sentence) consists of two subordinate constructions in the genitive absolute, surrounding the main clause οὐκ ἐνδέχεται καλῶς οὐδὲν αὐτῷ ἐξευρεῖν. The first construction stands for a conditional protasis, which develops the idea, expressed in the previous sentence, that in order to solve a problem, we

---

with appropriate premisses for a conclusion which is proved and inferred in the primary way (…) from the examination and combination of the things selected for each of the terms in the problem" (δείξας, πῶς καθ' ἕκαστον πρόβλημα τῶν οἰκείων τοῦ συμπεράσματος προτάσεων εὐπορήσομεν, ὃ προηγουμένως δείκνυται καὶ συνάγεται … ἐκ τῆς τῶν ἐκλελεγμένων περὶ ἑκάστου τῶν ὅρων τῶν ἐν τῷ προβλήματι ἐπιβλέψεώς τε καὶ συνθέσεως).

153 Cf. *An. Post.* I 7 and the commentary by Barnes, *Posterior Analytics*, pp. 130–132.

154 For Aristotle's discussion of problems in the context of demonstrative theory, cf. James G. Lennox, "Aristotelian Problems," in idem, *Aristotle's Philosophy of Biology: Studies in the Origins of Life Science* (Cambridge: Cambridge University Press, 2001), pp. 72–97.

need to have premisses appropriate to it. It says: "If the assumed premisses are not appropriate to the problem …" The main clause then appears to function as an apodosis. This is supported by the fact that it speaks of the impossibility of "finding" something: It surely makes sense to say that if our premisses are not appropriate to the problem, we cannot find anything we seek with respect to that problem (i.e. any conclusion). But the difficulty lies with the dative αὐτῷ, which seems to have no explicit referent.[155] Eduard Schwartz (followed by Stählin) marks a lacuna between καλῶς and οὐδὲν, suggesting that each word belongs to a different clause and that the referent of αὐτῷ has been omitted.[156] However, similar constructions in Galen suggest that the words οὐκ ἐνδέχεται καλῶς οὐδὲν ἐξευρεῖν are probably parts of the same phrase.[157]

Another difficulty is presented by the second subordinate clause; this might again express a condition or provide an explanation to the main clause. In the first case, it could be translated as follows: "… if the nature of the problem, which is also called the question, is unknown as a whole." In the second, we could replace 'if' with 'because'. The second ('explanatory') option is compatible with the forementioned interpretation of the first subordinate clause, but its weakness lies in the meaning it conveys: It is odd to suggest that we cannot draw a conclusion about a problem from inappropriate premisses *because* we do not know anything about the problem. Rather, our inability to draw the conclusion is already due to the fact that we do not have appropriate premisses. The first ('conditional') option would make sense if the object of ἐξευρεῖν were not the solution to the problem, but the appropriate premisses. The sentence would then mean that it is impossible to find any such premisses, if nothing about the problem is known.[158] But this is at odds with the view that the first

---

155 We might consider linking αὐτῷ to τῷ προβλήματι in 84,28 and interpret it as the dative of interest, but this would involve an awkward personification of 'the problem'. Another candidate is ὁ ἀποδεικτικός (84,21), but the last mention of him is too far away and in a different section.

156 This is Schwartz's proposal: οὐκ ἐνδέχεται καλῶς ⟨τὸν ἀποδεικτικὸν συλλογίζεσθαι· ἀλλὰ πάντως⟩ οὐδὲν ⟨οἷον τε⟩ αὐτῷ ἐξευρεῖν.

157 Cf. Galen, *Cris.* III 5 (184,11 Alexanderson/IX,729,1f. K.): οὐκ ἐνδέχεται περὶ σωτηρίας οὐδὲν ἐξευρεῖν. *UP* IV 13 (I,226,10 Helmreich/III,308 K.): οὔτ' ἐνδέχεται καλῶς εὑρεῖν χρείαν οὐδεμίαν (similarly ibid. IX 4/II,13,3f. Helmreich/III,700 K.). Cf. also *MM* III 2 (X,172,14f. K.): οὐκ ἐνδέχεται τἀληθὲς ἐν οὐδενὶ τῶν ὄντων ἐξευρεῖν.

158 For the discovery of premisses, cf. e.g. Galen, *PHP* II 3, 3 (CMG V,4,1,2: 108,26–28/V,219 K.), referring to *DD*: "The main point was that the suitable and appropriate premisses must be discovered from the very essence of the thing sought" (ἦν δὲ τὸ κεφάλαιον ὡς ἀπ' αὐτῆς χρὴ τῆς οὐσίας τοῦ ζητουμένου πράγματος ἐξευρίσκειν τὰ προσήκοντά τε καὶ οἰκεῖα λήμματα). Cf. also ibid. II 3, 13 (112,11–12/V,222 K.); *MM* I 5 (X,42,5–9 K.): "But discoveries,

subordinate clause is also the conditional of the main clause. Moreover, it is difficult to see how ἐξευρεῖν could have the appropriate premisses as their object.

The problem with αὐτῷ could be solved by changing αὐτῷ to αὐτῶν. Referring to λήμματα, αὐτῶν would then be taken as the genitive of origin, governed by ἐξευρεῖν ('… it is impossible to succeed in finding anything *from* these premisses', i.e. 'on their basis'). But even with this emendation, the role of the second subordinate clause remains obscure. I believe that something must have been omitted in the course of the textual transmission, perhaps another clause that mentioned the discovery of appropriate premisses ('… nor is it possible to find the appropriate ones, if …').

8, 6–9, 5

84,30 f. ἐν πᾶσιν οὖν τοῖς ζητουμένοις ἔστι τι προγινωσκόμενον. The particle οὖν marks a transition to a new theme and thus the sentence is an occasion to start a new paragraph; cf. below, 9, 6/85,14; 18, 2/91,6. Examples of τὰ ζητούμενα are given below: stones, plants, animals (9, 1/85,4), and the soul (9, 2/85,7). For previous knowledge, cf. below, 9, 1–5/85,1–13.

84,32 ὁρμητήριον. Originally denoting a military base (a 'starting-point' of military operations), the word is sometimes used by Galen to describe an observed fact or an 'indication' serving as both a point of departure and a reference-point of research.[159] In our text, the 'point of departure' corresponds to the agreed meaning of the proposed name (cf. above, 3, 3; 4, 1–2).

84,33 κριτήριον. Sextus distinguishes three ways in which the word κριτήριον is used with regard to logic: that "by which" (ὑφ' οὗ), that "through which" (δι' οὗ), and that "in virtue of which" (καθ' ὅ) a judgment about truth is made; *PH* II 16; *M.* VII 35–37. In our passage, κριτήριον is a standard 'in virtue of which' we make judgments about conclusions that seem to be right (τῶν εὑρῆσθαι δοκούντων), a standard credible by itself (cf. above, 4, 1–2). For this meaning

investigations, and demonstrations of the very essence of the thing … [are drawn] from scientific premisses, the manner of whose discovery was discussed in [*On Demonstration*]" (τὰς δὲ τῆς οὐσίας αὐτῆς τοῦ πράγματος εὑρέσεις τε καὶ ζητήσεις καὶ ἀποδείξεις … ἐκ τῶν ἐπιστημονικῶν λημμάτων, ὑπὲρ ὧν τοῦ τρόπου τῆς εὑρέσεως ἐν ἐκείνοις [*scil.* τοῖς περὶ τῆς ἀποδείξεως ὑπομνήμασιν] εἴρηται).

159 Cf. *Sem.* II 5, 42 (CMG V,3,1: 188,5/IV,634 K.), on anatomical observation: ὁρμητήριον … τῆς τῶν ζητουμένων εὑρέσεως. *MM* III 1 (X,158,18 K.), on indication: ἀρχὴ … καὶ οἱονεὶ ὁρμητήριόν … τῆς θεραπευτικῆς μεθόδου (cf. X,157,2 f. K.: τὴν ἀρχὴν τῆς … εὑρέσεως). *MM* XIV 17 (X,1005,5 K.).

of κριτήριον, cf. Galen, *PHP* IX 7, 3–4 (CMG V,4,1,2: 586,16–23/V,778 K.), who attempts to harmonize the views of the Stoics and the New Academy on these matters. In *Diff. Puls.* IV 2 (VIII,708,14 K.), Galen describes a definition based on the notion of the thing defined (ὃ κατὰ τὴν ἔννοιαν, *scil.* ὅρος) as a criterion of the definition of its essence (κριτήριον ... τοῦ κατὰ τὴν οὐσίαν). As he explains earlier, the notional definition "is believed by itself, as it expresses plainly apparent things" (ἐξ αὐτοῦ πιστεύεσθαι, φαινομένων ἐναργῶς πραγμάτων ἑρμηνείαν ἔχοντα); *Diff. Puls.* IV 2 (VIII,705,10 f. K.). As our text instructs the reader to turn that which is 'known in advance' into a criterion (χρὴ ποιεῖσθαι κτλ.), it is likely that it also points towards a definition of this sort.

85,1a πᾶσα γὰρ ζήτησις ἐκ προϋπαρχούσης εὑρίσκεται γνώσεως. There is a metonymic substitution of πάντα τὰ ζητούμενα in the subject. The sentence is an echo of Aristotle, *An. Post.* I 1, 71a1 f. (πᾶσα διδασκαλία καὶ πᾶσα μάθησις διανοητικὴ ἐκ προϋπαρχούσης γίνεται γνώσεως, cf. *EN* VI 3, 1139b26), but closer parallels (involving the notion of discovery) are found in Galen (*Inst. Log.* 1, 2/3,6–8 Kalbfleisch: ⟨δεῖ δὲ τῶν δι' ἀποδείξεως γιγνωσκομένων⟩ ἐκ προγιγνωσκομένων εἶναι τὴν εὕρεσιν) and Philoponus (*In An. Post.* I 1, 71a1/CAG XIII/3: 4,8 f.: πᾶσα εὕρεσις ἔκ τινος καὶ αὐτὴ προϋπαρχούσης γίνεται γνώσεως, ὥσπερ καὶ ἡ μάθησις); see also Simplicius, *In Phys.* VIII 5, 258b4 (CAG X: 1250,2–4): πᾶσα γὰρ μάθησις διανοητική ..., καὶ ζήτησις δηλονότι καὶ εὕρεσις, ἐκ προϋπαρχούσης γίνεται γνώσεως κτλ.

85,1b–13 This paragraph introduces a distinction between οὐσία (here probably best translated as 'substance') and its attributes, mentioning three kinds of attributes—works/activities, affections, and powers.[160] Since the attributes in question are useful for the inquiry concerning substance (cf. ζητεῖσθαι τὴν οὐσίαν in 85,7 f.), they must belong to the thing under inquiry in virtue of itself. On the other hand, we are told that some things are known solely in terms of their substance, without any knowledge of these attributes (85,2–5). Thus the text seems to imply a distinction between two kinds of attributes belonging to an object in virtue of itself, namely those that characterize it in

160 It is unclear if the text recognizes any distinction between 'works' (ἔργα) and 'activities' (ἐνέργειαι); if it does, it seems to regard ἐνέργειαι as instances of ἔργα; cf. 85,3 f. Cf. Galen, *Nat. Fac.* I 2 (105,13–106,3 Helmreich/II,6 f. K.), who points out that every ἐνέργεια can be called ἔργον, but not the other way round, as ἔργον may also refer to a product. However, even if our text recognizes this distinction, it probably uses the word ἔργον in the narrower sense of ἐνέργεια, as it is arguably in this sense only that ἔργον can be a *per se* attribute. I am grateful to Katerina Ierodiakonou for drawing my attention to this issue.

terms of its substance, and those that characterize it in terms of such things as works/activities, affections, or powers. Cf. Aristotle, *DA* I 1, 402a12–15, who distinguishes between the method of finding out what something is (τὸ τί ἐστι) and "demonstration concerning incidental properties" (τῶν κατὰ συμβεβηκὸς ἰδίων ἀπόδειξις); the properties being referred to, in the case of the soul, as its πάθη (cf. 402a7–10). Cf. also *DA* I 1, 402b16–403a1, where Aristotle says that the knowledge of 'what is' is useful for the understanding of things "incidental upon substances" (τῶν συμβεβηκότων ταῖς οὐσίαις) and, *vice versa*, the study of these properties contributes to the knowledge of 'what is'.[161] An interesting parallel is found in Galen, who argues that members of the same kind share one form (ἓν εἶδος) in virtue of which they are called by the name of that kind (e.g. 'dogs'), adding that "this is the case not only in regard to their substances (οὐ μόνον ταῖς οὐσίαις), but also in regard to the things that are incidental upon them, be they dispositions, activites, or affections (καὶ τοῖς συμβεβηκόσιν αὐτοῖς ἢ ὡς διαθέσεσιν, ἢ ὡς ἐνεργήμασιν, ἢ ὡς παθήμασιν)".[162] For the distinction between activities and affections, cf. Galen, *PHP* VI 1, 5–7 (CMG V,4,1,2: 360,22–362,2/V,507 K.); *Symp. Diff.* 1, 3 (CMG V,5,1: 200,18 f./VII,44 K.): ἡ μὲν γὰρ τοῦ ποιοῦντος κίνησις ἐνέργεια, πάθημα δὲ καὶ πάθος ἡ τοῦ διατιθεμένου πως ὑπ' αὐτοῦ. The typology of problems outlined in this passage is echoed below, 14, 4/88,25–30.

85,1 f. εἶναι δὲ τὴν γνῶσιν. The structure of the following sentence (until 85,10) consists of a series of constructions in the accusative and the infinitive, accompanied by the genitives-absolute, unruled by any verb. This is admittedly odd. Schwartz attempts to make things better by adding δυνατόν (*scil.* ἐστι) in 85,2; but the same problem reappears in 85,5–8 and, even if we convince ourselves that the infinitives in that clause (γιγνώσκεσθαι ... ἀγνοεῖσθαι δὲ καὶ ζητεῖσθαι) could still be governed by the element added by Schwartz, it is very hard to believe the same thing in the following clause (85,8–10). Thus, presumably, we have to put up with the fact that the syntax is elliptic, as it might suit the genre of excerpts. The most likely candidate for the unexpressed main verb is, in my view, λέγεται, φησί ("it is said", "he says") or something to that effect.

---

161 These properties seem to correspond to what Aristotle describes as '*per se* accidents' (τὰ καθ' αὐτὸ συμβεβηκότα), counting them among items shown in demonstrations; cf. *An. Post.* I 7, 75b1; I 22, 83b19–20; *Met.* Δ 30, 1025a30–32; *PA* I 3, 643a27–31. Cf. also Alexander, *In Met.* B 1 (CAG I: 176,19–30); *In Met.* K 1 (CAG I: 634,39–635,28).

162 *MM* II 7 (X,129,17–130,2 K.), trans. Hankinson, slightly modified. See also Hankinson's commentary *ad loc.*

85,4 λίθων, φυτῶν, ζῴων, ὧν τὰς ἐνεργείας ἀγνοοῦμεν. The exploration of stones, plants, and animals is an important part of pharmacology; cf. Galen, *SMT* VI prooem. (XI,791,15–792,4 K.) and the respective subjects of books VI–VIII (plants), IX (stones), and X–XI (animals). In this context, the research is concerned with their 'powers' (δύναμεις) or 'works' (ἔργα, ἐνέργειαι); see, for example, *SMT* VIII prooem. (XII,83,1–5 K.); IX 2 (XII,192,4–11 K.).

85,7f. ὡς τῆς ψυχῆς τὰς ἐπιθυμίας καὶ τὰ πάθη (*scil.* γιγνώσκεσθαι), ἀγνοεῖσθαι δὲ καὶ ζητεῖσθαι τὴν οὐσίαν. Inquiry about the substance of the soul is notoriously difficult; cf. Aristotle, *DA* I 1, 402a7–11. Our text does not indicate that it is an impossible task, but claims that it should start with known activities and affections of the soul, that is to say, with activities and affections whose existence is normally acknowledged and which are normally believed to belong to the soul. ἐπιθυμία is probably used in the general sense of 'desire', including desires pertaining to the rational part of the soul; cf. Galen, *QAM* 2 (35,16–36,8 Müller/IV,772 K.). A different question about the soul is proposed below, 14, 4/88,29f.

85,8–12 No example of the third kind of problems is given in our text (despite the opinion of Stählin, cf. below, on 88,30–33).

85,12f. The fourth kind of problems is presumably concerned with objects whose known activities are caused by unknown affections (the text hardly allows us to be more specific than that).

9, 6–9

85,14 ἡ μέθοδος τῆς εὑρέσεως. What follows until 13, 8/88,16 is a series of instructions about 'the method of discovery'. What is it a discovery of? It seems safe to assume that the discovery concerns that which is 'sought' in the problem, being an answer to the question proposed for inquiry (cf. 8, 6/84,32f.: τῶν εὑρῆσθαι δοκούντων). But we have seen that this answer cannot be merely an assertion (cf. 4, 1/81,25f.); if it is to be epistemically credible, it must have the form of an argument which meets the criteria of proof. Presumably, then, the 'method of discovery' is a method of finding such an argument.[163] As already

163 This was presumably also the topic of Galen's (lost) treatise *On Demonstrative Discovery* (περὶ τῆς ἀποδεικτικῆς εὑρέσεως), mentioned in *Lib. Prop.* 14, 22 (168,17f. Boudon-Millot/XIX,44 K.) and *MM* VII 5 (X,469,14 K.). For the Aristotelian background, cf. *An. Pr.* I 30, 46a24–26 (trans. Striker, slightly modified): "For if nothing that truly belongs to

mentioned, it involves such procedures as the definition of the problem and the discovery of premisses appropriate to it (cf. 8, 4/84,24–27).

85,15 γνωρίζειν τὰ προβλήματα. Cf. below, 9, 8/85,18–20.

85,15f. πολλάκις γοῦν ἐξαπατᾷ τὸ τῆς λέξεως σχῆμα καὶ συγχεῖ καὶ ταράττει τὴν διάνοιαν. In Aristotle's *Sophistical Refutations*, σχῆμα λέξεως is a name attached to a mode of producing a false impression in sophistical arguments, one of the six modes described as depending on λέξις; it is based on propositions that express things of one kind in a linguistic form suggestive of a different kind (e.g. a different gender or category). Cf. *Soph. El.* 4, 165b27; 166b10–19. Galen follows Aristotle's definition in his *Linguistic Sophisms* (*Soph.* 1/XIV,583,15–584,4 K.), but in *PHP* he employs the phrase σχῆμα λέξεως in a less rigorous manner, with regard to premisses formulated in a loosely ambiguous sense; see esp. *PHP* II 5, 26–30 (CMG V 4,1,2: 132,23–134,7/V,245f. K.), where the misleading element in the σχῆμα is the ambiguous preposition ἀπό. In our text, σχῆμα λέξεως is used in a similar connection, namely of a formulation which includes ambiguous elements (here the words ζῷον and τὸ κυούμενον). For the confusing effect of such formulations on the mind, cf. also Galen, *SMT* III 12 (XI,569,3f. K.): ἡ τῶν ὀνομάτων χρῆσις ταραχθεῖσα καὶ τὴν τῶν πραγμάτων ἐπιταράττει γνῶσιν. Cf. also below, 11, 2/86,23f.

85,17a ἐκ ποίας ἐστὶ διαφορᾶς, 'of what kind it is'; the subject of ἐστί presumably being the designated thing. Cf. below, 12, 5/87,14: ποίου κυνός;

85,17b εἰ ⟨ζῷον ἢ⟩ μὴ ζῷον τὸ κυούμενον. Stählin deletes μή, but Schwartz's correction is more likely both on textual grounds and because it provides the question with the proper Aristotelian form of a πρόβλημα (as opposed to πρότασις); cf. above, on 84,24f. In slightly different words (εἰ ζῷον τὸ κατὰ γαστρὸς ἢ οὐ ζῷον), the problem is mentioned in *Strom.* V (1) 5, 3 as an example of "convertible statements" (τὰ ἀντιστρέφοντα), explained as statements "that can equally be proposed by those who argue for the opposite thesis" (ἃ καὶ τοῖς τὸν ἐναντίον χειρίζουσι λόγον ἐπ' ἴσης ἔστιν εἰπεῖν); cf. Aulus Gellius, *Noct. Att.* IX 16. In the second century B.C., the same or a similar question was reportedly used by

---

the things has been left out in the collection of observations, we will be in a position to discover the demonstration and demonstrate anything that admits of demonstration (ἕξομεν περὶ ἅπαντος οὗ μὲν ἔστιν ἀπόδειξις, ταύτην εὑρεῖν καὶ ἀποδεικνύναι)." The method is expounded in *An. Pr.* I 27–30; cf. also the summary of Alexander, *In An. Pr.* I 31, 46a31–33 (CAG II/1: 333,12–18).

Carneades to illustrate his method of arguing persuasively on both sides of an issue.[164] The query whether the embryo is an animal has a long history reaching back to Aristotle (cf. esp. *GA* II 3) and beyond.[165] Various philosophical and medical views of the matter were transmitted in the doxographic literature.[166] There is an interesting testimony in Ps.Galen, *Def. Med.* 445 (XIX,452,13–17 K.), according to which Asclepiades of Bithynia (first cent. B.C.) was also occupied with the embryo question, giving it an ambiguous answer: Ἀσκληπιάδης δὲ οὔτε ζῶον, οὔτε μὴ ζῶον εἶπεν τὸ ἔμβρυον. Cf. also Ps.Galen, *An Ut.* 1 (XIX,158,3–5 K.): … τοῖς Ἀσκληπιάδαις καὶ τοῖς τούτων ἐκγόνοις ἡ τοῦ γιγνομένου καὶ συνισταμένου βρέφους ἐν μήτρᾳ ἀμφίβολος ἔννοια. For Asclepiades, see further below, on 87,31 f. Galen informs us that he had discussed the embryo question in *PHP* (in a part now lost) and, significantly, in *DD* (*UP* XV 5/II,357,24–28 Helmreich/IV,238 f. K.). See further below, on 87,4–7.

85,18–20 ἔχοντες γὰρ καὶ ζῴου τι νόημα καὶ κυουμένου ζητοῦμεν εἰ … τῇ κυουμένῃ οὐσίᾳ τό τε κινεῖσθαι δύνασθαι καὶ ἔτι τὸ αἰσθάνεσθαι ὑπάρχει. This is a summary of the argument developed in the subsequent paragraphs (10, 1–13, 8); the goal of the argument, insofar as it is preserved here, is not to solve the embryo question, but to instruct the readers, by means of this example, how to "recognize

164 Cf. Numenius, fr. 27 Des Places (= Eusebius, *Praep.* XIV 8, 4–5 Mras), trans. George Boys-Stones: "Carneades … did not make any concession however small, unless he could render his opponents impotent by allowing as plausible what he called positive or negative impressions of some particular thing being an animal or not (κατὰ τὰς ἀπὸ τοῦ πιθανοῦ λεγομένας αὐτῷ θετικάς τε καὶ ἀρνητικὰς φαντασίας τοῦ εἶναι τόδε τι ζῷον ἢ μὴ ζῷον εἶναι)." In his draft translation of the Numenius fragments, published on-line at academia.edu (2014, p. 21), Boys-Stones suggests that the words "of some particular thing being an animal or not" entered the text as a gloss. I find it more likely that something has been lost in Eusebius' report, e.g. οἷον after φαντασίας.—Unfortunately, we cannot be sure if Carneades' argument dealt with embryos or other subjects, as the question 'if X is an animal or not' could have taken various forms. See e.g. Ps.Plutarch, *Plac.* 874f5 f. = 'Aetius' I prooem. 3 (*Dox. Gr.* 274,6–7): ζητεῖ τις εἰ ζῷον ἢ μὴ ζῷον ὁ ἥλιος. Ps.Galen, *An Ut.* 1 (XIX,158 K.): ὡς γὰρ οἱ τοῦ ὅλου τὴν ἔρευναν ποιούμενοι οὐδὲν σαφῶς ὡρίσαντο πότερον ζῷον ἢ μή, τὸν αὐτὸν τρόπον καὶ οἱ τῆς τέχνης τῆς ἰατρικῆς προϊστάμενοι ἀόριστον τὸν ὑπὲρ τοῦ φυομένου παιδίου παρέδοσαν λόγον.

165 Cf. Marie-Hélène Congourdeau, *L'embryon et son âme dans les sources grecques* (*VIe siècle av. J.-C.–Ve siècle apr. J.-C.*) (Paris: Association des amis du Centre d'histoire et civilisation de Byzance, 2007).

166 Cf. Jaap Mansfeld, "Doxography and Dialectic: The *Sitz im Leben* of the 'Placita'," in *Aufstieg und Niedergang der römischen Welt*, II: Principat, Band 36.3, ed. W. Haase (Berlin/New York: de Gruyter, 1990), pp. 3056–3229, here 3184–3193.

the problems" (9, 6/85,15). First of all, each term of the problem is defined on the basis of the concept (νόημα) signified by its name, no doubt in a way that is agreed and epistemically credible (cf. above, 3, 1; 4, 2; 8, 6). These definitions articulate our previous knowledge about the investigated object while showing which attributes we need to find in that object in order to solve the problem; cf. 13, 8/88,13: τοῦτο δέ ἐστι τὸ ζητούμενον. For the investigation of these attributes, cf. below, 13, 2/87,26 f. For the implied definition of 'animal', cf. below, 10, 6/86,11 f.

85,20 f. ὥστε ἐνεργειῶν ἐστι καὶ παθῶν ἡ ζήτησις ἐπὶ προγινωσκομένης οὐσίας. Cf. above, 9, 1. In this case, 'activities' and 'affections' correspond to movement by impulse and to perception, respectively (cf. 9, 8; 11, 2; 13, 3–5.7).

10, 1–11, 1

85,23 f. εἰς διαφόρους χρήσεις ἠγμένον ... τοὔνομα. This seems to be a characteristically Galenic expression.[167]

85,24 f. Cf. above, 3, 3; 4, 1–2.

85,25 f. εἰ γὰρ ζῷον εἴποι καλεῖν ὅτιπερ ἂν αὐξάνῃ καὶ τρέφηται. This proposal is based on two assumptions: (a) 'living' is defined by nutrition and growth; (b) every living being is a ζῷον. For the first assumption, cf. Aristotle, *DA* II 1, 412a14 f. (ζωὴν δὲ λέγομεν τὴν δι' αὑτοῦ τροφήν τε καὶ αὔξησιν καὶ φθίσιν); for the second, cf. Plato, *Tim.* 77b1–3 (πᾶν γὰρ οὖν ὅτιπερ ἂν μετάσχῃ τοῦ ζῆν, ζῷον μὲν ἂν ἐν δίκῃ λέγοιτο ὀρθότατα). The proposal already insinuates the view, developed later in the argument, that the form of expression of the problem (cf. above, 9, 7/85,15 f.) may be analysed according to two meanings of the word ζῷον, a broader one including beings that live in the above-mentioned sense,[168] and a

---

167 Cf. *Symp. Diff.* 1, 2 (CMG V,5,1: 200,4–6/VII,43 K.): *παρὰ δὲ τὸ διακεῖσθαί πως τὸ τῆς διαθέσεως ὄνομα γέγονεν, εἰς ταύτην ἠγμένον τὴν χρῆσιν οὐχ ὑπὸ τῶν φιλοσόφων μόνων τῶν παλαιῶν, ἀλλὰ καὶ τῶν ἄλλων Ἑλλήνων.* Cf. *SMT* IX 2, 21 (XII,208,8 K.): εἰς χρῆσιν ἠγμένων [*scil.* τῶν λίθων]; ibid. XI 2, 11 (XII,377,3 f. K.): πρὸς χρῆσιν ἠγμένος [*scil.* ὁ σπόγγος].

168 In the *Topics*, Aristotle ascribes a similar definition of life ("a movement of a creature sustained by nutriment, congenitally present with it") to a certain Dionysius, probably his contemporary Dionysius of Chalcedon. Interestingly, Aristotle criticizes the definition for treating the word ζωή as if it applied to animals and plants synonymously (in the Aristotelian sense), whereas "life is generally understood to mean not one kind of thing only, but to be one thing in animals and another in plants" (*Top.* VI 10, 148a26–36); trans. Pickard-Cambridge. For the relation between the two passages, cf. Oliver Primavesi,

narrower one limited to beings that perceive and move by impulse. The second option is proposed below, 10, 6/86,10–12.

85,28 ἐπιδεικνύειν ἤδη χρή. Once the problem has been 'recognized' (cf. above, on 85,18–20), it is time to show whether the sought attribute holds of the object. Cf. below, 11, 2/86,23, where this phase of inquiry is described as ἡ διερεύνησις τοῦ πράγματος. See also below, on 87,12. However, with this definition of ζῷον, the attempted demonstration would miss the essence of the problem; cf. below, 13, 8/88,14–16.

85,29–86,1 Πλάτων γὰρ καὶ τὰ φυτὰ ζῷα καλεῖ "τοῦ τρίτου τῆς ψυχῆς εἴδους", τοῦ ἐπιθυμητικοῦ, μόνου μετέχοντα. Cf. Plato, *Tim.* 77b3–6. Ps.Galen (= Porphyry?), *Gaur.* 4, 7 (39,19–22 Kalbfleisch), quotes the same *Timaeus* passage as evidence that Plato includes plants among ζῷα. Plato was commonly regarded as a champion of the view that embryos are animals (cf. Ps.Plutarch, *Plac.* 907c1–3 = 'Aetius' V 15, 1/*Dox. Gr.* 425,15–17: Πλάτων ζῷον τὸ ἔμβρυον· καὶ γὰρ κινεῖσθαι ἐν τῇ γαστρὶ καὶ τρέφεσθαι), probably on the basis of such passages as *Tim.* 91b2–7, where he says that human seeds have ζωτικὴ ἐπιθυμία, and *Tim.* 91d2–5, where he describes them as ζῷα, nourished and formed in the womb. Cf. Ps.Galen, *Gaur.* 4, 1 (38,1–2 Kalbfleisch), on people who argue that if, according to Plato, the seeds have τὸ ἐπιθυμητικόν, then they must already have some perception and impulse. It was important for the critics of this view to show that Plato's understanding of ζῷον (and ἐπιθυμία in this connection) is such that it includes plants. Our text will allow for the possibility that embryos are animals, but not on the grounds that Plato calls them ζῷα or because they have τὸ ἐπιθυμητικόν (here presumably identified with the vegetative soul; see the next note).

86,1–5 Ἀριστοτέλης δὲ τῆς φυτικῆς τε καὶ θρεπτικῆς ψυχῆς μετέχειν οἴεται τὰ φυτά, ζῷα δ' ἤδη προσαγορεύειν οὐκ ἀξιοῖ … οὐ μὴν οἵ γε Στωϊκοὶ τὴν φυτικὴν δύναμιν ἤδη ψυχὴν ὀνομάζουσιν. For Aristotle's distinction between plants and animals, see e.g. *DA* II 2, 413a25–b4 (and references below, on 86,11 f.). The remark on the Stoics is superfluous and seems to be added mainly for the completeness of a conventional 'doxographic' pattern (but see below, on 86,12–18). For the Stoic view of plants, cf. *SVF* II,708–710.715. The term φυτικὴ ψυχή, common in Clement's time, is not found in Aristotle, but he does use the expression τὸ

"Dionysios der Dialektiker und Aristoteles über die Definition des Lebens," *Rheinisches Museum für Philologie* 135 (1992), pp. 246–261.

φυτικόν (cf. *EN* I 13, 1102a32), describing it as μέρος ψυχῆς (*EE* II 1, 1219b37; cf. *Mag. Mor.* I 4, 7). Our text seems to draw a parallel between the appetitive soul in the reference to Plato (86,1: τοῦ ἐπιθυμητικοῦ [τῆς ψυχῆς εἴδους] μόνου μετέχοντα [τὰ φυτά]) and Aristotle's vegetative soul.[169] The same parallel is drawn by Galen, in similarly phrased passages; cf. *PHP* VI 3, 7 (CMG V 4,1,2: 374,13–19/V,521 K.): κατὰ δὲ τὸ ἧπαρ ἐπιθυμητικὴν ἢ ὡς οἱ περὶ τὸν Ἀριστοτέλην, *θρεπτικὴν ἢ φυτικὴν* ἢ γεννητικήν (…) ὁ Πλάτων δ' ἀπὸ τοῦ πλήθους τῶν ἐπιθυμιῶν *ἐπιθυμητικήν*. οἱ δὲ Στωϊκοὶ *οὐδὲ ψυχὴν* ὅλως *ὀνομάζουσι* τὴν τὰ φυτὰ διοικοῦσαν, ἀλλὰ φύσιν. *Foet. Form.* 6, 31 (CMG V 3,3: 104,17–19/IV,700 K.): *φυτικὴν* μὲν ὑπὸ τῶν περὶ τὸν Ἀριστοτέλην καλουμένην, *ἐπιθυμητικὴν* δὲ ὑπὸ Πλάτωνος. ὑπὸ δὲ τῶν Στωϊκῶν *οὐδὲ ψυχὴν* ὅλως, ἀλλὰ φύσιν ἡγουμένων.[170] See also below, 10, 7–8.

86,5–10 Having acknowledged that he applies the name 'animal' to "everything that grows and is nourished" (cf. 10, 2/85,26), the opponent is now asked to admit that he uses the word in a sense broad enough to include plants. Indeed, it is difficult to escape this conclusion, unless one denies that plants have the capacity to grow and be nourished, or unless the definition of 'animals' in terms of these capacities is further qualified. If the opponent rejects it nevertheless (ἀποφήσαντος δὲ τοῦ προβαλόντος εἶναι ζῷα καὶ τὰ φυτά), it might be because he believes (or wishes to argue) that embryos are animals, but neither in the narrow nor in the broad sense of 'animals' mentioned above. A solution to this dilemma will be offered below, 13, 3–7.

86,9f. τὸ τρεφόμενόν τε καὶ αὐξανόμενον καὶ ζῷόν ἐστι καὶ οὐ ζῷον. The absurdity of the opponent's view is revealed by the implication that "that which is nourished and grows is an animal and is not an animal at the same time". This is a conjunction of two propositions, which might be taken to hold universally or in part of their subject. But they cannot both hold universally: The opponent, it is true, has been made to concede that 'everything that is nourished and grows is an animal'. But the case of a plant, which he does not believe to be an animal, forces him to admit at most that 'some things that are nourished and grow are not animals'. Should we then take it that the first predicate (ζῷον) belongs to all members of the subject class, but the second (οὐ ζῷον) only to some of them? Perhaps we should. Alternatively, we might think that the members to which they both belong are the same instances of the subject class, namely plants.[171]

---

169 For Aristotle's view on the difference between the two terms, cf. *EN* I 13, 1102b29: τὸ μὲν γὰρ φυτικὸν οὐδαμῶς κοινωνεῖ λόγου, τὸ δ' ἐπιθυμητικὸν καὶ ὅλως ὀρεκτικὸν μετέχει πως.

170 Cf. Mansfeld, *Doxography*, pp. 3187–3190.

171 I am sympathetic to Marwan Rashed's suggestion of adding φυτόν after αὐξανόμενον. But

86,11 f. εἰ αἰσθήσεώς τινος ἢ καὶ τῆς καθ' ὁρμὴν κινήσεως αὐτῷ μέτεστιν. Here a second option is proposed to the opponent. It could suffice for a being to partake (a) of "some sensory perception" or (b) of "movement by impulse" in order to be identified as an animal. Although (a) and (b) are presented as alternatives (as they are, even more strongly, below, 13, 4/87,32–88,1), it is hardly the case that they correspond to two independently sufficient conditions; rather, they describe the same condition from two different angles.[172] For Aristotle, the distinguishing feature of animals is perception (cf. e.g. *PA* III 4, 666a34; *GA* II 1, 732a12 f.; II 4, 741a9 f.: διαφέρει τὸ ζῷον τοῦ φυτοῦ αἰσθήσει) and, although perception is perhaps possible without movement (cf. *DA* II 2, 413b2–4), it is difficult to imagine how 'movement by impulse' could dispense with perception (cf. *DA* II 3, 414b1 f.: εἰ δὲ τὸ αἰσθητικόν, καὶ τὸ ὀρεκτικόν). Cf. also *Mag. Mor.* I 4, 9, where it is explained that the vegetative part of the soul does not have ὁρμή. For the vocabulary, cf. Galen, *Mot. Musc.* I 1 (IV,372,9–11 K.): *αἴσθησίν τε καὶ κίνησιν τὴν καθ' ὁρμήν*, οἷς τὸ ζῶον τοῦ μὴ ζώου διαφέρει. *Prop. Plac.* 13, 7 (CMG V 3,2: 108,12–14/IV,759 K.): ἀλλ' ἐὰν ἔρηταί μέ τις, τί πλεονεκτεῖ τῶν φυτῶν τὰ ζῶα, *τήν τε αἴσθησιν* λέγω καὶ *τὴν καθ' ὁρμὴν κίνησιν*.[173] Cf. also Ps.Galen, *Gaur.* 4, 1 (38,1 f. Kalbfleisch); Congourdeau, *L'embryon et son âme*, p. 291. The definition seems to go back to the Stoics.[174]

---

the same meaning could arguably be extracted from the transmitted text, if we take the expression τὸ τρεφόμενόν τε καὶ αὐξανόμενον in the sense of 'something that is nourished and grows'.

172 Cf. also above, 85,19 f.: τό τε κινεῖσθαι ... καὶ ἔτι τὸ αἰσθάνεσθαι. In our passage, it could even appear that the contrast is drawn not between perception and movement, but between perception on the one hand and perception *plus* movement (ἢ *καὶ* τῆς ... κινήσεως) on the other. But *καί* merely marks a 'balanced contrast', without much of a force of its own (cf. Smyth, *Greek Grammar* § 2862); the same applies below, 13, 4/87,32.

173 The reading is confirmed by the Vlatadon manuscript; cf. Véronique Boudon-Millot and Antoine Pietrobelli, "Galien ressuscité: édition princeps du texte grec du *De propriis placitis*," *Revue des études grecques* 118 (2005), pp. 168–213, here 187, ll. 2–4.

174 Cf. Galen, *PHP* II 3, 4 (CMG V,4,1,2: 110,1 f./V,219 K.), pointing out that the Stoics define τὸ ἡγεμονικόν as "the beginning of perception and impulse" (τὸ κατάρχον αἰσθήσεώς τε καὶ ὁρμῆς). Cf. SVF II 821 (from Arius Didymus): ἔχειν δὲ πᾶσαν ψυχὴν ἡγεμονικόν τι ἐν αὐτῇ, ὃ δὴ ζωὴ καὶ αἴσθησίς ἐστι καὶ ὁρμή. Cf. also Philo of Alexandria, *Quod Deus* 41–44 (= SVF II 458 part). For Aristotelian precedents, cf. e.g. *Mot. An.* 701a4–6 (trans. Nussbaum): "For the animal moves and progresses in virtue of desire or choice (ὀρέξει ἢ προαιρέσει), when some alteration has taken place in accordance with sense-perception or *phantasia* (κατὰ τὴν αἴσθησιν ἢ τὴν φαντασίαν)." Cf. LS I, pp. 321 f.

86,12–18 This is a similar account to the one already presented (10, 3–4/85,9–86,5), but it has a different function. Whereas the previous overview mainly serves the purpose of showing that the word ζῷον can be used in two different ways (the crucial point being the difference between Plato and Aristotle), this one focusses rather on the role of sense-perception in the definition of 'animal' (and the crucial point is the agreement between Aristotle and the Stoics); see below, on 86,16–18. The opponent is thereby encouraged to accept the second proposal offered in 10, 6/86,11 f.

86,15 οὐσία ἔμψυχος αἰσθητική. Although not found in Aristotle himself, this definition of ζῷον is widespread in the scholarly and doxographic literature of Clement's time; cf. Clement, fr. 38 (GCS 17: 219,28); Alexander, *In Met.α* 2, 994b20 (CAG I: 163,1 f.); *In Top.* V 4, 132a28 (CAG II/2: 391,6 f.); Sextus, *PH* II 224; Hippolytus, *Ref.* VII 18, 4; Ps.Galen, *Def. Med.* 28 (XIX,355,9 K.), etc.[175]

86,16–18 εἰ τοίνυν ἔμψυχον τὸ ζῷον ... δῆλον ὡς αἰσθητικὸν ἤδη τὸ ζῷον. According to the manuscript, the argument runs as follows: 'Animals are ensouled; souls are sentient natures; therefore, ensouled things are sentient.' But this is a remarkably bad argument, whose first premiss is irrelevant, as the second suffices for the conclusion to follow; and it misses the point of the discussion, which is concerned with the definition of 'animal'. Following Barnes, I read ζῷον instead of ἔμψυχον in 86,18.

11, 2–12, 3

86,20 f. ἔχει τὴν ἀπόκρισιν. Cf. above, 10, 2/85,28 f.

86,22 f. σαφὴς γίνεται ἡ διερεύνησις. Cf. above, on 85,18–20, and below, on 87,12. Cf. also above, 2, 2/80,18–20: οὐδὲ διερευνήσασθαι μέν, οὐχὶ δὲ ... δι' ἐρωτήσεως εἰς σαφήνειαν ἄγοντα τὸ ζητούμενον.

86,23 f. τῆς περὶ τοὔνομα ἀπάτης. Cf. above, 9, 7/85,15 f.

86,27 ἐριστικὸς ὑπάρχων γνωρισθείη. Cf. Aristotle, *Top.* VIII 11, 161b2–5 (trans. Pickard-Cambridge): "[F]or both he who asks contentious questions (ἐριστικῶς ἐρωτῶν) is a bad dialectician (φαύλως διαλέγεται), and also he who in answering

---

175 In *Top.* V 5, 135a16–19, Aristotle mentions οὐσία ἔμψυχος as an example of a 'property' (τὸ ἴδιον) of τὸ ζῷον; commenting on this passage, Alexander points out that αἰσθητική should be added (*In Top.* V 5, 135b16/CAG II/2: 405,2–5). Does this passage in the *Topics* draw on a definition of 'animal' common in Aristotle's time?

fails to grant the obvious answer (μὴ διδοὺς τὸ φαινόμενον) or to accept whatever question the questioner wishes to put." In our text, the 'eristic' nature of the opponent is revealed by the fact that he refuses to commit himself to any definition of the proposed word (thus preventing the possibility of establishing an agreed starting-point and leaving any statement about the problem open to attack).

86,28 f. ἑτέρου μὲν τοῦ κατ' ἐρώτησίν τε καὶ ἀπόκρισιν, ἑτέρου δὲ τοῦ κατὰ διέξοδον. 'Question and answer' is a traditional description of the dialectical method of argument, as opposed to 'exposition' (διέξοδος), which is more suitable to rhetoric; cf. DL VII 42; Alexander, *In Top.* I 1 (CAG II/2: 5,7–13). Cf. already Plato, *Protag.* 334c–338e. In our text, the phrase κατὰ διέξοδον refers more specifically to a didactic exposition; cf. Aristotle, *Soph. El.* 10, 171a38–b2, on the difference between the didactic and the dialectical argument: "[H]e who argues didactically should not ask questions but make things clear himself (μὴ ἐρωτᾶν ἀλλ' αὐτὸν δῆλα ποιεῖν), the other should merely ask questions." The exposition follows below, 12, 4–13, 5. Galen often uses the phrase κατὰ διέξοδον in reference to a detailed didactic account (distinguished from an outline or a summary of a teaching); cf. e.g. *Inst. Log.* 11, 2 (24,23–25,1 Kalbfleisch): ⟨τὸ⟩ νῦν γὰρ ὑπογραφή ἐστι τῆς λογικῆς θεωρίας, οὐ κατὰ διέξοδον διδασκαλία.

86,32 f. εἰ δὲ διακόπτειν ἐπιχειροίη τὴν ἐξέτασιν πυνθανόμενος, δῆλός ἐστιν οὐδὲ ἀκούειν βουλόμενος. It might seem odd that the opponent is not allowed to ask any questions in the course of the exposition, but the worry presumably is that such questions would disrupt the continuity of the argument and divert attention away from its subject. This, for example, is the reason why, in the treatise *On Bones*, Galen prefers to explain the meaning of the words he plans to use in his lecture before it starts.[176] In his book *On Crises*, Galen attributes the tendency to "thwart and interrupt the argument" to people "trained in the skill of contradiction", naming "the amazing Asclepiades" in this connection. For Galen, the problem with these interruptions is that they focus on the way we speak about something (in this instance, different kinds of 'crisis' and the manner of their discovery) rather than on the thing itself.[177] See also below, 12,

176 Galen, *Oss.* prooem. (II,734,1–5 K.): χρὴ δὲ ἴσως καὶ περὶ τῶν ἄλλων ὀνομάτων, οἷς χρησόμεθα κατὰ τὸν λόγον, ἤδη διελθεῖν, ὅπως μὴ μεταξὺ τῆς διηγήσεως χρωμένων αὐτοῖς ἢ ἀσαφὲς γένηται τὸ λεγόμενον, *ἢ τὸ συνεχὲς τῆς διδασκαλίας διακόπτηται*, τὸ παραπῖπτον ἀεὶ σαφηνίζειν προαιρουμένων ἡμῶν.

177 *Cris.* III 8 (188,10–14 Alexanderson/IX,735,1–6 K.): εἰ δὲ καὶ πόθεν εὕροιμεν αὐτὰς [*scil.* τὰς κρίσεις] ἐπιχειρούντων ἡμῶν διδάσκειν ἠσκηκώς τις ἀντιλογικὴν ἕξιν, οἵαν καὶ ὁ θαυμάσιος

7, on those "concerned with names". For Asclepiades, see above, on 85,17b, and below, on 87,31f.

87,1f. πρῶτον ἐρωτητέον ... ἐφ' ὅ τι φέρει πρᾶγμα τὸ ... ὄνομα. This is another phrase reminiscent of Galen's style; cf. *Nat. Fac.* I 2 (101,18–102,3 Helmreich/II,2 K.): ἀλλὰ *πρότερόν γε διελέσθαι τε χρὴ καὶ μηνῦσαι σαφῶς ἕκαστον τῶν ὀνομάτων*, οἷς χρησόμεθα κατὰ τόνδε τὸν λόγον, καὶ *ἐφ' ὅ τι φέρομεν πρᾶγμα*. *MM* I 4 (X,35,17–36,1 K.): ἐχρῆν ἀρχηγοὺς αἱρέσεως καθισταμένους αὐτοὺς ἐξηγήσασθαι *πρότερον ἐφ' ὅτου πράγματος ἕκαστον τῶν ὀνομάτων ἐπιφέρουσιν*. *MM* II 7 (X,144,3f. K.): ... *ἐρησόμεθα τί* δηλονότι *πρᾶγμα* μίαν ἰδέαν ἔχον, *ἐφ' οὗ* τοῦτο *τὸ ὄνομα φέρουσιν* ἅπαντες ἄνθρωποι, νόσον. *MM* IV 4 (X,267,14–16 K.): εἰ μὲν οὖν ἢ παλαιὸν ἢ παρά τινι τῶν Ἑλλήνων ἦν γεγραμμένον *τοὔνομα*, τάχ' ἂν ἴσως ἐξ ὧν ἐκεῖνοι γράφουσιν ἐνοήσαμεν *ἐφ' ὅτου πράγματος ἐπιφέρουσιν αὐτό*.

87,4–7 εἰ ... καὶ τὸ σπέρμα αὐτὸ ⟨τὸ⟩ καταβεβλημένον ... ἢ μόνα τὰ διηρθρωμένα τε καὶ ἤδη διαπεπλασμένα, τὰ ἔμβρυα καλούμενα. Like in the case of ζῷον, the opponent is invited to choose between two interpretations of the name by which the investigated object is called (τὸ κυούμενον/τὸ κατὰ γαστρός, words that have been used interchangeably). The first option is broader and includes "even the seed deposited (in the womb)".[178] The second option, in turn, includes only

---

Ἀσκληπιάδης ἤσκησεν, ἐμποδίζοι τε καὶ διακόπτοι τὸν λόγον, ἀναμιμνήσκειν χρὴ τοῦτον ὡς οὐ τὸ πρᾶγμ' αὐτὸ διαβάλλει τοῦτο πράττων ἀλλὰ τὸν ὑπὲρ αὐτοῦ λόγον.

178 Stählin is surely correct in excising καὶ τὰ ζῷα in 87,4 (presumably a gloss explaining the subject of διαπεπλασμένα); and Barnes (following Pohlenz) is probably right in suggesting that τὰ διαπεπλασμένα ἤδη should be excised as well: these words could have entered the text as a gloss, too (either prior to καὶ τὰ ζῷα or along with it), explicating the force of καί before τὸ σπέρμα on the basis of 87,6 (thus the second καί, excised by Pohlenz, should be retained). Even if we decide to keep the words τὰ διαπεπλασμένα ἤδη in the text, Bernays' addition of μή before διαπεπλασμένα, printed in Stählin's edition, should be rejected. Fregonara translates τὰ μὴ διαπεπλασμένα ἤδη as "ciò che non è ancora plasmato" ('that which has not been formed yet'). However, to accept this, we would have to suppose that ἤδη is used in the sense of ἔτι (unlike below, 87,6: τὰ ... ἤδη διαπεπλασμένα) and we would have to put up with the fact that the first option (presented in 87,4f.) excludes the developed stage of the embryo from the meaning of the 'the thing in the womb' (as Jonathan Barnes points out to me, "Bernays' μή produces not truth but absurdity"). I have considered the possibility that ἤδη might refer to a point at which the embryo, in the process of its formation, enters, as it were, the semantic field of the name in question: at the stage when the embryo is not formed yet, it may *already* be signified by that name. Cf. my "Galenus Christianus?" pp. 362f. But this is a forced interpretation, especially in light of the adverb's function in an almost identical phrase in 87,6.

those beings "that have been formed already"; the latter are further described as διηρθρωμένα ('articulated') and referred to as ἔμβρυα (this word is used for the first time in our text). Thus the question is whether the word 'the thing in the womb' covers all stages of the development of the embryo, including the stage of the deposited seed, or whether it refers only to the most advanced stage, when all of its organs are differentiated.[179]

For the embryo question, cf. above, on 85,17b. The vocabulary of our passage is reminiscent of Galen, *UP* XV 5 (II,357,24–28 Helmreich/IV 238–239 K.): "That the thing in the womb is already an animal, at least when all of its members are formed (ὅταν γε διαπεπλασμένον ἅπασιν ᾖ τοῖς μορίοις), we said in the treatises *On Demonstration* and *On the Doctrines of Hippocrates and Plato*." In our text, no such conclusion is made, but the distinctions proposed here (and below, 13, 3–5) provide us with the tools we need to argue in its favour.

87,7 f. περαίνειν ἤδη τὸ προκείμενον ἐφεξῆς καὶ διδάσκειν χρή. For the meaning of τὸ προκείμενον, cf. Galen, *MM* I 5 (X,45,5 f. K.): τὸ προκείμενον σκέμμα. Cf. also Galen, *PHP* III 8, 35 (CMG V,4,1,2: 232,3–12/V,357 f. K.): "… the man who really seeks the truth … as he approaches each particular problem … should study the premisses needed for proving it, which of them he should take … so as to bring the problem already to a conclusion (περαίνειν ἐξ αὐτῶν ἤδη τὸ προκείμενον)." After he has reached the conclusion, the researcher is expected to "teach" (διδάσκειν). Assuming that this teaching is concerned with the same subject as the conclusion, its task probably is to prove that the conclusion is true, in other words, that the middle term, discovered earlier in the discussion, holds of "the thing itself" (cf. below 88,13; 87,12 and note above, on 85,28).

---

179 For the underlying model of embryogenesis, cf. Galen, *Sem.* I 9, 2–9 (CMG V,3,1: 92,21–94,11/IV 542–543 K.), where the initial stage, called γονή, is one in which "the form of the seed is dominant" (ἡ τοῦ σπέρματος ἰδέα κρατεῖ); it is succeeded by the stage of κύημα, in which the main organs (the liver, the heart, and the brain) are "not differentiated and formed yet" (ἀδιάρθρωτα μὲν ἔτι καὶ ἄμορφα); next is the stage of ἔμβρυον, which further develops into παιδίον. Cf. Véronique Boudon-Millot, "La naissance de la vie dans la théorie médicale et philosophique de Galien," in L. Brisson, M.-H. Congourdeau, and J.-L. Solère (eds.), *L'embryon: formation et animation* (Paris: Vrin, 2008), pp. 79–94, here 85–87. In the translation, I follow the suggestion of Boudon-Millot, "La naissance," pp. 87 f., to render ἔμβρυον in this sense as 'fetus'.

12, 4–13, 8

87,9 εἰ δὲ ἡμᾶς λέγειν βούλοιτο ... ἀκουσάτω. Cf. above, 11, 4/86,29f. What follows until 13, 8 is an argument κατὰ διέξοδον.

87,12 ὑπὲρ αὐτῶν ἤδη τῶν πραγμάτων ἐσκοπούμην. Cf. above, 10, 2/85,28: ἐπιδεικνύειν ἤδη χρή. 11, 2/86,23: ἡ διερεύνησις τοῦ πράγματος.

87,13–16 The polysemy of the proposed name (τὸ κυούμενον/τὸ κατὰ γαστρός) is illustrated by the word κύων, a standard example of homonymy.[180]

87,15 ἀλλὰ καὶ Διογένους. Cf. Philo of Alexandria's list of the meanings of κύων: "... and besides the philosopher who came from the Cynic school, Aristippus and Diogenes" (*Plant.* 151). Cf. also Sextus, *M.* XI 29: "... and besides these the philosopher ..."

87,20–26 In these lines, the phonetic, the semantic, and the 'referential' aspects of the proposed word are distinguished in order to point out that the focus of research is on the nature of the investigated matter. Cf. above, 3, 2–4. The terminology is, for the most part, Stoic (but see below, on 87,24f.).

87,20 φωνὴ καὶ σῶμα καὶ ὂν καὶ τί. This is an echo of the Stoic view according to which the voice is a body because it acts.[181] Being a body, it is also 'existent', which indeed implies that it is 'something' as well, the latter being a more general category.[182]

87,24f. ἐκεῖνο μὲν ἀσώματόν τέ ἐστι καὶ λεκτὸν καὶ πρᾶγμα καὶ νόημα. The first three descriptions of meaning correspond to the Stoic usage.[183] For 'sayables',

---

180 Cf. Aristotle, *Soph. El.* 4, 166a16; Philo of Alexandria, *Plant.* 151; Athenaeus, *Deipn.* X 78 (II,485,3f. Kaibel); Galen, *MM* II 7 (X,128,16–129,4.9–15 K.); *Soph.* 1 (XIV,583,1–4 K.) and 2 (588,15f. K.); *Diff. Puls.* II 3 (VIII,573,11–15 K.); *Hipp. Off. Med.* III 25 (XVIIIb,869,15–870,1 K.); Sextus, *M.* XI 29.

181 Cf. DL VII 55–56: καὶ σῶμα δ' ἐστὶν ἡ φωνὴ κατὰ τοὺς Στωικούς ... πᾶν γὰρ τὸ ποιοῦν σῶμά ἐστι· ποιεῖ δὲ ἡ φωνὴ προσιοῦσα τοῖς ἀκούουσιν ἀπὸ τῶν φωνούντων.

182 Cf. Alexander, *In Top.* IV 1,121a10 (CAG II/2: 301,22–25 = SVF II 329, part).

183 Cf. esp. Sextus, *M.* VIII 12: σημαῖνον μὲν εἶναι τὴν φωνήν (...) σημαινόμενον δὲ *αὐτὸ τὸ πρᾶγμα* τὸ ὑπ' αὐτῆς δηλούμενον (...) τυγχάνον δὲ τὸ ἐκτὸς ὑποκείμενον (...). τούτων δὲ δύο μὲν εἶναι σώματα, καθάπερ τὴν φωνὴν καὶ τὸ τυγχάνον, ἓν δὲ *ἀσώματον*, ὥσπερ τὸ σημαινόμενον *πρᾶγμα*, καὶ *λεκτόν*. For 'sayables', cf. LS 33; Michael Frede, "The Stoic Notion of *lekton*," in S. Everson (ed.), *Language* (Companions to Ancient Thought 3; Cambridge: Cambridge University

cf. also below, 26, 4/96,23 f. However, the fourth description (meaning as νόημα) indicates that the author is using Stoic terminology against the background of a different model of signification, presumably the one outlined in Aristotle's *Int.* 1; cf. below, 23, 1.[184]

87,25 f. ἄλλη δέ τις ⟨ἂν⟩ εἴη τοῦ ζῴου φύσις. Since the beginning of the 'exposition' (12, 4), the meaning of the proposed predicate has not been discussed yet; instead, a more fundamental (and trivial) point has been established, namely that the subject is neither an utterance nor its meaning, for neither of these can be reasonably described as animals.

87,26 f. ἐναργῶς γὰρ ἐδείκνυτο τοῦ πράγματος αὐτοῦ τοῦ ζητουμένου, λέγω δὲ τοῦ ἐμβρύου, τῆς φύσεως ὁποία τίς ἐστιν. This sentence is presented as a confirmation of the preceding statement, based on information that has already been given to the reader. The information concerns the nature of the proposed subject and the confirmation rests on the fact that this nature is different from that of an utterance or a meaning.[185] Once this point has been clarified, the exposition proceeds to the question in what ways the proposed predicate could be attributed to this thing (87,27 f.: ἕτερον ὑπάρχον πρόβλημα κτλ.). Interestingly, the thing in question is described as τὸ ἔμβρυον, i.e. by a name reserved for 'the thing in the womb' in its most developed stage (cf. above, 12, 2/87,6 f.). It is unclear at what point the research object has been specified like this: in 12, 2, this solution was only proposed as one of two options in a dialectical argument. There is also a related question what passage is referred to by the words ἐναργῶς ἐδείκνυτο. Again, the only candidate in our text is the above-mentioned passage in 12, 2, but this outline of what ἔμβρυον means is arguably too sketchy to claim that it has shown anything ἐναργῶς. I suspect that the excerptor

---

Press, 1994), pp. 109–128; Jonathan Barnes in K. Algra et al. (eds.), *The Cambridge History of Hellenistic Philosophy*, pp. 197–213; Michele Alessandrelli, *Il problema del λεκτόν nello stoicismo antico: Origine e statuto di una nozione controversa*, Florence: Leo S. Olschki Editore, 2013. For πρᾶγμα as meaning, cf. Pierre Hadot, "Sur divers sens du mot *pragma* dans la tradition philosophique grecque," in P. Aubenque (ed.), *Concepts et catégories dans la pensée antique* (Paris: Vrin, 1980), pp. 309–319.

184 For the difference between νοήματα and λεκτά, cf. A.A. Long, "Language and Thought in Stoicism," in idem (ed.), *Problems in Stoicism* (London: The Athlone Press, 1996), pp. 75–113, here esp. 79–81.

185 We may wonder why it is necessary to confirm a point that, in itself, is already quite trivial; the answer probably lies partly with the polemical context, and partly with the didactic purpose of the discussion.

left something out, perhaps the contents of διδασκαλία mentioned in 12, 3/87,8. For ὁποία τίς ἐστιν, cf. above, 3, 4/81,23.

87,29f. εἰ τοῦτο λέγεις ζῷον τὸ δυνάμενον αἰσθέσθαι τε καὶ κινηθῆναι καθ' ὁρμήν. The speaker adopts the second definition proposed above, 10, 6/86,11f., adding the crucial element of 'capacity'. See already above, 9, 8/85,19f.

87,31f. δύναται γὰρ καὶ κοιμᾶσθαι ἢ μὴ παρόντων τῶν αἰσθητῶν μὴ αἰσθάνεσθαι. The capacities in terms of which a being is defined as an animal may not be actually used, either because the being is asleep or because no sensible objects are present. This qualification recalls the solution to the embryo question proposed by Asclepiades of Bithynia, who reportedly compared embryos to those who "have" senses but do not "use" them because they are asleep.[186] Cf. above, on 85,17b. It is possible that Asclepiades' solution is already (directly or indirectly) informed by Aristotle's distinction between potential and actual perception.[187] In any case, in our text, the 'Asclepiadean' solution is clearly developed in an Aristotelian framework. See further below, 13, 5/88,2–4.

87,32–88,1 τὸ δὲ δύνασθαι [L: δυνάμενον] ἤτοι ὁρμᾶν ἢ καὶ αἰσθάνεσθαι πεφυκὸς ζῴου γνώρισμα. Stählin (in the apparatus) proposes to change δυνάμενον to δύνασθαι and πεφυκός to πέφυκεν. But πεφυκός is probably correct; cf. Aristotle, *Top.* V 4, 133a8–11, who mentions the predicate τὸ αἰσθάνεσθαι πεφυκός as an example of a property (ἴδιον) of animals. In our text, the predicate is extended by ὁρμᾶν (in the sense equivalent to κινεῖσθαι καθ' ὁρμήν)[188] and by the qualification of these functions in terms of capacity (to which πεφυκός is attached). Τὸ δυνάμενον is more difficult. Perhaps it could be saved by regarding the phrase

---

186 Cf. Ps.Galen, *Def. Med.* 445 (XIX,452,13–17 K.): "But Asclepiades claimed neither that the embryo is an animal, nor that it is not an animal, but he said that it is similar to those who are sleeping: in the same way as they have senses, but do not use them, so too the thing in the womb (ὅνπερ γὰρ τρόπον ἐκεῖνοι μὲν ἔχουσιν τὰς αἰσθήσεις, οὐ χρῶνται δὲ αὐταῖς, οὕτωσὶ καὶ τὸ κατὰ γαστρός)."

187 Cf. Aristotle, *Top.* V 2, 129b33f. … τὸ αἰσθάνεσθαι πλείω σημαίνει, ἓν μὲν τὸ αἴσθησιν ἔχειν ἓν δὲ τὸ αἰσθήσει χρῆσθαι. *DA* II 5, 417a9–13: ἐπειδὴ δὲ τὸ αἰσθάνεσθαι λέγομεν διχῶς (τό τε γὰρ δυνάμει ἀκοῦον καὶ ὁρῶν ἀκούειν καὶ ὁρᾶν λέγομεν, κἂν τύχῃ καθεῦδον, καὶ τὸ ἤδη ἐνεργοῦν), διχῶς ἂν λέγοιτο καὶ ἡ αἴσθησις, ἡ μὲν ὡς δυνάμει, ἡ δὲ ὡς ἐνεργείᾳ. See also *GA* II 1, 735a4–11, where the distinction of various degrees of potentiality is applied to the question whether the seed has a soul.

188 Cf. e.g. Galen, *PHP* III 7, 25 (CMG V,4,1,2: 216,26/V,341 K.); IV 2, 17 (242,6/V,370 K.); IV 6, 37 (278,6/V,413 K.); *Loc. Aff.* VI 4 (VIII,402,16f. K.); *Hipp. Aph.* IV 1 (XVIIb,654,5). The first two passages are quotations from Chrysippus.

not as a description of a matter of fact, but as a quotation of a definition by which something is recognized as an animal ("The 'is naturally capable of ...' is a mark by which we recognize ..."). But this is a forced explanation and I prefer to adopt Stählin's conjecture.

88,2–4 πρῶτον μέν, εἰ τὸ κυούμενον ἤδη αἰσθάνεσθαι ... δύναται, ... ἕτερον δέ, εἰ τὸ κυούμενον αἰσθ⟨ήσ⟩εσθαι ποτὲ ... δύναται. The capacity to perform certain activities 'already' (ἤδη) is distinguished from the capacity to perform them in the future (ποτέ).[189] This distinction provides the basis for two qualifications of the definition of 'animal' proposed above (87,29 f.) and, in effect, two versions of the research question.[190] The second version will be rejected as trivial (88,5: οὐδεὶς ζητεῖ ἐναργὲς ὄν), whereas the first will be identified as the one worth looking into (88,13: τοῦτο ἐστι τὸ ζητούμενον). The same distinction is made in a similar context by the author of the Pseudo-Galenic treatise *To Gaurus on How Embryos are Ensouled* (possibly Porphyry). Reporting on the views of people who argue that embryos are animals potentially (δυνάμει μόνον), he distinguishes between the sort of potentiality (τὸ δυνάμει) which does not possess a certain δύναμις yet, but is capable of receiving it (ὃ μήπω δεδεγμένον τὴν δύναμιν οἷόν τ' ἦν ταύτην ἀνα⟨δ⟩έξασθαι),[191] and one that has already received it, but does not act accordingly (τὸ δὲ δεξάμενον, ὅταν μὴ ἐνεργῇ κατὰ ταύτην), like someone "who is occupied with something else or sleeps" (*Gaur.* 1, 2/33,14–18 Kalbfleisch). Later, the author of *Gaur.* takes pains to show that embryos are *not* animals potentially in the second sense of the word (*Gaur.* 13, 1–16, 9/52,3–58,16 Kalbfleisch). The background of the distinction is, of course, Aristotelian; see above, on 87,31 f., and, more generally, *DA* II 5, 417a21–418a6 and *Phys.* VIII 4, 255a33–b24.[192]

---

189 Stählin emends κινηθήσεσθαι δύναται ('is able to move in the future') to κινηθῆναι δυνήσεται ('will be able to move'); but it is hard to see why a scribe would make such an error. Changing αἰσθέσθαι to αἰσθήσεσθαι seems an easier option. Apparently, it is important for the argument that the capacity to do something in the future already belongs to the subject. Cf. below, on 88,9–13.

190 The 'qualification' is fully spelled out below, 13, 7/88,8–13: πάλιν τοῦτο διωρισάμεθα κτλ.

191 In *Gaur.*, this "capacity of receiving" is called 'suitability' (ἐπιτηδειότης); cf. Gwenaëlle Aubry, "Capacité et convenance: la notion d'epitêdeiotês dans la théorie porphyrienne de l'embryon," in L. Brisson, M.-H. Congourdeau, J.-L. Solère (eds.), *L'embryon: formation et animation* (Paris: Vrin, 2008), 139–155; see also below, 28, 4.

192 For the origin and use of the distinction in Aristotle's thought, cf. Stephen Menn, "The Origins of Aristotle's Concept of *Energeia*: *Energeia* and *Dynamis*," *Ancient Philosophy* 14/1 (1994), pp. 73–114.

88,5–16 The example of 'the method of discovery' introduced in 9, 6 is now concluded by a brief summary; cf. above, 8, 2–3.

88,6 μετελήφθη ... τοὔνομα εἰς λόγον. Cf. above, 3, 1/81,10 f.; 3, 3/81,18–20; 4, 2/82,2 f.; 84,25 f.

88,9–13 ἕτερον μὲν ... τὸ δυνάμει τοιοῦτον, ὃ ... ἔσται δέ ποτε τοιοῦτον, ἕτερον δὲ τὸ κατ' ἐνέργειαν ... τοιοῦτον, τούτου δὲ τὸ μὲν ἤδη ἐνεργοῦν κτλ. The two kinds of capacity mentioned above (13, 5/88,2–4) are now described from the perspective of a distinction between potency and act. Of course, the capacity to act 'already' is further actualized whenever it is used.

88,12 f. τὸ δὲ ἐνεργεῖν μὲν δυνάμενον, ἡσυχάζον δὲ ἢ κοιμώμενον. This is clearly the qualification (of the definition of animal) accepted by our author and precisely the one that does not apply to embryos according to Ps.Galen, *Gaur.* 13,1 (52,4–6 Kalbfleisch): ὅτι δὲ μη⟨δὲ⟩ δυνάμει ἔμψυχον τοῦ δυνάμει ἐξακουομένου ἐπὶ τοῦ ἀνειληφότος μὲν τὴν ἕξιν ἡσυχάζοντος δὲ καὶ μὴ ἐνεργοῦντος, δεικνύναι ἀκόλουθον. Cf. above, on 88,2–4.

88,13 τοῦτο δέ ἐστι τὸ ζητούμενον. Cf. Aristotle, *An. Post.* II 3, 90a35 f. (trans. Barnes): "It is plain then, that whatever is sought, it is a search for a middle term." Our text never identifies τὸ ζητούμενον as a middle term, but the discovery of the sought predicate provides precisely that.

88,14–16 Here is why the proposal offered in 10, 2 (85,25 f.) would fail; it misses the 'essence' of what it is for an embryo to be an animal.

88,15 τοῖς δ' ἄλλως συμβεβηκόσιν, 'incidental in another way', i.e. not in a way pertaining to the essence of the problem. Galen uses the same expression regarding causally irrelevant attributes of bodily parts.[193] For the relevant features as 'accidents', cf. above, on 85,1b–13.

---

193 Cf. *MM* II 3 (X,92,11–16 K.): "If someone does not know how to distinguish, in the field of things which naturally belong to the parts of the body (αὐτῶν τῶν κατὰ φύσιν ὑπαρχόντων τοῖς μορίοις), between the causes of activities and things incidental in another way (τά τ' ἄλλως συμβεβηκότα καὶ τὰ τῆς ἐνεργείας αἴτια), how is he likely to be able to isolate the disposition by which the activity is primarily harmed? For the things which are incidental to bodies in another way (τὰ μὲν γὰρ ἄλλως συμβεβηκότα τοῖς σώμασι) cannot damage an activity, even if they are altered thousands of times." *MM* II 5 (X,95,6–9): "... it is not possible to discover and demonstrate the number of diseases without knowing which of the things

14, 1–15, 1

88,17–33 This paragraph briefly summarizes what we have read so far (starting from chapter 2), while opening a discussion about another topic. It explicates the connection between the definition of demonstration reached in 3, 1–8, 3 and the subsequent discussion about the inquiry of problems. It concludes (abruptly) the longest coherent section of '*Strom.* VIII'.

88,17 f. κοινὴ ἐπὶ πᾶσι τοῖς εὑρίσκεσθαι λεγομένοις … τέτακται ἡ ἀπόδειξις, ἥτις ἐστὶ λόγος … πιστούμενος. Demonstration in the sense discussed above (cf. esp. 5, 1/82,12–14 and 7, 6/83,31–33) is "common", as it applies to everything said to be discovered (cf. above, 8, 6/84,32 f.: τῶν εὑρῆσθαι δοκούντων), i.e. to everything claimed to be the case about the thing sought (cf. above, 4, 1/81,25–82, 2).

88,20 τῷ μανθάνοντι. For demonstration as a teaching method, cf. above, 5, 3/82,17; 7, 7/84,1 f.

88,20 f. ἀρχὴ … τὸ πρὸς αἴσθησίν τε καὶ νόησιν ἐναργές. Cf. above, 7, 3/83,25; 7, 8/84,2 f.; 8, 1/84,12.

88,21–25 ἡ μὲν οὖν πρώτη ἀπόδειξις ἐκ τούτων σύγκειται, ἡ δ' ἐκ τῶν ἤδη φθασάντων ἀποδεδεῖχθαι κτλ. This distinction is mentioned for the first time in our text. It is also made by Alexander, *In Top.* I 1, 100a27–29 (CAG II/2: 16,30–18,5) and VIII 12, 162a37–162b2 (574,6 f.); *In Met.* Δ 3, 1014a37–b2 (CAG I: 356,29–34).

88,25–33 This classification of problems echoes the one made above, 9, 1–5. Unlike there, where four kinds of problems were distinguished, here it is said, oddly, that there are only three of them (88,26: τριῶν ὄντων, *scil.* εἰδῶν). In what follows, however, only two kinds will be mentioned (cf. below, on 15, 1/88,30–33).

88,25 f. τὸ μὲν οὖν πρῶτον εἶδος … ἐδείχθη. The first kind of problems was "exhibited" by the discussion of the embryo question; cf. above, 9, 9/85,20 f.

---

that belong to the parts of the body are causes of their activities and which are incidental to these parts in another way (τίνα μὲν τῶν ὑπαρχόντων τοῖς μορίοις αἴτια τῆς ἐνεργείας ἐστί, τίνα δ' ἄλλως τούτοις συμβέβηκεν)." (Trans. Hankinson, modified).

88,29 f. οἷον ἐν τίνι τοῦ σώματος μορίῳ τὸ ἡγεμονικόν ἐστι τῆς ψυχῆς. This is a different example from the one mentioned above, 9, 2/85,7 f., although both are concerned with the soul. The difference lies in the respective objects of inquiry: there, it was the substance of the soul (ζητεῖσθαι τὴν οὐσίαν), while here, it is the bodily location of the soul's ruling faculty. Even though in the second case, we are also ignorant of the substance of the investigated thing, it is not of immediate concern to us.

The question about the location of the ruling faculty is a doxographic *topos* (cf. Ps.Plutarch, *Plac.* 899a1–b4 = 'Aetius' IV 5, 1–10/*Dox. Gr.* 391 f.);[194] but it is also a matter of considerable importance for thinkers like Galen or Alexander of Aphrodisias. Galen discusses it extensively in *PHP*, providing (what he regards as) a demonstrative argument to the effect that the ruling faculty is located in the brain; the starting-point of the argument is the definition of the ruling part in terms of perception and voluntary movement.[195] Basing himself on the analysis of vision, Galen also addresses the question of the substance of the (ruling faculty of the) soul, leaving it open whether it is the *pneuma* in the brain, or something incorporeal which uses *pneuma* as its instrument; cf. *PHP* VII 7, 25 (CMG V,4,1,2: 474,22–27/V,643 K.).

88,30–33 ἡ δὲ αὐτὴ ἐπιχείρησις ... οὐδὲν ἄτοπον. William Lowth (in the 1715 Potter edition) marks this passage as the beginning of a new chapter and Reinhold Klotz (1832) follows him by giving it a new paragraph number. As noticed by von Arnim, *De octavo*, 10 f., this is clearly a mistake, as the passage belongs to the preceding discussion. It is less clear how precisely it belongs to it. It starts by saying that the same "reasoning of demonstration" applies in the case of "this problem, too" (κἀπὶ τούτου τοῦ προβλήματος). The most obvious referent of this is the problem just mentioned, namely "in which part of the body is the ruling faculty of the soul" (88,29 f.). However, in what follows, the text does not explain how demonstration is applied to this problem. Instead, it introduces a controversy about the number of ἀρχαί belonging to one animal, arguing that if the ἀρχαί are of different kinds, there is nothing absurd about there being more than one. Stählin thinks that this is an example of the third kind of problems mentioned above, 9, 3 4/85,8–12, i.e. of the situation when we have knowledge of some attributes and substances and ask which of these attributes belong to

---

194 Cf. the detailed treatment by Mansfeld, *Doxography*, pp. 3092–3108 *et passim*.

195 Cf. the summary of the whole discussion in *PHP* VIII 1, 1–24 (CMG V,4,1,2: 480,4–486,9/V,648–655 K.). Galen's methodology in dealing with the ἡγεμονικόν problem is analysed in particular by Tieleman, who also draws attention to *Strom.* VIII (4) 14, 4 in this connection (*Galen and Chrysippus*, pp. 20, 30 and 42 n. 16).

which substances.[196] This interpretation leads him to the view that τούτου τοῦ in 88,31 should be changed to τοῦ τρίτου [*scil.* προβλήματος].[197] But it is far from obvious how the controversy about the number of ἀρχαί in 88,31–33 exemplifies the third kind of problems.[198] Moreover, Stählin's reading does not mean "the third kind of problems" (cf. 88,27 f.: δευτέρα ... διαφορὰ προβλημάτων), but "the third problem". Finally, we have a reason to think that the controversy about the number of ἀρχαί could be relevant to the question about the location of the ἡγεμονικόν. The champions of the view that the ruling faculty of the soul is in the heart also argue to the effect that all faculties of the soul must be located in the same place. Cf. Galen, *PHP* VI 1, 1 (CMG V,4,1,2: 360,4–7/V,505 K.): "It was my purpose at the beginning to inquire about the δυνάμεις that govern us, whether they have the heart as their only source, as Aristotle and Theophrastus supposed, or whether it is better to posit three sources (τρεῖς ἀρχάς) for them."[199] Most likely, the two sentences preserved in 88,31–33 are related to the ἡγεμονικόν problem, not by representing the problem as such, but by introducing the controversy surrounding it. However, before the argument is developed, and before the third problem is even mentioned, the discussion is cut short.

(V) 15, 2–16, 3: Suspension of Judgement I

For the most part (15, 2–16, 1), this section is a polemic against scepticism, more specifically against the claim that "nothing is firm" and its practical implication that "we should suspend judgement about everything". The polemic is followed by a definition of αἵρεσις and δόγμα and a reflection on the role of the suspension of judgement for the 'dogmatists' (16, 2–3). There are not many threads connecting this section with the preceding ones, at least not in an explicit or plain manner. Demonstration is not mentioned and there is nothing particularly Aristotelian here either. On the other hand, it is not difficult to see how the theme of the 'chapter' *could* be relevant to the theory of demonstration, as the claim disputed in its first part constitutes a challenge to its principles and the partial concession to the sceptical standpoint in 16, 3 shows the limits of its

---

196 GCS 17, p. 88, note to line 31.

197 As noted in the apparatus, Stählin's correction is rejected by Barnes.

198 Suppose the problem is: 'Is it possible for there to be more than one ἀρχή of an animal?' If we substitute οὐσία for ἀρχή, then the task of finding what attributes belong to what substances already implies an affirmative answer to the problem.

199 Trans. De Lacy. Cf. Tieleman, *Galen and Chrysippus*, p. 30 n. 94, who suggests that *Strom.* VIII (5) 15, 1 deals with the number of psychic faculties.

application. Traces of the first concern may be discerned in the previous 'chapters' as well (cf. 1, 1–2; 4, 1; 9, 7), while the second concern is paralleled in Galen's approach to demonstration (see below, on 90,4f.).

While referring much of the material in *Strom.* VIII (including the next chapter) to one and the same *introductio dialectica*, von Arnim thinks that this 'chapter' comes from a different book, namely a Stoic one, and marks it as a Stoic fragment.[200] But this is a doubtful contention.[201] The only passage containing characteristically Stoic terminology (16, 2) seems to contain other doctrinal elements as well (see below, on 89,24–90,2) and the target of the polemic is probably construed with the help of post-Hellenistic sceptic sources (see below, on 89,2, 89,3 and 89,21–23).[202]

89,1 πρὸς τοὺς Πυρρωνίους. In L (352v8) these words are merged in the text (following after οὐδὲν ἄτοπον without punctuation) and they are left there in all editions before Stählin (printed as the beginning of the sentence and followed by εἴη, φασιν, ἡ ἐποχὴ κτλ.). But this is syntactically awkward and does not yield much sense either, and von Arnim is surely correct to suggest that the phrase should be separated.[203] It is less clear whether it originally functioned as a title of the chapter, as von Arnim proposes, or whether it entered the text as a marginal gloss.[204] On the one hand, the words correctly mark the beginning of a new thematic section and they partly correspond to its contents, too, insofar as the notion of ἐποχή introduced in 15, 2 and criticized until 16, 1 may be regarded as Pyrrhonian.[205] On the other hand, they are not well-suited to the remaining parts of the 'chapter' (16, 2–3). The second ('marginal gloss') option is supported

---

200 SVF II 121. Cf. von Arnim, *De octavo*, pp. 11f. and 16. Von Arnim's view is rejected by von Wedel, who finds it improbable that, in between purely logical fragments, Clement would suddenly start to use a different source in order to combat the sceptics (*Symbola*, p. 21).

201 Von Arnim, *De octavo*, p. 16, refers to Antipater's argument according to Cicero, *Ac.* II 109 (trans. Brittan): "Antipater was criticized … for saying that it was consistent for someone affirming that nothing was apprehensible (*ei qui adfirmaret nihil posse comprehendi*) to say that just that claim was apprehensible." But, though it might reflect a similar line of reasoning against the sceptics, the argument in our passage is not quite the same and the vocabulary depicting the sceptic point of view is quite different.

202 Cf. also the sceptical background to 16, 3 and 22, 1–4, discussed in the commentary.

203 *De octavo*, p. 11.

204 The latter is suggested by Le Boulluec, "Extraits d'œuvres de Clément d'Alexandrie. La transmission et le sens de leurs titres," in idem, *Alexandrie antique et chrétienne*, pp. 109–122, here 114.

205 The text itself seems to refer to the opponents addressed in 15, 2–16, 1 as οἱ ἐφεκτικοί (cf. below, on 90,4).

mainly by the contrast with another section in L introduced by a title, where the end of the preceding section and the title itself are clearly designated as such (GCS 90,7f. = L 353ʳ17f., see below, *ad loc.*). The position of the words πρὸς τοὺς πυρρωνίους in L indicates that in its *Vorlage* they did not have the same status as the other title, which, in turn, seems to indicate that they have a different origin. This also has a bearing on the textual presentation. Since we do not know who introduced the phrase or when it was introduced, it is perhaps risky to follow Dindorf (1869) and Stählin, who change the word πυρρωνίους (a Byzantine form witnessed, for example, in Photius) to πυρρωνείους (suitable to Clement's time). It is true that in L there are several instances of a substitution of this kind (cf. Stählin in GCS 17, p. xii, citing such examples as ἐπικούριος and πυθαγόριος), but in this particular case the form could equally well betray the hand of a Byzantine reader. Thus I prefer to retain the manuscript reading and mark the phrase as an addition (originating, presumably, as a marginal gloss).

15, 2–6

89,2 **εἴ φησιν ἡ ἐποχὴ βέβαιον εἶναι μηδέν.** The manuscript reading εἴη φασιν ἡ ἐποχὴ κτλ. is plausibly corrected by Stählin (the text could have been mutilated by someone trying to integrate the phrase πρὸς τοὺς πυρρωνίους into the syntax). For εἴ φησιν, cf. below, 89,6f.: εἰ μὲν ἀληθεύει ... εἰ δὲ μὴ ἀληθεύει, 89,11: εἰ ἔστιν ἀληθής. The section starts by assuming a condition about ἐποχή in order to show that, under this condition, ἐποχή refutes itself. The condition is that ἐποχή "says" something, namely that "nothing is firm". It is not difficult to see why the statement "nothing is firm" may be regarded as self-refuting. After all, what follows is a familiar line of argument against scepticism.[206] What is puzzling, however, is that the statement is attributed to ἐποχή. In its epistemological application, as developed in the Hellenistic debates between the Stoics and the Academic sceptics, the word ἐποχή designates the opposite of the Stoic συγκατάθεσις, namely the suspension of 'assent' to (i.e. a judgment about the truth of) a non-evident proposition.[207] Thus it refers to an action (or rather

206 Cf. Luca Castagnoli, *Ancient Self-Refutation: The Logic and History of the Self-Refutation Argument from Democritus to Augustine*, Cambridge: Cambridge University Press, 2010.

207 The earliest reports about this use of ἐποχή are associated with Arcesilaus; cf. Cicero, *Ac.* I 45; II 59; Sextus, *PH* I 132–133; DL IV 28. The Stoics could have used the word in the same sense, too, but there is little evidence of that; according to Cicero, Arcesilaus happily "agreed" with Zeno that the wise man has no opinions (*Ac.* II 66; cf. *Ac.* II 77 and I 41–42), but we never hear that Zeno employed the word ἐποχή or ἐπέχειν in this connection. Sextus once attributes the advice of "suspending judgement" to Chrysippus (*PH* II 253: οἱ περὶ τὸν

a 'non-action', a state of mind) described by the verb ἐπέχω, one that is also repeatedly mentioned in this section (cf. below, 15, 7–9; 16, 3). But how could a (non-)action be assumed to "say" something? Christiane von Wedel suggests that the word ἐποχή is used in two ways in our text: either (e.g. in 89,12) it refers to the act of suspending judgement, or (as in our lemma) it designates the 'ephectic' (i.e. sceptic) *doctrine*.[208] This seems to point in the right direction, but it cannot be right as such. If it were right, the meaning of one and the same grammatical subject would have to change not only within the same paragraph, but even within the same sentence. In 89,11–13, it is argued that if ἐποχή [a] is true, it will [b] start from itself (ἀφ' ἑαυτῆς ποιήσεται τὴν ἀρχήν) and [c] be an ἐποχή of itself. In 89,2 f. we read that if ἐποχή [a] says that nothing is firm, then, if it [b] starts from itself (ἀφ' ἑαυτῆς ἀρξαμένη), it will [c] first invalidate itself. In these parallel formulations ἐποχή is treated as an activity ([b] and [c]) endowed with a propositional content ([a]). As far as the activity is concerned, it is expressed by such verbs as ἀκυρόω ('invalidate') and ἀναιρέω ('eliminate'); cf. below, 89,3; 89,7.8–11. As regards the content, it is expressed by the assertion 'nothing is firm' (89,2).[209] All this indicates that in these passages the word ἐποχή does not refer to the state of suspension, nor does it designate merely a "doctrine"; rather, it covers a complex of views and practices associated with (leading towards) the suspension of judgement; more specifically, it covers an argumentative procedure in the course of which some (or, rather, any given) propositions are refuted, and a general belief about propositions, presupposed by this procedure. This complex of views and practices could correspond to what Sextus describes as ἡ σκεπτικὴ φιλοσοφία, ἡ σκεπτικὴ ἀγωγή, or simply as ἡ σκέψις.[210] The phrase ἡ ἐποχὴ φησίν could be explained along the lines of a passage where Sextus compares 'the sceptic persuasion' with the Cyrenaic one,

---

Χρύσιππον ... φασὶ δεῖν ἵστασθαι καὶ ἐπέχειν, re. the Sorites argument), but there the word ἐπέχειν likely paraphrases Chrysippus' notion of "being silent" (ἡσυχάζειν); cf. *M.* VII 416; Cicero, *Ac.* II 93 (= *SVF* II 275–277). For συγκατάθεσις, cf. below, on 90,2 f.

208 von Wedel, *Symbola*, p. 3: "... duobus modis scriptor voce ἐποχή utitur: modo doctrina ephectica est, modo actio cohibendi iudicii."

209 Cf. Castagnoli, *Self-Refutation*, p. 341: "Clement treats ἐποχή, quite peculiarly, as an equivalent to the claim that 'nothing is certain' or 'nothing is true'." However, the claim 'nothing is true' does not seem to be taken as an implication of ἐποχή. Cf. below, on 89,3–5.

210 Cf. e.g. *PH* I 5–7. Cf. also von Wedel, *Symbola*, p. 2, who notes that in our passage the word ἐποχή is used as if it were a name of the sceptic school (*nomen sectae scepticae*), "quam apud alios scriptores semper vel ἐφεκτικούς vel σκεπτικούς vel Πυρρωνείους vel σκέψιν nominatam inveniebam." The reference to σκέψις is crucial, I believe, but von Wedel does not seem to notice its particular relevance.

with regard to what each of them "says" (φησί).[211] Sextus also mentions that ἡ σκεπτικὴ ἀγωγή is called, among other names, 'suspensive' (ἐφεκτική, *PH* I 7 and 209), thus opening the floor to the same shorthand application of ἐποχή as in the case of σκέψις.

Variants of the view that 'nothing is firm' are associated with ἐποχή since the beginning of scepticism; cf. Cicero, *Ac.* I 45 (trans. Brittain): "[...] Arcesilaus used to deny that anything could be known ... For these reasons, he thought that we shouldn't assert or affirm anything, or approve it with assent ..."[212] However, the way it is articulated here is reminiscent more specifically of Aenesidemus' standpoint, as reported by Photius. Speaking of Aenesidemus' *Pyrrhonist Discourses*, Photius notes (with a distinct hint of irony) that the overall purpose of this book is "to affirm that nothing is firm so as to give rise to apprehension (βεβαιῶσαι ὅτι οὐδὲν βέβαιον εἰς κατάληψιν)".[213] Photius also says that, according to Aenesidemus, "he who philosophizes after the fashion of Pyrrho is ... wise insofar as he knows most of all this, namely that nothing is firmly apprehended by him (οὐδὲν αὐτῷ βεβαίως κατείληπται)."[214] Cf. already Philo of Alexandria, *Ebr.* 170: ... βέβαιον περὶ οὐδενὸς οὐδὲν ἂν ἔχοιμεν εἰπεῖν.[215]

---

211 *PH* I 215: φασὶ δέ τινες ὅτι ἡ Κυρηναϊκὴ ἀγωγὴ ἡ αὐτή ἐστι τῇ σκέψει, ἐπειδὴ κἀκείνη τὰ πάθη μόνα φησὶ καταλαμβάνεσθαι.

212 *negabat* [*scil. Arcesilas*] *esse quidquam quod sciri posset ... quibus de causis nihil oportere neque profiteri neque adfirmare quemquam neque adsensione approbare etc.* Cf. Carlos Lévy, *Cicero Academicus: Recherches sur les* Académiques *et sur la philosophie cicéronienne* (Rome: École française de Rome, 1992), pp. 262 f.: "Comme Socrate, Arcésilas s'appliquait à ruiner les fausses certitudes chez tous ceux qu'il rencontrait, et comme celui-ci dans certains dialogues platoniciens, il s'en tenait au constat d'aporie."

213 On Photius' irony cf. Karel Janáček, "Zur Interpretation des Photiosabschnittes über Ainesidemos" (1976), in idem, *Studien zu Sextus Empiricus, Diogenes Laertius und zur pyrrhonischen Skepsis*, edited by J. Janda and F. Karfík (Berlin/New York: De Gruyter, 2008), pp. 216–224, here 219 [95/96].

214 LS 71C1.3 = Photius, *Bibl.* 169b19 (Bekker).

215 As first recognized by Hans von Arnim, *Quellenstudien zu Philo von Alexandria* (Berlin: Weidmann, 1888), pp. 53–100, this passage likely goes back to Aenesidemus. Cf. Janáček, "Philon von Alexandreia und skeptische Tropen" (1982), in idem, *Studien*, pp. 265–282; Carlos Lévy, "Deux problèmes doxographiques chez Philon d'Alexandrie: Posidonius et Énésidème," in A. Brancacci (ed.), *Philosophy and Doxography in the Imperial Age* (Florence: Leo S. Olschki Editore, 2005), pp. 79–102; idem, "Philon d'Alexandrie est-il inutilisable pour connaître Énésidème? Étude méthodologique," *Philosophie Antique* 15 (2015), pp. 5–26. For possible echoes of Aenesidemus, cf. below, on 93,19; 93,21 f.; 93,22 f.; 93,24; 94,3 f.

89,2f. ἀφ' ἑαυτῆς ἀρξαμένη. The text further qualifies the condition under which ἐποχή refutes itself, namely that it applies its method to its own propositional content before anything else. Later it is argued that ἐποχή, if true, *will* start from itself; cf. below, on 89,11f.

89,3 ἀκυρώσει ἑαυτήν. The verb ἀκυρόω, 'deprive of authority', 'invalidate' (typically used of legal decrees[216]) does not belong to the common jargon of self-refutation.[217] It was probably introduced into the epistemological debate in the context of a polemic against dogmatism. Compare the Aenesidemean passage in Philo of Alexandria's *On Drunkeness*, where it is applied to mutually exclusive assertions: "Could anyone be such a crazy prattler as to assert firmly (φάναι παγίως) that the just, the wise, the beautiful, or the expedient is such and such? For whatever definition he gives, someone else, who has practised the contrary since childhood, will invalidate it (ἀκυρώσει)."[218]

89,3–5 ἢ τοίνυν δίδωσιν ἀληθές τι εἶναι καὶ οὐ περὶ πάντων ἐφεκτέον, ἢ ἐνίσταται μηδὲν εἶναι ἀληθὲς λέγουσα, καὶ δῆλον ὅτι οὐδ' αὐτὴ πρότερον ἀληθεύσει. Having proposed that 'nothing is firm', ἐποχή faces the question whether something is true. In response to this question, it may either "grant" that it is the case, or "object" that it is not.[219] Both of these options have destructive consequences for ἐποχή: the first, in showing that "we should not suspend judgement about everything"; the second, in being applicable to ἐποχή itself (the latter point is

---

216 Cf. e.g. Diodorus, *Bibl.* XII 17, 5 (ἀκυρῶσαι μὲν τὸν ὑπάρχοντα νόμον, βεβαιῶσαι δὲ τὸν διορθωθέντα); XII 18, 2 (ἀκυρώσας τὸν πρότερον νόμον); XVI 24, 2.5 (τὰ δόγματα τῶν Ἀμφικτυόνων ἀκυρώσει … τῆς δὲ προγονικῆς προστασίας ἀμφισβητῶν καὶ τὰς τῶν Ἀμφικτυόνων ἀδίκους ἀποφάσεις ἀκυρῶσαι βουλόμενος); XVI 29, 4 (ἀκυρῶσαι τὰς κρίσεις τῶν Ἀμφικτυόνων); Dionysius of Halicarnassus, *Ant. Rom.* VII 49, 2 (ἀκυρῶσαι τοὺς ἐπὶ τῇ πίστει τεθέντας νόμους). Cf. also Clement, *Paed.* II (10) 89, 1: τὸ δὴ διάταγμα τοῦ λόγου παντὶ διατηρητέον σθένει, καὶ οὐδὲν οὐδαμῶς παρανομητέον, οὐδὲ ἀκυρωτέον τὰς ἐντολάς.

217 For the Greek terminology of self-refutation see the General Index in Castagnoli, *Self-Refutation*, p. 393, s.v. self-refutation/jargon.

218 Philo, *Ebr.* 197: τίς οὕτως ἔκφρων ἐστὶ καὶ παράληρος, ὡς φάναι παγίως, ὅτι τὸ τοιόνδε ἐστὶ δίκαιον ἢ φρόνιμον ἢ καλὸν ἢ συμφέρον; ὃ γὰρ ἂν οὗτος ὁρίσῃ, τἀναντία μεμελετηκὼς ἐκ παίδων ἕτερος ἀκυρώσει. On the Aenesidemean background to this passage cf. above, note 215.

219 The verb ἐνίσταται ('object', as opposed to δίδωσιν) suggests that the claim 'nothing is true' is not understood as an implication of the propositional content of ἐποχή ('nothing is firm'). Rather, it is taken as one of two possible responses, on the part of the sceptic, to the proposition 'something is true'. But it is worrying that ἐποχή is not allowed to suspend judgement about this. More generosity is shown at the end of the debate; cf. below, 16, 1/89,21f.

developed below, 98,7–11). The first consequence bears with it an interesting implication. It will be destructive just in case ἐποχή already implies that "we *should* suspend judgement about everything": clearly, this practical implication is one of the beliefs that constitute the propositional content of ἐποχή. Cf. below, on 89,17 f.

It is a common strategy in the anti-sceptic polemic to make the opponent commit himself to a general negative view about the possibility of knowledge (the knowability of things or the truth of propositions) and then to argue that this view either exempts itself from itself or proclaims itself untrue. Cf. Cicero, *Ac.* II 29, 109; Aristocles in Eusebius, *Praep.* XIV 18, 5; Sextus, *M.* VII 440.[220] The argument goes back, at least, to Aristotle's polemic against the deniers of the principle of contradiction and their supposed claims that 'everything is true' or 'everything is false' (*Met.* Γ 8, 1012b13–22).[221]

89,6 f. εἰ μὲν ἀληθεύει, δίδωσιν ἄκουσά τι εἶναι ἀληθές. Here it is shown how ἐποχή might be forced to grant that something is true. In the previous sentence, an alternative was introduced, according to which ἐποχή either speaks the truth or not (ἤτοι ... ἀληθεύει ἢ οὐκ ἀληθεύει).[222] This alternative does not refer simply to a matter of fact about ἐποχή, but rather to alternative options for ἐποχή to consider. So much, at least, follows from our passage, which can only make sense if the protasis refers to something *conceded* by ἐποχή.

89,7–11 εἰ δὲ μὴ ἀληθεύει, ἀληθῆ ἀπολείπει ἅπερ ἀνελεῖν ἐβούλετο κτλ. The second option for ἐποχή is to insist that nothing is true, which would imply that ἐποχή is not true either (cf. above, 89,4 f.). Here a further consequence of this option is proposed: by conceding that it is not true, ἐποχή "admits" that things

220 For other references cf. Maria Lorenza Chiesara, *Aristocles of Messene: Testimonies and Fragments*, edited with translation and commentary (Oxford: Oxford University Press, 2001), p. 110 and n. 32.

221 Cf. Castagnoli, *Self-Refutation*, pp. 75–79, who aptly characterizes Aristotle's argument as a "dialectical silencer" of his opponents, rather than as an attempt to disprove their propositions. The same observation applies to our text whose main concern is not to prove that the propositional content of the ephectic persuasion is false, but rather to question its *credibility*. The thesis that "everything is false" (πάντα ψευδῆ) is already ascribed to a Greek author of the 5th century B.C. Xeniades of Corinth; cf. Sextus, *M.* VII 53, who criticizes the thesis from the sceptical point of view (*M.* VII 399; cf. Castagnoli, *Self-Refutation*, p. 117).

222 Cf. Cicero, *Ac.* II 95 (trans. Brittain): "A basic principle of dialectic is that anything asserted ... is either true or false."

it wanted to abolish are true. Most likely, this argument is not concerned with any proposition a sceptic might wish to refute (it would be silly to argue that if scepticism is refuted, all propositions refuted by the sceptics are true), but rather with one general proposition rejected by ἐποχή, namely that 'something is firm' or 'something is true'. A similar argument appears in Sextus, *M.* VIII 55 (trans. Bett): "… if everything is false, 'everything is false' will be also false, since it belongs to 'everything'. And if 'everything is false' is false, its opposite, 'not everything is false,' will be true."[223]

89,9f. ὡς ὁ ὄνειρος ὁ λέγων ψευδεῖς εἶναι πάντας τοὺς ὀνείρους. Here the self-refutation of ἐποχή is illustrated by an example of a dream saying that 'all dreams are false'. What does the example show? Suppose that a dream saying 'all dreams are false' were true; it would follow that all dreams are false but one dream is true, which is impossible. Suppose now that the dream were false; it would follow that not all dreams are false but one dream is false, which is possible. Hence, what the dream says cannot be true, but it can be false, which is presumably taken to mean that it is false. But if it is false that 'all dreams are false,' its opposite, 'not all dreams are false,' is true.[224]

89,10 ἑαυτῆς γὰρ ἀναιρετικὴ οὖσα τῶν ἄλλων γίνεται κυρωτική. Again, the "other things" confirmed by the self-refutation of ἐποχή can hardly be anything but the belief (or the fact) that something is firm or true (cf. above, on 89,7–11). Facing a similar objection concerning the sceptical argument against the existence of proof, Sextus retorts that even if the argument rejects itself, it does not thereby confirm the existence of proof: κἂν αὑτὸν δὲ ἐκβάλλῃ, οὐ διὰ τοῦτο κυροῦται τὸ εἶναι ἀπόδειξιν. Rather, having abolished every proof, the argument

223 Sextus, *M.* VIII 55: εἰ γὰρ πάντ' ἐστὶ ψευδῆ, ψεῦδος ἔσται καὶ τὸ "πάντ' ἐστὶ ψευδῆ", ἐκ πάντων ὑπάρχον. ψεύδους δὲ ὄντος τοῦ "πάντ' ἐστὶ ψευδῆ", τὸ ἀντικείμενον αὐτῷ ἀληθὲς ἔσται, τὸ "οὐ πάντ' ἐστὶ ψευδῆ".

224 Hermogenes of Tarsus (ca. 160–225) uses a similar example in his classification of questions 'without issue' (ἀσύστατα), i.e. such as do not provide a basis for an argument. One type of these questions involves an 'insoluble' case (τὸ ἄπορον) in which "no resolution or conclusion can be reached: e.g. Alexander is warned in a dream not to trust dreams; he takes advice. Whatever advice one gives will lead to the opposite result" (*On Issues* 6/33,3–7 Rabe, trans. Heath). This version is a variant of the Liar paradox, as reported by Cicero, *Ac.* II 95 (trans. Brittain): "If you say that you are lying and what you say is true, you're lying and saying something true." Our text would not wish the story to have the 'Liar paradox' consequence, as its aim is to show that the dream is false.

also 'brackets' itself: μετὰ τὸ … ἀνελεῖν καὶ ἑαυτὸν συμπεριγράφειν (*M.* VIII 480).[225] Our author does not deal with this particular answer of the sceptic, but he could have it in mind when insisting that ἐποχή should turn to itself *first* (see the next note).

89,11f. καὶ ὅλως εἰ ἔστιν ἀληθής, ἀφ' ἑαυτῆς ποιήσεται τὴν ἀρχήν. Earlier the consequences of ἐποχή's "speaking the truth" (εἰ μὲν ἀληθεύει) were considered, namely that it will have to grant that something is true (89,6f.). Here a more general point about ἐποχή is made, namely that if it is true (i.e. if it is considered true by the sceptics), it will make itself its first object. Unfortunately, the text does not give us any explanation of this inference, and the sceptic could argue that it is unfounded.[226] Presumably, the idea is that if the sceptic claims that 'nothing is firm', and if he is serious about that claim, he will be ready to face any objection from his opponent, the first objection coming to mind being the one from inconsistency.

89,12f. οὐκ ἄλλου τινὸς οὖσα ἐποχή, ἀλλ' ἑαυτῆς πρῶτον. The argument returns to the beginning (cf. 89,2f.). 'Being a suspension of itself' is an equivalent to 'invalidating itself' (89,3): the self-applicaton of ἐποχή is an application of the ephectic procedure to the propositional content of ἐποχή (and its practical implication that 'we should suspend judgement about everything'). See also below, 89,18f.

15, 7–16, 1

89,13–19 The argument now turns to the issue of pragmatic inconsistency: even while suspending judgement, the sceptic nevertheless apprehends certain things about himself (that he is a human being or that he suspends judgement), for otherwise he would not be able to dispute anything or claim that he suspends judgement. But the fact that he fails to do what he preaches, namely to suspend judgement about everything, questions the credibility of ἐποχή.

---

225 Cf. Castagnoli, *Self-Refutation*, pp. 251–307, esp. 290–293. On 'bracketing', see also below, on 93,25f.

226 Cf. Castagnoli, *Self-Refutation*, p. 341, commenting on this passage: "[Clement] does not provide any support for his contention concerning the priority of the self-application of ἐποχή, which is clearly polemically aimed at the Pyrrhonists' claim that their formulae and arguments eliminate themselves *after* eliminating the dogmatic doctrines which they target, or *together with* them." On the other hand, the sceptic too would have a hard time explaining why the self-application of ἐποχή should only be reserved for the end.

89,13 f. εἰ καταλαμβάνει τις ὅτι ἄνθρωπός ἐστιν ἢ ὅτι ἐπέχει, δῆλός ἐστι μὴ ἐπέχων. This is an analytically true statement with an indefinite subject (μὴ ἐπέχων is implied in καταλαμβάνει), but it is spelled out with specific regard to the sceptic (he who 'suspends judgement'). It is assumed that he 'apprehends' something, namely that he is a human being or that he suspends judgement, and this is taken both as evident (cf. δῆλος in 89,16) and as a breach of the ephectic rule of conduct (cf. above, on 89,3–5; below, 89,17 f.). The argument likely goes back to the Stoics, who regarded such propositions as 'I am conversing' (ἐμὲ διαλέγεσθαι) as evident (cf. Sextus, *M.* VIII 144: πρόδηλα) and who also seem to have appealed to self-reflective acknowledgments of a similar sort in their polemic against the sceptic notion of ἀκαταληψία.[227]

89,14 f. πῶς δ' ἂν καὶ τὴν ἀρχὴν εἰς τὴν ἀμφισβήτησιν ἀφίκετο περὶ πάντων ἐπέχων; 'To arrive at controversy' means to reach a point when opposite claims about

---

227 Evidence for this (uncontroversial) claim is surprisingly slim. There is a passage in Sextus, which, while mockingly developing the Stoic view that the Sage, who alone has firm knowledge, has not been found, seems to employ Stoic examples of self-evident propositions in order to draw the embarrassing conclusion that, according to the Stoics, even such things as these should not be known by the Stoics themselves. Sextus indicates that these examples (all self-reflective) are similar to those brought forward by the Stoics against the sceptics: "And since this is the case, the things said by the Stoics against the sceptics are available to be said, in turn, by the sceptics against them. For since among the inferior, according to them, are numbered Zeno and Cleanthes ..., then undoubtedly Zeno was ignorant as to whether he was contained in the universe or whether he himself contained the universe, and whether he was a man or a woman, and Cleanthes did not know whether he was a human being or some beast more crafty than Typhon" (*M.* VII 433; trans. Bett). Traces of the anti-sceptic appeal to self-reflective propositions, which likely go back to Stoic sources, can be discerned in Oenomaus of Gadara and Lucian of Samosata (both 2nd cent. A.D.); cf. Oenomaus *apud* Eusebius, *Praep.* VI 7, 10–12 (Mras), who describes self-perception as the "most credible" (πιστότερον) criterion by which we recognize that we are something, that we are animals, human beings, that we are presently conversing, etc.: ἆρά γέ τί ἐσμεν ἐγώ τε καὶ σύ; φαίης ἄν. τοῦτο δὲ ὁπόθεν ἴσμεν; τῷ ποτ' ἄρα τοῦτο εἰδέναι ἐκρίναμεν; ἢ οὐκ ἄλλο ἱκανὸν οὕτως ὡς ἡ συναίσθησίς τε καὶ ἀντίληψις ἡμῶν αὐτῶν; τί δ', ὅτι ζῷά ἐσμεν πῶς ποτ' ἄρα ἐξεύρομεν; πῶς δ' ὅτι καὶ ζῴων ... ἄνθρωποι ... πῶς δ' ὅτι διαλεγόμεθα ἐν τῷ παρόντι ἐγνώκαμεν;—Τί φής; ἆρ' οὐκ ὀρθῶς ἐκρίναμεν τὴν ἡμῶν αὐτῶν ἀντίληψιν τῷ πάντων ἐγγυτάτῳ πράγματι αὐτῷ;—Δηλονότι. οὐ γὰρ ἦν ἄλλο αὐτοῦ οὔτε ἀνώτερον οὔτε πρεσβύτερον οὔτε πιστότερον. On the Stoic background to this, cf. A.C. Lloyd, "*Nosce Teipsum and Conscientia*," *Archiv für Geschichte der Philosophie* 46 (1964) 188–200, esp. 199 f. The anti-sceptic context of such questions clearly transpires in Lucian, *Vit. Auct.* 27/Macleod II,48,14–17 (a customer asking Pyrrhias): Οὐδὲ ἡμεῖς ἄρα ἐσμέν τινες;—Οὐδὲ τοῦτο οἶδα.—Οὐδὲ ὅτι σύ τις ὢν τυγχάνεις;—Πολὺ μᾶλλον ἔτι τοῦτο ἀγνοῶ.

an issue are confronted. This confrontation takes place by means of arguments, whose aim is to establish or reject certain views. Cf. above, 4, 1/81,26: τὸ δόξαν. Our text seems to suggest that it is impossible to produce any argument *pro* or *contra* any view while suspending judgement about it. A similar objection against the sceptics is reported by Sextus, *PH* II 1–3.

89,15 f. πῶς δ' ἂν καὶ ἀπεκρίνετο πρὸς τὸ ἐρωτηθέν; In the end, the sceptic's answer to any question (e.g. 'Is P the case or not'?) is that he suspends judgement. But, according to our text, even this entails an exception to the sceptic rule. The word ἀποφαίνεται in 89,17 stands in a place where sceptics like Sextus would prefer to use such non-commital expressions as προφερόμαι or ἀπαγγέλω (cf. *PH* I 15).[228]

89,17 f. δεῖ πειθόμενον αὐτοῖς περὶ πάντων ἐπέχειν. Sextus distinguishes two ways in which the word τὸ πείθεσθαι may be used: "It means not resisting but simply following without strong inclination or adherence … and it sometimes means assenting to something by choice and, as it were, sympathy" (*PH* I 230, trans. Annas and Barnes). While admitting that the Academic sceptics did believe certain things in the strong sense, Sextus insists that the Pyrrhonians are different. On universal ἐποχή, cf. e.g. Cicero, *Ac.* I 45, on Arcesilaus (quoted above, on 89,2); II 78 (*omnium adsensionum retentio*); Sextus, *PH* I 205 (ἡ περὶ πάντων ἐποχή); I 232 (Ἀρκεσίλαος … περὶ πάντων ἐπέχει); Galen, *Opt. Doct.* 2, 3 (CMG V,1,1: 96,7–8/I,44 K.), on the 'older Academics': διὸ μηδ' ἀποφήνασθαι περὶ μηδενὸς ἠξίουν, ἀλλὰ περὶ πάντων ἐπέχειν.

89,19 εἴτε πειστέον αὐτῇ εἴτε καὶ μή. On the sceptic's (lack of) credibility, cf. Aristocles in Eusebius, *Praep.* XIV 18, 9.12–13.16; Sextus, *M.* VII 440.

89,20 f. εἰ τοῦτο αὐτό ἐστι τὸ ἀληθὲς τὸ μὴ εἰδέναι τὸ ἀληθές, [οὐδὲ τὴν ἀρχὴν] ἀληθές τι παρ' ἐκείνου δίδοται. As it stands in the manuscript, the text would have to be translated as follows: "… if this very [fact] is the truth, viz. that he does not know the truth, then it has not been granted by him that there is anything true at all" (or perhaps "… nothing true has been granted by him at

228 Cf. Castagnoli, *Self-Refutation*, p. 341, commenting on our passage: "[…] Sextus would reply that the Pyrrhonist's awareness and expression of his current mental state of ἐποχή are not the same as the dogmatic assent to, and assertion of, some unclear matter of fact, and therefore do not impair his ἐποχή in the least." But our text confronts (or pretends to confront) itself with a different kind of scepticism, which does not provide any means of distinguishing between clear and unclear matters of fact.

all"). For the meaning of ἀληθές τι ... δίδοται, cf. above, 89,3: δίδωσιν ἀληθές τι εἶναι, 89,6: δίδωσιν ... τι εἶναι ἀληθές, cf. also 89,22 f. But this does not make much sense in itself, nor does it go well with the following sentence. Von Arnim (*SVF* II, p. 37) suggests changing οὐδὲ to εὐθὺς, but it is perhaps more likely that the phrase οὐδὲ τὴν ἀρχὴν was introduced into the text by someone perplexed by the apparent admission that the sceptic might say something true.

89,21–23 εἰ δὲ καὶ τοῦτο ἀμφισβητήσιμον ἐρεῖ τὸ ἀγνοεῖν τἀληθές, κτλ. In the final sentence of the polemic, the sceptic is finally allowed to transcend the dilemma of whether he speaks the truth or not (cf. above, 89,5 f.) and to respond that he is not sure about that either. A similar response is already attributed to Arcesilaus; cf. Cicero, *Ac.* I 45 (trans. Brittan): "[...] Arcesilaus used to deny that anything could be known, not even the residual claim Socrates had allowed himself, i.e. the knowledge that he didn't know anything." Cf. also *Ac.* II 28, on Carneades; and Photius, *Bibl.* 170a11–14, on Aenesidemus (= LS 71C [8]). Our text recognizes and develops (with a bit of a grin) the regressive potential of this manoeuvre, arguing that however far it goes, there must be a point at which the sceptic believes to have found the truth. The description of the sceptic's attempt to escape the charge of self-refuting dogmatism (τὴν περὶ [τοῦ ἀληθοῦς] ἐποχὴν φαίνεται μὴ βεβαιῶν) is reminiscent of post-Hellenistic scepticism; cf. Sextus, *PH* I 4 (trans. Annas and Barnes): "... let us say that on none of the matters to be discussed do we affirm that things certainly are just as we say they are: rather, we report descriptively on each item according to how it appears to us at the time."[229] Cf. also LS 71A–D, on Aenesidemus.

16, 2–3

89,24–90,6 A Syriac translation of this passage is preserved in a collection of Patristic testimonies, dated to the 8th or 9th century (ms. Add. 14,533, fol. 137[a]).[230] It is introduced: "On what is αἵρεσις. Clement, *Stromateis*, end of book eight."[231] The most likely explanation of this reference is that in the manuscript

229 ... περὶ οὐδενὸς τῶν λεχθησομένων *διαβεβαιούμεθα* ὡς οὕτως ἔχοντος πάντως καθάπερ λέγομεν, *ἀλλὰ κατὰ τὸ νῦν φαινόμενον ἡμῖν* ἱστορικῶς ἀπαγγέλλομεν περὶ ἑκάστου.

230 Cf. William Wright, *Catalogue of the Syriac Mansucripts in the British Museum*, II (London 1871), p. 971. Zahn, *Forschungen*, p. 28, suggests that the collection is a translation of a Greek original dating to the end of the 6th century (*Forschungen*, p. 116).

231 MS Add. 14,153; fol. 137a, 1st col., ll. 16–31 (transcribed by Matteo Martelli):

ܕܡܢܐ ܗܝ ܗܪܣܝܣ ✤ ܕܩܠܝܡܝܣ ܣܛܪܘܡܛܣ ܣܘܦ ܡܐܡܪܐ ܕܬܡܝܢܝܐ ✤ ܗܪܣܝܣ
ܐܝܬܝܗ ܡܬܢܣܒܢܘܬܐ ܕܝܘܠܦܢܐ. ܐܘ ܐܝܟ ܐܢܫ. ܡܬܢܣܒܢܘܬܐ ܕܝܘܠܦܢܐ ܣܓܝܐܐ.

from which this passage was taken it had already been followed by a title of another section, presumably the same title as in L 353r,17 f. (= GCS 90,7 f.), which was understood as the beginning of another treatise. Cf. above, p. 8.

89,24–90,3 Our text starts to disintegrate into loosely connected segments. It is not clear what, if any, the function of these definitions of αἵρεσις, δόγμα, and κατάληψις in the argument of the 'chapter' is. If there is any connection with the preceding discussion, it is one of contrast: αἵρεσις is defined in terms of δόγματα, and δόγμα is a sort of κατάληψις, which, in turn, is defined as συγκατάθεσις, the opposite of ἐποχή. Thus αἵρεσις constitutes an alternative to the ephectic persuasion. Despite this thematic link, there does not seem to be any continuity in the polemical intent. As the parallels quoted below show, the definition of αἵρεσις would be approved by the sceptics and there is no trace of an attempt to turn it against them.[232]

89,24–90,2 πρόσκλισις δόγμασι πολλοῖς ἀκολουθίαν πρὸς ἄλληλα καὶ τὰ φαινόμενα περιέχουσι πρὸς τὸ εὖ ζῆν συντείνουσα. There is a close parallel to this definition of αἵρεσις in Sextus, *PH* I 16 (where it is also followed by a definition of

---

ܕܢܦܩܐ ܠܘܬ ܫܪ̈ܪܐ. ܘܗܠܝܢ ܕܡܬܚܙܝܢ ܐܚܝܕܐ. ܕܠܘܬ ܗܘ ܕܫܦܝܪ ܚܝܐ ܡܬܚܝܢ. ܘܠܗܕܐ
ܩܡ ܐܬܝܗܘܝ ܡܫܪ̈ܪܢܘܬܐ ܡܕܡ ܡܠܠܬܐ. ܡܫܪ̈ܪܢܘܬܐ ܕܝܢ ܡܩܒܠܢܘܬܐ. ܘܡܫܬܠܡܢܘܬܐ
ܕܐܬܪܚܩܢܘܬܐ. ܠܐ ܒܠܚܘܕ ܕܗܘܦܘܩܛܝܩܘ. ܐܠܐ ܘܦܪ ܡܠܦܢܐ ܕܡܕܡ ܠܒܗܘܢ ܡܟܒ.. ܐܘ ܡܛܠ
ܡܫܠܡܢܘܬܐ ܕܐܬܪܚܩܢܘܬܐ. ܐܘ ܡܛܠ ܠܐ ܠܚܡܘܬܐ ܕܡܣܒܪ̈ܢܐ. ܐܘ ܡܛܠ ܥܒܘܕܘܬ ܚܝܐ
ܕܡ̈ܠܐ.

According to Martelli (personal communication), the Greek original of the Syriac translation read περιέχουσα in 90,1 and probably συντείνοντα (rather than συντείνουσα) in 90,2. The Syriac transcription of the name of the sceptics (*hwpwqṭīqw*) confirms Sylburg's correction of the ms. ἐφεκτοί to ἐφεκτικοί in 90,4, although the Greek original of the transcription seems to have had a different prefix (ὑπ-).

232 Von Wedel, *Symbola*, pp. 4, 12 f. and 15, argues that Clement drew the definition of αἵρεσις from an anti-sceptic source whose intention was to show that scepticism does not qualify as a philosophical school. But in fact, this was precisely the point the sceptics wished to drive home (cf. below, ad 89,24–90,2). If we were to suppose a polemical intent behind this definition of αἵρεσις, it would have to be construed along different lines, namely along the lines of charging the sceptics with crypto-dogmatism and putting their views on a par with any fundamental beliefs of a dogmatist. However, although our text does argue that ἐποχή involves some (self-refuting) beliefs, it does not seem to indicate that these beliefs are δόγματα of sorts: no such ironic overtone is present and the distinction between the 'ephectic' and the 'dogmatist' points of view are respected (cf. 90,4).

δόγμα, cf. below, on 90,2f.). Other parallels are found in DL I 20, Ps.Galen, *Hist. Phil.* 7 (*Dox. Gr.* 604,2–4) and Ps.Galen, *Def. Med.* 12–13 (XIX,352,5–15 K.); see Table 1.[233] In Sextus and Diogenes, the definition of αἵρεσις is introduced in the context of a discussion about scepticism, specifically the question whether and in what sense it may be regarded as a 'school': Insofar as αἵρεσις is an inclination to δόγματα (defined as assent to something non-evident), scepticism is not a αἵρεσις, because it has no such δόγματα. On the other hand, if αἵρεσις is taken in the sense of a "persuasion congruent with some account according to appearance" (τὴν λόγῳ τινὶ κατὰ τὸ φαινόμενον ἀκολουθοῦσαν ἀγωγήν), the label is acceptable to the sceptic (Sextus, *PH* I 16–17; cf. DL I 20).[234]

The origin of the definition of αἵρεσις as an inclination to δόγματα is obscure; commenting on the version in Ps.Galen, *Hist. Phil.* 7, Diels traces it to a Stoic source (cf. *Doxographi Graeci*, pp. 246 and 251f.). Interestingly, this version defines αἵρεσις in terms of the mutual congruence of δόγματα (henceforth, Cδ) and/or a single goal (ἓν τέλος), but not of the congruence with φαινόμενα (henceforth, Cφ). The same is true of the first and the last in a series of five definitions of αἵρεσις in Ps.Galen, *Def. Med.* 12 (XIX,352,5–7.14–15 K.). We may call these definitions 'Group A'.[235] Definitions that include both the requirement of Cδ and of Cφ (ἀκολουθίαν πρὸς ἄλληλα καὶ τὰ φαινόμενα, ἀκολουθίαν ἑαυτοῖς

233 A source-critical analysis of these parallels is provided by von Wedel, *Symbola*, pp. 8–16. See below, p. 214 n. 237.

234 Here Sextus and Diogenes (the latter offering a more careful formulation: τὴν λόγῳ τινὶ κατὰ τὸ φαινόμενον ἀκολουθοῦσαν *ἢ δοκοῦσαν ἀκολουθεῖν*) clearly draw from the same source. On the relation between Sextus and Diogenes, cf. Jonathan Barnes, "Diogenes Laertius IX 61–116: The Philosophy of Pyrrhonism," in *Aufstieg und Niedergang der römischen Welt*, II: Principat, Band 36.6, ed. W. Haase (Berlin/New York: de Gruyter, 1992), pp. 4241–4301. The distinction between αἵρεσις and ἀγωγή might go back to Aenesidemus; cf. Chiesara, *Aristocles*, pp. 135f., commenting on Aristocles in Eusebius, *Praep.* XIV 18, 30 (Mras): διότι μὲν οὖν τὴν τοιαύτην εἴτε αἵρεσιν εἴτε *ἀγωγὴν λόγων* εἴτε ὅπῃ καὶ ὅπως ἐθέλει τις καλεῖν αὐτὴν οὐδεὶς ἂν εὖ φρονῶν ὀρθὴν εἶναι φαίη, δῆλον. See also Photius, *Bibl.* 170b2 Bekker; Sextus, *PH* I 210. The expression ἀγωγὴ λόγων is also used by Galen in a passage with sceptical undertones (*Loc. Aff.* III 3/VIII,143,11 K.). For ἀγωγή as a sceptic term of art, cf. also Galen, *Diff. Puls.* IV 2 (VIII,710,15–18 K.), on people influenced by "the persuasion of the sceptics and the so-called aporetics" (δεδιότες ἀποφήνασθαι περί τινος τῶν ἐκτὸς ὡς ὑπάρχοντος, ἐκ τῆς τῶν σκεπτικῶν καὶ ἀπορητικῶν ὀνομαζομένων ἀγωγῆς ἐπὶ τοῦτό γε ἡγμένοι). For Galen's attitude to scepticism, cf. below, pp. 216–217 n. 243.

235 It is hard to determine whether Diogenes' 'dogmatist' definition of αἵρεσις belongs to this group, as he fails to mention either appearances or a goal, referring to coherence only (DL I 20: πρόσκλισιν δόγμασιν ἀκολουθίαν ἔχουσιν).

TABLE 1

Clem. *Strom.* VIII 16, 2	Sextus, *PH* I 16	*DL* I 20	Ps.Galen, *Hist. Phil.* 7	Ps.Galen, *Def. Med.* 12 (XIX,352,5–7 K.)	Ps.Galen, *Def. Med.* 12–13 (XIX,352,12–353,2 K.)
αἵρεσίς ἐστι πρόσκλισις δογμάτων ἤ, ὥς τινες,	εἰ μὲν ⟨γάρ⟩ τις *αἵρεσιν* εἶναι λέγει	εἰ δὲ *αἵρεσιν* νοοῖμεν	κατ' ἔννοιαν δὲ τὴν *αἵρεσιν* [*scil.* ὑπολαμβάνουσί τινες] εἶναι	*αἵρεσίς* ἐστι	*αἵρεσίς* ἐστι
πρόσκλισις δόγμασι πολλοῖς ἀκολουθίαν πρὸς ἄλληλα καὶ τὰ φαινόμενα περιέχουσι	*πρόσκλισιν δόγμασι πολλοῖς ἀκολουθίαν ἔχουσι πρὸς ἄλληλά τε καὶ* ⟨τὰ⟩ *φαινόμενα*,	*πρόσκλισιν δόγμασιν ἀκολουθίαν ἔχουσιν*,	*πρόσκλισιν δογμάτων πολλῶν πρὸς ἄλληλα* συμφώνων ἢ	*πλήθους δογμάτων πρόσκλησις* τεχνικῶς συντεταγμένων καὶ	*πρόσκλησις πλήθους δογμάτων ἀκολουθίαν ἐχόντων ἑαυτοῖς καὶ τοῖς φαινομένοις*. ἢ … *πρόσκλησις ἐπὶ πλήθει δογμάτων ἀκολουθίαν ἐχόντων πρὸς ἄλληλα*
πρὸς τὸ εὖ ζῆν συντείνουσα.			πρὸς ἓν τέλος ἀναφερόντων	ἐφ' ἓν τέλος ἐχόντων τὴν ἀναφοράν.	ἐφ' ἓν τέλος.
καὶ τὸ μὲν *δόγμα* ἐστὶ κατάληψίς τις λογική, κατάληψις δὲ ἕξις καὶ *συγκατάθεσις* τῆς διανοίας	καὶ λέγει *δόγμα* πράγματι ἀδήλῳ *συγκατάθεσιν*,				*δόγμα* ἐστὶ … πράγματος συγκατάθεσις,
	φήσομεν μὴ ἔχειν αἵρεσιν [*scil.* ὁ σκεπτικός].	οὐκέτ' ἂν προσαγορεύοιτο αἵρεσις [*scil.* ἡ σκεπτική]			διὸ δὴ μᾶλλον ἡ λογικὴ αἵρεσις δογματικὴ κέκληται.

καὶ τοῖς φαινομένοις, etc.) usually lack a reference to a goal (Sextus, *PH* I 16; three versions in *Def. Med.* 12/XIX,352,7–13 K.); the only exception is our passage where, however, the goal is not described as τὸ τέλος, but more specifically as τὸ εὖ ζῆν (using a phrase πρὸς τὸ εὖ ζῆν συντείνουσα, which is probably of Stoic origin).[236] I call them 'Group B'. Despite their differences, it is likely, in my view, that all of these definions ultimately derive from one source, perhaps a Stoic author, for whom it was important to define αἵρεσις and who did so by pointing out 'inclination', Cδ, and 'one goal'.[237] While building upon this definition represented by Group A, Group B seems to be informed by someone who added the criterion of Cφ. The reason I think that Group A represents an earlier version is that the concepts of τέλος and the mutual coherence of beliefs belong to the standard repertoire of Stoic thought, whereas the requirement of congruence with phenomena does not.[238] The Group B definition is also

236 For the Stoic origin of the phrase, cf. SVF III 208 (Arius Didymus in Stobaeus), on the Stoic concept of virtue: συμφέρον [*scil.* τὴν ἀρετὴν προσαγορεύουσιν], φέρειν γὰρ τοιαῦτα ἃ συντείνει πρὸς τὸ εὖ ζῆν. A goal is also mentioned in Sextus, *PH* I 17, but not as a part of the 'dogmatist' definition of αἵρεσις, but as a part of a definition acceptable for the sceptic: the 'account' (λόγος) with which the sceptic persuasion 'congrues' is such that it shows "how it is possible to live in a way that seems to be right" (ὡς ἔστιν ὀρθῶς δοκεῖν ζῆν). Cf. von Wedel, *Symbola*, p. 13, who compares it with Clement's τὸ εὖ ζῆν.

237 Cf. von Wedel, *Symbola*, pp. 13 f., who traces the parallels in Clement, Sextus, and *Hist. Phil.* 7 to one (Stoic) source ('Y') dealing with the history of philosophy. At the same time, she postulates an intermediate anti-sceptic source for Clement ('*Clementis auctor*') and suggests that Sextus, apart from Y, was also influenced by a book written by a sceptic author ('X'), whom she identifies as Aenesidemus (see the discussion on pp. 8–16). But von Wedel's source-critical stemma fails to distinguish between definitions that do and those that do not include the requirement of Cφ. Moreover, Clement's anti-sceptic source could also be informed by the same source as Sextus (and Diogenes), i.e. by a sceptic author whose aim was to distinguish scepticism from a 'school'. This, I think, is quite likely in light of the preceding and (especially) the following context of our passage, which is clearly informed by a sceptic source of some kind.

238 The Group A definition reminds us of the definition of τέχνη as σύστημα ἐκ καταλήψεων συγγεγυμνασμένων πρός τι τέλος εὔχρηστον τῶν ἐν τῷ βίῳ, ascribed to Zeno (SVF I 73; cf. II 93–97). The requirement of Cφ played an important part in Epicurean methodology; cf. DL X 86–87 (from the *Letter to Pythocles*): οὐ γὰρ κατὰ ἀξιώματα κενὰ καὶ νομοθεσίας φυσιολογητέον, ἀλλ' ὡς τὰ φαινόμενα ἐκκαλεῖται. See also such expressions as τοῖς φαινομένοις (ταῖς αἰσθήσεσι) συμφωνία, τὸ σύμφωνον τοῖς φαινομένοις, etc. (X 86–87.93.96.112). For the notion of ἀκολουθία, see also DL X 104: μόνον ὁ μῦθος ἀπέστω· ἀπέσται δὲ ἐάν τις καλῶς τοῖς φαινομένοις ἀκολουθῶν περὶ τῶν ἀφανῶν σημειῶται. The phrase ἀκολουθεῖν τοῖς φαινομένοις is already found in Aristotle, *Met.* A 5, 986b31 (regarding sensible reality which 'forced' Parmenides to posit more than one principle).

reflected in Galen's criticism of Asclepiades, whose beliefs, in Galen's view, are congruent with his principles, but conflicting with appearances.[239]

90,2f. τὸ μὲν δόγμα ἐστὶ κατάληψίς τις λογική, κατάληψις δὲ ἕξις καὶ συγκατάθεσις τῆς διανοίας. Cf. Sextus, *PH* I 16 and Ps.Galen, *Def. Med.* 13 (XIX,352,16–353,2 K.), where δόγμα is defined as a kind of συγκατάθεσις. Cf. also Sextus, *PH* I 13.[240] Here our text fully adopts the Stoic vocabulary of knowledge. On κατάληψις and συγκατάθεσις, cf. texts in LS 41.[241] The qualification λογική seems to distinguish the kind of κατάληψις that counts as δόγμα from merely perceptual apprehension.[242] ἕξις καὶ συγκατάθεσις is a hendiadys, referring to the state in which the mind finds itself when assenting to something. On the semantics of δόγμα in Greek philosophical literature, cf. Jonathan Barnes, "The Beliefs of a Pyrrhonist," in M. Burnyeat and M. Frede (eds.), *The Original Sceptics: A Controversy* (Indianapolis: Hackett, 1997), pp. 58–91, here 67–78; on the Stoic usage, see pp. 71–73.

90,4f. οὐ μόνον οἱ ἐφεκτικοί, ἀλλὰ καὶ πᾶς δογματικὸς ἔν τισιν ἐπέχειν εἴωθεν. Again (as in 89,24), there is no syntactic connection between this and the preceding sentence. But there is continuity of topic, and it is likely that the sentence is extracted from a discussion which included both the anti-sceptic polemic (15, 2–16, 1) and the reflection on what constitutes a αἵρεσις (16, 2). Here, for the first time, the opponents are identified as οἱ ἐφεκτικοί, as opposed to οἱ δογματικοί. For this distinction, cf. DL I 16: τῶν δὲ φιλοσόφων οἱ μὲν γεγόνασι

---

239 Cf. *Nat. Fac.* I 14 (138,18–21 Helmreich/II,51 K.): τοῖς [γὰρ] Ἀσκληπιάδου γράμμασιν εἴ τις ἐπιμελῶς ὁμιλήσειε, τήν τε πρὸς τὰς ἀρχὰς ἀκολουθίαν τῶν τοιούτων δογμάτων ἀκριβῶς ἂν ἐκμάθοι καὶ τὴν πρὸς τὰ φαινόμενα μάχην, and further to 139,9 (II,52,14 K.). Cf. already *Nat. Fac.* I 14, 135,3–5 Helmreich (II,46,15–17 K.). In this connection, Galen sarcastically contrasts Asclepiades with Epicurus: ἀεὶ γὰρ τὸ μὲν ἀκόλουθον φυλάττειν βούλεται, τὸ δ' ἐναργῶς φαινόμενον ἀνατρέπειν ἔμπαλιν Ἐπικούρῳ (*Nat. Fac.* I 14, 135,9–11 Helmreich/II,47 K.). Similarly ibid., 138,21–139,1.7–9 (II,51f. K.). See the previous note.

240 The parallels are discussed by von Wedel, *Symbola*, pp. 17–21.

241 Cf. e.g. Michael Frede in Algra et al. (eds.), *The Cambridge History of Hellenistic Philosophy*, pp. 295–322; R.J. Hankinson in B. Inwood (ed.), *The Cambridge Companion to the Stoics* (Cambridge: Cambridge University Press, 2003), pp. 59–84.

242 Cf. Sextus, *PH* III 47: "Of the things which exist, they [i.e. the dogmatists] say, some are objects of perception and others objects of thought, some being apprehended by the mind and others by the senses (τὰ μὲν τῇ διανοίᾳ καταλαμβάνεσθαι, τὰ δὲ ταῖς αἰσθήσεσιν); and the senses are simply affected, while the mind proceeds from the apprehension of objects of perception to the apprehension of objects of thought (ἀπὸ τῆς τῶν αἰσθητῶν καταλήψεως ἐπὶ τὴν κατάληψιν τῶν νοητῶν ἰέναι)" (trans. Annas and Barnes, modified).

δογματικοί, οἱ δ' ἐφεκτικοί· δογματικοὶ μὲν ὅσοι περὶ τῶν πραγμάτων ἀποφαίνονται ὡς καταληπτῶν· ἐφεκτικοὶ δὲ ὅσοι ἐπέχουσι περὶ αὐτῶν ὡς ἀκαταλήπτων. The earliest witness to this designation of the sceptics is Aulus Gellius (ca. 125–ca. 180), who says that it is applied both to the Academics and the Pyrrhonians (*Noct. Att.* XI 5, 6): "Utrique [*scil.* Pyrrhonii et Academici] enim σκεπτικοί, ἐφεκτικοί, ἀπορητικοί dicuntur, quoniam utrique nihil adfirmant nihilque comprehendi putant."

There is no reason to suppose that our author abandons his reservations about scepticism. But, after rejecting the universal ἐποχή, he goes on to argue that even for a dogmatist thinker, συγκατάθεσις has its limits, such as may be usefully described by means of sceptic terminology. This combination of rejection and selective appropriation of scepticism is characteristic of Galen, who on the one hand disagrees with those (Academics and Pyrrhonians) who think that every assent is necessarily rash (ἅπασα συγκατάθεσις ἐξ ἀνάγκης ... ἐστι προπετής), but on the other hand agrees that with regard to certain questions (namely those in which we lack a clear criterion), "it is more appropriate to become sceptics, rather than to dogmatize rashly" (δίκαιον ἀπορητικοὺς γίνεσθαι μᾶλλον ... ἢ προπετῶς δογματίζειν).[243]

---

243 Cf. Galen, *Pecc. Dig.* 1, 3–4 (CMG V,4,1,1: 42,13–43,2/V,59,4–61,94 K.); *Mot. Musc.* II 5 (IV,441, 12–442,2 K.). Cf. also *PHP* V 4, 12 (CMG V,4,1,2: 315,30–33/V,457 K., trans. De Lacy, modified): "If the judgments should be more or less equal, as far as credence is concerned (εἰ μὲν ἰσάζοιέν πως ἀλλήλαις ... ὡς πρὸς πίστιν), we must withhold decision about the truth of the matter (ἐπέχειν ἀναγκαῖον ἡμᾶς ἔσται περὶ τῆς τοῦ πράγματος ὑπάρξεως), but if one of them should appear far more credible, we must assent to it and perform some act in accordance with our assent (συγκατατίθεσθαι μὲν καὶ πράττειν γέ τι κατὰ τὴν συγκατάθεσιν), not, however, without circumspection." For the importance of finding a middle way between universal ἐποχή and rash assent, see also *Art. Sang.* 5 (IV,720,15–721,6 K.): "But he who loves the truth neither rejects those things that are clearly known because of his ignorance of what is known, nor gives his assent to what is unknown because of his knowledge of what is clear (οὔτε τῶν ἐναργῶς γιγνωσκομένων ἀφίσταται διὰ τὴν τῶν ἀδήλων ἀγνωσίαν οὔτε τοῖς ἀγνώστοις συγκατατίθεται διὰ τὴν τῶν ἐναργῶν γνῶσιν). He who is not of this character either has doubts about what is known along with what is not clear, or because of what is known believes what is not clear. The first happens to the sceptics, the second to most of the dogmatists. How one may avoid both these pitfalls I have shown in another place." Trans. D.J. Furley and J.S. Wilkie in *Galen on Respiration and the Arteries* (Princeton, NJ: Princeton University Press, 1984), p. 165, slightly modified. On the avoidance of rash assent, see also *Loc. Aff.* III 3 (VIII,142 K.), re. the discovery of remedies: ὥσπερ δ' ἐν ἅπασι τοῖς ἄλλοις καθ' ὅλον τὸν βίον *ἐμαυτὸν ἀεὶ προπετοῦς συγκαταθέσεως ἐπέσχον*, οὕτω καὶ περὶ τούτων ἐζήτησα πολλῷ χρόνῳ κτλ. On Galen's attitude to scepticism, cf. Phillip De Lacy, "Galen's Response to Skepticism," in M. Marcovich (ed.), *Illinois Classical Studies* 16/1–2

90,5f. ἤτοι παρὰ γνώμης ἀσθένειαν ἢ παρὰ πραγμάτων ἀσάφειαν ἢ παρὰ τὴν τῶν λόγων ἰσοσθένειαν. As noticed by von Wedel, this passage provides a connecting link with the discussion about the causes of ἐποχή below, 22, 1–4.[244] The 'weakness of thought' is more fully explained below, 93,19–22; an equivalent expression is τὸ ἀβέβαιον τῆς διανοίας (93,28f.). The metaphor is presumably of Stoic origin; it refers to a state of mind also described as 'ignorance', which prevents the mind from distinguishing between evident and non-evident presentations, thus allowing it to assent to things of which it is not sure; cf. *SVF* III 548 (Arius Didymus in Stobaeus): τὴν γὰρ ἄγνοιαν μεταπτωτικὴν εἶναι συγκατάθεσιν καὶ ἀσθενῆ. μηδὲν δ' ὑπολαμβάνειν ἀσθενῶς [*scil.* φασι τὸν σοφόν], ἀλλὰ μᾶλλον ἀσφαλῶς καὶ βεβαίως, διὸ καὶ μηδὲ δοξάζειν τὸν σοφόν. Cf. Galen, on the Stoic notion of the 'weak assent': ἀσθενῆ δὲ λέγουσι συγκατάθεσιν, ὅταν μηδέπω πεπεικότες ὦμεν ἡμᾶς αὐτοὺς οὕτως ἀληθῆ τήνδε τινὰ δόξαν ὑπάρχειν, ὡς τὸ πέντε δακτύλους ἔχειν, εἰ τύχοι, καθ' ἑκατέραν χεῖρα καὶ ⟨τό⟩· τὰ δὶς δύο τέτταρα εἶναι (*Pecc. Dig.* 1, 3/CMG V,4,1,1: 41,16–19/V,59 K. = *SVF* III 172). The 'weakness of the mind' and the 'obscurity of things' are mentioned together by Philo of Alexandria in a passage that may be influenced by a sceptic source; cf. *Quaest. Gen.* II, fr. 54a: ἀνθρώποις μὲν γὰρ ἀσθενεῖς αἱ γνῶμαι καὶ ἀβέβαιοι ὡς τὰ πράγματα πολλῆς γέμοντα ἀδηλότητος. θεῷ δὲ οὐδὲν ἄδηλον οὐδὲ ἀκατάληπτον· ἰσχυρογνωμονέστατος γὰρ καὶ βεβαιότατος.

On ἀσάφεια παρὰ πραγμάτων, cf. Philo, *Jos.* 140: δεινὴ δ' ἀσάφεια καὶ πολὺ σκότος κατακέχυται τῶν πραγμάτων. Ibid. 131: τὴν δὲ τῶν ἐκτὸς ἀσάφειαν τίς οὐκ οἶδε;[245]

(1991), pp. 283–306; R.J. Hankinson, *The Skeptics: Arguments of the Philosophers* (London and New York: Routledge, 1995), pp. 131–134; idem, "Galen on the Limitations of Knowledge," in C. Gill, T. Whitmarsh, and J. Wilkins (eds.), *Galen and the World of Knowledge* (Cambridge: Cambridge University Press, 2009), pp. 206–242; R. Chiaradonna, "Galen and Middle Platonism," in ibid., pp. 243–260.

244 Cf. von Wedel, *Symbola*, pp. 7 and 21f. Von Wedel thinks that this partial concession to the sceptic point of view is Clement's own addition to his anti-sceptic source. But I do not find much support for this in Clement's extant writings. It is true that, according to Clement, "the soul was too weak to apprehend reality" (ἠσθένει πρὸς κατάληψιν τῶν ὄντων ἡ ψυχή), and that is why the Saviour was sent as a "divine teacher"; *Strom.* V (1) 7, 8. But this clearly applies to the situation before the incarnation of Christ. The idea that the sort of knowledge mediated by Christ is impossible to reach without him is never an occasion for Clement to employ sceptic terminology. Even after describing, in the fifth book of the *Stromateis*, an equivalent to the Platonist *via negativa* with regard to the first principle (i.e. God the Father), having said that the first principle transcends "place, time, name, and understanding", Clement allows for the possibility of knowing it through the Son; cf. *Strom.* V (11) 71, 2–5. As far as I can see, there is no place left for the suspension of judgement in Clement's thought.

245 The latter parallel is pointed out by Janáček, "Ainesidemos und Sextos Empeirikos" (1980),

A similar expression is found in Galen, *Diff. Puls.* III 1/VIII,637,9–14 K. (shortly after quoting Xenophanes, fr. 34 DK: δοκὸς δ' ἐπὶ πᾶσι τέτυκται): τοσαύτης γὰρ *οὔσης τῆς ἐν τοῖς πράγμασιν ἀσαφείας,* ὥστε εἰ καὶ τριπλασίονα βίον ζήσαιμεν οὗ νῦν ζῶμεν, οὐδὲ τότ' ἂν ἴσως ἁπάντων ἡμῖν ἀκριβῶς γνωσθέντων, ... πῶς οὐκ ἄτοπος ὁ περὶ τὰ μηδὲν προσήκοντα τρίβων τὸν χρόνον; See further below, on 93,22 f. and 94,3 f.

'Equipollence' (ἰσοσθένεια) is a 'term of art' of Pyrrhonean scepticism (cf. Sextus, *PH* I 8 *et passim*; DL II 73.110), which, in essence, designates the venerable Academic practice of producing "arguments of equal weight for the opposite sides of the same subject" (cf. Cicero, *Ac.* I 45, on Arcesilaus; cf. also DL IV 28).[246] The (post-Hellenistic) sceptics may have adopted the term from Epicurus (cf. DL X 32 = LS 16B[4]: "... neither can like sense refute like, because of their equal validity [διὰ τὴν ἰσοσθένειαν]"). It is not clear who did so, however; it is absent from our testimonies on Aenesidemus (but see *Anon. In Tht.* 61,24–28 = LS 71D[3], which seems to reflect an early phase of Pyrrhonism: οὐκ ἀποφαίνεται διὰ τὸ οἴεσθαι ἰσοκρατεῖς εἶναι τοὺς εἰς τὰ ἐναντία λόγους). Our text has already hinted at the concept above, 4, 1/81,26 f. See further below, 94,1.

(VI) 17, 1–21, 6: Division and Definition

The main focus of this chapter (consisting of a dense, thematically connected, but not entirely coherent collection of excerpts) is with division and definition as methods of arriving at the essence of the investigated object. There is continuity in topic and vocabulary with the preceding sections on demonstration, with definitions being regarded as the starting-points of demonstration, the objects of demonstration, and (in case they include explanations) as equivalent to demonstrations (cf. 17, 5–18, 1). The topic is treated in a wider context of investigation, which includes semantic analysis of the proposed name (17, 1.4), inquiry in terms of εἰ, τί, διότι questions (17, 2), identification of the categories (17, 2; 20, 1–2) and the differences (17, 4; 18, 2–7) appropriate to the investigated object, etc. The relation between the division of genera and the definition of

---

in idem, *Studien*, pp. 251–264, here 253 [617], who traces it back to Aenesidemus. As for *Jos.* 140, cf. Hans Joachim Krämer, *Platonismus und hellenistische Philosophie* (Berlin/New York: de Gruyter, 1971), p. 79 n. 299, who registers similar expressions in *Ebr.* 167 (πολὺ γὰρ σκότος τῶν ὄντων καὶ σωμάτων καὶ πραγμάτων κατακεχυμένον οὐκ ἐᾷ τὴν ἑκάστου φύσιν ἰδεῖν), but also in Cicero, *Ac.* I 44–45, II 61 and 122.

246 For the function of equipollence in the sceptical Academy, cf. Lévy, *Cicero Academicus*, pp. 260–263.

essence is explored and compared to the relation between analysis and synthesis (18, 4–7), and between matter and a craftsman (19, 1). The difference between species and parts is also explained (19, 3–8). Recalling the polemical context of the earlier discussion about demonstration, the text presents definitions as tools against "controversies" and "deception" (20, 4), but it also acknowledges that the capacity of definitions to grasp the essence is limited (21, 6). Added to definitions are differences corresponding to the Aristotelian 'properties', whose role is (unfortunately, very briefly) discussed in the last paragraph (21, 1–5). The doctrinal background of the whole chapter is essentially Aristotelian, mainly informed by the *Topics* and the second book of the *Posterior Analytics*; however, the exposition is also influenced by sources paralleled in the 'Middle Platonist' literature.

90,7f. αἱ τῶν ζητήσεων ἔφοδοι καὶ ἀρχαὶ περὶ ταῦτα καὶ ἐν τούτοις εἰσίν. In L, these words are written as a superscription of the same kind as στρωματεὺς ὄγδοος in L 346ᵛ and ἐκ τῶν Θεοδότου κτλ. in L 361ʳ (all three titles are centre-aligned and marked by an asterisk; see Plates 1–3 above, pp. 4–6). They are repeated (without εἰσίν) as a *subscriptio* in L 361ʳ, after the end of our 'eighth book' (33, 9/102,12), and thus refer to the whole section 17, 1–33, 9 (= chapters 6–9). We neither know who attached these words to our material nor when and why they were attached. Although they are presented as a title, it is unlikely that this was their original function (it would be odd to have deictic pronouns in a title). They rather seem to be a sentence extracted from a passage in which the topics signalled by the phrase περὶ ταῦτα καὶ ἐν τούτοις were introduced.[247] If this is the case, the sentence could have been picked up by the excerptor to summarize the contents of the subsequent sections. In fact, with the exception of the (apparently misplaced) chapter 7, it fits these contents remarkably well: Division and other methods discussed or mentioned in chapter 6 are indeed 'procedures' of investigation, while definitions (ch. 6), categories (ch. 8), and causes (ch. 9) may all be described as its 'principles'.

17, 1–18, 1

90,9f. προτακτέον δὴ καὶ τῶν ὅρων καὶ τῶν ἀποδείξεων καὶ διαιρέσεων ⟨τὸ⟩ ποσαχῶς λέγεται τὸ ζητούμενον. As is often the case in our 'chapter', the syntax is remiss or the text corrupt. It is not clear why definitions, demonstrations, and divisions are listed in that order and why διαιρέσεων lacks an article; but

247 Cf. e.g. Theophrastus, fr. 5 (Περὶ ἀνέμων), 1, 1 (Wimmer): αἱ γὰρ διαφοραὶ [*scil.* τῶν ἀνέμων] περὶ ταῦτα καὶ ἐν τούτοις, οἷον μέγεθος, μικρότης, ψυχρότης κτλ.

more disturbing is the fact that προτακτέον lacks an object. The last situation, I believe, calls for emendation, which could be achieved by placing an article before ποσαχῶς and taking the phrase τὸ ποσαχῶς κτλ. as an elliptic reference to a course of investigation already known to the reader (the article encapsulating τὸ διελεῖν or something of that sort).

The consecutive and resumptive δή indicates that, in its original context, the sentence resumed an earlier argument. The argument is not preserved here, but it was probably concerned with at least some of the topics mentioned in the first paragraph of the chapter. Starting from 18, 2, the chapter provides a relatively coherent account of division and definition. The aim of the first paragraph seems to be to prepare the reader for this topic by outlining a more general background to the investigative (and demonstrative) procedure and showing what role division and definition play in this regard. Apart from a disorderly manner of presentation, the paragraph is also extremely brief and dense, and deals with a number of difficult issues in a way that can be understood only by a reader who is already familiar with them. These features may be partly due to the excerptor and/or the copyist, but it is also possible that in the original text the reader was indeed expected to be already familiar with the topics and was only briefly reminded of them at the start of a new chapter. Thus, in our sentence, the reader is reminded what comes first in the order of investigation (for the notion of τάξις, cf. above, on 81,23 f.).

For the distinction of meanings of the proposed word, cf. Aristotle, *Top.* I 15 and 18 (106a1–107b37, 108a18–37). For the importance of making these distinctions before defining the thing in question, cf. Galen, *Diff. Puls.* II 3 (VIII,571,1–574,5 K.).[248] Cf. also the discussion about the meanings of the words 'animal' and 'embryo' above, 9, 7–13, 8.

90,10 f. τά τε ὁμώνυμα χωριστέον καὶ τὰ συνώνυμα εὐκρινῶς τακτέον κατὰ τὰς σημασίας. The manuscript reading (τά τε ὁμώνυμα) χειριστέον is defended by Stählin, who refers to the phrase τὸν λόγον χειρίζομαι in *Strom.* V (1) 5, 3; cf. also Sextus, *M.* I 131; I 165; VII 443; XI 21; Galen, *PHP* III 5, 18 (CMG V 4,1,2: 204,15/V,326 K.). But the meaning of χειρίζομαι in these contexts ('to avail oneself of', 'to employ') does not fit in with our sentence, nor does any other meaning of the verb found in the dictionary, and Sylburg's simple correction is probably right.[249]

---

248 Cf. Iwan von Müller, "Über Galens Werk vom wissenschaftlichen Beweis," in *Abhandlungen der königlich bayerischen Akademie der Wissenschaften, phil.-philol. Kl.* 20 (1897), pp. 403–478, here 445.

249 Cf. the use of χωρίζειν in Aristotle, *Top.* I 18, 108b6.

Both clauses, coordinated by τε ... καί, serve to explain what is entailed by an investigation of ποσαχῶς λέγεται τὸ ζητούμενον: Starting with the name by which the subject matter is called, one has to separate the subject matter from items having the same name, but a different definition (cf. below, 24, 7). As far as synonyms are concerned, our text probably uses the word in the (Aristotelian) sense of items having the same name and the same definition on account of the same genus (cf. below, 24, 2). When the proposed name is such that it could refer to a genus, the researcher may need to make sure whether the item referred to is in fact the genus as a whole or rather some of its (synonymously called) species; in doing so, he determines the σημασία of the proposed name.

90,11–13 εἰ τῶν πρὸς ἕτερα θεωρουμένων ἐστὶν τὸ προκείμενον ἢ καθ' αὑτὸ λαμβάνεται. According to Philo, *Ebr.* 187, "... each thing is incomprehensible by itself, while it seems to be known by being compared to something else (ἐξ ἑαυτοῦ μὲν γὰρ ἕκαστον ἀκατάληπτον, ἐκ δὲ τῆς πρὸς ἕτερον συγκρίσεως γνωρίζεσθαι δοκεῖ)." This agrees with the tenth Aenesidemean mode according to DL IX 87–88, which, after describing various attributes as known in relation to something else (κατὰ τὴν πρὸς ἄλλα σύμβλησιν, κατὰ τὴν ὡς πρὸς τὸ ἕτερον σχέσιν) and all things as known in relation to the mind (πρὸς τὴν διάνοιαν), concludes that nothing is known in itself (ἄγνωστα οὖν τὰ πρός τι καθ' ἑαυτά). In a more optimistic (and Aristotelian) manner, our text applies the same distinction to the inquiry of problems (for τὸ προκείμενον, cf. above, on 87,7 f.), maintaining that some belong to issues known in relation to something, while others are grasped by themselves. In light of a parallel distinction in 24, 1, where it is applied to the Aristotelian categories, it is likely that in our passage the dividing line is drawn between (a) problems concerned with entities belonging to the category of relation (such as a capacity or a cause of something; cf. Galen, *Nat. Fac.* I 4/107,8–14 Helmreich/II,9 K.) and (b) those dealing with issues falling under some of the remaining categories. See further below, on 24, 1/95,4 f.

90,13 f. εἰ ἔστι, τί ἐστι, τί αὐτῷ συμβέβηκεν ... εἰ ἔστι, τί ἐστι, διότι ἐστίν. At the beginning of book two of the *Posterior Analytics*, Aristotle distinguishes four kinds of research questions: (a) τὸ ὅτι, (b) τὸ διότι, (c) εἰ ἔστι, (d) τί ἐστι (*An. Post.* II 1, 89b24 f.). They correspond to the inquiry (a) if P is the case, (b) why it is the case, (c) if X exists, and (d) what it is.[250] In contrast, our text proposes two lines of investigation, describing them by means of two series of questions: in the second series, three of the Aristotelian questions (εἰ, τί, διότι) are picked

250 Cf. Barnes, *Posterior Analytics*, pp. 203 f.

up and arranged in a way that seems to be suggestive of the recommended order of inquiry. In the first series, the διότι question is replaced by an inquiry of attributes (τί αὐτῷ συμβέβηκεν). Now it is not entirely clear what the scope of the εἰ question is. On the one hand, it probably includes existence (cf. above, 3, 4: εἰ ὑπάρχει); on the other hand, it does not seem to be distinguished from the ὅτι question mentioned below, 90,27, where it must be concerned with the fact (cf. below, on 90,26 f. and 91,1 f., where εἰ picks up on the ὅτι of 90,27; cf. also on 90,28–91,3). The distinction between the two lines of research is also open to interpretation. With regard to the second line, we are told that it is a line of demonstration (17, 8/91,1 f.). The first line is left unexplained, but it could be explained in view of 9, 1, where the exploration of attributes of a known substance belongs to the scope of the demonstrative method as well (cf. above, on 85,1–13).

90,14–16 πρὸς δὲ τὴν τούτων θεωρίαν … συνυποβάλλει καὶ τὰ πρότερα καὶ †τὰ ἐπὶ τὰς† διαφορὰς [αἱ διαιρέσεις αὐτῶν]. As presented in the manuscript, this sentence is corrupt. There are several problems here, the most obvious one being the syntactically impossible end (τὰ ἐπὶ τὰς διαφορὰς αἱ διαιρέσεις αὐτῶν). Referring to Aristotle, *An. Post.* II 13, 96b25 (αἱ … διαιρέσεις αἱ κατὰ τὰς διαφοράς), Stählin proposes that we should read αἱ ἐπὶ τὰς διαφορὰς διαιρέσεις αὐτῶν. This provides us with four grammatical subjects whose common predicate is συνυποβάλλει. However, whatever this verb means, it is unlikely that it could function without a direct object. Thus Stählin (in the apparatus) suggests to change it to συμβάλλεται. This is attractive insofar as συμβάλλεται does not need a direct object and neatly connects with πρός τι in the sense of 'contribute to'.[251] But it is hard to explain why any scribe would mistake a perfectly intelligible word for an obscure one; the position of a predicate between two pairs of its subjects is also odd; and the meaning of the sentence, especially as regards its second part, remains unclear. But is there anything else we could do with the text?

If the word συνυποβάλλει is sound and if the sentence is complete,[252] it is likely that the nouns following it designate its objects. The first of them, τὰ πρότερα, probably refers to genera in terms of which things are defined (cf. e.g.

---

251 Stählin's solution is followed by Fregonara who translates the whole sentence as follows: "Allora, per questo tipo di esame contribuiscono insieme la conoscenza del particolare e dell' universale, i precedenti e le loro distinzioni secondo differenze."

252 Apart from Stählin, the first claim is challenged by Lowth (in the Potter edition) who suggests to change συνυποβάλλει to συνεπιβάλλει. Schwartz (in the Stählin edition) marks a lacuna after συνυποβάλλει.

Aristotle, *Top.* VI 4, 141b25–28). What follows must be somehow based on the idea that the genera are divided according to specific *differentiae*. It consists of two phrases: τὰ ἐπὶ τὰς διαφοράς and αἱ διαιρέσεις αὐτῶν. The second phrase likely refers to the genera and their division, but it is syntactically incompatible with the rest of the sentence. I assume that it entered the text as a marginal gloss, perhaps as an attempt to explain the phrase τὰ ἐπὶ τὰς διαφοράς. Indeed, this phrase is a puzzle: contrary to what is implied by Stählin (and made explicit in Fregonara's translation), ἐπὶ τὰς διαφοράς cannot mean 'according to *differentiae*', nor does it seem to convey anything intelligible in the context of the sentence.[253] I think it is likely that something has been omitted or distorted here. The most attractive solution at hand is to replace τὰ ἐπὶ τὰς with τὰς εἰδικάς; cf. below, 92,9; Galen, *San. Tu.* IV 2 (CMG V,4,2: 104,15/VI,236 K.); *Diff. Puls.* II 6 and 10 (VIII,601,16, 625,17 and 633,8 f. K.); *Soph.* 4 (XIV,597,8–12 K.), etc. As far as the verb συνυποβάλλει is concerned, its meaning could be derived from the attested use of ὑποβάλλω in the sense of 'provide'.[254] The idea, presumably, is that the knowledge of the particular and that of the universal provides that which is necessary for the study of the above-mentioned questions, namely the genus and the *differentiae*. This is further explained in 90,16–21.

90,16 f. ἡ μὲν οὖν ἐπαγωγὴ φέρει τὸ καθόλου καὶ τὸν ὁρισμόν. Cf. Aristotle, *Top.* I 12, 105a13 f.: ἐπαγωγὴ δὲ ἡ ἀπὸ τῶν καθ' ἕκαστα ἐπὶ τὸ καθόλου ἔφοδος. The ὁρισμός reached by induction must be a universal proposition that something is the case, derived from particular instances, e.g. "the skilled man is the best at his particular task" (*Top.* 105a15 f., trans. Pickard-Cambridge). Cf. below, 17, 5/90,22–24, on two kinds of definition, and 17, 6–7/90,24–27, on induction.

90,18 τὸ τί ἦν εἶναι καὶ τὸ ἄτομον. For τὸ τί ἦν εἶναι, cf. W.D. Ross, *Aristotle's Metaphysics*, vol. I (Oxford: Clarendon Press, 1924), p. 127; Barnes, *Posterior Analytics*, p. 174. Cf. below, 91,7 f. Τὸ ἄτομον refers to the unsplittable species, i.e. the essence of the subject matter, discovered through division; cf. below, 18, 2–7 and note on 91,16 f.

---

253 This is recognized by Stählin who, in his 'Nachträge und Berichtigungen' to the 1909 edition, proposes to change τὰ ἐπὶ τὰς διαφοράς to αἱ κατὰ τὰς διαφοράς (Clemens Alexandrinus, Bd. IV: Register, GCS 39, 1936, p. LXIX, *ad* 90,16).

254 cf. Galen, *Loc. Aff.* V 8 (VIII,366,11–13 K.): πολλάκις γὰρ ἡ μὲν τύχη μεγάλας εὐδοκιμήσεως ἀφορμὰς ὑποβάλλει, χρῆσθαι δ' αὐταῖς οἱ πολλοὶ δι' ἀμαθίαν οὐκ ἴσασιν. Cf. the use of φέρει, συναύξουσι, and παρίσταται in a similar sense below, 90,17.20.21.

90,18 f. ἡ δὲ θεωρία τοῦ ποσαχῶς τὴν λῆψιν τοῦ οἰκείου. Here τὸ οἰκεῖον is probably a difference appropriate to the thing in question; cf. Aristotle, *Top.* I 18, 108b4–6 (trans. Pickard-Cambridge): "[The discovery of differences] … helps us in recognizing what a thing is, because we usually distinguish the account that is proper to the substance of each particular thing by means of the *differentiae* that are appropriate to it (ταῖς περὶ ἕκαστον οἰκείαις διαφοραῖς)." For 'appropriate differences', cf. also *Met.* B 3, 998b25; Galen, *PHP* IX 6, 56 (CMG V 4,1,2: 584,11/V,775 K.). For the distinction of meanings, cf. above, 17, 1.

90,19 αἱ δὲ διαπορήσεις τὰς κατ' αὐτὰς διαφοράς. Again, the background is presumably Aristotle's *Topics*, where the examination of ποσαχῶς λέγεται is followed by the investigation of differences; cf. *Top.* I 16, 107b38–108a6 and 18, 108a38–108b6. The text seems to link this investigation with the Aristotelian method of 'going through the difficulties'; cf. e.g. *Top.* I 2, 101a34–36 (trans. Smith): "[This study] is useful in relation to the philosophical sciences because if we have the ability to go through the difficulties on either side (πρὸς ἀμφότερα διαπορῆσαι) we shall more readily discern the true as well as the false in any subject."

91,20 τὴν θεωρίαν συναύξουσι. For the notion of 'adding to the study', cf. Galen, *Hipp. Nat. Hom.* II 6 (CMG V,9,1: 70,5–7/XV,136 K.): "And what shall we say concerning those [doctors] who have added most to the study of anatomy after these (περὶ τῶν μετ' αὐτοὺς ἐπὶ πλεῖστον αὐξησάντων τὴν ἀνατομικὴν θεωρίαν), like Herophilus and Eudemus …?"

90,22 τὸ κεφαλαιωθὲν ἐκ τῆς διαιρέσεως ὅρος γίνεται. The idea of definition as a summary of the outcomes of division is further explained below, 18, 2–7 (cf. esp. on 91,7 f.); another word expressing the relation between the two procedures is σύνθεσις (cf. 18, 4 and 19, 1/91,28 f.: ἐπισυντιθείς). There is an interesting parallel to this definition of ὅρος in the Suda lexicon (*s.v.* ἀνάλυσις): ὅρος ἐστι λόγος κατὰ ἀνάλυσιν ἀπαρτιζόντως ἐκφερόμενος τοῦ ὁριστικοῦ καὶ κεφαλαιωδῶς. The word κεφαλαιωδῶς, 'summarily' (governed by ἐκφερόμενος and co-ordinate with ἀπαρτιζόντως), probably expresses the view that ὅρος sums up what is discovered about the defined object through ἀνάλυσις.[255] The same definition is at the background to a difficult passage in Alexander's commentary on the *Topics* (CAG II/2: 42,27–43,2): οἱ δὲ λέγοντες *ὅρον εἶναι λόγον κατὰ ἀνάλυσιν ἀπαρτιζόν-*

255 Presumably, ὁριστικοῦ should read ὁριστοῦ. I take it that the genitive is governed by ἀνάλυσιν. Cf. the parallel in Alexander quoted below.

τως ἐκφερόμενον, ἀνάλυσιν μὲν λέγοντες τὴν ἐξάπλωσιν τοῦ ὁριστοῦ καὶ κεφαλαιωδῶς, ἀπαρτιζόντως δὲ τὸ μήτε ὑπερβάλλειν μήτε ἐνδεῖν κτλ. Here the word κεφαλαιωδῶς seems to be misplaced and it is tempting to think that it originally belonged to the definition of ὅρος, too (it could have been there in the same position as in the Suda, perhaps even along with, and after, τοῦ ὁριστοῦ). The context in Alexander indicates that the definition is of a Stoic origin; this is supported by another parallel ascribed to Antipater.[256]

90,22–24 καὶ πρότερος τῆς διαιρέσεως καὶ ὕστερος ὁ διορισμός κτλ. The first kind of definition seems to apply to universal propositions that serve as starting-points of a dialectical argument or proof; one way to arrive at such definitions is by means of induction; cf. above, on 90,16. The second kind applies to disputed problems; it shows the essence of the subject matter on the way of providing a proof. Cf. above, on 84,25 f.

90,25 ἐκ τοῦ καθ' ἕκαστα κεφαλαιοῦται τὸ καθόλου. For the vocabulary, cf. Sextus, *M.* VII 224, reporting on Peripatetic epistemology: ἡ συγκεφαλαίωσις τῶν ἐπὶ μέρους εἰς τὸ καθόλου ἔννοια καλεῖται. Cf. also Galen, *Hipp. Off. Med.* I 3 (XVIIIb,652,1–3 K.): ὀνομάζουσι δὲ τὴν συναρίθμησιν (*scil.* τῶν αἰσθήσεων) οὐχ οὕτως μόνον, ἀλλὰ καὶ συγκεφαλαίωσιν. Alexander, *In Met.* A 1 (CAG I: 4,28–5,2): ἡ μὲν γὰρ ἐμπειρία τῶν πολλῶν τε καὶ καθ' ἕκαστα ἤγουν αἰσθητῶν ... εἰς μίαν περίληψιν καὶ γνῶσίν ἐστι συστολή, ἡ δὲ τέχνη τῶν πολλῶν ἐμπειριῶν συγκεφαλαίωσίς ἐστιν.

90,26 f. ἡ ... ἐπαγωγὴ οὐ τὸ τί ἐστι δείκνυσιν, ἀλλ' ὅτι ἐστὶν ἢ οὐκ ἔστιν. Cf. Aristotle, *An. Post.* II 7, 92a38–b1: οὐ γὰρ τί ἐστι δείκνυσιν (*scil.* ἐπαγωγή), ἀλλ' ὅτι ἢ ἔστιν ἢ οὐκ ἔστιν.

90,28–91,3 ὅ τε διορισμὸς ... τὸ τί ἐστι διδάσκει, οὐχὶ δὲ τὸ εἰ ἔστιν, ἥ τε ἀπόδειξις τὰ τρία, τό τε εἰ ἔστιν καὶ τὸ τί ἐστιν καὶ τὸ διὰ τί ἐστιν, σαφηνίζει. The contrast between definition and demonstration seems to be based on the aporematic chapters in *An. Post.* II 3–7, where it is shown why definitions cannot be used as proofs. Cf. e.g. *An. Post.* II 3, 91a1 f.: "Definitions show what something is, whereas demonstrations show that this is or is not true of that." Ibid. II 7, 92b19 f.: "It is clear too that, in the current methods of definition, definers do

---

256 Cf. DL VII 60: ὅρος δέ ἐστιν, ὥς φησιν Ἀντίπατρος ἐν τῷ πρώτῳ Περὶ ὅρων, λόγος κατ' ἀνάλυσιν ἀπαρτιζόντως ἐκφερόμενος. For κεφαλαιωδῶς, cf. also Epictetus, *Diss.* II 12, 9: κακῶς ὡρίσω· οὐ γὰρ ἀντακολουθεῖ τῷ κεφαλαιώδει τὸ ὁρικόν.

not prove that anything exists." The triad of questions belonging to the agenda of proof (εἰ, τί, διὰ τί) reflects, in turn, Aristotle's discussion in *An. Post.* II 8–12.

91,3 εἰσὶ δὲ ἔνιοι καὶ τῆς αἰτίας ἐμπεριεκτικοὶ ὅροι. Cf. *An. Post.* II 10, 93b38f.: … ἄλλος δ' ἐστὶν ὅρος λόγος ὁ δηλῶν διὰ τί ἔστιν. As Barnes points out, "this (…) is in fact the *only* genuine sort of definition which Aristotle's science will admit" (*Posterior Analytics*, p. 224).

91,4–6 ἐπεὶ δὲ ἐπίστασθαι ἔστιν ὅταν ἴδωμεν τὴν αἰτίαν, αἰτίαι δὲ τέσσαρες … τετραχῶς ἔσται ὁ διορισμός. Like ὅρος in the previous sentence, the word διορισμός in this one likely refers to a conclusion of a demonstrative argument (cf. Aristotle, *An. Post.* II 10, 94a13f.: τῆς τοῦ τί ἐστι ἀποδείξεως συμπέρασμα). The four sorts of definition mentioned here differ according to the sort of explanation provided by the middle term. Cf. *An. Post.* II 11, 94a20–24: ἐπεὶ δὲ ἐπίστασθαι οἰόμεθα ὅταν εἰδῶμεν τὴν αἰτίαν, αἰτίαι δὲ τέτταρες … πᾶσαι αὗται διὰ τοῦ μέσου δείκνυνται. This thought, discussed by Aristotle in *An. Post.* II 11–12, is not developed any further in our text (though it could be seen as a connecting link between this chapter and the section on causes; cf. esp. 28, 2/98,3–5).

18, 2–7

91,6–26 This paragraph is thematically linked to *An. Post.* II 13, a chapter whose aim is to show "how we should hunt out the items predicated in what something is" (96a22f.). The correspondences noted above with the preceding chapters of *An. Post.* II make it likely that this affinity is not accidental. However, the way the topic is presented here is largely independent from Aristotle's exposition, incorporating elements from other Aristotelian writings as well as later sources, paralleled, most conspicuously, in Alcinous and Galen.

91,6 ληπτέον οὖν πρῶτον τὸ γένος, ἐν ᾧ ἐστι τῶν ἐπάνω τὰ ἐγγύτατα. The genus in terms of which an object is defined should contain universals that are as near to the object as possible. Plainly, these universals are species into which the genus in question is divided. Cf. below, 18, 5/91,13: τὰ ἐνυπάρχοντα αὐτῷ [*scil.* τῷ γένει] εἴδη, 18, 7/91,23f.: τὰ προσεχῆ τῶν … εἰδῶν τῷ ἀνθρώπῳ, and note on 91,17–23. This instruction is not mentioned in *An. Post.* II 13, but it could be derived from *Top.* VI 5, 143a15–28; cf. 143a20–22 (trans. Pickard-Cambridge): "[F]or the man who puts [the object defined] into the nearest [genus] has stated all the higher genera (ὁ γὰρ εἰς τὸ ἐγγυτάτω θεὶς πάντα τὰ ἐπάνω εἴρηκεν), seeing that all the higher genera are predicated of the lower." As Galen reports, there was a "considerable discussion as to what genus should be placed first of all in a definition—the highest (…), or the proximate, or perhaps one of the

intermediates ..." (*Diff. Puls.* IV 7/VIII,734,11–735,1 K.).[257] For the function of οὖν, cf. above, on 84,30 f.

91,7 τὴν προσεχῆ διαφοράν. Cf. Aristotle, *An. Post.* II 13, 96b35–97a6, who uses the expression πρωτὴ διαφορά to the same effect (97a1): "The first difference of animal is that into which every animal falls; and similarly for everything else (...) If you continue in this way you can know that nothing has been omitted ..." (trans. Barnes). For the vocabulary, cf. Alcinous, *Did.* 5, 3 (H 157,7): τέμνειν [*scil.* τὸ γένος] κατὰ τὰς προσεχεῖς διαφορὰς κτλ. Galen, *Diff. Puls.* IV 8 (VIII,736,12 ff. K.): οὐ μόνον τὸ γένος ... ἀλλὰ καὶ τὴν ... προσεχῆ διαφορὰν κτλ.

91,7 f. ἡ ... συνέχεια τῶν διαφορῶν ... τὸ τί ἦν εἶναι ἐκπληροῖ. Cf. Aristotle, *PA* I 3, 643b33 f.: "... the continuity of the differences derived from the genus according to its division (ἡ συνέχεια ... τῶν ἀπὸ τοῦ γένους κατὰ τὴν διαίρεσιν διαφορῶν) means just this, that the whole is a single thing."[258] It is important for Aristotle that this continuity is not merely a 'conjuction' (σύνδεσμος), but a unity corresponding to one of the senses of being (i.e. the categories); cf. *An. Post.* II 10, 93b35–37; *PA* I 3, 643b17 ff.; *Met.* Z 4, 1030b7–13;[259] cf. also *Met.* I 1, 1052a19–25. Cf. below, 20, 2/92,20–23. For ἐκπληροῖ, cf. Barnes in Porphyry, *Introduction*, pp. 179 f. (on *Isag.* 10,10).

91,9 f. οὐκ ἀνάγκη δὲ πάσας ... τὰς διαφοράς, ἀλλὰ τὰς εἰδοποιούσας. Cf. Aristotle, *An. Post.* II 13, 97a6 f.: "If you are defining and dividing you do not need to know everything there is." For the vocabulary, cf. Aristotle, *Top.* VI 6, 143b8 f.: πᾶσα γὰρ εἰδοποιὸς διαφορὰ μετὰ τοῦ γένους εἶδος ποιεῖ. Cf. also Galen, *MM* I 3 (X,23,10–16 K.): "For ... not every difference that is conjoined with the genus contributes something towards the creation of the species, but only that from the appropriate division of the genus. Only these are species-forming differences: all the others are superfluous (αὗται γάρ εἰσιν εἰδοποιοὶ μόναι τῶν διαφορῶν, αἱ δ' ἄλλαι πᾶσαι περιτταί)."[260]

91,10–12 ἡ γεωμετρικὴ ἀνάλυσις καὶ σύνθεσις τῇ διαλεκτικῇ διαιρέσει καὶ ὁρισμῷ ἔοικεν κτλ. There is no doubt that analysis and synthesis were practised in

257 Trans. Jonathan Barnes, *Porphyry: Introduction*, translated with an introduction and commentary (Oxford: Clarendon Press, 2013), p. 112.

258 Trans. Lennox, modified.

259 For other references, cf. Barnes, *Posterior Analytics*, p. 217 (*ad* 92b30).

260 Trans. Hankinson, modified.

ancient geometry, but their precise function is disputed.[261] According to a widespread interpretation, whose elements, at least, go back to the classical period, analysis is a heuristic method applied to problems or theorems in order to find some agreed ἀρχαί from which they could be solved or proved. Synthesis, then, is a process of proving a theorem or solving a problem (e.g. constructing a proposed figure) from these ἀρχαί.[262] For analysis as a way towards principles, cf. also above, 8, 1/84,9–13. While explaining the relation between division and definition, our text indicates that it is similar to the one between analysis and synthesis. How is it similar? According to our passage, the point of convergence between division and analysis is that both are procedures by which "we ascend to what is more simple and primitive" (91,11 f.). Later we learn that this ascent consists in dividing the genus of the investigated object (τὸ ζητούμενον) into its "inherent" species (91,12 f.), until we reach "the simplest species" (τὸ ἁπλούστατον εἶδος), which includes nothing but the object in question (91,21–23).[263] This is followed by a 'synthesis' of the specific *differentiae* discovered through division (91,25 f.); cf. also above, 17, 4/90,22; 18, 2/91,7 f., on the "summary" of the *differentiae*. At first, the analogy seems to work poorly, as the division of a genus is not a reversal of the process by which we construct definitions. But in fact, the item likened to the object of analysis is not the genus, but the *name* of the thing sought, or rather the concept associated therewith. Cf. below, 19, 1/91,27–29: ἡ διαίρεσις ... τὴν ἁπλότητα τοῦ ὀνόματος ἀναζητοῦσα. The background is Aristotelian; cf. esp. *Phys.* I 1, 184a21–184b14, on definitions arrived at by a division of a compounded whole signified by a given name.[264] Cf. also Galen, *MM* II 7: "And the decomposition of each name into its constituent definition proceeds in this way."[265] The idea that definition is based on 'analysis' is also present in

---

261 I have been educated on this matter, which exceeds my competence, by Fabio Acerbi; see his studies "The Language of the 'Givens': Its Forms and Its Use as a Deductive Tool in Greek Mathematics," *Archive for History of Exact Sciences* 65/2 (2011), pp. 119–153, and "The Geometry of Burning Mirrors in Classical Antiquity: Analysis, Heuristic, Projections, Lemmatic Fragmentation," *Archive for History of Exact Sciences* 65/5 (2011), pp. 471–497, here esp. 481 f.

262 For the philosophical view of analysis, whose traces can already be discerned in Aristotle and Plato, cf. esp. Stephen Menn, "Plato and the Method of Analysis," *Phronesis* 47/3 (2002), pp. 193–223.

263 For the implied notion of genera as compounds, see below, on 91,13 f.

264 Cf. Robert Bolton, "Aristotle's Method in Natural Science: Physics I," in L. Judson (ed.), *Aristotle's Physics: A Collection of Essays* (Oxford: Clarendon Press, 1991), pp. 1–29.

265 *MM* II 7 (X,149,17–150,1 K.): καὶ τό γε διαλύειν ἕκαστον τῶν ὀνομάτων εἰς λόγον οὕτω γίνεται. Trans. Hankinson. For this passage, cf. above, on 83,26 f.

the Stoic definition of ὅρος mentioned above (on 90,22); there analysis is equivalent to an ἐξάπλωσις ('explanation') of the *definiendum*.[266] Cf. also Galen, *Ars. Med.* Ia 1.4, where the 'analysed' object is not a name but a notional definition, whose analysis (also described as διαίρεσις or ἐξάπλωσις) leads to the definition of essence.[267]

91,13 f. εἰς τὰ ἐνυπάρχοντα αὐτῷ [*scil.* τῷ γένει] εἴδη … εἰς τὰ ἐμφαινόμενα εἴδη. It is a surprising thing to say that species are 'inherent' and 'mirrored' in their genus.[268] It is reminiscent of the way Aristotle describes the division of quantity into parts, rather than of genus into species; cf. *Met.* Δ 13, 1020a7 f.: ποσὸν λέγεται τὸ διαιρετὸν εἰς ἐνυπάρχοντα ὧν ἑκάτερον ἢ ἕκαστον ἕν τι καὶ τόδε τι πέφυκεν εἶναι, and Alexander's commentary *ad loc.*: τὸ δὲ γένος οὐχ οὕτως εἰς τὰ εἴδη διαιρεῖται ὡς εἰς ἐνυπάρχοντα· οὐ γὰρ ἐν τῷ γένει οἱ τῶν εἰδῶν ἐνυπάρχουσι λόγοι, εἰς οὓς διαιρεῖται (CAG I: 396,9 f.). But our author is aware of the difference between the two kinds of division (cf. below, 19, 3–6) and even notes that "genus is inherent in the species" (19, 7/92,11 f.). Clearly the language of our passage is still influenced by the analogy between division and analysis, where genera "appear to be compounds" (cf. 91,15 f.: τὰ σύνθετα δοκοῦντα εἶναι γένη).[269] There is an interesting parallel in Philo, *Opif.* 76 (trans. Runia): "Most excellently, after

266 For ἐξάπλωσις, cf. also Troilus the 'Sophist', *Prolegomena* in *Rhetores Graeci*, ed. C. Walz, Vol. VI (Stuttgart/Tübingen: Cotta, 1834), p. 50,2–5: ὅρος … ἐστὶ λόγος σύντομος, δηλωτικὸς τῆς ἑκάστου φύσεως· λόγος μὲν πρὸς ἀντιδιαστολὴν τοῦ ὀνόματος· τὸ γὰρ ὄνομα συστολή ἐστι τοῦ λόγου· ὁ δὲ λόγος ἐξάπλωσίς ἐστι τοῦ ὀνόματος.

267 *Ars. Med.* Ia 1.4 (274,6–9; 275,8–15 Boudon/I,305 f. K.): καλεῖν δὲ ἔξεστι τὴν τοιαύτην διδασκαλίαν (*scil.* ὅρου διάλυσιν), οὐ μόνον ὅρου διάλυσιν, ἀλλὰ καὶ διάπτυξιν, ὥς τινες ὠνόμασαν, *ἢ ἀνάλυσιν, ἢ διαίρεσιν*, ὡς ἕτεροί τινες, ἢ *ἐξάπλωσιν*, ἢ ἐξήγησιν, ὡς ἄλλοι … εὐμνημόνευτα γὰρ ἱκανῶς ἐστι τὰ ἐξ ὅρου διαλύσεως ἅπαντα, διὰ τὸ περιέχειν ὅλης τῆς τέχνης ἐν ἑαυτῷ τὰ κεφάλαια τὸν ἄριστον ὅρον, ὅνπέρ καὶ οὐσιώδη τινες ὀνομάζουσιν, *ἀντιδιαιρούμενοι τοῖς ἐννοηματικοῖς προσαγορευομένοις*· ἐκεῖνοι μὲν γὰρ ἀπὸ τῶν συμβεβηκότων οἷς ὁρίζονται πράγμασιν, οὗτοι δὲ ἀπὸ τῆς οὐσίας αὐτῆς συνίστανται.

268 Aristotle too sometimes uses these verbs to convey two aspects of the same relation; cf. esp. *DA* II 2, 413a13–16: οὐ γὰρ μόνον τὸ ὅτι δεῖ τὸν ὁριστικὸν λόγον δηλοῦν, ὥσπερ οἱ πλεῖστοι τῶν ὅρων λέγουσιν, ἀλλὰ καὶ τὴν αἰτίαν *ἐνυπάρχειν καὶ ἐμφαίνεσθαι*. Cf. also *Pol.* I 5, 1254a28–32.

269 The view of genera as compounds could be influenced by a difficult passage in *An. Post.* II 13, 96b15–25, where Aristotle speaks of items "compounded from the unsplittable [species]" (96b21: τοῖς γὰρ συντιθεμένοις ἐκ τῶν ἀτόμων κτλ.). According to a traditional interpretation (described and disputed by W.D. Ross, *Aristotle's Prior and Posterior Analytics*, Oxford: Clarendon Press, 1957, pp. 657 f.), these items are genera, whose attributes are inductively inferred from the definitions of the species of which the genera 'consist'; cf. Themistius, *In An. Post.* (CAG V/1: 57,8 f.): συντίθεται δὲ ἐκ μὲν τῶν καθ' ἕκαστον ὁ τοῦ εἴδους ὁρισμός, ἐκ δὲ τῶν εἰδῶν ὁ τοῦ γένους. Philoponus, *In An. Post.* (CAG XIII/3: 402,8 f.): εἴπομεν

[God] had called the genus human being, he separated its species and stated that it was created 'male and female', even though the individual had not yet taken shape. This is because the most proximate of the species are present in the genus and become apparent as if in a mirror (τὰ προσεχέστατα τῶν εἰδῶν ἐνυπάρχει τῷ γένει καὶ ὥσπερ ἐν κατόπτρῳ διαφαίνεται) to observers with sharp vision."

91,16 f. ἐπὶ τὸ ζητούμενον καὶ μηκέτι τομὴν ἐπιδεχόμενον κτλ. The end-point of division is the unsplittable species; cf. Aristotle, *Top.* II 2, 109b15 f.; III 6, 120a34 f.; *An. Post.* II 13, 96b15 f.; *PA* I 3, 643a7–20; *Met.* Δ 10, 1018b5 f.; Z 8, 1034a8; I 9, 1058b9 f.; K 1, 1059b34–38. This constitutes the essence of the investigated object; cf. Aristotle, *Met.* Z 11, 1038a15–30. Cf. above, 90,18. For the vocabulary, cf. Galen, *PHP* IX 5, 13 (CMG V,4,1,2: 566,20 f./V,753 K.): ἀπὸ μὲν τοῦ πρώτου καὶ γενικωτάτου καταβαίνοντας *ἐπὶ τὰ μηκέτι τομὴν δεχόμενα κτλ.*

91,17–23 θνητὸν καὶ ἀθάνατον … χερσαῖον καὶ ἔνυδρον … πτηνὸν καὶ πεζόν … λογικὸν καὶ ἄλογον. This list of differences, arrived at by the division of ζῷον, provides for a definition of man as ζῷον θνητόν, χερσαῖον, πεζόν, λογικόν (91,26). In order to illustrate the earlier points about division, each of these differences should be 'species-forming' (cf. 18, 3/91,9 f.), the first pair should be 'proximate' to the genus (cf. 18, 2/91,7), and the genus should contain species that are 'nearest' to the defined object (cf. 18, 2/91,7 f.). Clearly the text is trying to meet these conditions. Interestingly, the 'nearest species', also described as "proximate to the investigated object" (91,20; cf. below, 91,23 f.; 21, 4/93,8 f.), are arranged in an order of higher and subordinate items (i.e. some being more 'proximate' than others), but the status of γένος is reserved for the first item in the definition only.

The first pair of differences distinguishes (the species of) man from the heavenly bodies; the second from mortal creatures living in the water; the third from earthly creatures who have wings; and the fourth from non-rational earthly pedestrians. The implied classification is similar to the one outlined in Plato's

---

γὰρ τὸ γένος ὅλον εἶναι ἐκ μερῶν τῶν εἰδῶν συγκείμενον. Eustratius, *In An. Post.* (CAG XXI/1: 193,4–6): συντιθέμενα τὰ γένη φησίν, ἄτομα δὲ τὰ εἰδικώτατα, ἃ οὐκ ἔστιν ἔτι εἰς εἴδη τεμεῖν. λέγει οὖν ἐκ τῶν εἰδῶν ὡς μερῶν τὰ γένη ὡς ὅλα συντίθεσθαι. When explaining the ὅλον τι in *An. Post.* II 13, 96b15 as τὸ γένος, whose 'parts' are the unsplittable species, Philoponus refers to Porphyry, *Isag.* (CAG IV/1: 8,1 f.): ὅλον γάρ τι τὸ γένος, τὸ δὲ ἄτομον μέρος, τὸ δὲ εἶδος καὶ ὅλον καὶ μέρος κτλ. For the description of species as 'parts' of their genus, cf. Barnes, *Porphyry: Introduction*, pp. 148–150 and 339–342. Cf. also ibid., p. 114, on the widespread use of περιέχειν ('to contain') in this connection.

Timaeus 39e10–40a2: εἰσὶν δὴ τέτταρες [*scil.* ἰδέαι], μία μὲν οὐράνιον θεῶν γένος, ἄλλη δὲ πτηνὸν καὶ ἀεροπόρον, τρίτη δὲ ἔνυδρον εἶδος, πεζὸν δὲ καὶ χερσαῖον τέταρτον. However, unlike Plato, who puts pedestrian and earthly creatures on the same level, our text regards the flying animals as a subclass of those living on earth.[270] As far as the fourth pair is concerned (λογικὸν καὶ ἄλογον), it is surprising to see it used as a specific difference on the level of pedestrian earthlings, as, according to common wisdom, the possession of reason distinguishes human beings not only from pedestrian earthlings, but from *all* mortal animals.[271] One would expect the predicate δίπουν instead.[272] On θνητόν and λογικόν, see further below, on 92,33 and 93,12.

91,23 f. τὰ προσεχῆ τῶν … εἰδῶν τῷ ἀνθρώπῳ. Cf. above, 91,20; below, 21, 4/93,8 f.

91,25 ἐπισυνθέντες εἰς ἕνα λόγον. Cf. above, on 90,22 and 91,10–12.

19, 1–8

91,27 f. ὕλης μὲν τάξιν ἐπέχει ἡ διαίρεσις τῷ ὅρῳ … τεχνίτου δὲ καὶ δημιουργοῦ ὁ ὅρος. After the geometrical simile (91,10–12), the relation of division to definition is now compared to one between matter and a craftsman. The point of comparison is that division produces the components of definition, but does not suffice to put them together (cf. above, 90,22; 91,25). When saying that it "seeks for the simplicity of the name" of the defined object (τὴν ἁπλότητα τοῦ ὀνόματος ἀναζητοῦσα), our author presumably hints at the idea of matter "desiring the form" (cf. Aristotle, *Phys.* I 9, 192a16–23: ἐφίεσθαι τὸ εἶδος). The simplicity of the name probably corresponds to the end-point of division (91,21: τὸ ἁπλού-

270 For other parallels, cf. Jaap Mansfeld, *Heresiography in Context: Hippolytus'* Elenchos *as a Source for Greek Philosophy* (PhA 56; Leiden/New York/Cologne: Brill, 1992), p. 81.

271 Cf. Alexander, *Mant.* (Suppl. Arist. II/1: 169,11–13): "The appropriate divisive differences of something do not extend further than that which they divide (οὐκ εἰσὶν ἐπὶ πλέον ἐκείνου, ὃ διαιροῦσιν)—e.g. none of the differences which divide animal or holds of anything which is not an animal." Trans. Barnes, *Porphyry: Introduction*, p. 182. Moreover, the predicate λογικός, used here as a subdivision of πεζόν, normally applies not only to humans, but also to gods; cf. Galen, *Protr.* 9, 2 (CMG V,1,1: 132,8 f./I,21 K.): τὸ δὴ τῶν ἀνθρώπων γένος, ὦ παῖδες, ἐπικοινωνεῖ θεοῖς τε καὶ τοῖς ἀλόγοις ζῴοις, τοῖς μέν, καθ' ὅσον λογικόν ἐστι, τοῖς δέ, καθ' ὅσον θνητόν. Cf. idem, *MM* I 3 (X,24,3–5 K.): εἰ δ' ἤτοι λογικὸν καὶ ἀθάνατον, ἢ λογικὸν καὶ θνητόν, εἴη ἂν τὸ μὲν εἴδει θεός, τὸ δὲ ἄνθρωπος. Thus it is usually treated as a proximate species of ζῷον alongside with (on the same level as) θνητόν. Cf. Barnes, *Porphyry: Introduction*, pp. 183 f.

272 Cf. Ps.Plato, *Def.* 415a11; Aristotle, *Top.* I 7, 103a27; V 3, 132a2; *An. Post.* II 5, 92a1; 6, 92a29 f.; II 13, 96b31 f.; *Met.* Z 12, 1037b12.

στατον εἶδος), which enables us to replace the name of the object by a definition of its essence; see parallels above, on 91,10–12.

The phrase τάξιν τίνος ἐπέχειν (τινί) refers to a relative position of something as analogous to a relative position of something else. Cf. e.g. SVF II,583: ἐπεὶ γὰρ ἡ γῆ κέντρου τάξιν ἐπέχει, ὥσπερ ἀπὸ κέντρου κύκλος γίνεται, οὕτω καὶ ἀπὸ τῆς γῆς εἰκὸς ἔξω περιφέρειαν γεγονέναι. For the expression ὕλης τάξιν ἐπέχειν, cf. e.g. *Strom.* IV (6) 39, 3; Alcinous, *Did.* 27, 2 (H 180,13 f.). Cf. also below, 21, 1/92,31 f.; 28, 2/98,5 f.

91,28 f. ἐπισυντιθεὶς καὶ κατασκευάζων ⟨τὸν λόγον⟩. The verbs can hardly dispense with an object. Cf. above, 91,25: ἐπισυνθέντες εἰς ἕνα λόγον τὸν ὅρον … ἀποδίδομεν.

91,30 οὔτ' αὐτῶν τῶν πραγμάτων οὔτε τῶν ἰδεῶν οἱ ὅροι. In *Met.* Z 15, Aristotle argues that neither sensible particulars nor 'ideas' (as understood by the Platonists) can be defined. Our passage seems to reflect this argument, without necessarily commiting its author to a Platonist view of separate forms.[273] Alternatively, the word ἰδέα could be used in the sense of an entity underlying the definition of species, as opposed to the concept we have of that species; cf. Galen, *MM* II 7 (X,131 K., trans. Hankinson): "[T]here is a particular object that underlies each separate type of thing signified (καθ' ἕκαστον οὖν τῶν σημαινομένων ἕν τι πρᾶγμά ἐστιν ὑποκείμενον), but there is not necessarily one thing that underlies each word. (…) In the cases in which not only is something signified by the word, but there is an object that underlies it, *there are as many forms of objects as there are types of signification* (ὅσαπερ ἂν ᾖ τὰ σημαινόμενα, τοσαῦται καὶ αἱ τῶν πραγμάτων ὑπάρχουσιν ἰδέαι)." See further below, on 95,2 f.

91,31 f. τούτων τῶν διανοιῶν τοὺς ἑρμηνευτικοὺς λόγους ⟨ὅρους⟩ εἶναί φαμεν. Cf. below, 20, 5/92,27–30. The underlying epistemology is that of Aristotle's *Int.* 1; cf. above, on 87,24 f. and below, 23, 1.

92,1–3 ἡ μέν τις εἰς εἴδη … ὡς γένος, ἡ δέ τις εἰς μέρη ὡς ὅλον, ἡ δὲ εἰς τὰ συμβεβηκότα. These and other kinds of διαίρεσις are listed by Alcinous, *Did.* 5, 2

273 *Pace* Mansfeld, *Heresiography*, p. 63 and n. 15, who argues, partly on the basis of this passage, that "the dominant view" of *Strom.* VIII is "without doubt a Platonist one". But the background to the use of 'ideas' in this passage is uknown; it could be as dialectical as in *Met.* Z 15.

(H 156,34–44) and Galen, *PHP* IX 9, 43–46 (CMG V 4,1,2: 608,12–29/V,804 K.). Cf. also Sextus, *PH* II 213; Alexander, *In Top*. VIII 1, 157a6 (CAG II/2: 532,19–21); Ps.Galen, *Hist. Phil.* 14.[274] For εἰς τὰ συμβεβηκότα, cf. below, on 92,4f.

92,3f. ἡ μὲν οὖν τοῦ ὅλου εἰς τὰ μέρη ... κατὰ μέγεθος ἐπινοεῖται. In *Met.* B 3, 999a3, Aristotle draws a contrast between a divison "according to quantity" and "according to species" (κατὰ τὸ ποσὸν ἢ κατ' εἶδος).[275] While implying a similar distinction, our text speaks of 'magnitude' rather than 'quantity', the word μέγεθος being probably used in the more narrow sense of continuous quantity (τὸ συνεχές); cf. Aristotle, *Cat.* 6, 4b20–5a14 and Galen, *Morb. Diff.* 8 (VI,867,14–868,5 K.).[276] Cf. also Galen, *PHP* IX 9, 43 (CMG V,4,1,2: 608,13f./V,804f. K.): ... ὅταν ὅλον τι συνεχὲς ὄν ⟨εἰς⟩ τὰ μόρια τέμνηται. The problem with this kind of division is explained by Galen in *PHP* VIII 2, 8: "With reference to the division of bodies in magnitude (τῆς κατὰ μέγεθος τομῆς τῶν σωμάτων), the geometricians have pointed out that there is no end to it; the section always contains a magnitude smaller than itself."[277]

92,4f. ἡ δὲ εἰς τὰ συμβεβηκότα οὐδέποτε ὅλη δύναται διαληφθῆναι, εἴ γε καὶ οὐσίαν ἑκάστῳ δεῖ πάντως τῶν ὄντων ὑπάρχειν. Alcinous' list of divisions includes "division of subjects according to accidents (ἡ δὲ ὑποκειμένων εἰς συμβεβηκότα), as when we say that of men some are good, others bad, and others middling" (*Did.*

---

274 For these and other parallels, cf. Mansfeld, *Heresiography*, pp. 81 and 229f.; Barnes, *Porphyry: Introduction*, pp. 341f.

275 Mansfeld, *Heresiography*, p. 81, commenting on our passage, refers (among other parallels) to Alexander, *In Met.* B 3, 999a1–6 (*CAG* I: 208,10–18) and Boethius, *De divis.* 879b. Surprisingly, Mansfeld does not mention the background passage in Aristotle himself. He also overlooks the distinction between τὸ μέγεθος and τὸ ποσόν, thus incorrectly describing Boethius' *divisio secundum quantitate* as an "exact parallel" to our passage.

276 "Futhermore, it is clear to everyone, I imagine, that both genera, that pertaining to number and that pertaining to magnitude, can be subsumed under another higher category, that of quantity (ὡς ἀμφοῖν τοῖν γενοῖν, τοῦ τε κατ' ἀριθμὸν καὶ τοῦ κατὰ μέγεθος, ἀνωτέρω τάττειν ἑτέραν ἐγχωρεῖ κατηγορίαν τὴν τοῦ ποσοῦ). That is to say, there is a discrete form of quantity (τὸ διωρισμένον) which is also called quantity in the proper sense (ἰδικῶς), and a continuous form (τὸ συνεχές) which is also called size (πηλίκον). But in the present [treatise] it seemed to me clearer to divide things thus—that is to classify one difference of quantity as number and the other as magnitude (κατὰ μὲν τὸν ἀριθμὸν τάξαντι τὴν ἑτέραν διαφορὰν τοῦ ποσοῦ, κατὰ δὲ τὸ μέγεθος τὴν ἑτέραν)." (Trans. Ian Johnston in *Galen on Diseases and Symptoms*, Cambridge: Cambridge University Press, 2006, p. 150, modified).

277 CMG V,4,1,2: 492,5–7/V,663 K.; trans. De Lacy.

5, 2/H 156,39–41).[278] While also speaking of division εἰς τὰ συμβεβηκότα, our text seems to have in mind accidents belonging to the same subject. It argues that no subject can be fully explained by its division into accidents, as there must always remain something to which the accidents belong. Cf. Galen, *PHP* IX 9, 45: "But they do not say that substances are 'divided' into the powers in them (οὐ μὴν τάς γε οὐσίας διαιρεῖσθαι … εἰς τὰς ἐν αὐταῖς δυνάμεις), but that each substance, being undivided, has some activity in accordance with the powers in it."[279]

The phrase οὐδέποτε ὅλη δύναται διαληφθῆναι could be translated 'can never be divided as a whole', the verb διαλαμβάνω having the additional force of making something distinct or clear. Neither of these meanings (nor any other meaning of the verb found in the dictionary) tallies with the subject of the phrase (διαίρεσις), but they do fit well with the object of division. Arguably, the sentence involves a metonymical shift of reference, not unlike the one witnessed above, 9, 1/85,1.

92,9 εἰδικὰς διαφοράς. The manuscript reads ἰδικάς. According to LSJ, the word ἰδικός is used in two ways: as a 'late' form of εἰδικός and as an equivalent of ἴδιος. In our passage, of course, the first meaning applies, the adjective being derived from εἶδος. It is difficult to determine which layer of transmission is responsible for the spelling we have in L; in the editions of the second century literature, εἰδικός is more common, although the forms may vary within the same author[280] and even the same edition.[281] It is reasonable to suppose that where such variation occurs, ἰδικός is not the original form. I assume that our passage originally read εἰδικάς as well; the point has a bearing on the solution to a difficult passage above, 17, 3/90,16.

92,9 f. τὸ εἶδος ἀεὶ ἔν τινι μέρει θεωρεῖται κτλ. Speaking of the division of genera, our text goes on to explain the difference between species and parts. Although

---

278 Trans. John Dillon, *Alcinous: The Handbook of Platonism* (Oxford: Clarendon Press, 1993), pp. 8 f., modified.

279 CMG V,4,1,2: 608,22–24/V,804 K.; trans. De Lacy.

280 For example, with regard to this particular phrase, there are numerous instances of the form εἰδικαὶ διαφοραί in Kühn's edition of Galen (some are mentioned above, on 90,14–16); but there is an exception in *Hipp. Off.* III 37 (XVIIIb,921,5 K.): διαφορᾶς … οὔσης ἰδικῆς, ἣν ἄρτι διῆλθον κτλ.

281 Cf. Alexander, *In Top.* VI 1 (CAG II/2: 420,3): διδάξας ὁ Ἀριστοτέλης περὶ τῶν γενικῶν καὶ ἰδικῶν προβλημάτων κτλ., as opposed to ibid. VII 1 (CAG II/2: 497,11 f.): ἀλλὰ καὶ εἰ τὰ εἰδικὰ προβλήματα ὑπὸ τὰ γένη ἐστὶν κτλ.

species is always observed in some part of the genus, the fact of X's being a part of Y does not qualify X as the species of Y. The logic of the argument demands that species are regarded as 'parts' of their genera; for this view, cf. above, on 91,13 f. However, the expression ἔν τινι μέρει θεωρεῖται seems to suggest a distinction between species and the parts of the genera in which they are observed. There are two options that account for this distinction: Either the 'parts' in question are specific *differentiae*, or they are individual representatives of the species. For the latter (more likely) view, cf. Porphyry, *Isag.* (CAG IV/1: 8,1–3; trans. Barnes): "For a genus is a sort of a whole, an individual a part, and a species both a whole and a part—but a part of one thing and a whole not of other items, but in other items (for a whole is in the parts)."[282] For the distinction between species and parts, cf. Cicero, *Top.* 30–31; Mansfeld, *Heresiography*, pp. 82 f.; Barnes in Porphyry, *Introduction*, pp. 339–342.

92,10 f. ἡ γὰρ χεὶρ μέρος μέν ἐστι τοῦ ἀνθρώπου, εἶδος δὲ οὐκ ἔστιν. Cf. Cicero, *Top.* 30 (trans. Reinhardt): "In a partition there are, as it were, limbs, like the head of a body, the shoulders, the hands, the sides, the lower legs, the feet and so on; in a division there are species which the Greeks call εἴδη." For the hand as an example of a part, cf. e.g. Aristotle, *Cat.* 15, 15b23; Sextus, *M.* XI 24.

92,11–13 καὶ τὸ μὲν γένος ἐν τοῖς εἴδεσιν ἐνυπάρχει, τὸ γὰρ ζῷον καὶ ἐν τῷ ἀνθρώπῳ καὶ ἐν τῷ βοΐ. A genus is 'in' its species in the sense that it is a part of their definition; cf. Aristotle, *Int.* 11, 21a17 f.: ἐνυπάρχει γὰρ ἐν τῷ ἀνθρώπῳ τὸ δίπουν καὶ τὸ ζῷον. Cf. also *Phys.* IV 3, 210a18: τὸ γένος ἐν τῷ εἴδει, and *Met.* Δ 25, 1023b22–24: διὸ τὸ γένος τοῦ εἴδους καὶ μέρος λέγεται.[283]

92,14 κυριώτερον τὸ εἶδος τοῦ μέρους. Species are more important than parts "for the purpose of finding the definition of a substance, no doubt".[284]

92,14 f. ὅσα τοῦ γένους κατηγορεῖται, ταῦτα πάντα καὶ τοῦ εἴδους κατηγορηθήσεται. Cf. Aristotle, *Top.* IV 2, 122b9: ἀνάγκη γὰρ τοὺς τῶν γενῶν λόγους κατηγορεῖσθαι τοῦ εἴδους καὶ τῶν μετεχόντων τοῦ εἴδους.

282 ὅλον γάρ τι τὸ γένος, τὸ δὲ ἄτομον μέρος, τὸ δὲ εἶδος καὶ ὅλον καὶ μέρος, ἀλλὰ μέρος μὲν ἄλλου, ὅλον δὲ οὐκ ἄλλου ἀλλ' ἐν ἄλλοις· ἐν γὰρ τοῖς μέρεσι τὸ ὅλον.

283 For other references, see Barnes, *Porphyry: Introduction*, p. 339 and n. 33.

284 Mansfeld, *Heresiography*, p. 83.

20, 1–5

92,17 f. ἄριστον … εἰς δύο διαιρεῖν εἴδη τὸ γένος, εἰ δὲ μή, εἰς τρία. This reflects a Platonic view according to which "we must always split into the least possible number" (Plato, *Polit.* 287c4 f.).[285] Contrast the Aristotelian critique of 'dichotomizers' (οἱ διχοτομοῦντες) in *PA* I 2, 642b5–3, 643b9.[286]

92,18–20 γενικώτερον μὲν [διαιρούμενα χαρακτηρίζεται] τῷ τε ταὐτῷ καὶ θατέρῳ … χαρακτηρίζεται. The phrase διαιρούμενα χαρακτηρίζεται is repeated twice in the sentence and Stählin (in the apparatus) suggests that on the first occasion it could be deleted. But the case for the correction is strong enough to introduce it into the text. The opposition γενικώτερον μὲν/ἔπειτα δὲ can only apply in a meaningful way to the levels of description of the species (χαρακτηρίζεται), not to the levels of their division (διαιρούμενα). It makes sense to say that, on a more general level, species are characterized by the 'same' and the 'diverse', i.e. by that which constitutes their genus and their specific differences (cf. above, 19, 5/92,7–9); but it does not make sense to say that they are characterized by the same and the diverse *when divided* on a more general level. The less general level, in turn, corresponds to the description of the species when we think of them as belonging to one of the Aristotelian categories, thus *dividing* them accordingly (cf. below, 92,20–23). For the description of the categories as τὰ γενικῶς σημαινόμενα, cf. Aristotle, *Cat.* 4, 1b26; below 23, 6/94,25.

92,21 f. ὥσπερ ὅταν λέγωμεν· τῶν ὄντων τὰ μὲν σώματά ἐστι, τὰ δὲ ἀσώματα. The predicates σώματα/ἀσώματα split the genus τὰ ὄντα into two species according to the category of substance. Galen offers a similar example of a predication about substance (*Inst. Log.* 2, 1/5,4 f. Kalbfleisch): ὁ ἀὴρ σῶμά ἐστιν· ὁ ἀὴρ οὐκ ἔστι σῶμα.

92,22 ⟨ἢ κεῖσθαι ἢ ἔχειν⟩. Aristotle provides the list of *ten* categories on two occasions only (*Cat.* 4, 1b26 f. and *Top.* I 9, 103b22 f.), otherwise referring to them summarily or selectively; the categories of κεῖσθαι and ἔχειν are among the least often mentioned ones. But it is likely that in our text all ten categories were originally listed; cf. below, 21, 4/93,8, on "ten categories" as something already known and 23, 6/94,25–95,1, where they are enumerated in full.

---

285 Cf. Barnes, *Porphyry: Introduction*, pp. 132 f., for parallels; cf. also Mansfeld, *Heresiography*, pp. 83 f., quoting Cicero's critique of (Epicurean) tripartite division in *Fin.* II 26.

286 For details, cf. the references in Mansfeld, *Heresiography*, p. 330; cf. also James G. Lennox, *Aristotle: On the Parts of Animals* I–IV, translated with a commentary (Oxford: Clarendon Press, 2001), pp. 152 f.

92,23–30 This section is a summarizing statement about definition, evaluating the topic from the point of view of its use; cf. above, 8, 3; 14, 1–15, 1.

92,26 f. τὰς πολλὰς ἀμφισβητήσεις ... καὶ τὰς ἀπάτας. Cf. above, 9, 7/85, 15–17; 11, 2–3/86,22–27.

92,28 f. ἡ δὲ τῆς διανοίας ἑρμηνεία ὅρος ἐστίν. Cf. above, 19, 2/91,30–32.

21, 1–6

92,31 f. ἡ διαφορὰ σημείου τάξιν ἐπέχουσα τῷ ὅρῳ. We have already heard about specific differences and their role within definitions (cf. above, 18, 2/91,6–10); in what follows (21, 1–5), the text discusses another sort of difference, one that is not a part of definitions, but is *added* to them (προσλαμβάνεται, cf. below, 92,32: προστεθέν; 93,1: προσαγόμενα; 93,13: προσπαραληπτέον). The relation of this difference to a definition is analogous to the relation of a sign to the signified object (for the phrase τάξιν τίνος ἐπέχειν τινί, cf. above, on 91,27 f.); the point of this comparison is brought out more clearly below, 22, 2/93,1 f.: a difference added to a definition of X, which uniquely belongs to X but does not show its essence, may be regarded not only as a sign of X, but also of its definition.

92,32 τὸ γελαστικὸν προστεθέν. In *PA* III 10, 673a8, Aristotle notes that man "is the only one of the animals that laughs" (μόνον γελᾶν τῶν ζῴων ἄνθρωπον); later this becomes one of the stock examples of a 'property' (τὸ ἴδιον). Cf. e.g. Lucian, *Vit. Auct.* 26 (Macleod II,48,1 f.); Clement, *Paed.* II (5) 46, 2; Sextus, *PH* II 211; Alexander, *In Top.* I 5, 102a18–30 (CAG II/2: 45,21–24).[287]

92,33 [λογικὸν] θνητὸν χερσαῖον πεζὸν ⟨λογικὸν⟩ γελαστικόν. The text shows how a διαφορά is added to a definition, building upon the example produced above, 18, 6–7/91,17–23. In light of that example, and the way it was arrived at, it is difficult to account for the order of differences given here in the manuscript. It is true that ζῷον λογικὸν θνητόν is a standard definition of man, sometimes presented as an alternative to the (Aristotelian) definition ζῶον πεζὸν δίπουν (cf. Galen, *Diff. Puls.* IV 8/VIII,739,11 f. K.; *MM* I 3/X,24,3–7). But its combination with the predicates χερσαῖον πεζόν can only make sense if λογικός serves as a specific difference on the level of πεζόν (as it does above, 18, 7/91,22 f.), not if it is regarded as a proximate difference of ζῷον. The latter option would render the other predicates irrelevant and would be confusing from the didactic point of view.

---

287 For other references, see Barnes, *Porphyry: Introduction*, p. 208 n. 22.

I think it is likely that the order of predicates was changed in the course of the textual transmission, perhaps under the influence of the 'standard' definition of man mentioned above. Cf. also below, on 93,12.

93,1 f. σημεῖα γάρ ἐστι … οἰκείων πραγμάτων, τὴν δὲ φύσιν αὐτῶν τῶν πραγμάτων οὐκ ἐνδείκνυται. Unlike definitions, the additional differences do not indicate the essence of the things they belong to; nevertheless, they function as signs of the things they belong to, since they do not belong to anything else. Cf. Alexander, *In Top.* I 5, 101b39 (CAG II/2: 43,2–4): λόγος γὰρ καὶ τὸ 'ζῷον γελαστικόν', καὶ ἀπερίττως τε καὶ ἀνελλιπῶς σημαίνει τὸν ἄνθρωπον. See further below, on 93,3–5.

93,3 ἰδίου ἀπόδοσίν φασιν εἶναι τὴν διαφοράν. Chrysippus reportedly described definition (ὅρος) as ἰδίου ἀπόδοσις.[288] While possibly hinting at this description, our text nevertheless draws a distinction between definition and the sort of difference it has in mind. Whoever is meant by φασίν, the reference is an occasion to connect the διαφορά in question with the Aristotelian notion of τὸ ἴδιον.

93,3–5 καὶ καθ' ὃ τῶν ἄλλων ἁπάντων διαφέρει … ὃ μόνῳ ὑπάρχει καὶ ἀντικατηγορεῖται τοῦ πράγματος. Picking up on the previous sentence, the text goes on to explain that the object of which a διαφορά is predicated (τὸ ἔχον τὴν διαφοράν) differs from all other objects in respect of the feature just mentioned, namely its ἴδιον, further characterized by Aristotle's own words; cf. *Top.* I 5, 102a18f.: "A property is something which does not indicate the essence of a thing, but yet *belongs to that thing alone and is counterpredicated of it* (ἴδιον δ' ἐστὶν ὃ μὴ δηλοῖ μὲν τὸ τί ἦν εἶναι, μόνῳ δ' ὑπάρχει καὶ ἀντικατηγορεῖται τοῦ πράγματος)."[289]

Stählin changes ὃ to ἥ, thus transferring Aristotle's description of ἴδιον to our text's διαφορά. With this reading, the text would say that every object having a διαφορά (in the sense of the Aristotelian ἴδιον) differs from all other objects; but this is not an enlightening statement. In addition, Stählin (following the punctuation of L and modern editions) suggests that the sentence continues with ἐν τοῖς ὅροις, thus giving the impression that, according to our text, ἡ διαφορά (in the sense of τὸ ἴδιον) is predicated "in definitions". But this would be surprising not only with regard to what Aristotle has to say about the subject,

---

288 SVF II 226; cf. also Alexander, *In Top.* I 5, 101b39 (CAG II/2: 43,2), criticizing the Stoic definition of definition: … οὐδὲν ἂν λέγοιεν τὸν ὅρον διαφέρειν τῆς τοῦ ἰδίου ἀποδόσεως.

289 I take it that καθ' ὃ in 93,3 is either analeptic, referring back to ἰδίου on the same line, or proleptic, anticipating ὃ μόνῳ ὑπάρχει κτλ. But the difference between the two options is not great, as in the first case the object of καθ' ὃ is further explicated by ὃ μόνῳ ὑπάρχει κτλ., and in the second, the backward reference to ἰδίου is implied by the context.

but also with regard to our own text's description of διαφορά as a 'sign' of the defined object (or of the object's definition), which is *added to* the definition (cf. above, on 92,31 f.). It is more likely, in my view, that the aim of the lines 93,3–6 is precisely the opposite, namely to draw a contrast between predications in terms of ἴδια (93,3–5) on the one hand and 'definitions' (93,5 f.) on the other. Cf. also the opposition ἐν τοῖς ἰδίοις/ἐν τοῖς ὅροις in Aristotle, *Top.* V 2, 130a30 f. and 130b13–14.25–28. However, since the two clauses that ought to represent the two poles of this contrast are not syntactically connected in this (or any other) way, I suspect that something is missing between them.

93,5 f. ἐν τοῖς ὅροις ἀνάγκη τὸ πρῶτον γένος ὡς ἀρχηγὸν καὶ ὑποστατὸν [δεῖ] παραλαμβάνειν. Adding iota to ἀνάγκη, Sylburg allows us to preserve δεῖ in 93,6. The emphatic phrase ἀνάγκῃ δεῖ could be designed to underline the contrast between predications in terms of properties and definitions (see the previous note). But I find it more likely that δεῖ found its way into the text as a gloss; cf. above, 92,18 and below, 93,17, for interventions of a similar sort. For the genus as the first constituent of definitions, cf. Aristotle, *Met.* Δ 28, 1024b4 f.: ἐν τοῖς λόγοις τὸ πρῶτον ἐνυπάρχον, ὃ λέγεται ἐν τῷ τί ἐστι, τοῦτο γένος. Cf. also Galen, *Diff. Feb.* I 1 (VII,274,8–10 K.), on "the first and most important genus of differences being taken from the essence of the object under division (τοῦ πρώτου τε καὶ κυριωτάτου γένους τῶν διαφορῶν ἀπὸ τῆς τοῦ τεμνομένου πράγματος οὐσίας λαμβανομένου)".

93,7 f. τῶν εἰδῶν ... διὰ τῶν δέκα κατηγοριῶν. Cf. above, 20, 2/92,20–23.

93,8 f. τῶν προσεχῶν εἰδῶν. Cf. above, on 91,17–23.

93,12 ἄνθρωπός ἐστι τὸ ζῷον ⟨θνητὸν λογικόν⟩ [γελαστικόν]. As it stands in the manuscript (ἄνθρωπός ἐστι τὸ ζῷον γελαστικόν), the text makes little sense; it is supposed to be an example of a definition, which includes a genus and "the two most necessary species" (93,10 f.). Further, the species are supposed to be such that they "grasp the essence and designate the nature of the object" (93,9 f.). But the predicate we have here, γελαστικόν, is but one and, moreover, of the kind that "does not indicate the nature" of its object (cf. 92,32–93,2). By contrast, in the next sentence (93,12–14), we are told that something else should be employed *in addition* (προσπαραληπτέον), namely something that uniquely belongs to the object (τὸ ἐξαιρέτως συμβεβηκός, ἡ ἰδία ἀρετή, τὸ ἴδιον ἔργον, etc.), i.e. its 'property'. It is likely, in my view, that in the course of the textual transmission something was omitted from the text, namely two predicates corresponding to the essential attributes of 'man', and that this loss was later

compensated by another predicate, presumably one originally written on the margin of the next sentence. It is also likely that the lost predicates can be found in the definition of man arrived at above, 91,26; they probably correspond to the first and the last of those predicates, i.e. θνητόν and λογικόν (the first one being a proximate difference of the genus and the last one specifying the unsplittable species); see further above, on 91,17–23 and 92,33.

93,12–14 τό τε ἐξαιρέτως συμβεβηκὸς τῷ ὁριζομένῳ κτλ. Cf. above, 21,1–2/92,32–93,3 and the previous note.

93,14 f. ἐξηγητικὸς ... ὢν τῆς τοῦ πράγματος οὐσίας. For the vocabulary, cf. Galen, *MM* I 9 (X,74,11 K.): *σαφῶς ἐξηγησάμενοι τὴν οὐσίαν* (τοῦ συμπτώματος); *UP* V 9 (I,276,17 Helmreich/III,377 K.): τὴν οὐσίαν ἐξηγήσασθαι (τοῦ σώματος τῶν κύστεων); *SMT* V 14 (XI,752,7 f. K.): τὴν οὐσίαν ἐξήγημαι (τῶν τοιούτων φαρμάκων).

93,15–17 περιλαβεῖν μὲν ἀκριβῶς τὴν φύσιν τοῦ πράγματος ἀδυνατεῖ, διὰ δὲ τῶν κυριωτάτων εἰδῶν τὴν δήλωσιν τῆς οὐσίας ποιεῖται. For the thought and the vocabulary cf. Galen, *Diff. Puls.* IV 2 (VIII,703,16–18 K.): "If definition is to explain precisely the whole essence of an object, it should omit none of its most important [attributes] (τὸν ὁρισμὸν οὖν εἴπερ ὅλην *τὴν οὐσίαν ἀκριβῶς τοῦ πράγματος* μέλλει *δηλώσειν*, οὐδὲν αὐτῆς *τῶν κυριωτάτων* παραλείπειν προσήκει)." Nevertheless, according to Galen, definitions do not succeed in expressing the essence of the defined object (in this case, the pulse) precisely, because they cannot contain all its essential attributes: "But all this cannot be contained in a definition (οὐ μὴν ἅπαντά γε ταῦτα *δυνατὸν* ὅρῳ *περιλαβεῖν*). People fond of definitions do not want them to extend to such a length. Therefore, no definition of pulses is essential in a precise manner (οὐδέ ἐστι πᾶς ὅρος ... οὐσιώδης *ἀκριβῶς*), lest it be very long" (ibid. IV 2/VIII,708,5–8 K.).

93,17 f. σχεδὸν ἐν ποιότητι [ὁ ὅρος] τὴν οὐσίαν ἔχει. For the idea that definitions grasp essential 'qualities', cf. Aristotle, *Met.* Δ 14, 1020a33 f.: τὸ ποιὸν λέγεται ἕνα μὲν τρόπον ἡ διαφορὰ τῆς οὐσίας, οἷον ποιόν τι ἄνθρωπος ζῷον ὅτι δίπουν κτλ. Cf. ibid. Δ 28, 1024b4–6; *Top.* IV 6, 128a26 f. and e.g. Alexander, *In Top.* VI 1, 139a28–31 (CAG II/2: 421,16 f.): ἐν γὰρ τῷ τί ἐστι κατηγορεῖται τὸ γένος, αἱ δὲ διαφοραὶ ὡς ποιότητες μᾶλλον. In his polemic against Archigenes, Galen criticizes the use of the word 'qualities' in the broad sense of generic differences; cf. *Diff. Puls.* II 4 (VIII,580,9–13 K.): "But to call the first generic differences of pulses 'qualities' is also negligent. A difference in terms of great and small is not one of quality, unless you wish to claim that three cubits are a quality, too." See further ibid. II 10 (VIII,632,5–634,6 K.).

The repetition of the subject (cf. 93,14 f.) is probably due to an incorporation of a marginal gloss; cf. the (erroneous) repetition of διαιρούμενα χαρακτηρίζεται above, 92,18.

(VII) 22, 1–4: Suspension of Judgement II

This section is thematically linked to the discussion of ἐποχή above, 15, 2–16, 3, especially to its last bit (16, 3), where the main reasons why even a dogmatist sometimes suspends judgement are sketched. Here the same point, namely the causes of ἐποχή, is developed more fully, in a similar manner and with similar vocabulary. But it would be rash to suppose that it simply picks up on the argument made in 16, 3. Whereas there the perspective of ἐποχή is (moderately) dogmatist, here it seems to be (sophisticatedly but uncompromisingly) sceptic. To all appearances, the passage is an excerpt or a paraphrase taken from a sceptic source where ἐποχή was discussed along the lines of the Aenesidemean 'modes', i.e. the ways of arguing that lead us to the suspension of judgement.[290] This excerpt or paraphrase was probably made by the same author who accepts the limited validity of ἐποχή in 16, 3, not in order to embrace scepticism, but rather to represent the sceptic's position with his own words. Why the section is preserved between chapters 6 and 8 in our manuscript is a mystery.

93,19 τὰ ποιητικὰ τῆς ἐποχῆς αἴτια … τὰ ἀνωτάτω. In their expositions of the 'modes of suspension' (οἱ τρόποι τῆς ἐποχῆς), neither Sextus nor Diogenes nor Aristocles use the word αἴτια.[291] But it is used, in the relevant context, by Philo of Alexandria, who introduces the Aenesidemean modes as explanations (αἴτια) why "we should not be able to say anything firm about anything" (βέβαιον περὶ οὐδενὸς οὐδὲν ἂν ἔχοιμεν εἰπεῖν); cf. *Ebr.* 170–171. The qualification of the αἴτια as τὰ ἀνωτάτω in our passage seems to suggest that, in its original context, a less general account of the explanations was provided as well, of which the passage

290 Several versions of 'the modes of suspension' (or, alternatively, of 'inapprehensibility': ἀκαταληψία) are known from antiquity, the earliest (preserved in Philo of Alexandria, Aristocles in Eusebius, Sextus, Dogenes Laertius, and a Byzantine *Commentary on Metaphysics* ascribed to Herennius) going back to Aenesidemus (1st cent. B.C.). Translation of the principle sources, along with a philosophical commentary, is provided in Julia Annas and Jonathan Barnes, *The Modes of Scepticism: Ancient Texts and Modern Interpretations* (Cambridge: Cambridge University Press, 1985). For the ten modes of Aenesidemus, see also Hankinson, *The Skeptics*, pp. 139–162.

291 For the terminology of the modes, cf. Annas and Barnes, *The Modes of Scepticism*, p. 21.

we have here is a sort of summary. Compare the summary of the ten modes in Sextus, *PH* I 38 (trans. Annas and Barnes): "Superordinate to these (τούτων δὲ ἐπαναβεβηκότες) are three modes: that deriving from the subject judging; that deriving from the object judged; that combined from both."

93,20 τὸ πολύτροπον καὶ ἄστατον τῆς ἀνθρωπίνης γνώμης. Again, the language is reminiscent of Philo's account in *On Drunkenness*, where, however, similar expressions refer not to judgement, but to phenomena and imagination; cf. *Ebr.* 170: ἅτε *μὴ ἑστῶτος* τοῦ φανέντος, ἀλλὰ *πολυτρόποις* καὶ πολυμόρφοις χρωμένου ταῖς μεταβολαῖς. 180: γίνεται δ' οὐχ ἥκιστα τὸ περὶ τὰς φαντασίας *ἄστατον*. Another parallel is found in Sextus, *M.* VIII 473, who cites τὸ πολύτροπον τῆς ἀνθρωπίνης διανοίας as a reason why the sceptic does not even assert that his argument against proof is true.

93,21f. τῆς διαφωνίας ἤτοι τῆς ἀλλήλων πρὸς ἀλλήλους ἢ τῆς ἑαυτῶν πρὸς ἑαυτούς. The concept of disagreement (with oneself and with one another) underlies the first five Aenesidemean modes according to DL IX 79–84 (or the first four and the tenth, according to Sextus, *PH* I 36–39).[292] The corresponding passages in Philo are *Ebr.* 171–180 and 193–202. For the vocabulary, cf. DL IX 95: καὶ ὁ μὲν ἄνθρωπος καὶ πρὸς αὑτὸν διαφωνεῖ καὶ πρὸς τοὺς ἄλλους, ὡς δῆλον ἐκ τῶν διαφόρων νόμων καὶ ἐθῶν (this echoes the fifth mode according to DL IX 83–84; cf. Philo, *Ebr.* 193–197).[293] Cf. also Philo, *Ebr.* 175: οὐ γὰρ μόνον ἄλλοτε ἄλλως τὰ αὐτὰ κρίνουσιν [*scil.* οἱ ἄνθρωποι], ἀλλὰ καὶ ἑτέρως ἕτεροι κτλ. See further below, on 94,3.

93,22f. ἡ ἐν τοῖς οὖσι διαφωνία. The idea that there is disagreement "in things" (as opposed to disagreement about them) is rare, but it is also found in DL IX 76: *τῶν μὲν γὰρ πραγμάτων διαφωνούντων* τῶν δὲ λόγων ἰσοσθενούντων ἀγνωσία τῆς ἀληθείας ἐπακολουθεῖ. As noted by Janáček,[294] the Sextan equivalent is ἡ ἐν τοῖς πράγμασιν ἀνωμαλία. Cf. *PH* I 12, where Sextus says that "the anomaly in things" induces "the men of talent" to investigate "what in things is true and what false" (τί τε ἀληθές ἐστιν ἐν τοῖς πράγμασι καὶ τί ψεῦδος); see further below, on 94,3f.

---

292 Among the five modes ascribed to Agrippa, διαφωνία characterizes the first one (cf. Sextus, *PH* I 164–165; DL IX 88: ἀπὸ τῆς διαφωνίας).

293 Cf. Janáček, "Ainesidemos," in idem, *Studien*, p. 253.

294 "Ainesidemos," in idem, *Studien*, pp. 253f.

93,23 εἰκότως ἐμποιητικὴ … τῆς ἐποχῆς. For the use of εἰκότως in this context, cf. Sextus, *PH* I 87: εἰ δὲ τὰ αὐτὰ διαφόρως κινεῖ παρὰ τὴν διαφορὰν τῶν ἀνθρώπων, εἰσάγοιτ' ἂν εἰκότως καὶ κατὰ τοῦτο ἡ ἐποχή. I 134: διόπερ εἰκότως ἂν καὶ οὗτος ὁ τρόπος εἰς ἐποχὴν ἡμᾶς περιάγοι.

93,23–27 μήτε γὰρ πάσαις ταῖς φαντασίαις πιστεύειν … μήτε πάσαις ἀπιστεῖν … μήτε τισὶ μὲν πιστεύειν, τισὶ δὲ ἀπιστεῖν. For the structure of the argument, cf. Sextus, *M.* VII 389: οὔτε δὲ πᾶσαν [*scil.* φαντασίαν] ἀληθῆ λεκτέον εἶναι οὔτε ψευδῆ οὔτε τινὰ μὲν ἀληθῆ τινὰ δὲ ψευδῆ.[295] A detailed discussion of these three options (in that order) continues until the end of *M.* VII, in the context of a refutation of the criterion of truth. Cf. the summary in VII 439: εἰ μήτε πᾶσαι αἱ φαντασίαι εἰσὶ πισταὶ μήτε πᾶσαι ἄπιστοι, μήτε τινὲς μὲν πισταὶ τινὲς δὲ ἄπιστοι, οὐκ ἂν εἴη κριτήριον τῆς ἀληθείας ἡ φαντασία.

93,24 διὰ τὴν μάχην. Cf. Sextus, *M.* VII 392: εἰ δέ ἐστί τί τινι μαχόμενον, οὐ πᾶσα φαντασία ἐστὶν ἀληθής. The word μάχη often appears in Sextus in this and similar connections (see Janáček's Index), the closest parallel being *PH* I 44 (the exposition of the Aenesidemean modes), where he speaks of "the conflict of presentations" (μάχη τῶν φαντασιῶν). One possible source of this conflict is vividly described by Philo, *Ebr.* 196 (a section on differences in customs and laws): "But if people are not only at slight variance with one another, but also in utter discord (τοῖς ὅλοις ἀπᾳδόντων), to the extent that they oppose and fight one another (ὡς ἀντιστατεῖν καὶ διαμάχεσθαι), it follows that presentations occurring to them (τὰς προσπιπτούσας … φαντασίας) will differ too and their judgements will be in mutual conflict (ἀλλήλαις πεπολεμῶσθαι)." Cf. ibid. 180, on "conflicting and discordant dreams" (μαχόμενα καὶ ἀσύμφωνα ὀνείρατα).

93,25f. διὰ τὸ καὶ τὴν λέγουσαν πάσας ἀπίστους ὑπάρχειν ἐξ ἁπασῶν οὖσαν συμπεριγράφεσθαι πάσαις. While discussing the option that all presentations are false, Sextus argues that by affirming this, we are already determining that something is true, namely this very affirmation (*M.* VII 399; cf. above, 89,6f. and 20f.). Here our text takes a step further to say that the presentation according to which no presentation is credible is 'bracketed' (i.e. cancelled, see below) along with all of them and cannot be claimed to be credible either. This step is not designed to refute scepticism, but rather to save it from inconsistency. It

295 The parallel is noted by von Arnim, *De octavo*, pp. 15f. Cf. also Janáček, "Ainesidemos," in idem, *Studien*, pp. 253f.

is not explicitly used by Sextus in the parallel discussion (but see *M.* VII 398, mentioned below), but it is known to him, of course.[296] It also seems to be presupposed by the polemic above, 89,21–23.

For the form and vocabulary of the argument, cf. Sextus, *M.* VII 397 (trans. Bett): "For if everything is true, everything will become clear to us, and if this is so, everything's being unclear to us will also be sound and true—since it is one of 'everything' (ἓν ἐκ τῶν πάντων καθεστώς)." Later Sextus says that we should not say that "all presentations are false" either, for "analogous reasons" (398). Cf. also *M.* VIII 55, quoted above, on 89,7–11, and *PH* II 86.[297] The metaphor of 'bracketing' (περιγράφω), probably derived from the practice of textual criticism, is also known to Sextus.[298] Our passage seems to be the earliest evidence of this usage.[299]

93,27a διὰ τὴν ἰσότητα. Cf. Sextus, *M.* VII 402–403: "For presentations come about from unreal things as well as from real ones. And an indication of their indistinguishability is their being found equally plain and striking (τὸ ἐπ' ἴσης ταύτας ἐναργεῖς καὶ πληκτικὰς εὑρίσκεσθαι), while an indication of their being equally striking and plain is the fact that the corresponding actions are connected with them."[300]

93,27b κατήχθημεν εἰς ἐποχήν. Janáček lists the following verbs used by Sextus in a similar sense in connection with (εἰς) ἐποχήν: εἰσάγω, περιάγω, συνάγω,

296 Cf. Castagnoli, *Self-Refutation*, pp. 251–307.

297 All these parallels are noted by Janáček, "Ainesidemos," in idem, *Studien*, p. 254.

298 Cf. Castagnoli, *Self-Refutation*, pp. 251–307, esp. 254 f. and 262–272. As for the precise meaning of the term, Castagnoli (p. 268) points out, on the basis of papyrological evidence, that "[the] verb περιγράφειν could denote, at least starting from the first century AD, that specific kind of cancellation practiced on a text by copyists, correctors and philologists which we call 'expunction' or 'deletion'" and argues that it may have been used by Sextus against the background of this technical meaning. Castagnoli credits this brilliant idea to Walter Cavini.

299 Cf. Castagnoli, *Self-Refutation*, pp. 254 n. 11 and 351 f. n. 137, who, while considering the possibility that the metaphor could have been first adopted for sceptic use by Sextus, points out that the relative chronology of Clement and Sextus is uncertain. But, irrespective of chronology, it is unlikely that Clement would know this metaphor from Sextus, as there is no evidence of his acquaintance with Sextus' work. To all appearances, the kind of scepticism Clement combats in 15, 2–16, 1 and mediates in 22, 1–4 is of a pre-Sextan variety and the occurrence of the word συμπεριγράφεσθαι in this context allows us to argue safely, I think, that it belonged to the pre-Sextan phase of scepticism, too.

300 Trans. Bett, slightly modified.

συνεισάγω, καθίσταμαι, καταλήγω, καταντῶ, περιίστημι.[301] The verb used in our passage is arguably chosen for its connotation of the ship's landing; cf. LSJ, s.v. κατάγω, 4.a.

93,28 τῶν ἀρχικωτάτων τῆς ἐποχῆς ⟨αἰτίων⟩. The object of the sentence cannot be directly linked to ἀρχικωτάτων, and so a noun qualified by this adjective must have been omitted from the text. For the choice of αἰτίων, cf. above, 93,19. Cf. also Sextus, *M.* IX 5: τὸ μὲν πρῶτον καὶ ἀρχικώτατον αἴτιον κτλ.

93,28 f. τὸ ... ἀβέβαιον τῆς διανοίας. Cf. above, 90,5 f. and 93,20. On 'infirmity', cf. above, 89,2; Philo, *Ebr.* 170, quoted above, on 93,19.

93,30 f. δικαστηρίων τε καὶ βουλευτηρίων καὶ ἐκκλησιῶν. This is a standard list (and order) of the key institutions of the *polis*, going back to the classical age; cf. Plato, *Gorg.* 452e2 f.; Aristotle, *Pol.* III 11, 1282a34–37; Arius Didymus in Stobaeus, *Anth.* II 7, 26.85; Philo of Alexandria, *Abr.* 20, etc.

93,32 αἱρέσεως ⟨καὶ φυγῆς⟩. In support of this emendation, Stählin refers to Philo, *Ebr.* 169 and 171; but since κακά in 93,31 can hardly be conceived as a possible object of αἵρεσις, the addition of φυγή is very likely anyway.

93,32–94,1 διανοίας ... μετοκλαζούσης. Presumably a Homeric allusion; cf. *Il.* XIII 279–281 (trans. Lattimore): "The skin of the coward changes colour one way and another,/and the heart inside him has no control to make him sit steady,/but he shifts his weight from one foot to another (μετοκλάζει καὶ ἐπ' ἀμφοτέρους πόδας ἵζει) ..." For ἰσοσθένεια, cf. above, on 90,5 f.

94,2 καὶ αἱ συντάξεις καὶ αἱ πραγματεῖαι. These words seem to have entered the text as a gloss specifying the genre of the βιβλία in question. The semantic distinction between the two terms, which may both refer to scientific treatises, is hard to determine.[302]

---

301 Janáček, "Ainesidemos," in idem, *Studien*, p. 253.

302 A possible distinction is this: Whereas the word σύνταξις is used primarily in view of the arrangement of a treatise (signalling that it consists of several coordinated parts), πραγματεῖα refers to the way it handles its subject. Cf. Arnold Schumrick, *Observationes ad rem librariam pertinentes. De* syntaxis, syntagma, pragmateia, hypomnēma *vocabulis*, Marburg diss. 1909. For πραγματεῖα, cf. the case study by Philip van der Eijk, "Galen and the Scientific Treatise: a Case Study of *Mixtures*," in M. Asper (ed.), *Writing Science: Medical and Mathematical Authorship in Ancient Greece* (Berlin/Boston: de Gruyter, 2013), pp. 145–175.

94,3 f. τῶν διαφωνούντων ἐν τοῖς δόγμασι … τὴν ἐν τοῖς οὖσιν ἀλήθειαν γινώσκειν. Again, the vocabulary is reminiscent of Philo, *Ebr.* 198: "… even the multitude of the so-called philosophers, pretending to seek for the clarity and infallibility in things (τὸ ἐν τοῖς οὖσι σαφὲς καὶ ἀψευδές), is divided into groups and clubs and holds discordant beliefs (δόγματα ἀσύμφωνα), often even opposite ones, not only about one accidental matter, but about nearly all things, small and great, in which controversies are implicated (ἐν οἷς αἱ ζητήσεις συνίστανται)." See also Photius, *Bibl.* 169b21–26 (Bekker)/LS 71C (2): "Consequently, he [i.e. Aenesidemus] says, neither the Pyrrhonists nor the others know the truth in things (τὴν ἐν τοῖς οὖσιν ἀλήθειαν); but the philosophers of other schools, as well as being ignorant in general, … are also ignorant of the very fact that they have apprehension of none of the things of which they think that they have gained apprehension (οὐδὲν αὐτοῖς τῶν δοξάντων εἰς κατάληψιν ἐληλυθέναι κατείληπται)."[303] The expression ἡ ἐν τοῖς οὖσιν ἀλήθεια is also used by Sextus, in the context of a discussion about the criterion; cf. *M.* VII 89, 114, and 435.[304] For similar phrases, which seem to be a hallmark of the Aenesidemean style, see further above, on 90,5 f. and 93,22 f.

(VIII) 23, 1–24, 9: Categories

This section picks up on chapter 6, where the Aristotelian categories were introduced as the general meanings of predicates in definitions (20, 1–2; 21, 4). Here, the categories become the focus of attention, being presented as elementary concepts as well as "elements of things in matter"; this expression presumably refers to their role as the genera of being (23, 4.6). The discussion is prefaced by a conceptual scheme derived from Aristotle's *De interpretatione* and it also includes a classification of 'onymies' that goes beyond Aristotle and clearly reflects a commentary tradition on the *Categories*. There is continuity between this section and the preceding material on demonstration, marked, among other things, by the interest in the categories as the elements of "everything sought" (23, 3) and by the connections with chapter 6, mentioned above. Again, the doctrinal background is a second-century variety of Aristotelianism, with most conspicuous parallels found in Galen (cf. 23,

303 Trans. Long and Sedley, slightly modified.

304 Detailed analysis of these parallels is provided by Janáček, "Ainesidemos," in idem, *Studien*, pp. 255–258. Cf. also idem, "Zur Interpretation des Photiosabschnittes," in *Studien*, p. 222 [98].

6/95,1–3 and 24, 1/95,4 f.), Alexander (cf. 23, 6/95,1–3), and especially Aspasius (cf. 23, 1/94,7 f. and 24, 9/95,24–26).

23, 1–6

94,5–12 Cf. Aristotle, *Int.* 1, 16a3–8. In the 1st cent. B.C., the Aristotelian authorship of *Int.* was disputed by Andronicus of Rhodes (on the basis of *Int.* 1, 16a6–9). It was later defended by Alexander of Aphrodisias and before him by Aspasius (2nd cent. A.D.), who is the first known commentator on that writing.[305] Between Aspasius and Alexander, Galen also regarded *Int.* as Aristotle's work and wrote a commentary on it.[306] The fact that our text frames the exposition of the categories with the distinctions made in *Int.* 1 suggests that it draws on sources siding with the same view about its authorship. Cf. also above, on 87,24 f.; 91,31 f.; 92,11–13.

94,5 f. τά τε ὀνόματα σύμβολα ὄντα τῶν νοημάτων. Aristotle describes "the items in voice" (τὰ ἐν τῇ φωνῇ), i.e. utterances, as "symbols of affections in the soul" (τῶν ἐν τῇ ψυχῇ παθημάτων σύμβολα); *Int.* 1, 16a3 f. Our text paraphrases this, by narrowing τὰ ἐν τῇ φωνῇ to ὀνόματα (cf. *Int.* 2, 16a19–21: "ὄνομα is a voice significant by convention, without time, none of whose parts is significant in separation") and by interpreting τὰ ἐν τῇ ψυχῇ παθήματα as νοήματα, i.e. concepts (cf. *Int.* 1, 16a9 f.: ἐν τῇ ψυχῇ ... νόημα, *DA* III 6, 430a26–28).

94,6 f. τῶν νοημάτων κατὰ τὸ προηγούμενον, κατ' ἐπακολούθημα δὲ καὶ τῶν ὑποκειμένων. This distinction between the primary and the secondary reference is not found in Aristotle, but it is known to the commentators, who attribute it to the 'ancients'; cf. Dexippus, *In Cat.* I 3 (CAG IV/2: 9,22–25): μόνα δὲ σημαινόμενα οἱ ἀρχαῖοι λέγουσι τὰ νοήματα, ἐπειδὴ περὶ τῶν πραγμάτων ἐστὶ ταῦτα καὶ ἀπὸ τῶν πραγμάτων. προηγουμένως μὲν τὰ νοήματα, κατὰ δεύτερον δὲ λόγον καὶ τὰ πράγματα σημαίνεται. Cf. ibid. 7,1–2: προηγουμένως σημαίνεται τὰ νοήματα, κατὰ συμβεβηκὸς δὲ καὶ τὰ πράγματα. A parallel in Simplicius suggests that it goes back to Boethus of Sidon (1st cent. B.C.); cf. Simplicius, *In Cat.* 1, 1a16 (CAG VIII: 41,28 f.): ὁ δὲ Βόηθος μόνα λεγόμενα καὶ σημαινόμενα τὰς νοήσεις εἶναί φησι παρὰ τοῖς ἀρχαίοις.[307]

94,7 f. τὰ νοήματα ὁμοιώματα καὶ ἐκτυπώματα τῶν ὑποκειμένων. For ὁμοιώματα, cf. Aristotle, *Int.* 1, 16a7. The description of concepts as 'imprints' seems to be

---

305 Cf. Moraux, *Aristotelismus* I, pp. 117–119; II, p. 231.

306 *Lib. Prop.* 14, 11 and 17, 1 (166,8 f. and 171,8 Boudon-Millot/XIX,41.47 K.).

307 Cf. Moraux, *Aristotelismus* II, pp. 336 f.

influenced by the Stoic theory of concept formation.[308] There is an interesting parallel in Aspasius, who reportedly explained that Aristotle's 'affections in the soul' are caused by sensible objects.[309]

94,8–11 ἅπασι καὶ τὰ νοήματα τὰ αὐτά ἐστι … οὐκέτι δὲ καὶ τὰ ὀνόματα κτλ. Cf. *Int.* 1, 16a5–8.

94,13 ὡρισμένα γὰρ χρὴ εἶναι τὰ στοιχεῖα. Grammar reduces names to twenty-four "universal elements" (94,12 f.), because the elements (of speech) must be limited. Aristotle argues that the elements of speech (τὰ τῆς φωνῆς στοιχεῖα) are limited in kind, but not in number; for if they were limited in number, there would be just as many letters in the world as there are these elements; cf. *Met.* B 4, 1000a1–4; B 6, 1002b17–21. Our text does not apply such a distinction to the elements, claiming simply that they must be limited. There is no doubt that, according to our author, the elements limit the sounds of speech in kind. But when saying that the elements themselves are limited, he seems to be referring to their limited number. The example is used in a similar way by Galen, *On Medical Experience* 3, 88.[310]

94,13 f. τῶν γὰρ καθ' ἕκαστα ἀπείρων ὄντων μὴ εἶναι ἐπιστήμην. Cf. Aristotle, *Met.* B 3, 999a26–28.

94,14–16 ἴδιον δὲ ἐπιστήμης … εἰς τὰ καθόλου ἀνάγεται. This passage is quoted by John of Damascus in the *Sacra Parallela* (fr. 278 Holl).[311]

---

308 For the Stoic theory, cf. R.J. Hankinson, "Stoic Epistemology," in Inwood (ed.), *The Cambridge Companion to the Stoics*, pp. 61–64. For the metaphor of 'imprint', cf. Eusebius, *Praep.* XV 20, 2 (= SVF I 141); Sextus, *M.* VII 228–230; 251; *DL* VII 50. See already Plato, *Theaet.* 191d6 and 194b5; Aristotle, *Mem.* 1, 450a27–32.

309 Cf. Boethius, *In Arist. De interp.* 41,16–19 (Meiser²): *arbitratur* [*scil. Aspasius*] *Aristotelem passiones animae non de rebus incorporalibus, sed de his tantum quae sensibus capi possunt passiones animae dixisse*. Cf. Moraux, *Aristotelismus*, II, pp. 231 f.

310 "… let them (the Empiricists) know that the sounds of speech, though endless in number, could not be retained and comprehended by mere memory, but that a wise man grasped them and limited them, because, having reflected upon them and examined them, he discovered that the principles and the elements of which these sounds are composed—I mean the letters—are twenty-four in number according to Greek reckoning" (trans. Walzer).

311 *Fragmente vornicänischer Kirchenväter aus den Sacra parallela*, hrsg. Karl Holl (Leipzig: J.C. Hinrichs, 1899), p. 110.

94,15 καθολικοῖς ... θεωρήμασι καὶ ὁρισμοῖς. In L, theorems on which knowledge is based are described as "universal" (καθολικά) and "limited" (ὡρισμένα). The first qualification indicates that the theorems in question must be universal propositions.[312] The second echoes the view that the elements of speech must be limited (94,13: ὡρισμένα ... τὰ στοιχεῖα). The first qualification is unproblematic, but the point of the latter is hard to explain. A variant preserved by Damascene, which reads ὁρισμοῖς instead of ὡρισμένοις, makes better sense. Presumably, 'definitions' are added in order to include the (undemonstrated) principles among the foundations of knowledge (cf. above, on 83,27–29). The error can be explained by attraction to ὡρισμένα in 94,13.

94,16 f. ἡ δὲ τῶν φιλοσόφων πραγματεία κτλ. By excluding grammar from the philosopher's agenda, our text possibly distances itself from the Stoic interpretation of Aristotle's *Categories* as an (inadequate) inquiry about words (λέξεις).[313]

94,18 f. ὑφ' ἃ πᾶν τὸ ζητούμενον ὑπάγεται. For πᾶν τὸ ζητούμενον, cf. above, 3, 2/81,17; 8, 4/84,24; 17, 1/90,10. The relevance of the doctrine of the categories to the practice of inquiry and demonstration is explained by Galen, *Inst. Log.* 12, 1 (26,17–20 Kalbfleisch): χρώμεθα δ' αὐτοῖς [*scil.* categorical syllogisms] ἐν ταῖς ἀποδείξεσιν ἐν αἷς ὑπὲρ ἑνὸς τῶν ὄντων ἐστὶ ζήτησις ὁπηλίκον ἐστὶν ἢ ὁποῖον ἢ ποῦ κείμενόν (ἐστιν ἢ) τι τοιοῦτον τῶν κατὰ τὰς ἄλλας κατηγορίας.[314] Cf. above, 17, 2/90,11–12 and 20, 1–2/92,18–23. In our passage, the categories are described as 'elements' both with regard to the concepts we have of the investigated objects and with regard to the objects themselves (τούτων in 94,18 referring to both τὰ νοήματα and τὰ ὑποκείμενα). See further below, on 95,1.

---

312 The association between theorems and knowledge (in a very broad sense) is pointed out by Alexander, *In Top.* I 11, 104b1–5 (CAG II/2: 74,9–11), who defines a theorem as "an account that brings about knowledge" (λόγος γνώσεως αἴτιος). The expression 'universal theorem' (τὸ καθόλου/καθολικὸν θεώρημα) is very common in Galen; cf. e.g. *Hipp. Epid.* VI, II 39 (CMG V,10,2,2: 106,6–8/XVIIa,978 f. K.): πολλὰ δ' ἐν τοῖς τῶν Ἐπιδημιῶν γέγραπται μὲν ἐπί τινων ἑωραμένα κατὰ μέρος, ὑποπέπτωκε δὲ καθολικοῖς θεωρήμασιν ἤτοι διηνεκέσιν ἢ ὡς τὸ πολὺ γινομένοις ("Many things written in the *Epidemics* have been observed in some particular cases, but fall under universal theorems that hold either invariably or for the most part").

313 Cf. Moraux, *Aristotelismus*, II, pp. 585–601.

314 On this passage, cf. Benjamin Morison, "Les *Catégories* d'Aristote comme introduction à la logique," in O. Bruun and L. Corti (eds.), *Les Catégories et leur histoire* (Paris: Vrin, 2005), pp. 103–119, here 118.

94,20f. εἰ δὲ πάντα διαφύγοι, μηδαμῇ εἶναι. Cf. Aristotle, *Met.* Δ 7, 1017a22–24 (trans. Ross): "Those things are said in their own right to be (καθ' αὑτὰ … εἶναι) that are indicated by the figures of predication (ὅσαπερ σημαίνει τὰ σχήματα τῆς κατηγορίας); for the senses of 'being' are just as many as these figures." The phrase μηδαμῇ εἶναι is probably an allusion to Parmenides, fr. 7,1 DK, in a version quoted by Alexander, *In Met.* N 2 (CAG I: 805,20f.): οὐ γὰρ μήποτε τοῦτο *μηδαμῇ εἶναι* μὴ ἐόντα (μηδαμῇ replacing the presumably original δαμῇ, attested elsewhere). Alexander must have seen this version in his copy of *Metaphysics*, where the verse is mentioned in the context of a polemic against the (Academic) view that even that-which-is-not is (N 2, 1089a2–5). It is likely that our passage reflects this discussion in some way. Cf. also Galen, *Inst. Log.* 17, 6 (44,3f. Kalbfleisch): "… when someone says that things-that-are-not are or that things-that-are are not, all Greeks think that he is speaking falsely."

94,21–23 τῶν λεγομένων τὰ μὲν ⟨μετὰ συμπλοκῆς … τὰ δὲ⟩ ἄνευ συμπλοκῆς … ὡς τὸ "τρέχει". The text now turns to Aristotle's *Categories*, starting with a paraphrase of *Cat.* 2, 1a16–19. As preserved in L, however, it does not make much sense and the corrections printed here and in 94,24 in Stählin's edition are the best available.

94,23f. ὅσα λόγον οὐκ ἀποτελεῖ οὐδὲ τἀληθὲς ἢ τὸ ψεῦδος ἔχει. Examples mentioned in *Cat.* 2, 1a19 (ἄνθρωπος, βοῦς, τρέχει, νικᾷ) are summarized in view of *Cat.* 4, 2a8–10: τῶν δὲ κατὰ μηδεμίαν συμπλοκὴν λεγομένων *οὐδὲν οὔτε ἀληθὲς οὔτε ψεῦδός* ἐστιν, οἷον ἄνθρωπος, λευκόν, τρέχει, νικᾷ.

94,24–95,1 τῶν δὲ ⟨μὴ⟩ μετὰ συμπλοκῆς λεγομένων τὰ μὲν οὐσίαν σημαίνει, τὰ δὲ ποιόν, τὰ δὲ ποσόν κτλ. Cf. *Cat.* 4, 1b25–27. It has been observed that, unlike in most of Aristotle's lists, quality precedes quantity in this passage.[315] This could be interesting insofar as the order of these two categories in the *Categories* has been criticized by some Platonist readers as being at variance with their ontological views.[316] But it is unlikely that this debate affects our passage in

315 Cf. Michael Frede, "Les *Catégories* d'Aristote et les Pères de l'Église grecs," in O. Bruun and L. Corti (eds.), *Les Catégories*, pp. 135–173, here 144.

316 Cf. Moraux, *Aristotelismus*, II, pp. 522f.; Riccardo Chiaradonna, "Autour d'Eudore: Les débuts de l'exégèse des Catégories dans le Moyen Platonisme," in M. Bonazzi and J. Opsomer, *The Origins of the Platonic System: Platonisms of the Early Empire and Their Philosophical Contexts* (Leuven: Peeters, 2009), pp. 89–111, here 97–99. Cf. Simplicius, *In Cat.* (CAG VIII:206,3–207,26), referring to Eudorus and Pseudo-Archytas; cf. also Philo of Alex-

any way. As Galen's example shows, the order of these (and other) categories may vary within the writings of the same author.[317] This is also true of our text; cf. above, 20, 2/92,20–23, where the order of the two categories is reversed. Moreover, Aristotle himself sometimes gives precedence to quality; cf. *An. Post.* I 22, 83a21–23; *Met.* Δ 7, 1017a25–27; *Met.* N 2, 1089a7–9; for the last reference, cf. above, on 94,20 f.

95,1 ἃ δὴ καὶ στοιχεῖα τῶν ὄντων φαμὲν τῶν ἐν ὕλῃ. This clause picks up on 94,18 f., where it is mentioned that some 'elements' of concepts and 'underlying things' have been discovered. Having enumerated ten items signified by uncombined expressions, the text says that these very items are "the elements of things in matter". How shall we understand this? The role of the 'elements' has been explained before: it enables us to reach knowledge of concepts and things by reducing them to universal propositions (94,13–19). Now we are told that these 'elements' are the ten predications discussed in the *Categories*. Thus, in order to gain knowledge of a 'thing in matter', we must reduce it to a proposition whose predicate (a) holds of all members of the class to which the thing belongs and (b) falls under one of the ten 'categories'. This predicate, then, is not only something we mean when speaking of the thing (class) in question, but also something that is the case with regard to that thing (class). Thus the ten 'categories' are the kinds of items the 'things in matter' can be; in other words, they are the 'elementary' kinds of being. When we look for a definition, we normally do not proceed as far as these 'elements' themselves, but the first genus in terms of which anything is defined should presumably always be reducible to one of them.[318]

95,1 f. καὶ μετὰ τὰς ἀρχάς. Although there might be a case for taking this phrase as a predicate coordinate with στοιχεῖα, it is more natural to regard it as a

---

andria, *Decal.* 30; Hippolytus, *Ref.* VI 24, 2 (231,7 Marcovich); Mansfeld, *Heresiography*, p. 68.

317 Cf. e.g. *Diff. Puls.* II 10 (VIII,632,17 f. K.): τὸ μὲν ... οὐσία, τὸ δὲ ποιόν, τὸ δὲ ποσόν, τὸ δὲ πρός τί, as opposed to *Inst. Log.* 12, 1 (26,19 f. Kalbfleisch): ὑπὲρ ἑνὸς τῶν ὄντων ἐστὶ ζήτησις ὁπηλίκον ἐστὶν ἢ ὁποῖον ἢ ποῦ κείμενόν.

318 Cf. a fragment of Alexander's commentary on *An. Post.* I 22, preserved in Parisinus 1843, fol. 190r, and quoted in CAG II/1, p. xix: τοῦ μὲν γὰρ τινὸς ἀνθρώπου προσεχῶς μὲν ὁ ἄνθρωπος ἐν τῷ τί ἐστι κατηγορεῖται, τούτου δὲ τὸ ζῷον πεζὸν δίπουν, καὶ τούτου τὸ ζῷον πεζόν, καὶ τούτου τὸ ζῷον, καὶ τούτου ἡ ἔμψυχος οὐσία, ἧς ἡ οὐσία· καὶ οὐδὲ ταύτης ἐστὶν ἐξωτέρω τῶν ἐν τῷ τί ἐστι τοῦ τινὸς ἀνθρώπου κατηγορουμένων, ὃ κοινότερον κατηγορεῖται. ὁ δ' αὐτὸς λόγος καὶ ἐπὶ τῶν ἄλλων ἐννέα γενῶν. Cf. also Galen, *MM* II 7 (X,151,14–152,1 K.), quoted above, on 83,26 f.

second qualification of τῶν ὄντων, coordinate with ἐν ὕλῃ.[319] The expression μετὰ τὰς ἀρχάς is appropriate for an object of knowledge described in relation to principles on the basis of which it is known (see the next note). Thus it is appropriate for 'the things in matter', insofar as they are objects of knowledge.

95,2f. ἔστι γὰρ λόγῳ θεωρητὰ ταῦτα, τὰ δὲ ἄυλα νῷ μόνῳ ληπτά ἐστι. The contrast between ταῦτα and τὰ ἄυλα supports the view that ταῦτα refers to 'the things in matter'.[320] These things must be particulars taken as objects of knowledge, i.e. limited in kind. As long as they are limited in this way, they are capable of being known λόγῳ. The predicate λόγῳ θεωρητά is introduced as an explanation (γάρ), the point explained being either the content of φαμέν in the preceding clause, or the qualification of these entities as μετὰ τὰς ἀρχάς. The latter option is more likely, I think. The contrast between things that are λόγῳ θεωρητά and those that are νῷ μόνῳ ληπτά does indeed explain why the former are μετὰ τὰς ἀρχάς, if λόγος is an argument by which we demonstrate something, whereas νοῦς is an act by which we grasp the undemonstrated principles on which any such argument must ultimately depend.[321] To put it plainly, if something is capable

319 I have rehearsed the former interpretation in my "Categories in *Stromata* VIII," *Elenchos* 33/2 (2012), pp. 206–209, where I took the 'elements of things in matter' as referring to enmattered forms that become objects of knowledge on the basis of conceptual analysis (distinguishing such principles as 'matter' and 'form'). However, as pointed out by Dmitry Biriukov, "О платонизме в философско-богословской системе Климента Александрийского" (On Platonism in the Theological and Philosophical Doctrine of Clement of Alexandria) in *Plato Collection* II (Supplement to the Bulletin of the Russian Christian Academy for the Humanities 14; Moscow/Saint Petersburg: RGGU-RHGA, 2013), pp. 276–483, here 287f., this explanation was not entirely convincing. In her commentary on our passage, Anna Zhyrkova, "Reconstructing Clement of Alexandria's Doctrine of the Categories," in J.J. Finamore and R.M. Berchman (eds.), *Conversations Platonic and Neoplatonic: Intellect, Soul, and Nature. Papers from the 6th Annual Conference of the International Society for Platonic Studies* (Sankt Augustin: Academia Verlag, 2010), pp. 145–154, also construes μετὰ τὰς ἀρχάς as a predicate coordinate with στοιχεῖα, but interprets the phrase differently. Explaining the 'principles' as the principles of proof and the 'elements' as the most universal concepts, Zhyrkova thinks that the 'principles' precede the 'elements' (pp. 150f.). But if the most universal concepts are said to be "after" the principles of proof, how shall we understand the relation between the two? Shall we take our text as saying that the most universal concepts ('substance', 'quality', etc.) can be proved?

320 For a different view, see my "Categories," pp. 206–209.

321 Cf. already Aristotle, *EN* VI 11, 1143a36–1143b1: καὶ γὰρ τῶν πρώτων ὅρων καὶ τῶν ἐσχάτων νοῦς ἐστὶ καὶ οὐ λόγος, καὶ ὁ μὲν κατὰ τὰς ἀποδείξεις τῶν ἀκινήτων ὅρων καὶ πρώτων κτλ. Cf.

of being known by λόγος in this sense, there must be some principles on the basis of which it can be known, and it is with regard to these principles that any such object is described as μετὰ τὰς ἀρχάς.

If the contrast between λόγος and νοῦς serves to explain why 'the things in matter' are μετὰ τὰς ἀρχάς, it is likely that the objects of νοῦς in our passage correspond to these ἀρχαί. Now these objects are called τὰ ἄυλα, in contradistinction to the objects of λόγος. We have seen that the objects of λόγος must be particulars limited in kind, which is equivalent to saying that they are kinds of particulars. These kinds are 'in matter', but they are known on the basis of something immaterial, graspable by the intellect. Unfortunately, the text tells us too little of these immaterial entities to allow for a convincing account of their status, content, and function. There is an interesting passage in Alexander's *De anima*, where a distinction is made between ἐπιστήμη, defined as "demonstrative syllogism or knowledge of the cause of the thing known", and νοῦς, explained as "the capacity to grasp the immaterial forms (τῶν ἀύλων εἰδῶν ληπτική), or the capacity by which we make true assertions about undemonstrated principles (περὶ τὰς ἀναποδείκτους ἀρχάς)".[322] Alexander also draws a contrast between immaterial forms and forms in matter (τὰ ἔνυλα), which are intelligible in a secondary sense only (insofar as they are separated from matter and actualized as concepts).[323] It is conceivable that our author shares some ontological and epistemological views with Alexander; however, if he does, he does not tell us about it.

The immaterial things, taken as the principles of understanding, possibly include the ten kinds of predicates. This is neither said nor necessarily implied in our text, but it would explain why these principles are mentioned here at all.[324]

---

ibid. VI 8, 1142a25 f.: ὁ μὲν γὰρ νοῦς τῶν ὅρων, ὧν οὐκ ἔστι λόγος κτλ. See also *Mag. Mor.* I 34, 13, on the distinction between νοῦς (περὶ τὰς ἀρχὰς τῶν νοητῶν καὶ τῶν ὄντων) and ἐπιστήμη (περὶ τὰ μετὰ τὰς ἀρχὰς μετ' ἀποδείξεως ὄντα).

322 Alexander, DA (Suppl. Arist. II/1: 66,16–19).

323 Cf. Alexander, DA (Suppl. Arist. II/1: 87,24–90,2). A Platonist version of the distinction is found in Alcinous, *Did.* 4,7 (H 155,39–41); cf. Mansfeld, *Heresiography*, p. 63 n. 15. Cf. also *Strom.* V (14) 94, 2, where Clement ascribes to Plato a view that God created the 'perceptible species' (τὰ εἴδη τὰ αἰσθητά) according to the 'intelligible genera' (τὰ γένη τὰ νοητά); cf. Plato, *Tim.* 30c–d; 39e; Philo of Alexandria, *Opif.* 16.

324 The author possibly includes the categories among the 'simple' objects of intellection mentioned above, 7, 4/83,27.

95,3 κατὰ τὴν πρώτην ἐπιβολήν. In Greek philosophical literature (starting with Epicurus), the word ἐπιβολή often describes an activity by which a capacity of cognition (perception, imagination, or thought) is applied to its object.[325] κατὰ τὴν πρώτην ἐπιβολήν may then qualify a cognition that grasps its object immediately, "at first application" of the capacity concerned.[326] This is also the case in our passage, where the immaterial entities are said to be grasped by the intellect like that. The point presumably is to draw a contrast between this kind of cognition, taken as an immediate grasp of self-evident objects, and knowledge mediated by an argument (cf. above, 6, 7–7, 5 and note on 83,27–29). Cf. Galen, *MM* I 4, speaking of axioms as undemonstrated phenomena, subjected to intellection κατὰ πρώτην ἐπιβολήν.[327]

24, 1–9

95,4 ὑπὸ τὰς δέκα κατηγορίας ὑποτασσομένων. Cf. above, 23, 3/94,18f.: ὑφ' ἃ πᾶν τὸ ζητούμενον ὑπάγεται.

95,4f. τὰ μὲν καθ' αὑτὰ λέγεται … τὰ δὲ πρός τι. Some predications belong to the investigated object in virtue of itself, some in virtue of its relation to something else. This distinction is not found in Aristotle, for whom things predicated in any one of the categories belong to the object in virtue of itself (though predications of substance do so primarily, while the other ones in the secondary sense only).[328] But it is known to Simplicius, who ascribes it to the Old Academy (οἱ περὶ Ξενοκράτη) and to Andronicus, "who think they can cover everything by [the categories of] 'in itself' and 'in relation to something', and thus so large a number of genera [*scil.* as distinguished by Aristotle] is, in their view, superfluous."[329] However, when applying this distinction, Andronicus appears to have drawn the dividing line between the category of substance

325 For Epicurus, cf. DL X 31.51.69; *Sent.* 24 (76,13f. Usener).

326 Cf. e.g. Galen, *Dig. Puls.* III 1/VIII,883,17 K. (κατὰ τὴν πρώτην ἐπιβολὴν τῆς ὄψεως); ib. 888,9f. (κατ' αὐτὴν τὴν πρώτην ἐπιβολὴν τῆς ἁφῆς); *Inst. Od.* 1/CMG Suppl. V: 34,9/II,857 K. (κατὰ τὴν πρώτην ἐπιβολὴν τῆς διανοίας).

327 *MM* I 4 (X,36,10–18 K.), quoted above, on 83,26f. For later parallels, cf. a definition of νοῦς in an anonymous commentary on Hermogenes in H. Rabe (ed.), *Rhetores Graeci XIV: Prolegomenon Sylloge* (Leipzig: Teubner, 1931), pp. 228,21f.: νοῦς μέν ἐστιν ἡ πρώτη ἐπιβολὴ τοῦ πράγματος, διάνοια δὲ ἡ διάκρισις κτλ. Cf. also John Philoponus, *In DA* (CAG XV,79,24): ἀδιαίρετος γὰρ οὗτος (*scil.* νοῦς) καὶ πάντη ἑαυτῷ ὅμοιος, ἅμα τῇ πρώτῃ ἐπιβολῇ ἀχρόνως νοῶν τὰ πράγματα.

328 Cf. Aristotle, *Met.* Δ 7, 1017a22–30 (quoted above, on 94,20f.); Z 4, 1030a17–1030b13.

329 Simplicius, *In Cat.* 4, 1b25 (CAG VIII: 63,22f.): οἱ γὰρ περὶ Ξενοκράτη καὶ Ἀνδρόνικον πάντα τῷ καθ' αὑτὸ καὶ τῷ πρός τι περιλαμβάνειν δοκοῦσιν, ὥστε περιττὸν εἶναι κατ' αὐτοὺς τὸ τοσοῦ-

and the others conceived as its accidents.[330] In contrast, our text reserves the designation of πρός τι for the category of relation, while describing the remaining nine categories as καθ' αὑτά predications. This version, too, is known to Simplicius (who presents it as his own view),[331] and, closer to our text, it is used by Galen.[332]

95,6–9 The description of συνώνυμα closely follows Aristotle, *Cat.* 1, 1a6–12. On the definition of ζῷον, cf. above, 86,15 f.

95,9–11 ἑτερώνυμα ... περὶ τὸ αὐτὸ ὑποκείμενον ἐν διαφόροις ἐστὶν ὀνόμασιν κτλ. The class of heteronyms is not found in Aristotle. According to Boethus (reported by Simplicius), it was (first?) used by Speusippus, but not in the same sense as in our text.[333] The earliest witness of heteronyms in our sense (namely as items having different names and definitions, but corresponding to the same object) is Alexander, *In Top.* V 4, 133b15 (CAG II/2: 398,1–5); cf also

---

τον τῶν γενῶν πλῆθος. The division of being according to these two 'categories' presumably goes back to the Old Academy; cf. Hans J. Krämer, *Platonismus und hellenistische Philosophie* (Berlin/New York: de Gruyter, 1972), pp. 81–96; for Clement, see p. 91 and n. 362. Cf. also Witt, *Albinus*, p. 38 and n. 5.

330 Cf. Simplicius, *In Cat.* (CAG VIII: 63,24–26): ἄλλοι δὲ εἰς οὐσίαν καὶ συμβεβηκὸς διατέμνουσιν· καὶ οὗτοι δὲ ταὐτόν πως δοκοῦσι τοῖς προτέροις λέγειν τὰ συμβεβηκότα πρός τι λέγουσιν, ὡς ἄλλων ἀεὶ ὄντων, καὶ τὴν οὐσίαν καθ' αὑτό. For the scope of Andronicus' 'Hauptgruppen', cf. Moraux, *Aristotelismus*, I, p. 103; Mansfeld, *Heresiography*, pp. 59 f. and n. 8. For a detailed discussion of Andronicus' view, cf. Michael Griffin, *Aristotle's* Categories *in the Early Roman Empire* (Oxford: Oxford University Press, 2015), pp. 48–54.

331 *In Cat.* (CAG VIII: 174,16): καὶ ῥητέον ὅτι ἐν τῷ καθ' αὑτὸ αἱ ἐννέα κατηγορίαι θεωροῦνται. The context is a polemic against Eudorus' criticism of Aristotle's alleged failure to treat the 'category' of καθ' αὑτό in the *Categories*. Cf. Moraux, *Aristotelismus*, II, pp. 521 f. and n. 48; Mansfeld, *Heresiography*, pp. 59 f. n. 8.; Griffin, *Aristotle's Categories*, pp. 82–87.

332 Cf. *Dig. Puls.* II 2 (VIII,839,13–16 K.): ἆρ' οὖν τῶν καθ' ἑαυτὰ λεγομένων ἐστὶ τὸ μέγα καὶ τὸ μικρόν; ἀλλ' εἰ τοῦτο, πάντως δήπουθεν ἢ οὐσίας ἐστὶ δηλωτικά, ἢ ποιότητος, ἤ τινος τῶν κατὰ τὰς ἄλλας κατηγορίας λεγομένων.

333 Speusippus reportedly distinguished two classes of names, 'tautonyms' (ταὐτώνυμα) and 'heteronyms' (ἑτερώνυμα), further dividing the latter class into three groups: (1) 'heteronyms proper' (τὰ ἰδίως ἑτερώνυμα), (2) polyonyms, and (3) paronyms. He defined the 'heteronyms proper' as items that are "different as far as names, things, and definitions are concerned, such as the art of grammar, human being, and wood" (Simplicius, *In Cat.*/CAG VIII: 38,20–39,2). Thus, what Speusippus calls 'heteronyms proper', corresponds to the ἕτερα below, 95,11–14. Cf. Jean Pépin, "Clément d'Alexandrie, les *Catégories* d'Aristote et le fragment 60 d'Héraclite," in idem, *De la philosophie ancienne à la théologie patristique* (London: Variorum, 1986), IV, pp. 271–284, here 275.

idem, *In Met.* Γ 2, 1003b22 (CAG I: 247,22–24), where the following examples are mentioned: τὸ ἀμερές and τὸ ἐλάχιστον, σπέρμα and καρπός, ἀνάβασις and κατάβασις.[334] Cf. also Simplicius, *In Cat.*1a1 (CAG VIII: 22,30–33); Ammonius, *In Cat.* 1a1 (CAG IV/4: 16,26–29).[335]

Although the items "subordinated to the categories" should include both concepts and underlying objects (cf. above, 94,16–21), the description of ἑτερώνυμα in our passage can only apply to concepts.

95,10f. ὁδὸς γὰρ ἡ αὐτή, ἤτοι εἰς τὸ ἄνω ἢ εἰς τὸ κάτω. 'Ascent' and 'descent' are examples of heteronyms mentioned by other sources; in all of them, the underlying object is the ladder.[336] Our text's ὁδός could be based on Heraclitus' fr. 60 DK (ὁδὸς ἄνω κάτω μία καὶ ὡυτή), or perhaps on Hippocrates, *Alim.* 45 (ὁδὸς ἄνω κάτω μίη).[337] In light of the Galen hypothesis, the latter option is more attractive, as Galen often quotes *Alim.* in his writings and even wrote a commentary on it.[338] In a text published in 1576 by Giovanni Battista Rasario, which claims to contain this commentary (probably a forgery composed by Rasario himself),[339] there is an interesting passage in which the two ways (ὁδὸς ἄνω and κάτω) in *Alim.* 45 are explained as bodily vessels (ἀγγεῖα), "differing only in relation (τῇ σχέσει μόνῃ διαφέροντα), just like ladders making both for ascent and descent (ὥσπερ καὶ κλίμακες τῇ τε ἀναβάσει καὶ τῇ καταβάσει ὑπηρετοῦσι)."[340]

334 Both passages are noted by Pépin, "Clément," p. 277 and n. 3, pp. 281f. and n. 1.

335 These parallels are noted by Ernst, *De Clementis*, pp. 49f. Cf. also Pépin, "Clément," p. 278; Concetta Luna, *Commentaire et notes*, in I. Hadot (ed.), *Simplicius: Commentaire sur les Catégories*, III (PhA 51; Leiden: Brill, 1990), pp. 52f. and n. 30.

336 Cf. Alexander, *In Top.* V 4, 133b15 (CAG II/2: 398,3–5): … ὡς ἐπὶ ἀναβάσεως ἔχει καὶ καταβάσεως, ὅταν κατὰ ἑνός, τῆς κλίμακος, καὶ ἄμφω τὰ ὀνόματα λέγηται· τὸ γὰρ ὑποκείμενον ἕν ἐστιν, αὐτὴ ἡ κλίμαξ. Similarly Simplicius, *In Cat.*1a1 (CAG VIII: 22,32f.); Ammonius, *In Cat.* 1a1 (CAG IV/4: 16,27–29).

337 The first option is discussed by Pépin, "Clément," pp. 279–284.

338 For quotations, cf. e.g. *Opt. Corp. Const.* 4 (IV,746,7 K.); *San. Tu.* I 15 (CMG V,4,2: 36,32–37,1/VI,80 K.); *Alim. Fac.* I 1 (VI,467,17f. K.); *Comp. Med. Gen.* I 17 (XIII,446,12f. K.); *Hipp. Epid. VI*, III 1 (XVIIb,6,13f. K.); *Hipp. Aph.* II 11 and 18 (XVIIb,467,7 and 488,18f. K.). For the commentary, cf. *Lib. Prop.* 9, 12 (161,14f. Boudon-Millot/XIX,36 K.). The commentary is lost, but some sections are preserved in an Arabic translation; cf. Ivan Garofalo, "Il commento di Galeno al *De alimento* e gli estratti di Ibn Riḍwān," *Galeno* 6 (2012), pp. 123–164.

339 Cf. Karl Deichgräber, *Pseudhippokrates: Über die Nährung* (Mainz: Akademie der Wissenschaften und der Literatur, 1973), pp. 12f. n. 12, whose opinion is now generally accepted.

340 Ps.Galen, *Hipp. Alim.* IV 24 (XV,411 K.). The passage is also mentioned by Pépin, "Clément," p. 281 and n. 4.

What makes this passage interesting is that it links the Hippocratean ὁδὸς ἄνω κάτω with a traditional example of heteronyms (in the non-Speusippean sense), while also explaining it in a way that comes close to our description of heteronyms. Due to the obscure origins of *Hipp. Alim.*, it is difficult to appreciate the value of this parallel, however.[341]

95,11–14 τὸ δ' ἄλλο εἶδος τῶν ἑτερωνύμων … ἕτερα δὴ λεκτέον, οὐχ ἑτερώνυμα. This definition of the ἕτερα coresponds to the Speusippean ἑτερώνυμα.[342] For the distinction between the two, cf. Simplicius, *In Cat.* 1a1 (CAG VIII: 22,30–33); Ammonius, *In Cat.* 1a1 (CAG IV/4: 16,24–26).

95,14f. πολυώνυμα … ἄορ, ξίφος, φάσγανον. The definition of polyonyms (unknown to Aristotle) goes back to Speusippus.[343] Out of the four examples listed by Speusippus (ἄορ, ξίφος, μάχαιρα, φάσγανον), our text mentions just three; the same triplet (in the order ξίφος, ἄορ, φάσγανον) is mentioned by Galen, *Diff. Puls.* II 3 (VIII,574,11 f. K.).[344] The concept of polyonyms is also adopted by Alexander, whose examples include φάσγανον and μάχαιρα.[345]

95,15f. παρώνυμα δ' ἐστὶ τὰ παρά τι ἕτερον ὠνομασμένα κτλ. This is a simplified version of Aristotle's *Cat.* 1, 1a12–15, whence also comes the example ἀπὸ τῆς ἀνδρείας ὁ ἀνδρεῖος.

95,16–18 τὰ δὲ ὁμώνυμα … οἷον ἄνθρωπος τό τε ζῷον καὶ γεγραμμένος. This is a paraphrase of Aristotle, *Cat.* 1, 1a1–3, but the example is different. Aristotle's example (οἷον ζῷον ὅ τε ἄνθρωπος καὶ τὸ γεγραμμένον) is based on the polysemy

---

341 As far as I know, no source-critical study of Ps.Gal. *Hipp. Alim.* is available. For some passages, cf. Tommaso Raiola, "Alle origini di un falso galenico: Il *Commento* al *De alimento* e una citazione di Sabino," *Aion* 32 (2010), pp. 101–110. It seems clear, however, that Rasario, before composing his Pseudo-Galenica, searched through a number of sources, including Galen himself and authors quoting him, to find suitable Galenic passages. Cf. Ivan Garofalo, "Il falso commento di Galeno al *De humoribus* e un saggio di edizione del vero," in S. Fortuna, I. Garofalo, A. Lami, and A. Roselli (eds.), *Sulla tradizione indiretta dei testi medici greci* (Pisa/Rome: Fabrizio Serra, 2009), pp. 201–218, here 203. I am grateful to Christina Savino for advice on this issue.

342 Cf. above, note 333.

343 Cf. Simplicius, *In Cat.* 1a12 (CAG VIII: 38,24–26); Ernst, *De Clementis*, p. 50; Witt, *Albinus*, p. 38.

344 Cf. Mansfeld, *Heresiography*, p. 64 n. 16. Galen mentions the word πολυώνυμα in *Syn. Puls.* 2 (IX,435,1 K.).

345 Cf. esp. Alexander, *In Met.* Γ 2, 1003b22–32 (CAG I: 247,27–29).

of ζῷον, which can mean either 'animal' or 'picture' (cf. LSJ s.v. ζῷον II.). In contrast, our text distinguishes homonymous meanings of the word ἄνθρωπος, which can either refer to an animal (i.e. man as a living being), or to a man on the picture.[346] Cf. below, on 95,20 f. Aristotle himself provides similar examples of homonymy (based on a distinction between an object and its image) elsewhere; cf. *DA* II 1, 412b20–22; *Mem.* 1, 450b20–25.

95,18–26 The classification of homonyms in this passage is loosely based on Aristototle, *EN* I 6, 1096b26–28: ἀλλὰ πῶς δὴ [*scil.* τὸ ἀγαθόν] λέγεται; οὐ γὰρ ἔοικε τοῖς γε ἀπὸ τύχης ὁμωνύμοις. ἀλλ᾽ ἆρά γε τῷ ἀφ᾽ ἑνὸς εἶναι ἢ πρὸς ἓν ἅπαντα συντελεῖν, ἢ μᾶλλον κατ᾽ ἀναλογίαν;[347] It may be usefully compared with an equally 'scholastic', but different treatment of homonyms in Porphyry's *In Cat.* (CAG IV/1: 65,12–67,32).[348]

95,19 f. τὰ μὲν ἀπὸ τύχης … τὰ δὲ ἀπὸ διανοίας. Cf. Porphyry's *In Cat.* (CAG IV/1: 65,18 f.), re. ὁμωνύμων τρόποι: ὁ μέν ἐστιν ἀπὸ τύχης, ὁ δὲ ἀπὸ διανοίας. Porphyry's examples of homonyms by coincidence are Alexander, the son of Priam, and Alexander, the Macedonian king (65,22–24). For the Ajax example, cf. ibid. 64,9–21, where it is pointed out, however, that various men called Ajax, though each having a different 'description' (ὑπογραφή), nevertheless have the same definition.

95,20 f. τὰ μὲν καθ᾽ ὁμοιότητα κτλ. Cf. above, on 95,16–18. The same kind (and example) of homonyms is mentioned by Porphyry, *In Cat.* (CAG IV: 65,25–30).

95,21 f. τὰ δὲ κατὰ ἀναλογίαν, ὡς "πόδες Ἴδης" καὶ οἱ ἡμέτεροι πόδες κτλ. This group of homonyms, already mentioned by Aristotle (cf. above, on 95,18–26), is also known to Porphyry, whose examples are the first items in the relations between the monad and the numbers, the point and the line, the source and the rivers, and the heart and the animals; these items may in all cases be called

346 Both variants are known to (and combined by) the commentators. Cf. Themistius, *In An. Post.* II 13 (CAG V/1: 57,13 f.): ζῷον γὰρ ὁμωνύμως τὸ γεγραμμένον, ὅτι καὶ ἄνθρωπος ὁμωνύμως ὁ γεγραμμένος. Porphyry, *In Cat.* (CAG IV: 65,25–30), as opposed to ibid. 66,23–28. Cf. also the scholion on Dionysius Thrax, *Ars gram.* in A. Hilgard, ed., *Scholia in Dionysii Thracis Artem grammaticam* (Grammatici Graeci 3; Leipzig: Teubner, 1901), pp. 135,31–136,1: ὁμώνυμα γὰρ λέγονται ἐκεῖνα, ὧν τὸ ὄνομα μόνον κοινόν ἐστιν, ὁ δὲ ὁρισμὸς διάφορος, ὥσπερ ὁ ἄνθρωπος ὁ ζῶν καὶ ὁ γεγραμμένος κτλ.

347 Cf. Pépin, "Clément," p. 275.

348 For an earlier attempt to make this comparison, cf. my "Categories," pp. 213–216.

ἀρχή (*In Cat.*/CAG IV/1: 65,31–66,2). Porphyry later makes a distinction between analogy and metaphor (unlike analogy, metaphor is something's own name transferred to something else) and notes that "many authors, among them Atticus, have erred in counting metaphor and analogy together as a single kind of homonymy, and have confused metaphor with analogy". Porphyry's examples of metaphor that should not be confused with analogy include the Homeric description of the slopes of a mountain as its 'feet' (cf. *Il.* XX 59).[349] This is the example of homonyms from analogy used in our text, which, therefore, seems to represent the tradition rejected by Porphyry and followed, among others, by the Platonist scholar Atticus.[350]

95,22–24 τὰ δὲ κατ' ἐνέργειαν, ὡς ποὺς πλοίου ... καὶ ποὺς ὁ ἡμέτερος κτλ. This group is not on Porphyry's list. A remote parallel is found in Ammonius, *In Cat.* 1a5 (CAG IV/4: 22,7 f.), where διὰ τὴν τῆς ἐνεργείας ὁμοιότητα is a subgroup of homonyms according to similarity. 'The foot of a ship' is again a Homeric expression (*Od.* X 32), referring to the ship's rudder. It is mentioned by Porphyry as another expression based on metaphor, not homonymy; cf. *In Cat.* (CAG IV/1: 67,13–26).

95,24 f. λέγεται ὁμώνυμα ἀπὸ τοῦ αὐτοῦ καὶ πρὸς τὸ αὐτό. This description of homonyms is based on Aristotle's phrase τῷ ἀφ' ἑνὸς εἶναι ἢ πρὸς ἓν ἅπαντα συντελεῖν (quoted above, on 95,18–26). In Porphyry's classification, two groups of homonyms are distinguished on the same basis, namely (a) "when different things get their designation from some one thing (ἀπό τινος ἑνός)", and (b) "when different things that all seek the same goal (πρὸς ἓν σπεύδοντα τέλος) receive a common designation from that goal" (*In Cat.*/CAG IV/1: 66,2–15). Porphyry notes that some interpreters link these two groups together, "referring to the whole class as homonyms deriving from and relative to a single source (τὸ ὅλον τοῦτο καλέσαντες ἀφ' ἑνὸς καὶ πρὸς ἕν)" (66,15–17). This is clearly the case in our text, too. Moreover, unlike Porphyry, our text does not present this as another subgroup of homonyms ἀπὸ διανοίας, but rather as an additional characteristic of homonyms, whose scope of generality is not specified.

---

349 *In Cat.* (CAG IV/1: 66,34–67,31).

350 We do not know whether Atticus, a Platonist who considered the doctrine of the categories irrelevant for the understanding of Plato's views (fr. 2, Des Places), ever composed a commentary on *Categories*, but it seems rather unlikely; cf. George Karamanolis, *Plato and Aristotle in Agreement? Platonists on Aristotle from Antiochus to Porphyry* (Oxford: Oxford Univerity Press, 2006), p. 177 and n. 81.

95,25 ὡς [ἀπὸ τοῦ ἰατροῦ] τὸ βιβλίον καὶ τὸ σμιλίον ἰατρικά. As preserved in L, the text is not sound. Stählin deletes ἀπό, but it seems a better option to follow Potter and extract the whole phrase ἀπὸ τοῦ ἰατροῦ as a scribal note. The information that the medical book *belongs to* the doctor is irrelevant and the origin of the whole phrase (which simply echoes the explanation of ἀπὸ τοῦ αὐτοῦ given in 95,26) is easier to explain than the intrusion of an isolated preposition. For the meaning of the example, see the next note.

95,26 ἀπό τε τοῦ χρωμένου ἰατροῦ καὶ πρὸς τὸν αὐτὸν λόγον τὸν ἰατρικόν. A scalpel and a book both receive their name ('medical') from the same thing, namely the medical man who uses them. This is the meaning of ἀπὸ τοῦ αὐτοῦ which may be translated as "after the same thing". Even though they are called after the same thing, each of them nevertheless has a different definition. A similar example of homonyms ἀπό τινος ἑνός is provided by Porphyry: A book, a drug, and a scalpel are called 'medical' (ἰατρικόν), each of them having this name from one thing, namely the art of medicine, but each for a different reason. Therefore, even though the name is always derived from the same thing, in each case it has a different definition (*In Cat.*/CAG IV/1: 66,2–10).

As far as the expression πρὸς τὸν αὐτὸν λόγον τὸν ἰατρικόν is concerned, it has no equivalent in Porphyry's classification: his group of homonyms πρὸς ἓν is represented by things that are called 'healthy' in view of their common goal, namely health (66,12–15). This tallies well with Aristotle's πρὸς ἓν ἅπαντα συντελεῖν in *EN* I 6 (1096b27 f.), which does indeed suggest that the object of πρός is a goal. By contrast, in our text, the item in view of which these things are called 'medical' is not a goal, but ἰατρικὸς λόγος. Whatever this expression means, it is clear that the phrase πρὸς τὸ αὐτό is understood in a generally relative sense. This reminds us of other passages in Aristotle, where items predicated πρὸς ἕν constitute a group of their own, distinct from synonyms or homonyms, as they have different definitions, but these definitions are variously related to the same thing, as, for example, such items as patient's body, operation, and an instrument are all related to the same art of medicine.[351] Basing themselves on these passages, some authors mentioned by Porphyry

---

351 Cf. esp. *Met.* Γ 2, 1003a33–1003b6 and Z 4, 1030a35–1030b2 (trans. Ross): "… we use the word [viz. 'to be'] neither homonymously nor in the same sense, but just as we apply the word 'medical' (τὸ ἰατρικόν) when there is a *reference* to one and the same thing, not *meaning* one and the same thing (τῷ πρὸς τὸ αὐτὸ μὲν καὶ ἕν, οὐ τὸ αὐτὸ δὲ καὶ ἕν), nor yet speaking homonymously; for a patient and an operation and an instrument (σκεῦος) are called medical neither homonymously nor in virtue of one thing, but with reference to one thing," namely the art of medicine (cf. Γ 2, 1003b1).

claim that the items said ἀφ' ἑνὸς καὶ πρὸς ἕν are not homonyms at all, but "instead they put them between homonyms and synonyms, because different things are all called 'medical' by participation in the same account (διὰ τὸ λόγου μετοχῇ τοῦ αὐτοῦ), and similarly the things that are called 'healthy' are all so called because they preserve the same health".[352] The 'account' in which these items participate is presumably an account of the art of medicine, in which they participate by being related to it in their respective ways.

However, our text follows yet a different tradition than the people mentioned by Porphyry, the main difference being that it uses the phrase ἀπὸ τοῦ αὐτοῦ καὶ πρὸς τὸ αὐτό as a characteristic of homonyms (or some kinds of them), not of a separate group. An interesting parallel may be found in Aspasius' *Commentary on Nicomachean Ethics*.[353] Commenting upon the opening chapters of *EN* VIII, where various meanings of the word 'friendship' are discussed, Aspasius notes that 'friendship', like 'being', is among things that are "said in many ways", and, more narrowly, those that are called "after the same thing in view of the same thing" (ἀπὸ τοῦ αὐτοῦ πρὸς τὸ αὐτὸ [λεγόμενα]). As far as friendship is concerned, various things are called 'friendship' "according to the similarity" (κατὰ τὴν ὁμοιότητα) with the friendship of good people, all friendships having their name "from" the latter (ἀπὸ ταύτης τοὔνομα ἔχουσι).[354] This indicates that Aspasius is acquainted with homonyms 'according to similarity', taking them as a kind of a more general class of homonyms called ἀπὸ τοῦ αὐτοῦ πρὸς τὸ αὐτό. A few pages later, while commenting upon Aristotle's definition of friendship as "reciprocal good-will" (*EN* VIII, 1155b33f.), Aspasius asks whether one definition covers every friendship, for if it did, all the various friendships would not be homonyms. In this connection, he makes a distinction between a "precise definition" (ὁρισμὸς ἀκριβής) and an account that is merely an "outline" (ὑπογραφή), and explains that the latter kind of account may be applied to "things said in many ways", i.e. to the homonyms: "It is clear, as we stated above, for which things said in many ways we can take one account (ἕνα λαβεῖν λόγον)—for things that are not too far apart, whenever these many things are called after one thing (ὅταν ἀφ' ἑνὸς τὰ πολλὰ λέγηται), for example, a medical book and an instrument are called after a first instance, the medical man (ἀπὸ πρώτου λέγεται τοῦ ἰατρικοῦ ἀνθρώπου), it is possible to take in outline the single account of the 'medical' pertaining to all these (ἔστι τύπῳ λαβεῖν ἕνα πάντων λόγον ἰατρικόν), and say [this] about everything called in whatever way after medicine (πᾶν τὸ

352 Porphyry, *In Cat.* (CAG IV/1: 66,17–21).

353 The following passage is an abridged version of an argument developed in my "Categories," pp. 216–219.

354 Aspasius, *In Eth. Nic.* (CAG XIX/1: 161,11–16).

ἀπὸ ἰατρικῆς). And the same holds of friendship, too."[355] Aspasius' claim that homonymous things, though not having the same definition, may nevertheless be grasped by the same account sheds light on Clement's statement that a medical book and a scalpel are homonymous not only "after the medical man that uses them", but also "in view of the same λόγος ἰατρικός". This expression presumably means the same thing as in Aspasius, namely an account of the name ἰατρικός, which does not give a definition of things thus called, but helps us to describe them in an 'outline'. As far as I am aware, these remarks in Aspasius' commentary constitute the closest parallel to Clement's account of homonymy ἀπὸ τοῦ αὐτοῦ καὶ πρὸς τὸ αὐτό in extant literature.[356]

(IX) 25, 1–33, 9: Causes

This section is about causes. It is framed by two partly overlapping classifications (25, 1–4; 32, 7–9), whose common elements include the distinction between procatarctic and synectic causes. Apart from presenting the notion of a cause as that which does something (25, 5, continued in 27, 3–28, 1), it also deals with the question of the corporeality of causes and effects (26, 1–27, 2), the 'suitability' of the patient (28, 2–6), the relativity of causes (29, 1–6), reciprocal causation (30, 1–5), multiple causation (31, 1–32, 3), and other topics. The text partly consists of reports of other people's views, the views being sometimes confronted with one another (cf. esp. 26, 1–5; 31, 1–2); but the author seems to have his own views, too (cf. e.g. 26, 4/96,25: ὅπερ καὶ μᾶλλον, 28, 5/98,15: καταγέλαστον δὲ τὸ λέγειν, 31, 4/100,13: τῷ ὄντι δέ), sometimes presenting them as solutions to the reported controversies (26, 5–27, 2; 31, 4–5). Often the status and function of the various doctrines presented here are not elucidated, however, and the resulting impression is of an incoherent and occasionally inconsistent assemblage of material. But this may be a false impression cre-

355 Aspasius, *In Eth. Nic.* (CAG XIX/1: 164,1–8): εἰ δέ ἐστιν ὁρισμὸς εἷς πάσης φιλίας, οὐκ ἂν ὁμώνυμοι εἶεν αἱ φιλίαι, ὥσπερ δοκεῖ αὐτῷ. ἢ ἔστι καὶ πολλαχῶς τινων λεγομένων ἕνα λαβεῖν λόγον, οὐκ ἀκριβῆ μὲν ὁρισμόν, ὑπογραφὴν δὲ μᾶλλον. τίνων δὲ πολλαχῶς λεγομένων ἔφαμεν ἐγχωρεῖν ἕνα λαβεῖν λόγον, οὐκ ἄδηλον, ὅτι τῶν μὴ πολὺ διεστώτων, ἀλλ' ὅταν ἀφ' ἑνὸς τὰ πολλὰ λέγηται, ὥσπερ τὸ ἰατρικὸν βιβλίον καὶ ὄργανον ἀπὸ πρώτου λέγεται τοῦ ἰατρικοῦ ἀνθρώπου, ἔστι τύπῳ λαβεῖν ἕνα πάντων λόγον ἰατρικόν, λέγοντας πᾶν τὸ ἀπὸ ἰατρικῆς ὁπωσοῦν ὄν· καὶ ἐπὶ τῆς φιλίας δὲ ταὐτόν πως ἔστι.

356 For Aspasius' account of homonymy, cf. Enrico Berti, "Amicizia e «focal meaning»," in A. Alberti—R.W. Sharples (eds.), *Aspasius: The Earliest Extant Commentary on Aristotle's Ethics* (Berlin/New York: de Gruyter, 1999), pp. 176–190.

ated by the excerptor, who seems to have been more interested in the doctrines themselves than in the continuity of the argument.

For the most part, the reported views are anonymous, but on one occasion they are attributed to the Stoics and to Aristotle, respectively (26, 4). The theoretical outlook of the author is influenced by the Stoics, but sometimes the Stoic tenets are developed with the help of Aristotelian tools (cf. e.g. 25, 5) and there is a tendency to harmonize Stoic views with Aristotle (cf. esp. 26, 4–27, 2; 28, 2–6). Some examples and distinctions are likely to be of medical origin (cf. esp. 28, 7; 30, 1; 32, 4–7), including a passage where the author clearly presents his own opinion (31, 4–5). The text includes glosses of a Christian reader (28, 5/98,14 f.; 30, 4/99,25) and it also seems to include at least one attempt to adapt the original argument to the purpose of Christian thought (29, 3–6; cf. also 30, 5/99,32–100,2).

Apart from the author's interest in controversies (and their solutions), there are few signs of continuity with the preceding parts of the eighth book. One possible link is the role of the concept of ὁμολογούμενον in the argument (26, 4/97,3). The relevance of the discussion to the demonstrative theory could be indicated by a passage where causes are classified as objects of knowledge (32, 4–6). Also, the philosophical profile of the source seems to be broadly similar to the previous parts. The most significant parallels are found in Cicero, Seneca, Sextus, the medical doxography, and Galen.

25, 1–5

95,27–96,5 This passage is listed as SVF II 346. Though containing some Stoic elements, it is probably not derived from a Stoic source, however; see comments below.

95,27 f. τὰ μὲν προκαταρκτικά, τὰ δὲ συνεκτικά, τὰ δὲ συνεργά, τὰ δὲ ὧν οὐκ ἄνευ. In the following lines, only the first two kinds of causes will be explained (95,28–96,2), the third and the fourth being merely illustrated by an example (96,4 f.). Other (partly overlapping) classifications are offered below, 28, 2/98,3–5; 32, 4–6/101,4–12; 32, 7/101,13–16.

95,28–31 προκαταρκτικὰ μὲν τὰ πρώτως ἀφορμὴν παρεχόμενα ... οὐ μὴν κατηναγκασμένως. The concept of the procatarctic causes was likely introduced by the Stoics in the context of a discussion about fate and responsibility.[357] The name

357 Cf. Bobzien, "Chrysippus' Theory of Causes," in K. Ierodiakonou (ed.), *Topics in Stoic Philosophy* (Oxford: Clarendon Press, 1999), pp. 196–242, here 218–222.

is derived from προκατάρχω, possibly in the sense of 'to begin [something] preliminarily' (i.e. to initiate something before another causal factor comes in), or perhaps 'to precede the beginning [of something]'.[358] The definition given here (τὰ πρώτως ἀφορμὴν παρεχόμενα κτλ.) could go back to the Stoics, too, as their aim was to distinguish a class of causes that provide a stimulus to action, but do not suffice to produce action on their own.[359]

The efficiency of the procatarctic cause is illustrated by an example of intemperate people and the way that beauty (τὸ κάλλος) is a cause to them (τοῖς ἀκολάστοις) of love (τοῦ ἔρωτος):[360] when seen, beauty creates in them an erotic condition "only", "not necessarily" (μόνον, οὐ μὴν κατηναγκασμένως). This presumably means that (a) when seen by intemperate people, beauty becomes an efficient cause to them of a condition that may turn into love, but is not

358 Cf. Bobzien, "Chrysippus' Theory," pp. 220 f. Apart from these two options, Bobzien proposes a third: "That which commences [*intr.*] beforehand." English equivalents used by modern scholars include 'preliminary' (LS), 'antecedent' (Hankinson) and 'initiating' (Sharples). The last option is the most attractive of the three, but it is rarely used. 'Antecedent' is confusing, as Cicero uses the expression *causa antecedens* in a more general sense; cf. *De fato* 34.40–42 (80,7 f.; 84,23 f.; 86,6.17 f.; 88,2 Sharples). As far as the form 'procatarctic' is concerned, *The Oxford English Dictionary* registers it in the medical literature as early as the 17th century. Cf. e.g. Gideon Harvey, *Morbus Anglicus* XXVIII 155: "Those Procatarcticks that required a larger comment, as love, grief, &c. we have discoursed of in particular Chapt." John Mason Good, *The Study of Medicine*, II (London: Baldwin, Cradock, and Joy, 1822), while mentioning the "procatarctic" causes, uses the equivalent "occasional" (p. 40 *et passim*).

359 Cf. Brad Inwood, *Ethics and Human Action in Early Stoicism* (Oxford: Clarendon Press, 1985), pp. 45 f. It should be mentioned, however, that the word ἀφορμή is not used in a recognizably Stoic fashion. In the Stoic sources, ἀφορμή typically refers to one of the following things: a kind of impulse (ὁρμή) which commands us not to do something (cf. Inwood, *Ethics*, pp. 226 f.), or a natural inclination to virtue (cf. SVF II 214; 228, 264; Inwood, *Ethics*, p. 45). Neither of these meanings is appropriate here; rather, the word is used in the non-technical sense of 'that which gives rise to something', an 'occasion'; cf. e.g. Isocrates, *Or.* VII (*Areop.*) 32 (146b): τοῖς δ' εἰς τὰς ἄλλας ἐργασίας ἀφορμὴν παρέχοντες. Philo of Alexandria, *Flac.* 40: ἵνα μὴ παρέχῃ … ἀφορμὴν εἰς ὕβριν κτλ. On the other hand, non-technical use of the word is attested in the Stoic doxography of DL (VII 76; cf. Sextus, *M.* II 70) and in Epictetus (cf. e.g. *Diss.* I 7, 20; III 10, 15).

360 Hankinson, *Cause and Explanation*, p. 247, links the words τοῖς ἀκολάστοις τοῦ ἔρωτος together and translates "to those intemperate in love"; but it is more likely that the combination of the dative and the genitive in the description of a causal process has the same function here as in other passages in this chapter, namely to signify the object to which something is caused, and the effect caused to that object, respectively; cf. below, 26, 2; 30, 1–5. For the implied notion of ἔρως as a kind of ἐπιθυμία, cf. below, note 362.

love yet, and that (b) by being an efficient cause of this condition, it does not compel these people to love.[361]

The description of (the vision of) beauty as a stimulus of ἔρως recalls the Stoic definition of ἔρως as 'an effort to produce friendship through the appearance of beauty' (ἐπιβολὴ φιλοποιΐας διὰ κάλλος ἐμφαινόμενον).[362] The example fits in with the aim of the Stoics to make a distinction between those factors in the causal process that are not in our power (such as the presence of a φαντασία) and those that are (such as assent to that φαντασία and the ensuing impulse to action), and it is possible that it was already used by them in this context.[363] There is an intriguing parallel in Cicero's account of causes in *Top.* 59.[364] Having distinguished a class of causes without which an effect cannot take place, Cicero mentions one type of them, namely those that "provide a kind of preliminary to efficient action and carry with them certain assisting factors, albeit not necessary ones". He then adduces the following example: "The

361 It seems from the context that οὐ μὴν κατηναγκασμένως should qualify the coming about of ἔρως, rather than the formation of ἐρωτικὴ διάθεσις. Cf. Görler, "Hauptursachen," p. 259, arguing that οὐ κατηναγκασμένως "bezieht sich selbstverständlich auf eine mögliche Tat des ἀκόλαστος, nicht auf τὴν διάθεσιν ἐμποιεῖ"; similarly Hankinson, *Cause and Explanation*, p. 247. Yet it is difficult to extract this meaning from our passage. Perhaps something is missing after οὐ μὴν κατηναγκασμένως, or—as I am inclined to think—the phrase has been transposed from a position after εἰς τὸ γίγνεσθαί τι in 95,29.

362 SVF III 395 (from Stobaeus); trans. B. Inwood and L.P. Gerson in *Hellenistic Philosophy: Introductory Readings* (Indianapolis: Hackett, 1997, 2nd edn), p. 231. Similar definitions are found in SVF III 396–397. For the meaning of ἐπιβολή, cf. Christopher Gill, "Stoic Erôs—Is There Such a Thing?" in E. Sanders, C. Thumiger, C. Carey, and N.J. Lowe (eds.), *Erôs in Ancient Greece* (Oxford: Oxford University Press, 2013), pp. 143–161, here 147 (with references). The association of ἔρως with intemperate people in our passage suggests that the word is used in the sense of an irrational affection or, as Galen puts it, "a runaway motion of the desiderative power" (κίνησις ἔκφορος τῆς ἐπιθυμητικῆς δυνάμεως); *PHP* IV 1, 16 (CMG V,4,1,2: 238,16 f./V,365 K.). Galen attributes this understanding of ἔρως to Chrysippus; cf. *PHP* IV 1, 17 (238,22 f./V,366 K.), re. Chrysippus' Περὶ ψυχῆς: τὸν μὲν ἔρωτα τῆς ἐπιθυμητικῆς εἶναι δυνάμεως [ἔγραψε]. Cf. DL VII 113 (SVF III 396), where ἔρως is described as a sort of ἐπιθυμία (i.e. ἄλογος ὄρεξις), "which is not concerned with virtuous things (οὐχ ἢ περὶ σπουδαίους ἐστίν)." Contrast DL VII 130 (SVF III 716), re. Chrysippus' Περὶ ἔρωτος: καὶ μὴ εἶναι ἐπίμεμπτον αὐτόν [*scil.* τὸν ἔρωτα]; SVF III 717 (from Stobaeus): τὸν δὲ ἔρωτα οὔτε ἐπιθυμίαν εἶναι οὔτε τινὸς φαύλου πράγματος. For ἔρως as an irrational affection, cf. also Galen, *PHP* IV 1, 12 (CMG V,4,1,2: 236,26/V,364 K.); IV 6, 39 (278,16–18/V,413 K.); *Aff. Dig.* 6, 2.7–9 (CMG V,4,1,1: 19,15–20 and 20,10–31/V,27–30 K.).

363 Cf. Inwood, *Ethics and Human Action*, p. 54, on presentations (φαντασίαι) as necessary but not sufficient conditions for action.

364 The parallel is noted by Witt, *Albinus*, pp. 39–41. Cf. also Görler, *Hauptursachen*, pp. 263 f.

meeting had brought the cause for love, the love the cause for the crime."[365] Commentators have pointed out similarities with the description of procatarctic causes in our passage, and it is indeed likely that Cicero had this kind of causes in mind.[366] The example used by Cicero, involving the sequence of meeting, love, and crime, is probably based on the story of Medea, mentioned by Cicero in *Fat.* 35 and alluded to in *Top.* 61.[367] It is quite likely that our example, involving an intemperate person, originates within the same context of a discussion where the Medea story served as a paradigm of causal sequence; cf. below, 27, 4–5/97,17–22. Cicero suggests that this context is Stoic[368] and it could go back to Chrysippus, whose interest in Euripides' *Medea* is well attested.[369]

---

365 *Top.* 59: *alia autem praecursionem quandam adhibent ad efficiendum et quaedam afferunt per se adiuvantia, etsi non necessaria, ut: Amori congressio causam attulerat, amor flagitio.* I quote according to the edition and translation of Tobias Reinhardt, *Cicero's Topica*, Oxford: Oxford University Press, 2003.

366 Cf. Reinhardt, *Cicero's Topica*, p. 326. According to Seneca, a similar model of causation was used by Posidonius; cf. *Ep.* 87,31 (trans. Inwood): "[Posidonius] says that riches *are* the cause of the bad outcomes, not because riches themselves *do* anything but because they instigate people to action (*non quia ipsae faciunt aliquid, sed quia facturos inritant*). For there is a difference between the efficient cause (which must do harm immediately) and the antecedent cause (*alia est enim causa efficiens, quae protinus necessest noceat, alia praecedens*). Riches have this antecedent causality." Note the implication that antecedent causes do not bring about their effects necessarily; this corresponds both to our passage and the parallel in Cicero, *Top.* 59; cf. also *Fat.* 44. The distinction between the 'efficient' cause (probably translating the Greek ποιοῦν or ποιητικόν, cf. below, 25, 3/96,1; 27, 2–3/97,9–15; 28, 2/98,3 and Seneca, *Ep.* 65,14) and the 'antecedent' cause (here the Greek word is a matter of guess) also seems to be reflected in Posidonius, fr. 95 EK (Stobaeus, *Anth.* I 13, 1c): αἴτιον δ' ἐστί τινος, δι' ὃ ἐκεῖνο, ἢ τὸ πρῶτον ποιοῦν, ἢ τὸ ἀρχηγὸν ποιήσεως ("a cause is of something, because of which that thing is, either the primary producer or the instigator of the production"; trans. Hankinson, *Cause and Explanation*, p. 328); for δι' ὃ as a category covering both the efficient and (broadly speaking) antecedent causes, cf. below, 27, 3–5/97,13–22 (there, however, only the efficient cause is worthy of the name 'cause').

367 Cf. Duhot, *Causalité*, p. 202. The relevance of the Medea story to the present discussion is brought out especially by the final sentence of *Fat.* 35 (trans. Rackham): "It was not the case that those events brought the cause of love (*non erat ut eae res causam adferrent amoris*)."

368 Cf. *Top.* 59: "It is from this type of causes, depending on one another from all eternity, that the Stoics fashion their chain of fate."

369 Cf. DL VII 180; Galen, *PHP* III 3, 13–22 and IV 6, 19–27 (CMG V,4,1,2: 188,16–190,15 and 274,5–23 / V 306–308 and 408–410 K.). The Chrysippean passage discussed by Galen concerns the decision-scene in *Medea* 1078 f., taken as a starting-point for a debate about affections and their relation to the reasoning faculty of the soul. Cf. Christopher Gill, "Did Chrysippus Understand Medea?" *Phronesis* 28/2 (1983), pp. 136–149; Teun Tieleman, *Chrysippus' On*

But it is likely that Cicero's source is a more recent author, Posidonius of Apamea being a plausible candidate.[370]

However, the role of the 'erotic condition' in our passage is not easily explained in the context of the Stoic theory of action. At first sight, it seems to correspond to the role of φαντασία ὁρμητική, i.e. the sort of presentation that sets the soul to motion.[371] But it is very odd for a Stoic to describe φαντασία as διάθεσις, a term normally reserved for "an enduring state which additionally does not admit of degrees", such as virtue.[372] Most likely, the expression refers not to a presentation (as understood by the Stoics), but rather to some transient condition of the soul (or the body) brought about by the vision of beauty. This usage could be labelled as Aristotelian,[373] but the description of the procatarctic cause as an external factor that provides for some διάθεσις has closer parallels in medical literature, particularly Galen; cf. e.g. *Diff. Feb.* I 3 (VII,282,7–9 K.) and I 4; *Caus. Puls.* I 1 (IX,2,11–13 K.); *MM* IV 3 (X,242–249 K.).[374] Cf. below, 30, 1/99,13–16. Galen also informs us that morbid symptoms of love were treated under the heading of procatarctic causes in the medical tradition.[375] On procatarctic causes, see further below, 31, 4/100,13–16; 33, 1/101,17–19; 33, 9/102,10–12.

Affections: Reconstruction and Interpretation (PhA 94; Leiden/Boston: Brill, 2003), pp. 170–178. In his criticism of the Chrysippean views of these matters, Galen appeals to Posidonius' work Περὶ παθῶν (cf. *PHP* IV 3–7, *passim*); for Galen's use of Posidonius as an ally against Chrysippus, cf. Tieleman, *Chrysippus' On Affections*, pp. 198–287.

370 Cf. above, note 366. Cicero refers to Posidonius in *Fat.* 5–7 (speaking of him as a "master"), the book referred to being presumably his lost treatise *On Fate* (cf. DL VII 149). For Posidonius as a possible source of Cicero's *Fat.* 44, *Top.* 59, and other passages mentioned above (esp. *Fat.* 34–36), cf. Susanne Bobzien, *Determinism and Freedom in Stoic Philosophy* (Oxford: Clarendon Press, 1998), pp. 322–324.

371 Cf. SVF III 169 (from Stobaeus): τὸ δὲ κινοῦν τὴν ὁρμὴν οὐδὲν ἕτερον εἶναι λέγουσιν ἀλλ' ἢ φαντασίαν ὁρμητικὴν τοῦ καθήκοντος αὐτόθεν, τὴν δὲ ὁρμὴν εἶναι φορὰν ψυχῆς ἐπί τι κατὰ τὸ γένος.

372 The quotation is from LS I, p. 376; cf. Simplicius, *In Cat.* 8b26 (CAG VIII: 237,25–238,32) = SVF II 393.

373 Cf. *Cat.* 8, 8b35–9a1 (trans. Ackrill): "It is what are easily changed and quickly changing that we call conditions (διαθέσεις), e.g. hotness and chill and sickness and health and the like. For a man is in a certain condition in virtue of these but he changes quickly from hot to cold and from being healthy to being sick." In contrast, says Simplicius, the Stoics described such states as "unstable health" as ἕξεις (CAG VIII: 238,10).

374 Cf. Hankinson, "Galen's Theory of Causation," in *Aufstieg und Niedergang der römischen Welt*, II: Principat, Band 37.2, ed. W. Haase (Berlin/New York: de Gruyter, 1994), pp. 1757–1774, here 1766 f. See also below, on 102,10–12.

375 Cf. Galen, *Hipp. Prog.* I 4 (XVIIIb,18,15–19,1 K.): τοὺς δ' ἤτοι καταλεπτυνομένους ἢ ἀχροοῦντας

95,31–96,2 συνεκτικὰ δὲ ἅπερ συνωνύμως καὶ αὐτοτελῆ καλεῖται κτλ. In the Chrysippean argument, as presented by Plutarch, the procatarctic causes were contrasted with the 'self-complete' ones (αὐτοτελῆ).[376] Our passage explains the name αὐτοτελής as follows: "they bring about their effect (ἀποτέλεσμα) sufficiently by themselves (αὐταρκῶς δι' αὐτῶν)" (96,1 f.). There is debate whether the Stoics used the concept only in order to make their point about the procatarctic causes, namely that they are *not* sufficient to produce their effect, or whether they admitted that self-complete causes also exist.[377] It is worth noting in this connection that Cicero, while speaking of causes that "by their own force bring about with certainty the result that is subject to this force", cites the following examples: fire and wisdom (*Top.* 58–59).[378] The example of fire is

---

ἢ ἀγρυπνοῦντας ἢ καὶ πυρέξαντας ἐπὶ προφάσεσιν ἐρωτικαῖς ἐν ἐκείνῳ τοῦ λόγου τῷ κεφαλαίῳ περιλαμβάνουσιν οἱ παλαιοί, καθὸ περὶ τῶν προκαταρχόντων αἰτίων διεξέρχονται. For the expression ἐρωτικὴ διάθεσις, cf. also Galen, *Hipp. Epid. VI*, V 6 (XVIIb,255,18–256,3 K.): τούτων δ' ἀτοπώτερα λέγουσιν οἱ τὰ τοιαῦτα πάθη φάσκοντες ἐν τῷ γήρᾳ λύεσθαι, τῆς ψυχῆς μὲν οὖν ὀργάς, φιλαργυρίας, δεισιδαιμονίας, τοῦ δὲ σώματος ἐρωτικὰς διαθέσεις.

376 Cf. Plutarch, *Stoic. Repug.* 1056b (= SVF II 997); Bobzien, "Chrysippus' Theory of Causes," pp. 218–222. The most common English equivalent of αὐτοτελής (used, for example, by Sharples and Bobzien) is 'self-sufficient'; this is slightly misleading, though, as the English prefix (in this case) is reflexive, whereas the Greek one (in this case) is not: As used here, the word αὐτοτελής seems to render the idea that the cause in question is complete in virtue of itself; in other words, that it does not need anything else to produce its effect. 'Self-complete' is Bobzien's rendering of αὐτοτελής in the context of Stoic logic (e.g. in *The Cambridge Companion to the Stoics*) and one that seems preferable to me in the present connection, too. Other proposals include 'perfect' (Frede and Hankinson, cf. Cicero, *Fat.* 41) and 'complete' (LS).

377 Cf. Bobzien, "Chrysippus' Theory," pp. 236–241.

378 In his much-debated report on the cylinder example, going back to Chrysippus, Cicero makes a distinction between factors that provide "the beginning of motion" (*principium motionis*) to an object (like someone pushing a cylinder), and those by which the object actually moves (like the force and nature of a cylinder rolling down a hill); *De fato* 41–43 (part of SVF II 974; LS 62C). Earlier, Cicero makes a notoriously puzzling distinction between two pairs of causes: *perfectae et principales* vs. *adiuvantes et proximae*. It is an uncontroversial thing to say that the adjective *perfectae* translates the Greek αὐτοτελῆ. It is less clear whether any factor mentioned in the cylinder example corresponds to this kind of cause. But if the cause mentioned in *Top.* 58 that "by its own force brings about with certainty the result that is subject to this force" (*quod vi sua id quod sub eam vim subiectum est certe efficit*) is of the 'self-complete' variety (which appears quite likely), it is tempting to think that the factors involved in the actual movement of the cylinder, namely its 'force' (*vis*) and 'nature' (*natura*), correspond to this description, too; cf. Reinhardt, *Cicero's Topica*, p. 325; Michael Frede, "The Original Notion of a Cause," in idem, *Essays in Ancient*

also found in Sextus, who draws a contrast between 'self-complete' causes and those that "need the affected matter as a collaborator" (συνεργοῦ … δεῖται τῆς πασχούσης ὕλης), and argues that fire does not belong in the former group (*M.* IX 237–238.241–243); this seems to imply a belief, on the part of Sextus' opponents, that fire is a self-complete cause.[379] For the example of wisdom, cf. SVF I 89.[380] On self-complete causes, see also below, on 101,19 f.

However, the first name applied to this kind of causes in our passage is συνεκτικά. The name is derived from συνέχω, 'to contain', and thus the causes are such that they 'contain' or 'hold together' something.[381] Again, the concept was likely introduced by the Stoics, with regard to that on account of which bodies are coherent (συνέχεται) in a certain way, i.e. primarily the *pneuma*.[382] But, as Galen points out, a 'synectic cause' in this sense is a cause of existence (ὑπάρξεως), rather than of a 'coming about' of something (γενέσεως τινος).[383]

Philosophy (Minneapolis: University of Minnesota Press, 1987), pp. 139–141. As far as the cylinder's 'nature' is concerned, namely its 'rollability' (*volubilitas*), it seems to have a different status before and after the object is set to motion, the former being perhaps referred to as *causa principalis*, the latter as *causa perfecta*. Cf. Hankinson, *Cause and Explanation*, pp. 247–249, whose account of *causae perfectae et principales* I find wholly convincing.

379 Cf. also *M.* VIII 198–199. While criticizing this (Stoic?) view, Sextus employs the concept of auxiliary causes in a sense that seems to be alien to the Stoics. Cf. below, on 98,5–9. For a possible Stoic answer, cf. Reinhardt, *Cicero's Topica*, p. 325: "Of course fire, in order to set something on fire, needs some prerequisites. The Stoics and Cicero made a distinction between passive prerequisites or background conditions and active causes. Accordingly, a cause would be deemed 'perfect', if it required no second *active* cause to become efficient" (italics mine). Cf. also Frede, "Notion of a Cause," pp. 141 f.

380 SVF I 89 (from Stobaeus, citing Zeno): αἴτιόν ἐστι δι' ὃ γίγνεταί τι, οἷον διὰ τὴν φρόνησιν γίνεται τὸ φρονεῖν καὶ διὰ τὴν ψυχὴν γίνεται τὸ ζῆν καὶ διὰ τὴν σωφροσύνην γίνεται τὸ σωφρονεῖν. ἀδύνατον γὰρ εἶναι σωφροσύνης περί τινα οὔσης μὴ σωφρονεῖν ἢ ψυχῆς μὴ ζῆν ἢ φρονήσεως μὴ φρονεῖν. The passage is noted by Reinhardt, *Cicero's Topica*, p. 328.

381 Proposed English translations of the term include 'containing' (the most common equivalent), 'sustaining' (LS), and 'cohesive' (Bobzien); cf. Hankinson, *Cause and Explanation*, p. 240 n. 6.

382 Cf. Plutarch, *Stoic. Repug.* 1053f (quoting Chrysippus = SVF II 449): ὑπὸ τούτων γὰρ συνέχεται τὰ σώματα. For the Stoic origin of the synectic causes and the original meaning of the term, cf. esp. Galen, *CC* (= *De Causis Contentivis*) 1, 1–5 (CMG Supp. Or. II: 50–53 = LS 55F). Bobzien, "Chrysippus' Theory," p. 228, notes that "the term συνεκτιὸν αἴτιον, which is standard in later taxonomies of causes, is not attested for Chrysippus, or for the early Stoics in general. Rather, the cohesive function of *pneuma* is talked about in various forms of the verb συνέχειν". It is possible that the very term συνεκτιὸν αἴτιον was only introduced by the later Stoics, e.g. Posidonius; cf. Galen, *CC* 2, 1 (CMG Supp. Or. II: 55,5–7).

383 *Syn. Puls.* 9 (IX,458,8–14 K.) = SVF II 356; LS 55H. Cf. Bobzien, "Chrysippus' Theory,"

In contrast, Galen claims, causes are primarily causes of something's coming about.[384] We have no testimony to the effect that the Stoics identified the self-complete and synectic causes and it is likely, in my view, that this identification presupposes a different, non-Stoic use of the term 'synectic cause'.[385] Again, it is Galen who informs us of this other use, developed by doctors (and adopted by Galen himself), according to which the name was applied in a loose sense (καταχρώμενοι τῇ προσηγορίᾳ) to causes that bring about some change, for example to conditions of the body in relation to (the coming about of) symptoms.[386] Galen mentions two characteristics belonging to the concept of the synectic cause in this sense: (a) something comes about by it, and (b) that which comes about by it ceases along with it.[387] This corresponds to the definition of the synectic cause given below, 33, 1/101,17–19 (see note *ad loc.*).

The word συνωνύμως in 95,31 (and below, 33,2/101,19) is used in the sense of a different name of the same item; this is originally a Stoic usage (widely adopted

---

p. 229, who points out that Galen's distinction between existence and something's coming about in this context is "roughly equivalent to the Stoic [distinction] between σχέσεις and κινήσεις".

384 Cf. *Syn. Puls.* 9 (IX,458,14 f. K.): δέδεικται γὰρ οὔτ' ἄλλου τινὸς αἴτιον πρότερον οὐδὲν, ὅτι μὴ γενέσεως. *CC* 7, 2 (CMG Supp. Or. II: 63,29–31).

385 Bobzien, "Chrysippus' Theory," argues that Chrysippus could not have understood 'synectic' and 'self-complete' as "two alternative ways of expressing membership of a cause in a particular class", and even though the synectic causes in the original sense could have been regarded as 'self-complete' (namely as the causes of an object's being that object, or being in the state of being that object), this was merely a coincidence of conceptually independent properties (p. 230). This supposed (philosophically possible, but textually unattested) coincidence could help us explain why modern scholars have sometimes identified the *causae perfectae et principales* in (Cicero's report on) the cylinder analogy with the synectic causes (in the early Stoic sense of συνέχω); on this, see Bobzien, "Chrysippus' Theory," pp. 230–233. But we may doubt if it played any role in the identification of the synectic and self-complete causes in Greek philosophy itself. Apparently the only passage where this identification is made is the present one (and, of course, below, 101,19 f.; see note *ad loc.*); here, however, both names clearly refer to causes of something's coming about (cf. e.g. below, 28, 7/98,19 f.; 33, 2–3/101,19–21). It would seem that this identification already presupposes the understanding of the synectic cause as a cause of change, which brings about its effect by its own force. Taken in this way, it could easily be identified with the self-complete causes spoken of by the Stoics, insofar as these too were regarded as causes of change.

386 Cf. *Syn. Puls.* 9 (IX,458,8–14 K.); *CC* 8, 9–12; 10, 2–3 (CMG Supp. Or. II: 67–69.71).

387 *Adv. Jul.* 8, 22 (CMG V,10,3: 70,3–6/XVIIIa,298 f. K. = SVF II 355): καὶ μὴν οὐδ' ἄλλην τινὰ ἔννοιαν εἰπεῖν ἔχει [*scil.* ὁ Ἰουλιανός] τοῦ συνέχοντος αἰτίου παρὰ τὸ γίνεσθαί τι πρὸς αὐτοῦ καὶ παύεσθαι σὺν αὐτῷ, πλὴν εἰ κἀνταῦθα πάλιν ἐξαίφνης ἑαυτὸν εἶναί φησι Στωϊκόν, ὡς ἐν ἄλλοις ἐποίησεν.

in late Hellenistic literature), which differs from the Aristotelian view of συνώνυμα explained above, 24, 2/95,7–9 (and probably implied in 17, 1/90,11).[388] This inconsistency could be taken as a sign that the present passage derives from a different source than the material in chapters 6 and 8; or the source is the same, but includes quotations from other sources; or the source is the same, but uses the word in two ways. Finally, the phrasing of the sentence could be due to Clement himself.[389]

96,2f. ἑξῆς δὲ πάντα τὰ αἴτια ἐπὶ τοῦ μανθάνοντος δεικτέον. Instead of giving a definition of the remaining two kinds of causes, the author informs us of his plan to exhibit "all the causes" with the example of a pupil. This plan, sketched briefly in the following lines (96,3–5), could be Clement's own. Cf. below, on 96,4f.

96,3f. ὁ μὲν πατὴρ αἴτιόν ἐστι προκαταρκτικὸν … ὁ διδάσκαλος δὲ συνεκτικόν. How does the father serve the role of a procatarctic cause of his offspring's μάθησις? Perhaps by sending him to a teacher.[390] The teacher then fits the description of the synectic cause insofar as: (a) he is regarded as the only agent in the process of learning and (b) the effect of his teaching activity is understood as something that ceases along with this activity (cf. below, 33, 1/101,17–19). Both conditions are conceivable against the background of Aristotle, *Phys.* III 3, 202a13–b22, where teacher and student represent the moving cause (τὸ κινητικόν or τὸ κινοῦν) and that which is moved by it (τὸ κινητόν or τὸ κινούμενον), respectively; their movement, insofar as it is the change of the patient, is described as μάθησις.

96,4f. ἡ δὲ τοῦ μανθάνοντος φύσις συνεργὸν αἴτιον. The definition of the auxiliary cause exemplified by the student's nature has not been given. In the classification of causes below, 32, 7–33, 9, the auxiliary cause is described as a factor that, by accruing to another, independently effective cause, aids it in such a way that the effect of that other cause is intensified (cf. below, on 33, 7/102,1f.);

---

388 For the Stoic use of συνώνυμα, cf. Simplicius, *In Cat.* (CAG VIII: 36,8–11), who explains that it is equivalent to the Speusippean πολυώνυμα (cf. above, 95,14f.). Cf. already Dionysius Thrax, *Ars gram.* I 12 (36,5f. Uhlig): συνώνυμον δέ ἐστι τὸ ἐν διαφόροις ὀνόμασι τὸ αὐτὸ δηλοῦν, οἷον ἄορ ξίφος μάχαιρα σπάθη φάσγανον.

389 Cf. *Strom.* IV (14) 96, 2 and V (12) 79, 3, where Clement uses the word συνωνύμως in the Stoic sense. In *Strom.* I (20) 98, 1, a passage surrounded by close parallels with our chapter, Clement follows the Aristotelian usage, however. Cf. below, on 100,12f.

390 Cf. e.g. Plato, *Laches* 178a–181d. I owe this suggestion to David Sedley.

the crucial implication of this description is that the cause to which the auxiliary factor accrues is capable of producing the same effect even without it, only with less intensity (cf. below, 33, 8–9 and note on 102,4f.). But this cannot be the meaning of 'auxiliary' here. Elsewhere (28, 3–5/98,5–14), the name συνεργός is attached to a different item, namely a recipient of an activity, which undergoes that "to which it is naturally disposed" (εἰς ὃ πέφυκεν): Although it does not produce any effect on its own, nevertheless it contributes to the production of an effect by its 'suitability' (ἐπιτηδειότης); in that case, the effect of the (active) cause cannot be produced without this (passive) contribution. Cf. below, on 98,8f. Out of these two meanings of the word 'auxiliary', the example of the student's nature is clearly based on the latter.

There is an interesting parallel to this passage in *Strom.* II 6, 25, 4–26, 2, where Clement uses examples of suitability to illustrate his point about the receptivity (παραδοχή) or obedience (εὐπειθεία) of a pupil (τοῦ μανθάνοντος): the 'suitability' of the student is presented as a factor contributing to the success of (divine) teaching, i.e. to the pupil's learning (μάθησις). One example used in this connection is earth, which "being fertile, helps (συνεργεῖ) with regard to the dissemination of the seeds".[391] Cf. also below, 30, 2/99,19f.

96,5 ὁ δὲ χρόνος τῶν ὧν οὐκ ἄνευ λόγον ἐπέχει. The concept of the prerequisites goes back to Plato; cf. esp. *Phaedo* 99a–b, where Socrates points out that prerequisites are not causes. In *Top.* 59, Cicero distinguishes two groups of prerequisites, described as items without which an effect cannot be brought about (*sine quibus effici non potest*): some of them are "dormant, of no active efficacy, somehow inert", while others "provide a kind of preliminary to the efficient action". The latter group likely corresponds to the procatarctic causes (cf. above, on 95,28–31); examples of the first group include "place, time, matter, tools, and other things like that". This agrees with our text's account of prerequisites in this passage and (as far as matter is concerned) below, 28, 3–6. Time and place are often described as prerequisites; cf. Seneca, *Ep.* 65, 11; Philo of Alexandria, *Decal.* 30. In Galen, *MM* III 10 (X,224,5–14 K.), time and place are examples of items "coming up from the outside and having the status of prerequisites (τὸν ὧν οὐκ ἄνευ λόγον ἔχουσαι)". In *Symp. Diff.* 1, 12 (CMG V,5,1: 206,11–14/VII,47f. K.), Galen makes a distinction between causes that "contribute something to the

391 *Strom.* II (6) 26, 1: συνεργεῖ οὖν καὶ γῆ γόνιμος ὑπάρχουσα πρὸς τὴν τῶν σπερμάτων καταβολήν. Other examples include straw (which will burn easier when it is dry and "ready to receive the burning power"), pieces of iron attracted by a loadstone (διὰ συγγένειαν), and a heap of chaff moved by amber; however, Clement describes these recipients not as 'auxiliary', but as 'joint-causes' (συναίτια) (26, 2).

completion of what is brought about (εἰσφέρεταί τινα τῷ γενομένῳ συντέλειαν)" and those that "do not contribute anything, but are not separated from those that contribute (μὴ χωριζόμενα τῶν εἰσφερομένων)"; the latter, he says, "have the status of prerequisites" (τὸν ὧν οὐκ ἄνευ λόγον ἐπέχει). Cf. also Galen, *CP* (= *De causis procatarcticis*) 7, 89 (CMG Suppl. II: 21,30–35): "Rightly, then, those who are concerned with the truth say that these things have the status of prerequisites (*rationem eorum sine quibus non … habere*) and are not causes of bringing about what is made."[392] For the phrase λόγον τίνος ἐπέχειν, cf. above, 91,27 f., and below, 98,5 f. and 12. On prerequisites, see further below, on 98,5–9.

96,6–10 This passage is listed as SVF II 344; however, Stoic elements are developed by means of a distinction between potency and act, which is alien to the Stoics; cf. von Wedel, *Symbola*, p. 23; Duhot, *Causalité*, p. 215, and comments below.

96,6 αἴτιον δὲ κυρίως λέγεται τὸ παρεκτικόν τινος ἐνεργητικῶς. Cf. Cicero, *Fat.* 34 (trans. Sharples): "[C]ause should not be understood in such a way that what precedes each thing is the cause for that thing, but what precedes each thing and brings it about (*quod cuique efficienter antecedat*)." Sextus, *PH* III 14 (trans. Annas and Barnes): "A cause would seem in general, according to [the dogmatists], to be that because of which, by being active, the effect comes about (δι' ὃ ἐνεργοῦν γίνεται τὸ ἀποτέλεσμα)." Cf. also below, 27, 6/97,24 f. The notion of a cause as that which does something is characteristic of Stoic thought; cf. Seneca, *Ep.* 65, 4: "The Stoic view is that there is one cause, that which does something (*id quod facit*)."[393] Our text further develops this view by employing a distinction between the *capacity* to bring about an effect, and the *activity* by which this is done. It is only in the latter case (namely when producing an effect ἐνεργητικῶς) that something is described as a cause "properly" (κυρίως); but, as explained below (96,8–10), "bringing about something" is used in both ways. There is a close parallel to this discussion in Galen, *CAM* 14, 3 (CMG V,1,3: 98,5–8/I,272 K.): "Causes properly deserve this name when they are active (κυριώτατα μὲν, ὅταν ἐνεργῶσι, ταύτης τῆς προσηγορίας ἀξιοῦται). Nevertheless, they are often called causes even when they are not producing anything yet, solely

392 Trans. Hankinson, *Galen: On Antecedent Causes* (Cambridge: Cambridge University Press, 1998), p. 101 (slightly modified).

393 Cf. Frede, "Notion of Cause," pp. 126–128. There is a Platonic precedent to this view; cf. Plato, *Hipp. Maj.* 296e8 f., 297a4 f.: οὐ τὸ αἴτιον ποιοῦν ἐφάνη; Πάνυ γε. *Philebus* 26e6–8: Οὐκοῦν ἡ τοῦ ποιοῦντος φύσις οὐδὲν πλὴν ὀνόματι τῆς αἰτίας διαφέρει, τὸ δὲ ποιοῦν καὶ τὸ αἴτιον ὀρθῶς ἂν εἴη λεγόμενον ἕν;

on the grounds of being able to act (εἰ καὶ μηδέπω μηδὲν ἐνεργεῖ, κατ' αὐτὸ τὸ δύνασθαι μόνον), for example when indigestion is called a cause of an illness, even though it does not make one ill yet."

96,9 f. **καὶ τὸ ἤδη ἐνεργοῦν καὶ τὸ μηδέπω μέν κτλ.** The distinction between the potential and the actual causes goes back to Aristotle, *Phys.* II 3, 195b3–6. See also above, on 88,2–4.

26, 1–27, 2

96,11–14 This passage is listed as SVF II 345. It is the beginning of a section dealing with the question of the corporeality of causes and (especially) their effects; the discussion continues until 27, 2/97,13.

96,11 **οἳ μὲν οὖν σωμάτων, οἳ δ' ἀσωμάτων φασὶν εἶναι τὰ αἴτια.** It is not clear if the alternative in this sentence (bodies/incorporeals) concerns causes or their effects:[394] The corporeality of causes will be discussed immediately below, 96,11–17, but the (in)corporeality of effects, first introduced in 96,16, is the main theme afterwards. The context rather speaks in favour of the effects; cf. the function of the genitive below, 96,15 (αἰτίαν εἶναι τοῦ τέμνειν, ἐνεργείας οὔσης καὶ ἀσωμάτου); 96,19 (τὸ δὲ οὗ ἐστιν αἴτιον); 99,13 (ἀλλήλων ... τὰ αἴτια ... αἴτια); for the brachylogy, cf. below, 96,23 f. (λεκτῶν ... τὰ αἴτια).[395]

The 'doxography' in this paragraph seems to be influenced by a source whose aim was to contrast, for polemical reasons, opposite views about causes; cf. Sextus, *M.* IX 210 (trans. Bett): "... if there is cause, either body is cause of body (σῶμα σώματος), or incorporeal of incorporeal (ἀσώματον ἀσωμάτου), or body of incorporeal, or incorporeal of body." A parallel in Galen suggests that arguments against causes in *M.* IX 210–217 go back to the Alexandrian doctor

394 The genitive case can be taken to mean "causes belong among bodies/incorporeals" or "causes are causes of bodies/incorporeals".

395 Most readers with whom I have discussed this passage think that the genitives are partitive and refer to causes. There are three main reasons why I think otherwise: (1) There are similar formulations in 96,23 f. and 99,13, where the genitives certainly refer to effects; (2) There is no other evidence of a taxonomic genitive in our text, apart from 98,28 f. (τὸ αἴτιον τῶν πρός τι); but this is a common periphrasis for calling something a relative, and one in which the genitive is duly introduced by an article; (3) If the first division of opinions (bodies x incorporeals) concerns causes, it is not clear why it is mentioned before the second division (bodies κυρίως x incorporeals κυρίως), being merely a less sophisticated version of the same alternative. Cf. also Michael Frede, "Notion of Cause," p. 135, who takes our passage as saying that "some philosophers assume that causes are causes of bodies".

Herophilus (4th–3rd cent. B.C.); cf. Galen's report in *CP* 16, 199–200, concluded by a quotation from Herophilus: "Either a corporeal cause is the cause of something corporeal, or an incorporeal cause of something incorporeal, or one of the other possibilities distinguished in the process of division. But clearly none of these is the case: so it is evident at all events that there are no causes."[396] The aim of our text seems to be to resolve this issue; cf. 26, 3–26, 5, for arguments in favour of the view that effects are incorporeal. Cf. also above, on 85,17b, on the polemical background to the embryo discussion.

96,11–13 οἳ δὲ τὸ μὲν σῶμα κυρίως αἴτιον … τὸ δὲ ἀσώματον καταχρηστικῶς κτλ. The Stoics famously characterized causes as bodies; cf. SVF I 89 and II 336, from Stobaeus (= LS 55A); Sextus, *M.* IX 211 (= SVF II 341/LS 55B). Cf. also Seneca, *Ep.* 117, 2 (trans. Inwood): "Whatever does something is a body." We do not know who made the distinction between the more and less proper use of the word αἴτιον, however.

96,13–17 ἄλλοι δ' ἔμπαλιν ἀναστρέφουσι, τὰ μὲν ἀσώματα κυρίως αἴτια λέγοντες κτλ. The argument about the incorporeality of causes is based on the Stoic analysis of causation hinted at below, 26, 4/96,23 f., according to which effects are incorporeal predicates. People referred to in this passage paraphrase this view by saying that effects are 'activities' (ἐνέργειαι) and, as such, are incorporeal (96,16). This is also the opinion of our author (cf. below, 96,21 f.). Unlike the Stoics, however, the people referred to here think that causes are also, properly speaking, activities (and thus incorporeal). For example, in the case of cutting a body with a knife, the cut (τομή) is someone's activity that brings about the effect of 'cutting' to the knife and 'being cut' to the body. It is only in a loose sense that a knife (a body) may be described as a cause. Our author does not say whether he accepts this analysis or not; nevertheless, when speaking of causation in his own terms, he always refers to the (corporeal) agents as causes.

The description of causes as activities has a parallel in Galen's discussion of synectic causes: "There are those who say that the heating of the head by the sun is the synectic cause of the warmth that results in it and that the cut is the synectic cause of the wound." Galen retorts that what these people describe as causes is actually that by which the bodies in question are affected.[397]

---

396 Trans. Hankinson, *Galen: On Antecedent Causes*, pp. 147 f. Hankinson plausibly suggests that Herophilus' 'scepticism' could be influenced by his contacts with the Megarian school, hinted at by an anecdote about Herophilus and Diodorus Cronus, preserved in Sextus, *PH* II 245 (ibid. pp. 278 f.). See also below, on 97,3–7.

397 *CC* 9, 6–7, Arabic version (CMG Supp. Or. II: 68–71, trans. Lyons). For the idea of a cause as

96,18–21 τὸ "τινῶν ἐστιν αἴτιον" λέγεται τριχῶς, τὸ μὲν ὅ … τὸ δὲ οὗ … τὸ δὲ ᾧ κτλ. The phrase λέγεται τριχῶς, lit. 'is said in thee ways', should probably be understood in the sense that every predication of the form 'is the cause of Y' implies the notion of three items. The items are the agent, the effect, and the object. Henceforth, these three items will be expressed in the nominative, the genitive, and the dative, respectively. The example goes back to Aristotle (cf. below, on 98,3–5), and the form of expressing causal relations, while typically Stoic, has Aristotelian precedents, too.[398] There is an interesting parallel in Seneca, *Ep.* 65, 3 (trans. Inwood): "A statue had matter, to yield to the artisan, and an artisan, to give shape to the matter (*qui materiae daret faciem*). So in the case of the statue the material was the bronze and the cause (*causa*) was the workman. The same state of affairs holds for all things—they consist of that which becomes and that which makes (*ex eo … quod fit, et ex eo, quod facit*)." Cf. also below, 29, 1–2/98,25–30.

96,21–97,1 τὸ γίνεσθαι οὖν καὶ τὸ τέμνεσθαι … οὗ πάλιν πτῶσίς ἐστι τὸ ναῦν γίνεσθαι. This passage is listed as Archedemus, fr. 7 (SVF III, p. 262 f.), and LS 55C. Referring as it does to Stoic thinkers (see the next note), caution is at place regarding the extent to which it can be attributed to a Stoic source. The first sentence (τὸ γίνεσθαι οὖν κτλ.) is an inference from the preceding passage where causation is analysed with the help of an Aristotelian example. The description of effects as activities (something's coming about) and as incorporeal items corresponds to the Stoic doctrine (cf. below, on 96,23 f. and 96,25–97,1); however, here it is presented as the view of our author, who is not always bound by Stoic distinctions and terminology (cf. above, on 96,6–10). Thus we cannot be sure, for example, if the word ἐνέργειαι in 96,22 (used in the same connection above, 96,16) goes back to a Stoic source. For the doctrinal perspective of 96,25–97,1, cf. below, *ad loc.* For the effect as the 'coming about' of something, cf. above, on 95,31–96,2. For the example of cutting, cf. above, 96,7 f. and below, 96,26.

96,23 f. κατηγορημάτων ἤ, ὥς τινες, λεκτῶν … τὰ αἴτια. The conclusion reached in the previous sentence is now associated with the description of effects as

---

something incorporeal, cf. also Ps.Galen, *Med. Def.* 154 (XIX,392 K.): "Cause is that which does something in the body, while being itself incorporeal" (αἴτιόν ἐστιν ὃ ποιοῦν τι ἐν τῷ σώματι καὶ αὐτὸ ἀσώματόν ἐστι). Cf. Jutta Kollesch, *Untersuchungen zu den pseudogalenischen Definitiones medicae* (Berlin: Akademie-Verlag, 1973), pp. 120 f., who identifies it as a Peripatetic view. Another version of the view is mentioned in 'Aetius' I 11, 3 (*Dox. Gr.* 310,1–4).

398 Cf. e.g. *An. Post.* II 17, *passim*; *EN* III 5, 1114b1–4.

predicates, or alternatively as 'sayables'.[399] The latter name is explained by a reference to Cleanthes and Archedemus, who are said to have called predicates 'sayables' (λεκτά). This reference, listed as SVF I 48, is the only testimony we have of the employment of the concept of λεκτά before Chrysippus. Cleanthes (3rd cent. B.C.) may have treated the topic in his book *On Predicates* (Περὶ κατηγορημάτων) mentioned in DL VII 175. We know nothing about the context in which he did so, however. Our passage does not say that Cleanthes used λεκτά of predicates *only*,[400] nor does it imply that he used the word in connection with causes. A plausible way of explaining why he is mentioned by name, however, is that he was the first to describe predicates as λεκτά.[401] Archedemus of Tarsus was a Stoic philosopher of the 2nd century B.C., who could have discussed λεκτά in his treatise *On Utterance* (Περὶ φωνῆς).[402] For the Stoic background to the view of effects as predicates, see the next note. The Stoics regarded predicates as incorporeal precisely because they thought of them as λεκτά, i.e. things uttered about an object, but distinct from every item involved in the predication that counts as a body, because it does something (the object, its properties, the utterance itself).[403]

The elliptic structure κατηγορημάτων ἤ ... λεκτῶν ... τὰ αἴτια does not need to be complemented by adding αἴτια as a predicate, as it mirrors the form of 96,11: οἳ μὲν οὖν σωμάτων, οἳ δ' ἀσωμάτων φασὶν εἶναι τὰ αἴτια. Cf. also below, 99,13: ἀλλήλων οὐκ ἔστι τὰ αἴτια κτλ.

96,25–97,1 ἤ, ὅπερ καὶ μᾶλλον, τὰ μὲν κατηγορημάτων ... τὰ δ' ἀξιωμάτων κτλ. The notion of effects as predicates, mentioned above, is developed by an exam-

---

399 For 'sayables', see references above, on 87,24f.

400 Cf. Jonathan Barnes in Algra et al. (eds.), *The Cambridge History of Hellenistic Philosophy*, p. 210.

401 Cf. Alessandrelli, *Il problema del λεκτόν*, pp. 43–46.

402 On Archedemus, cf. Barnes in Algra et al. (eds.), *The Cambridge History of Hellenistic Philosophy*, pp. 71f. The fact that Archedemus is mentioned along with Cleanthes in our passage is difficult to explain. He was hardly the first Stoic after Cleanthes to include predicates among λεκτά. Perhaps he was the first one to limit the scope of λεκτά explicitly to predicates, in contrast to the description of predicates as 'incomplete' λεκτά; cf. DL VII 64 (LS 33G): "A predicate is what is asserted of something, or a state of affairs attachable to something or some things, as Apollodorus [viz. of Seleucia, Archedemus' contemporary] says, or a deficient sayable (λεκτὸν ἐλλιπές) attachable to a nominative case for generating a proposition." Alternatively, the testimony could be (directly or indirectly) based on Archedemus' analysis of causation in which he spoke of effects as λεκτά, while mentioning Cleanthes in the same connection.

403 Cf. LS 33A–H.

ple of a verbal predicate (τέμνεται), also expressed in a substantival form (τὸ τέμνεσθαι).[404] The author indicates earlier (96,23f.) that this notion of effects is Stoic and the example used here seems to go back to Stoic sources, too; cf. Sextus, *M.* IX 211 (SVF II 341/LS 55A): "... the Stoics say that every cause is a body which functions as cause to a body of something incorporeal—for example, the scalpel (a body) as cause to flesh (a body) of the incorporeal predicate 'being cut' (τοῦ τέμνεσθαι κατηγορήματος), or again fire (a body) as cause to wood (a body) of the incorporeal predicate 'being burnt'" (trans. Bett).[405] Against the view reported in 96,23f., describing effects as predicates without qualification, the author now proposes an alternative option, according to which some causes are causes of predicates, while others of propositions. This option, preferred by Clement or his source (ὅπερ καὶ μᾶλλον),[406] is not paralleled in Stoic testimonies, but it is based on the Stoic distinction between the 'deficient' and the 'self-complete' λεκτά, the former group including κατηγορήματα (e.g. 'writes') and the latter ἀξιώματα (in the sense of propositions, e.g. 'Socrates writes').[407] Whatever the doctrinal background to this, more differentiated view of effects, our author seems to use it for reasons determined by the context of the present discussion: It enables him to include all items expressed in a substantival form into the description of effects, without abandoning the view of effects as something incorporeal; this will serve as a bridge to the 'Aristotelian' notion of effects as προσηγορίαι (i.e. items referred to by nouns), discussed below, 97,1–7 (see the next note).[408]

---

404 In L, both forms of the predicate are substantival; it reads: οἷον τοῦ τέμνεσθαι, οὗ πτῶσις τὸ τέμνεσθαι. That this is a mistake (probably on the part of a scribe) is confirmed by an analogous, but grammatically impossible rendering of the second example: ὡς τοῦ ναῦς γίνεσθαι κτλ. Both verbs have been corrected already by Gentien Hervet.

405 According to the doxographic tradition, the view of effects as predicates goes back to Zeno; cf. Stobaeus, *Anth.* I 13, 1c (Wachsmuth I,138,15f./SVF I 89/LS 55A): καὶ τὸ μὲν αἴτιον σῶμα, οὗ δὲ αἴτιον κατηγόρημα. Cf. also Posidonius, fr. 95 EK (Stobaeus, *Anth.* I 13, 1c): ... καὶ τὸ μὲν αἴτιον ὂν καὶ σῶμα, οὗ δὲ αἴτιον οὔτε ὂν οὔτε σῶμα, ἀλλὰ συμβεβηκὸς καὶ κατηγόρημα.

406 On more than one occasion, Clement uses the phrase ὅπερ καὶ μᾶλλον to endorse one of the alternative views he refers to; cf. *Strom.* II (15) 68, 1; V (14) 130, 4; VII (16) 104, 6; *Exc.* 13, 4.

407 Cf. DL VII 63.

408 Alessandrelli, *Il problema del λεκτόν*, pp. 37, 40f. and 48, suggests that the more differentiated option is introduced by a Stoic author (possibly Archedemus) in response to the Peripatetic critique of the notion of effects as incorporeal predicates; according to this critique, the Stoic notion fails to explain how new bodies, such as a ship, can be produced. Whatever the background to the distinction between the causing of predicates and of propositions in our passage, it is admittedly tempting to think that its original aim is to

It is presumably for the same reason that the concept of 'case' (πτῶσις) is introduced in our passage. It is notoriously difficult to determine the precise meaning of this term in Stoic thought. 'Case' is sometimes taken to refer to an inflectible (i.e. substantival) form of expression, which may include nouns, but also substantival clauses.[409] Alternatively, it has been suggested that the Stoics originally conceived of cases not as linguistic forms, but rather as individual and common qualities of an object, signified by proper names and nouns.[410] Finally, it has been argued that cases, like κατηγορήματα, are components of meaning ('deficient' sorts of λεκτά), distinct from both linguistic forms and (corporeal) qualities.[411] When we reflect on our text in light of these interpretations, we find that it points not in one, but in two directions. In 96,26 and 97,1, πτῶσις likely refers to a form of expression. By introducing the distinction between a verbal and a substantival predicate, the text cannot mean that these are two sorts of λεκτά; rather, they are two sorts of expression of the same κατηγόρημα (taken as an incomplete λεκτόν), one of them being described as a πτῶσις of the other. The same, of course, applies in the case of ἀξίωμα. But if πτῶσις serves to distinguish one sort of expression from another, it is fair to conclude that it is regarded as a form of expression.

But why is it important that the same (deficient or self-complete) λεκτόν may be expressed in two ways? It does not seem to make much difference to the Stoics mentioned by Sextus, *M.* IX 211 (quoted above), who use the substantival form (τοῦ τέμνεσθαι) only. Again, the solution probably lies in the role that the distinction between the two forms plays in the context of the present argument. By showing that effects (whether conceived of as predicates or propositions)

---

distinguish between the causing of new effects to existing bodies on the one hand, and the causing of substantial generation on the other (David Sedley has kindly proposed this interpretation to me). However, the role the distinction plays in the context of the present argument is yet another matter.

409 Cf. LS I, p. 201; Alessandrelli, *Il problema del λεκτόν*, pp. 137–154. This view essentially goes back to Eduard Zeller, *Die Philosophie der Griechen* III/1 (Leipzig: Reisland, [5]1923), p. 90 n. 2.

410 Cf. Andreas Schubert, *Untersuchungen zur stoischen Bedeutungslehre* (Vandenhoeck & Ruprecht: Göttingen, 1994), pp. 79–109; Michael Frede, "The Stoic Notion of a Grammatical Case," *Bulletin of the Institute of Classical Studies* 39/1 (1994), pp. 13–24; idem, "The Stoic Notion of *lekton*," pp. 121–128. A crucial text suggesting this option is DL VII 58.

411 Cf. Richard Gaskin, "The Stoics on Cases, Predicates, and the Unity of Proposition," in R. Sorabji (ed.), *Aristotle and After* (London: University of London, 1997), pp. 91–108, here 94–101; Jean-Baptiste Gourinat, *La dialectique des stoïciens* (Paris: Vrin, 2000), pp. 122–129. For earlier references, see Schubert, *Untersuchungen*, p. 79 n. 40.

can be expressed in a substantival form, the author creates a point of contact with the view of effects discussed below, according to which effects are items referred to by nouns. While discussing this view, the author again uses the word πτῶσις, not in the sense of a linguistic form in which a λεκτόν may be expressed, but rather in the sense of something incorporeal that the items referred to by nouns 'obtain' (cf. below, 97,3.6f.); this something is obviously a λεκτόν. By interpreting the 'Aristotelian' view of effects in this manner, the author establishes a connection between the two theories: Once the reader has acknowledged that (incorporeal) effects can be expressed in a substantival form, it should not be difficult to accept that such expressions as 'house' or 'ship' (referring to effects in the 'Aristotelian' view) actually signify something incorporeal. It is slightly disappointing that the word πτῶσις is used in two different meanings in the course of the argument, but I could not find a more charitable interpretation of this passage.

97,1f. Ἀριστοτέλης δὲ προσηγοριῶν, οἷον τῶν τοιούτων κτλ. The word προσηγορία is used by the Stoics of that part of speech which signifies a common quality ('man', 'horse'), as opposed to a (proper) name, signifying an individual quality (DL VII 58). Normally, it is translated as 'noun' (or, to underline its technical usage, as 'appellation'). However, in our passage, it does not refer to nouns, but rather to something designated by them: a house, a ship, etc.[412] There is a close parallel to this in Sextus, *PH* III 14, where effects are also described as nouns, in contrast to the notion of effects as predicates.[413] There, as in our passage, items referred to by the word προσηγορία are probably the qualities of a given object that enable us to call it in a certain way (e.g. 'horse'); these qualities, according to the Stoics, are corporeal; thus the notion of effects as 'nouns' probably amounts to the view that causes are causes of corporeal qualities (as opposed to the view of effects as incorporeal λεκτά).

97,3–7 ἡ πτῶσις δὲ ἀσώματος εἶναι ὁμολογεῖται … οὐδὲ γὰρ τὴν οἰκίαν λέγομεν σῶμα οὖσαν, ἀλλὰ τὴν πτῶσιν ἀσώματον οὖσαν, ἧς οἰκία τυγχάνει. Commenting upon the Aristotelian view of effects mentioned above, the text makes a distinction between items referred to by nouns and something we *say* when speaking about these items (we should presumably think of a speech that describes them as effects of a cause). The items in question are regarded as corporeal objects;

412 Cf. Frede, "Notion of a Case," p. 16; Gaskin, "Stoics on Cases," p. 101 n. 26.

413 *PH* III 14: οἱ μὲν προσηγοριῶν αἴτιον εἶναι τὸ αἴτιον φάσκοντες, οἷον τῆς χύσεως, οἱ δὲ κατηγορημάτων, οἷον τοῦ χεῖσθαι. In SVF I 65 (= LS 30A), the word προσηγορία is arguably used in the sense of a corporeal object, too; cf. below, note 415.

when speaking of them, however, we do not say these objects themselves (they do not go through our mouth), but we express something incorporeal that they obtain. When explaining what it is, the text once again picks up the notion of πτῶσις, taking it as something agreed to be incorporeal.

Whatever the doctrinal background to this passage, it is clearly informed by the Stoic view that when referring to an object, speech signifies something incorporeal about it, also described as τὸ λεκτόν.[414] The notion of an object 'obtaining' its case (ἧς … τυγχάνει) seems to be linked to the Stoic habit of speaking of objects as τὸ τυγχάνον ('that which obtains').[415] And there are good reasons to think that (at least some) Stoics conceived of cases as (deficient) λεκτά.[416] But the purpose for which the concept of πτῶσις is employed here, namely to represent the effects signified by nouns (thus replacing the corporeal candidates for this role), does not seem to have any parallel in the Stoic sources. The agenda of the whole passage seems to be driven by an attempt to overcome the disagreement between the Stoic and the Aristotelian view of effects, introduced above, 26,1/96,11. The basis on which it is done, while largely influenced by the Stoic theory, seems to be as doctrinally independent here as everywhere else in the chapter.

The sophism to which the concept of πτῶσις provides a solution is listed as LS 33O. Another version of it is quoted in DL VII 187 (= SVF II 279, LS 37R) among a number of puzzles (some of them probably going back to the Megarian school) already discussed by Chrysippus. It is based on the ambiguity of the phrase 'to say something', which may refer to one of three things: (a) an utterance, (b) a meaning signified by it (here identified as πτῶσις), or (c) the object referred to. The solution proposed by our text may be reconstructed as follows:

414 Cf. esp. Sextus, *M.* VIII 11–12.75.

415 See the previous note and other references in Frede, "Notion of a Case," p. 19. A passage in Simplicius, *In Cat.* 8, 8b25 (CAG VIII: 209,13), suggests that the items 'obtained' by the objects are cases: τὰς πτώσεις τευκτὰς ἀπὸ τοῦ τυγχάνεσθαι (*scil.* οἱ ἀπὸ τῆς Στοᾶς ἐκάλουν). See also SVF I 65 (= LS 30A, from Stobaeus, *Anth.* I 12, 3/Wachsmuth I,137,3–6): Ταύτας [*scil.* τὰς ἰδέας] δὲ οἱ Στωικοὶ φιλόσοφοί φασιν ἀνυπάρκτους εἶναι καὶ τῶν μὲν ἐννοημάτων μετέχειν ἡμᾶς, *τῶν δὲ πτώσεων ἃς δὴ προσηγορίας καλοῦσι τυγχάνειν*. This much-disputed fragment should likely be translated as follows: "The Stoic philosophers say that these [ideas] do not exist and that we participate in concepts, whereas *those items they refer to as 'nouns' obtain cases.*" Here the "items referred to as nouns" are probably corporeal objects. For the syntactic construction of the sentence, cf. Andreas Graeser, "The Stoic Theory of Meaning," in J. Rist (ed.), *The Stoics* (Berkeley/Los Angeles: University of California Press, 1978), pp. 77–100, here 85; Gaskin, "Stoics on Cases," p. 101 n. 26, quoting Hülser's translation in FDS 316.

416 See references above, note 411.

The premiss "whatever you say, goes through your mouth" pertains to (a), the premiss "but you say 'house'" pertains to (b), and the conclusion "therefore house goes through your mouth" pertains to (c).[417] The crucial distinction here is that between the corporeal object and the incorporeal πτῶσις that the object obtains. A similar distinction seems to be applied by Galen to another puzzle quoted by DL in the same passage: "If someone is in Megara, he is not in Athens; but man is in Megara; therefore, it is not the case that man is in Athens."[418] Reporting on this puzzle, Simplicius suggests that the solution lies in the distinction between a particular man and man in general.[419] Having probably the same solution of the same puzzle in mind, Galen suggests that the ambiguous use of the word ἄνθρωπος may be resolved by distinguishing between man as a 'being' (οὐσία) and man as πτῶσις.[420]

97,8 f. καὶ τὸν οἰκοδόμον οἰκοδομεῖν λέγομεν … οὕτως φαμὲν χλαμύδα ὑφαίνεσθαι. The distinction between corporeal objects and their incorporeal πτῶσις, introduced in the face of the Aristotelian view of effects (97,1–7), is now further explained by reminding the reader of situations when we speak of an object that has not come about yet. Picking up on the example used in the

417 I fail to see why Clement's solution should contain a "muddle", as Long and Sedley put it (LS I, p. 201), or why "it is not quite free of confusion", as Gaskin contends ("Stoics on Cases," pp. 96 f. n. 14). It is perfectly fine for our author to accept the first premiss of the sophism, as long as it refers to an utterance. The core of his response lies in the interpretation of the object of λέγειν in the second premiss as πτῶσις (in the sense of an incorporeal λεκτόν), rather than the corporeal object referred to. In contrast to what is sometimes claimed (LS I, p. 201; Gourinat, *La dialectique*, p. 124 and n. 2), the response does not imply that the incorporeal πτῶσις goes through our mouth (this is already shown by Graeser, "Theory of Meaning," p. 85 f.).

418 DL VII 187.

419 Simplicius, *In Cat.* 5, 3b10–23 (CAG VIII: 105, 13 f.): ὁ γὰρ ἄνθρωπος οὐ τίς ἐστιν· οὐ γάρ ἐστιν τὶς ὁ κοινός.

420 Galen, *Soph.* 4 (XIV,595,14–16 K.): τρίτη δὲ [*scil.* ἀμφιβολία] παρὰ τὴν ἐν τοῖς συνθέτοις ὁμωνυμίαν, οἷον 'ἄνθρωπός ἐστιν'· ἀμφίβολος γὰρ ὁ λόγος, εἴτε τὴν οὐσίαν εἴτε τὴν πτῶσιν εἶναι σημαίνει. The passage is included in SVF II 153 and LS 37Q. That Galen's solution refers to the puzzle quoted above is suggested by Frede, "Notion of a Case," p. 21; cf. also R.B. Edlow, *Galen on Language and Ambiguity* (PhA 31; Leiden: Brill, 1977), pp. 65 f. There is some controversy about the meaning of οὐσία and πτῶσις in this passage; cf. Gaskin, "Stoics on Cases," p. 99 f. n. 21. I take it that οὐσία is used in the Aristotelian sense of a primary substance (cf. e.g. Galen, *PHP* VIII 2, 6/CMG V,4,1,2: 490,29 f./V,662 K., where Dion is an example of οὐσία ἄτομος), while πτῶσις refers to something we mean when we say 'man'.

preceding lines, the text points out that we speak of a house-builder building a house (οἰκοδομεῖν) with reference to something that will come about in the future (κατὰ τὴν ἐπὶ τὸ γενησόμενον ἀναφοράν). The item of which we speak in this way is obviously the object of the building activity, namely a house, just as we speak of a 'mantle' with reference to something that is being weaved yet. The conclusion is not explicit, but it is supposed to be plain: If the item of which we speak has not come about yet, surely it cannot be corporeal.[421]

97,9–11 τὸ γὰρ ποιοῦν ἐνεργείας δηλωτικὸν ὑπάρχει, οὐκ ἔστι δὲ ἑτέρου μὲν τὸ ποιοῦν, ἑτέρου δὲ τὸ αἴτιον κτλ. This sentence is introduced as an explanation (γάρ); the point explained presumably is the claim that when speaking about a house's being built or a mantle's being weaved, we are referring to something that will come about in the future (see the previous note). The explanation starts by a concession to an implied objection, namely that what an agent does is an activity, i.e. something coming about now, rather than in the future.[422] The text acknowledges that speaking of X as an agent implies that X does something ("'agent' indicates an activity"),[423] but argues that what the agent does is not the activity itself, but rather something caused by the activity, e.g. a house or a mantle.[424]

---

421 Commenting on this passage, von Wedel, *Symbola*, pp. 26f., suggests that it launches a new point ("Novi aliquid incipit Clemens"); in her attempt to reconstruct the supposed continuity with the following argument (where the identity of 'that which makes X' and 'the cause of X' is emphasized), she proposes that something is missing after ὑφαίνεσθαι in 97,9, namely a clause in which the agent of 'weaving' was mentioned. Stählin agrees on all counts, printing the sentence as the beginning of a new paragraph; indicating a lacuna in 97,9; and proposing the addition of, e.g., καὶ αἴτιός ἐστιν τῆς χλαμύδος καὶ τῆς οἰκίας ὁ ποιῶν. But it is likely that the sentence is a continuation of the preceding argument and no correction is necessary.

422 The interpretation of the first clause as a concession is, in my view, very likely in light of the overall argument and the parallel quoted in the commentary. The ground on which it stands would be even firmer if we had μὲν after ἐνεργείας. In fact, δὲ in the next clause (97,10) raises the suspicion that it could have been the case in an earlier version of the text.

423 For the meaning of the phrase, cf. a similar formulation below, 33, 3/101,21 ("If 'cause' is indicative of a self-complete activity ..."), where δηλωτικὸν ἐστι is equivalent to σημαίνει in 101,22.

424 Thus it is not a good idea to delete τῆς χλαμύδος καὶ τῆς οἰκίας in 97,11, as proposed by Stählin; again, Stählin's presentation of the text is probably influenced by von Wedel, who thinks that these words were erroneously added by Clement to his source (cf. *Symbola*, p. 27).

There is a close parallel to this argument in Galen, *Thras.* 27 (67,21–68,5.11–15 Helmreich/V,854 f. K.), voiced against people who fail to distinguish between an activity by which something is done (τὸ ποιεῖν τι, ἐνέργεια τοῦ τέλους) and the object or goal of this activity. Examples illustrating this difference include those of house-building as opposed to a house, weaving as opposed to a garment, and ship-building as opposed to a ship: "It is not the building of a house that is the goal of house-building, but the house itself; not the weaving of a garment, or the construction of a ship, or a pallet, or any individual article, but the article itself that is constructed" (trans. Singer).[425] Galen also suggests that the goal of each art precedes it as an object of desire; cf. *Thras.* 29 (71,10–13 Helmreich/V,859 K.): "No one would wish to compose an art of medicine without first having a desire for health (μὴ προποθέσας ὑγιείας), any more than one would wish to compose an art of building without a prior need for a house (εἰ μὴ κἀνταῦθα οἰκίας ὠρέχθη), or an art of weaving without the need of a garment." This arguably sheds light on our text's view of effects referred to by nouns as something incorporeal.

97,11 καθ' ὃ γὰρ αἴτιος ... κατὰ τοῦτο καὶ ποιητικός ἐστι τοῦ γίνεσθαι. The predicate αἴτιος indicates that the (unexpressed) grammatical subject should be masculine; rather than returning to οἰκοδόμος, mentioned in 97,8, we should perhaps think of the indefinite pronoun τις. The sentence explains why, in cases like building or weaving, that which is done by an agent is the same item as that which is caused by a cause. The reason is that, generally, when X is the cause of the coming about of P, then, in the same respect, X also makes P come about. The focus of this argument still pertains to the incorporeality of effects (items caused by an activity are its objects before they come about), but the identification of a cause as that which does something opens a new topic.

27, 3–28,1

97,13–22 Τὸ αὐτὸ ἄρα αἴτιον καὶ ποιητικόν ... οὐ πάντα τῆς τεκνοκτονίας αἴτια τυγχάνει, μόνη δὲ ἡ Μήδεια. This section is listed as SVF II 347. Its first sentence is formally a conclusion of the preceding discussion, but in fact, it starts a new theme, shifting the attention from effects to causes.[426] The text returns to the view of a cause as that which does something, introduced above, 25,

---

425 *Thras.* 27 (67,24–68,4 Helmreich/V,854 K.): οὐ γὰρ οἰκοδομεῖν οἰκίαν αὐτὸ δὴ τοῦτο τέλος ἐστὶ τῆς οἰκοδομικῆς, ἀλλ' οἰκία, καθάπερ οὐδ' ὑφαίνειν ἐσθῆτα καὶ ναῦν συμπήττειν καὶ σκίμποδα καὶ τῶν ἄλλων ἕκαστον, ἀλλ' αὐτὸ τὸ δημιουργηθέν.

426 The words καὶ δι' ὃ in 97,13, repeated in the next line, are inappropriate here and probably entered the text due to a scribal error.

5/96,6–10; for the Stoic background to this view, see above, on 96,6. Causes in this proper sense are further distinguished from items that do not qualify as causes, but fit a more general description of that 'on account of which' (δι' ὅ) something comes about.[427] These items do not have a name here, but they are illustrated by examples taken from the story of Medea, where they function as conditions for something to happen, namely for Medea's killing her children (97,17–21). Notwithstanding the fact that this outcome would not have happened without these conditions, the only factor deserving to be called a 'cause' in the story is Medea herself (97,21f.), plainly because it is she who commited the crime.[428]

Elements of the Medea story (on the basis of its rendering by Ennius) are used in a similar way by Cicero, *Fat.* 35; cf. also *Top.* 61; for the context cf. above, on 95,28–31. It is tempting to think that our text reflects Cicero's Greek source; however, it is hard to say whether it draws from this source immediately or how closely it follows it.

97,23–98,2 τὸ μὴ κωλῦον ἀνενέργητόν ἐστιν … οὐκ ἂν οὖν εἴη αἴτιον τὸ μὴ κωλῦον. This passage closely follows on the preceding one, developing another consequence of the view of a cause as that which does something: Factors that do not prevent P do not count as causes of P, for the following reasons: (1) X is a cause insofar as it does something; but not preventing is not doing something; hence, insofar as X does not prevent P, it is not a cause of P. (2) If X is a cause of P, then P is because of the presence of X; hence, insofar as P is because of the absence of X (namely because X capable of preventing P does not do so), X is not a cause of P. Compare a different view in Aristotle, *Phys.* II 3, 195a12–14; *Met.* Δ 2, 1013b12–16: ἄμφω δέ, καὶ ἡ παρουσία καὶ ἡ στέρησις, αἴτια ὡς κινοῦντα. For the notion of presence, cf. below, 33, 1/101,18f.

There is a close parallel to this passage in *Strom.* I (17) 82, 3–5 (= SVF II 353); cf. above, pp. 57–59. In L, the words διὰ τοῦτο γοῦν … οὐ πάρεστιν (97,25f.) are placed after τὸ μὴ κωλῦον in 98,2. Von Arnim's proposal to move them after τοῦ γινομένου makes more sense and corresponds to almost a verbatim parallel in *Strom.* I (17) 82, 4–5 (GCS 15: 53,24f.).[429]

427 For δι' ὅ, cf. Posidonius, fr. 95 EK, quoted above, p. 266 n. 366.

428 The idea of a string of conditions, without which Medea would not have commited her crime, is probably influenced by the initial lines of Euripides' *Medea*.

429 Cf. Von Arnim, *De octavo*, p. 13.

28, 2–7

98,3–5 τετραχῶς τὸ αἴτιον λέγεται κτλ. Nothwithstanding the view of a cause as that which does something, presented above (25, 5 and 27, 3–28, 1), here the text introduces the Aristotelian classification of causes, using the traditional example of a statue; cf. Aristotle, *Phys.* II 3, 194b24 f., 195a33–35, 195b6–12; *Met.* Δ 2, 1013b6–9. Aristotle does not use this example to illustrate all four causes, however, but only the efficient cause (the sculptor or the art of sculpture) and the material one (bronze). A full account of Aristotelian causes based on this example is found in Seneca, *Ep.* 65, 4–6 (after the passage quoted above, on 96,18–21). Cf. also Alexander, *Fat.* 3 (Suppl. Arist. II/2: 167,2–12). The following discussion shows that our author continues to regard the effects of these causes as activities (98,6: τὸ γενέσθαι τὸν ἀνδριάντα), rather than as (corporeal) objects; cf. above, 26, 2–3/96,18–22. The four causes are also mentioned above, 18, 1/91,4–6.

98,5–9 τῶν ὧν οὐκ ἄνευ λόγον ὁ χαλκὸς ἐπέχει … καὶ ὁμοίως ἐστὶν αἴτιον … αἴτιον δὲ οὐχ ἁπλῶς … συνεργὸν δέ. The view of matter as a cause is qualified in light of the fact that matter has the status of prerequisites. Unlike synectic causes, the text argues, prerequisites are not causes without qualification (ἁπλῶς), but are instead (merely) auxiliary. Here, against the background of the Aristotelian classification, the text re-employs the terminology introduced above, 25, 1–3, focussing on the distinction between synectic causes (cf. above, on 95,31–96,2) and prerequisites (cf. above, on 96,5). Surprisingly, the prerequisites are now characterized as 'auxiliary', i.e. by a name reserved for a different class of causes above (25, 1/95,27 f.), as well as below (33, 1–9), where the description of the auxiliary causes is incompatible with that of the prerequisites (cf. esp. 102,1 f. and 8–10). As mentioned earlier (on 96,4 f.), it is impossible to avoid the conclusion that the word συνεργός is used in (at least) two different meanings in our text.

For the use of συνεργός in the present passage, cf. Sextus, *M.* IX 237–245, where matter is described as an auxiliary factor that is also a cause, because the effect it helps to produce cannot occur without it. A similar view is expressed by Galen, *CP* 7, 75–76 (trans. Hankinson): "For whenever you remove the patient, out of which the thing which is created is created, you destroy the agentive force, by which it comes to be, and these contributory factors (*quae conferunt*) are rightly called its causes. All men call those things which of their own nature contribute to something's coming to be 'causes'." However, unlike our passage, Galen makes a further distinction between such passive factors as matter, which contribute something of their own nature (76–77), and those that "have the status of prerequisites (*rationem habent eorum sine quibus non*), cutting

off and preventing a thing's creation if they are not present, but contributing nothing essentially by their presence (*ipsa autem nichil in essendo presentia cooperantia*)" (84).[430] Galen's examples of the latter kind are place and the surrounding air (78). Cf. above, on 96,5.[431]

Whereas the material cause from the Aristotelian classification is characterized as a prerequisite and an auxiliary factor, the efficient cause is represented by the synectic in this passage.[432] Unlike prerequisites, the synectic cause is regarded as a cause 'without qualification' (cf. 98,8). This agrees with the notion of a cause as that which does something; being applied to causes "in the proper sense" (cf. above, 25, 5/96,6), it allows for a less rigorous use of the word 'cause' (provided it is further qualified by such adjectives as 'auxiliary'; cf. below, 98,13 f.). But the concept of synectic causes implied here (and below, 28, 7; 31, 5) seems to be in conflict with their description as 'self-complete' above (25, 3/95,31–96,2) and below (33, 2/101,19 f.). This, at least, is how Sextus sees it, when he says that "if there is any cause, it is a cause of something either self-completely—that is, using its own power—or it needs the affected matter as a collaborator for this purpose" (*M.* IX 237; trans. Bett, modified). Galen agrees that these are two mutually exclusive descriptions, but (unlike Sextus, *M.* IX 245) rejects only the first one: "… no cause is of itself complete (*per se perfecta est*): just as the material requires the efficient cause, so the efficient cause plainly requires suitable material" (*CP* 13, 165, trans. Hankinson).[433] Thus, to all appearances, we have a serious inconsistency in the understanding of the synectic cause in our text; for a possible solution, cf. below, on 101,19 f.

98,9 f. **πᾶν τὸ ἐνεργοῦν παρέχει τὸ ἀποτέλεσμα μετ' ἐπιτηδειότητος τοῦ πάσχοντος.** Here it is claimed that the causal process *always* involves an agent and a patient. The contribution of the patient is described as its suitability (ἐπιτηδειότης). This term, in the sense of an inherent capacity for acting or being affected in a certain way, was probably introduced to the philosophical vocabulary by Philo

430 Here *cooperantia* is likely an equivalent of the Greek συνεργοῦντα. Cf. the retroversion by Kurt Bardong, *CP* 7, 84 (CMG Suppl. II: 20,33).

431 Galen's distinction just mentioned could shed light on the distinction between the auxiliary causes and prerequisites above, 25, 1/95,27 f. It also roughly corresponds to the examples of the two kinds of causes in 25, 4/96,4 f.

432 This need not (and probably does not) mean that these two classes are co-extensive; our author seems to regard both the synectic and the procatarctic causes as efficient; cf. below, 98, 7 and 31, 4–5.

433 The expression *per se perfecta* likely corresponds to the Greek αὐτοτελές (cf. Hankinson, *Galen: On Antecedent Causes*, pp. 254 f.).

the Logician.[434] It is not attested by the Stoics.[435] For our passage, cf. Sextus, *M.* IX 243, who uses the concept of suitability to combat the view that fire is a self-complete cause of burning: "If [fire brings about burning] together with the suitable condition (σὺν τῇ ἐπιτηδειότητι) of the burning wood, where do we take the right to say that *it* is the cause of the burning, as opposed to the suitable condition of the wood?" (trans. Bett); cf. also *M.* VIII 199. Galen employs it in a polemical context, too; cf. *Diff. Feb.* I 6/VII,290,11–14 K. (χρὴ γὰρ μεμνῆσθαι … ὡς οὐδὲν τῶν αἰτίων ἄνευ τῆς τοῦ πάσχοντος ἐπιτηδειότητος ἐνεργεῖν πέφυκεν) and *CP* 8, 107/CMG Suppl. II: 26,8 (where *aptitudo* certainly translates ἐπιτηδειότης); in both passages, the target of Galen's polemic is the (Erasistratean) view according to which the antecedent conditions traditionally regarded as causes of fever (e.g. heat or chilling) cannot be counted as causes at all, as they do not bring about the same effect in everyone. While criticising the notion of a cause implied by this view, Galen resorts to arguments similar to Sextus' critique of self-complete causes.[436] See further below, 32, 1.3.

98,10–12 διατίθησι μὲν γὰρ τὸ αἴτιον, πάσχει δὲ ἕκαστον εἰς ὃ πέφυκέν τι, παρεκτικῆς τῆς ἐπιτηδειότητος κτλ. The verb describing the activity of a cause (διατίθησι) suggests that the effect brought about by this cause is a διάθεσις of the object. The διάθεσις is something that the object endures (πάσχει), but it is not determined solely by the agent force, as it actualizes something to which the object is naturally disposed (εἰς ὃ πέφυκέν τι). This disposition to receive a specific διάθεσις is the object's ἐπιτηδειότης. Cf. above, 95,29–31, where causation is described in a similar way ('suitability' corresponding to the object's ἀκολασία).[437]

---

434 Cf. Alexander, *In An. Pr.* (CAG II/1: 184,6–10); Simplicius, *In Cat.* (CAG VIII: 195,32–196,2).

435 For the history of the term ἐπιτηδειότης, cf. Robert B. Todd, "*Epitedeiotes* in Philosophical Literature: Towards an Analysis," in *Acta Classica* 15 (1972), pp. 25–35. With respect to Alexander and the later tradition, Todd's conclusions are refined by Aubry, "Capacité et convenance," who shows that it came to be used in the sense of Aristotle's first potentiality (namely of a capacity to receive certain ἕξις); cf. above, on 88,2–4 and below, on 98,10–14.

436 Cf. Hankinson, "Galen's Theory of Causation," pp. 1760–1764; *Cause and Explanation*, pp. 303–306.

437 There is a comparable use of ἐπιτηδειότης in Galen, *Temp.* I 8 (32,14–19 Helmreich/I,560 K.): "Here, again, we must make at least the following distinction with regard to names: 'hot mixture' in activity (ἐνεργείᾳ) is something else than 'hot mixture' in capacity (δυνάμει). We say that something is such-and-such in capacity, when it is not yet what we say, but it may easily become that, as it has a natural condition suitable to that effect (φυσικήν τιν' ἐπιτηδειότητα κεκτημένοις εἰς τὸ γενέσθαι)." For Galen, mixture is a διάθεσις of the body.

98,13 f. οὐκ ἔστιν δὲ αἰτία, ἀλλὰ συνεργός, ἐπεὶ πᾶν αἴτιον ἐν τῷ δρᾶν νοεῖται. Cf. above, 25, 5/96,6; 27, 3.6/97,13.24 f.; 28, 3/98,8 f.

98,14 f. αὐτὴν δὲ οὐκ ἂν ποιοίη ἡ γῆ, ὥστε οὐδὲ αἰτία ἂν εἴη ἑαυτῆς. This statement (which interrupts the continuity of thought in 98,13–18) is likely Clement's gloss, indicating how the discussion could be useful for the interpretation of Gen 1:1. Cf. below, 29, 3–6/99,1.7–10; 30, 4/99,25.

98,15–18 καταγέλαστον δὲ τὸ λέγειν μὴ τὸ πῦρ αἴτιον τῆς καύσεως, ἀλλὰ τὰ ξύλα κτλ. In *M.* IX 244, while rejecting the view that the status of a cause belongs to the agent force only, Sextus argues that wood is the cause of burning just as the fire is: "[I]t is most absurd of a man to say that the fire is the cause of the burning and the wood is not the cause." Unlike Sextus' opponents, our text acknowledges that suitability is a causal factor (cf. above, 98,7 f.); nevertheless, unlike Sextus, it also combats the view that the causal status of suitability is equal to that of the agent force (98,12–14). To emphasize this point, it reduces the claim of an equal status to an absurd notion of inequality, according to which suitability is the *only* cause involved in the process. It is tempting to think that this reduction mirrors, in a polemically reversed manner, the argument used by Sextus.

98,19–22 τὸ συνεκτικὸν αἴτιον οὐ δεῖται χρόνου ... τῶν προκαταρκτικῶν τὰ μὲν χρόνου δεῖται ... τὰ δὲ οὐ δεῖται, ὡς ἡ πτῶσις τοῦ κατάγματος. Here we have another example of the synectic cause: a cautery.[438] It shows something that, according to our text, is true of the synectic causes in general, namely that they bring about their effect immediately (for the expression "without time", cf. below, 98,22–24). In this regard, they differ from procatarctic causes that need time to bring about their effect. The text does not provide an example, but we may safely think of such cases as cold, heat, fatigue, indigestion, or drunkenness, mentioned below, 31, 4/100,15. Interestingly, some procatarctic causes do not need time either, e.g. a fall bringing about a fracture.[439] As noted by Duhot, the background to the whole discussion is plainly medical.[440] For time as a necessary condition of the efficiency of the procatarctic causes, cf. Galen, *Caus. Morb.*

---

438 For other examples, cf. above, 25, 4/96,3 f. and below, 31, 5/100,18 f.

439 Why is the status of a branding-iron different from that of a fall? Our passage seems to imply that the effect of a branding-iron (namely pain) ceases when the cause is removed (cf. below, 33, 1/101,18 f.).

440 *Causalité*, p. 221.

2 (VII,8,4–9,14 K.).[441] For the example of a fall as an immediate cause of a fracture (ὡς πτῶμα κατάγματος), cf. Ps.Dioscorides, *De iis quae virus ejaculantur* (= Περὶ ἰοβόλων), *Praef.* (ed. Sprengel/MGO XXVI/2: 54).[442]

98,22–24 μή τι οὖν οὐ κατὰ στέρησιν χρόνου ἄχρονα λέγεται ... ἀλλὰ κατὰ μείωσιν, ὡς καὶ τὸ ἐξαίφνης κτλ. Here the word ἄχρονος refers to what Aristotle describes as 'imperceptible time'; cf. *Phys.* IV 13, 222b14 f.: τὸ δ' ἐξαίφνης τὸ ἐν ἀναισθήτῳ χρόνῳ διὰ μικρότητα ἐκστάν. Cf. Simplicius, *In Phys.* IV 13, 222b14 (CAG IX: 753,4): ... ὡς δοκεῖν ἄχρονον εἶναι ὅπερ ἐξαίφνης λέγομεν, and Alexander's commentary quoted ibid., 753,12–28. Cf. also Philo, *Agr.* 176 (ἐξαίφνης ... καί, εἰ οἷόν τε τοῦτ' εἰπεῖν, ἀχρόνως); Galen, *Dig. Puls.* III 1/VIII,885,6 f. K. (πᾶν ... ὅ τι ἐν ἐλάττονι τοῦ πρώτου πρὸς αἴσθησιν χρόνῳ γίγνεται, τοῦτ' ἄχρονον εἶναι φαντάζεται); *PHP* II 5, 36 (CMG V,4,1,2: 134,26 f./V,247 K.: οὐδείς ἐστι χρόνος αἰσθητὸς ἐν τῷ μεταξὺ τοῦ τρωθῆναί τε τὸ μόριον αἰσθέσθαι τε τὸ ζῷον), etc.

29, 1–30, 5

98,25–28 πᾶν αἴτιον ὡς αἴτιον διανοίᾳ ληπτὸν τυγχάνει, ἐπεὶ τινὸς καὶ πρός τινι νοεῖται κτλ. Here starts a new thematic section concerned with the relativity of causes. Various aspects of this topic are discussed until 32, 3/101,3. Causes are conceived in two relations: as the causes 'of something' (τινός), i.e. their effect, and 'to something' (πρός τινι), i.e. the object acted upon.[443] Cf. above,

441 Cf. e.g. 8,18 f.: χρόνου δεῖται ... τὸ μέλλον ποιήσειν ὁπωσοῦν. For this Galenic passage, see also below, on 98,25–28 and 100,20–22.

442 The passage is noted by von Wedel, *Symbola*, p. 30. Interestingly, Pseudo-Dioscorides mentions 'fall' not as a procatarctic cause of a fracture, but as an example of a cause that is "in one respect preparatory and in another respect synectic" (κατὰ μέν τι παρασκευαστικόν, κατὰ δέ τι συνεκτικόν), as it "occurs together with its effect and separates itself" (συνεφιστάμενον τῷ ἀποτελεσματικῷ [?] χωριζόμενόν τε). (There is a problem concerning the meaning of τῷ ἀποτελεσματικῷ. This word should mean 'productive', thus referring to something producing an effect, i.e. to a cause; but the context demands that τὸ συνεφιστάμενον also refers to a cause. It is tempting to think that we should read τῷ ἀποτελέσματι instead.) In this passage, the immediate effect of a cause is taken as a mark of its 'synectic' ingredient, whereas the fact that the impact lasts after the cause is removed (this being the point of χωριζόμενόν τε) is linked to its 'preparatory' character. Cf. the distinction between the procatarctic and synectic causes made earlier in the same text (p. 52): τινὰ μέν ἐστι τὰ προκατάρξαντα, ἃ ποιήσαντα πάθος χωρίζεται (...), τὰ δὲ συνεκτικά, διὰ τὸ ποιεῖν τὸ πάθος καὶ παραμένειν.

443 In 98,26, L reads πρός τι, thus producing a pair of relations τινὸς καὶ πρός τι. But, as 98,27 shows, this is a scribal error (cf. von Wedel, *Symbola*, p. 30). The category of πρός τι (mentioned in 98,29) likely includes both relations distinguished in 98,25–28.

26, 2/96,18–21; Sextus, *M.* IX 207. Stählin inserts διπλῇ before διανοίᾳ in 98,25, thus suggesting that these two relations are grasped in acts of mind described as διπλῇ διάνοια (cf. below, 98,29 f.: δυεῖν ἐπιβάλλομεν). However, this is a strange phrase to use in this sense and, in any case, the text can be defended as it is. Presupposing the view that relations are grasped by the mind, it says that "every cause is grasped by the mind as a cause". A similar view is used by the sceptics as an argument against the existence of causes; cf. Sextus, *M.* IX 208: "But things in relation to something are only conceived, and do not exist ... therefore cause too will be only conceived, and will not exist" (trans. Bett).[444] But our text does not claim that relations are grasped by the mind *only*, i.e. that there is no reality behind our understanding of a cause as a cause. Compare the distinction made by Galen in his discussion of powers (δύναμις) in the sense of 'active causes' (αἰτία δραστική): ἥ γε νόησις ἐν τῷ πρός τὶ τῆς δυνάμεως, ἡ θ' ὕπαρξις ἐν τῇ πρὸς ἕτερον σχέσει (*Plen.* 3/VII,524,13–15; 525,6 f. K.).

For the examples of a knife and a fire, cf. above, 25, 5; 26, 1–3; 28, 6. For the notion of suitability, cf. above, on 98,9 f. Galen uses the same examples in the same connection; cf. *Caus. Morb.* 2 (VII,8,4–9,14 K.). For *adamas* (i.e. presumably a diamond) as an object that cannot be burned, cf. *Strom.* VII (11), 67, 8; cf. Theophrastus, *Lap.* 19; Hero, *Pneum.* I prooem. (6,15 Schmidt); Pliny, *Nat. Hist.* XXXVII 57 (V,406,1–3 Mayhoff), etc. Cf. also Galen, *Caus. Morb.* 2 (VII,9,7 f. K.): τῷ ξίφει τέμνειν οὐκ ἂν οὐδὲ τοὺς λίθους ἐπιχειρίσειας, μὴ ὅτι γε ἀδάμαντα.

98,28–30 τὸ αἴτιον τῶν "πρός τι" ... ὥστε δυεῖν ἐπιβάλλομεν κτλ. The author sums up the preceding distinctions by saying that causes belong to the category of relation. Cf. Sextus, *M.* IX 239–240, who presents it as a doctrine of the 'dogmatists'; cf. Aristotle, *Met.* Δ 15, 1020b28–30 and 1021a14–19; Galen, *Ars. Med.* 23 (344,18 Boudon/I,366 K.: πᾶν αἴτιον ἐν τῷ πρός τι); *Nat. Fac.* I 4 (107,10–14 Helmreich/II,9 K.); *QAM* 2 (33,22 f. Müller/IV,769 K.). Thus, when we think of something as a cause, we focus our mind on that to which it is related as a cause, and we do this in the two respects mentioned above: as that 'of which' and that 'to which' it is a cause. A similar view is expressed by Galen, *CC* 9, 7: "As for the terms wound, fracture, severing and tear, each contains an indication of two notions, one being that of the cut itself and the other that of the body in which it occurs."[445] For 'focussing', cf. above, on 95,3.

---

444 τὰ δέ γε πρός τι ἐπινοεῖται μόνον ἀλλ' οὐχ ὑπάρχει (...) καὶ τὸ αἴτιον ἄρα ἐπινοηθήσεται μόνον, οὐχ ὑπάρξει δέ. Cf. the parallel in DL IX 97.

445 CMG Supp. Or. II: 71, trans. Lyons.

99,1 ὁ αὐτὸς καὶ περὶ τοῦ δημιουργοῦ καὶ ποιητοῦ λόγος καὶ πατρός. According to this sentence, an argument (λόγος) about a 'craftsman', a 'creator', and a 'father' is the same (ὁ αὐτός). How shall we understand this? 'Father' is mentioned in the following passage as an illustration of the principle that nothing is its own cause (99,2.6.11.12). 'Craftsman' and 'creator' are not mentioned, but it is likely that they should be taken in connection with the discussion of the cause of matter below, 99,7–10, i.e. as referring (in a Platonic and Christian manner) to a divine agent. 'Father', of course, could also be used in this sense, but the position of πατρός in the sentence suggests that it denotes a different object. Thus we seem to have two objects referred to by this sentence, one being a divine agent (a demiurge) and the other a father, and we are told that the argument about both of them is the same. This could be taken in two ways: It could mean that in both cases the argument is the same as with causes in general, namely that they are conceived in relation to something else; thus ὁ αὐτός would point back to the preceding lines about the relativity of causes (98,28 f.). However, the aim of the following discussion (29, 3–6/99,1–12) is not to apply any λόγος about the relativity of causes to a specific case (or cases), but rather to argue in favour of another (implied) principle, namely that 'nothing is its own cause'.[446] The argument is based, among other things, on the example of a father, which is presumably why it is described as an argument about a father (περὶ τοῦ πατρὸς λόγος or Aπ). The case of the divine agent is different. If the father serves as an illustrative example of a general statement (it is clear to everyone that no one can be his own father), the argument about the demiurge (Aδ) seems to have another aim, namely to show that matter cannot create itself, and so it must be created by the demiurge (cf. below, on 99,7–10). Thus, apart from making a general point about causes, the passage also tries to show how the same argument could be used to solve a cosmological problem. I believe it is this connection between the two arguments (the applicability of Aπ to Aδ) that makes them "the same" in our passage.[447] I find it very likely that the sentence was introduced by Clement to indicate a way in which the discussion in the following lines could be useful for his purpose.

99,2 οὐκ ἔστιν αὐτό τι ἑαυτοῦ αἴτιον οὐδὲ ἑαυτοῦ τις πατήρ. This sentence (syntactically unconnected with the preceding one) articulates an implication of the principle according to which causes are causes in relation to something

446 The centrality of this thesis is highlighted by its position at the beginning and at the end of the passage (99,2 and 99,12).

447 If this is correct, the first and third καί should presumably be understood in the sense of "both ... and".

else (κατὰ τὴν πρὸς ἕτερον σχέσιν, 98,28f.). The same point is already made by Plato with regard to the causes of generation; cf. *Hipp. Maj.* 297c: οὐκοῦν καὶ τόδε καλῶς λέγω, ὅτι οὔτε ὁ πατὴρ ὑός ἐστιν, οὔτε ὁ ὑὸς πατήρ; (…) οὐδέ γε τὸ αἴτιον γιγνόμενόν ἐστιν, οὐδὲ τὸ γιγνόμενον αὖ αἴτιον. *Phil.* 27a: ἄλλο ἄρα καὶ οὐ ταὐτὸν αἰτία τ' ἐστὶ καὶ τὸ δουλεῦον εἰς γένεσιν αἰτίᾳ.

99,3f. τό γε μὴν αἴτιον … διατίθησι, τὸ … γενόμενον … διατίθεται. Cf. above, 28, 4/98,10–12.

99,4f. οὐ δύναται δὲ τὸ αὐτὸ πρὸς ἑαυτῷ λαμβανόμενον ἐνεργεῖν ἅμα καὶ διατίθεσθαι. A similar point is made by Galen in the context of a discussion about the relativity of causes. After noting that an active cause is understood and exists in relation to something else (a passage quoted above, on 98,25–28), he rejects as 'absurd' a contrasting view, according to which anything could act upon itself: ποιεῖν δ' εἰς ἑαυτὸ λέγειν ὁτιοῦν, ἢ ἐνεργεῖν εἰς ἑαυτὸ, παρὰ τὴν ἔννοιάν ἐστιν (*Plen.* 3/VII,525,7f. K.).[448]

99,6 τὸ αἴτιον τοῦ ὑπ' αὐτοῦ γινομένου προχρονεῖ κατὰ τὴν οὐσίαν. Cf. Plato, *Phil.* 27a: ἆρ' οὖν ἡγεῖται μὲν τὸ ποιοῦν ἀεὶ κατὰ φύσιν, τὸ δὲ ποιούμενον ἐπακολουθεῖ γιγνόμενον ἐκείνῳ; Unlike Plato, however, our text limits the priority of causes to a priority in time. Cf. also Aristotle, *Met.* Θ 8, 1049b23–27, on the chronological priority of efficient causes.

99,7–10 οὐ δύναται δὲ τὸ αὐτὸ … καὶ προχρονεῖν τῇ ὕλῃ … ἅμα καὶ ὑστερεῖν καὶ ὑστεροχρονεῖν κτλ. The principle of chronological priority is now used for an argument that nothing can be a cause of matter and a product of matter at the same time: for this would entail that one and the same thing is both prior and posterior to matter, which is impossible. Most likely, the aim of the argument is to show that matter cannot be its own cause; cf. above, 28, 5/98,14f. By implication, it is an "argument about a craftsman and a demiurge" (cf. above, on 99,1), which seems to be drafted in such a way as to serve Clement's own philosophical interests.

99,10–12 διαφέρει τε τὸ εἶναι τοῦ γίνεσθαι … οὐθὲν οὖν ἐστιν ἑαυτοῦ αἴτιον. How does the distinction between being and becoming pertain to the present discussion? We have seen that causes are causes of something's coming about (cf.

---

448 The argument is directed against the notion of a synectic cause, which, apart from sustaining other things, also sustains itself; cf. *CC* 6 (CMG Suppl. Or. II: 61–63).

above, 96,18–21); this is a starting-point of another argument against the possibility of anything's being its own cause: If X were the cause of itself, it would be the cause of its becoming itself, i.e., of becoming something that it already is. But it is impossible for the same item to be (εἶναι) and to become (γίνεσθαι) P at the same time and in the same respect. Hence, nothing can be the cause of itself.

99,13–24 This passage is listed as SVF II 349 (= LS 55D).

99,13 ἀλλήλων οὐκ ἔστι τὰ αἴτια, ἀλλήλοις δὲ αἴτια. Here is another difference from Aristotle (cf. *Phys.* II 3, 195a8–11; Met. Δ 2, 1013b9–11), consequent to the one already noted: causes are causes of something's coming about to an object, not of objects themselves; cf. above, 26, 2–4/96,18–97,2.[449] Thus they are causes 'to' (*dat.*), not 'of' (*gen.*), one another; cf. above, 29, 1/98,25–28.

99,13–16 ἡ γὰρ σπληνικὴ διάθεσις ... οὐ πυρετοῦ αἴτιος κτλ. Here, as above (95,30), διάθεσις is used in the sense attested in medical literature (cf. above, on 95,28–31). For the condition of the spleen that brings about fever, cf. Galen, *Diff. Feb.* II 6 (VII,346,6–8 K.): "Often it also happens that the condition of the liver or the spleen brings about fever (ἡ κατὰ τὸ ἧπαρ ἢ τὸν σπλῆνα διάθεσις ἤνεγκε πυρετὸν), when an inflammation of sorts or a blockage produces the aforementioned affections."

99,16 f. οὕτως καὶ αἱ ἀρεταὶ ... διὰ τὴν ἀντακολουθίαν. Virtues are causes to one another of 'not being separated' (τοῦ μὴ χωρίζεσθαι). The example is based on the Stoic doctrine of the 'inter-entailment' of virtues; cf. SVF III 295–304, esp. DL VII 125; Plutarch, *Stoic. Repug.* 1046e8–10 (LS 61F); Galen, *Opt. Med.* 3, 10 (291,10–13 Boudon-Millot/I,61 K.). On the 'inseparability' of virtues (ἀχωρίστους εἶναι), cf. SVF III 280 (LS 61D); III 305, etc. Cf. also *Strom.* II (9) 45, 1; II (18) 80, 2–3.

99,17 f. οἱ ἐπὶ τῆς ψαλίδος λίθοι ... τοῦ μένειν κατηγορήματος. The example of stones in a vault as inter-related items seems to go back to Stoic sources; cf. Varro, *Ling. Lat.* X 59 (= SVF II 155), referring to Chrysippus.[450] Galen uses a

449 Cf. Ernst, *De Clementis*, p. 39.

450 Varro, *Ling. Lat.* X 59 (trans. Taylor) "Sometimes the one can be identified from the other and sometimes vice versa, as Chrysippus writes, just as a father can often be recognized from the son and a son from the father, and as in arches the two sides mutually support each other, the right no less than the left (*neque minus in fornicibus propter sinistram dextra stat quam propter dextram sinistra*)." The parallel is noted by Ernst, *De Clementis*, p. 40.

similar image ("stones lying next to each other") to depict the idea of a causal chain (στίχον αἰτιῶν).[451] For effects as predicates, cf. above, 26, 4/96,23–97,1.

99,19f. καὶ ὁ διδάσκαλος δὲ καὶ ὁ μανθάνων ... τοῦ προκόπτειν κατηγορήματος. This example could be Clement's own contribution; cf. *Strom.* I (1) 12, 3, quoted above, on 80,24f.

99,21f. ὡς ὁ ἔμπορος καὶ ὁ κάπηλος ἀλλήλοις εἰσὶν αἴτιοι τοῦ κερδαίνειν. An example of a mutually advantageous exchange between a wholesaler (ἔμπορος) and a retailer (κάπηλος); for the distinction, cf. Plato, *Resp.* II, 371d5–7.[452]

99,22f. καθάπερ ἡ μάχαιρα καὶ ἡ σάρξ. A recurrent example, now used for the last time; for other instances, cf. above, on 98,25–28.

99,25 τὸ "ὀφθαλμὸν ἀντὶ ὀφθαλμοῦ" καὶ "ψυχὴν ἀντὶ ψυχῆς". Quotations from Lev 24:20 (cf. Matt 5:38) and 24:18. A gloss of a Christian reader, probably Clement, which seems to be related to the following discussion (99,26–100,3); cf. above, 98,14f., and below, on 99,32–100,2. It is placed in the middle of a continuous exposition (30, 1–5).

99,26–31 ὁ μὲν γὰρ πλήξας τινὰ θανασίμως κτλ. Another example to illustrate the thesis framing the discussion (99,13; below, 100,2f.): People who kill one another are causes of death to one another, but the causes of their respective deaths are not what each of them is a cause of to the other. The situation on which the example is based is presumably that of two wrestlers killing each other at the stadium; cf. above, 98,17f. A similar motif is used, in a different context, by Galen.[453] Interestingly, the author first describes the effect by a

---

451 *Symp. Diff.* 1, 13 (CMG V,5,1: 206,15–208,1/VII,48 K.): τούτων οὖν οὕτως ἐχόντων ἐνδέχεταί τινα στοῖχον αἰτιῶν γενέσθαι πολλάκις ἄλληλα διαδεχομένων, ὡς εἰ καὶ ψηφίδων ἐφεξῆς ἀλλήλαις κειμένων πλειόνων κινήσειέ τις τὴν πρώτην, αὕτη δὲ τὴν δευτέραν κἀκείνη τὴν τρίτην καὶ οὕτω κατὰ τὸ ἑξῆς ἑκάστη τὴν μετ' αὐτήν.

452 For a detailed discussion of these terms, cf. Charles M. Reed, *Maritime Traders in the Ancient Greek World* (Cambridge: Cambridge University Press, 2003), pp. 6–14.

453 Cf. Galen's mocking speech in CP 11, 146–147 (trans. Hankinson): "Gentleman of the jury: you have seen this man fighting in the stadiums many times, raining heavy blows on his opponents, and receiving similar treatment from them in turn; but in spite of this, neither has he died at the hands of one of them, nor have any of them died after a beating at his hands. No, because a beating cannot cause death: for if it could, then he who was beaten more would die more rapidly; but things do not happen like this. This man, in

noun (τοῦ θανάτου), correcting himself afterwards to preserve continuity with the preceding argument. This seems to show that, while sharing the view of effects as predicates, he does not attach much importance to the linguistic form of their expression.

99,32–100,2 ὁ ἀδικήσας … αἴτιος … τοῦ ἀδικηθέντος, ὁ δὲ … νόμος οὐκ ἀδικήματος κτλ. This is a more complex example, as it explains retaliation as something brought about by a different subject (the law) than the one to which the injury has been commited (another person). So it does not really illustrate the case of causes being causes to each other. Moreover, the law can hardly be taken as a cause in the sense demanded by our text, namely as that which does something. The author could be thinking of situations where retaliation ordained by the law is in fact executed by the victim (*lex talionis*); however, if it were significant for the argument, he would have said so. Instead, much attention is paid to the difference among effects: One is an act of injustice; another, brought about in response (on the basis of the law), is an act of just retribution and education. Perhaps the point is to illustrate merely the negative part of the thesis by showing that crime and punishment are not related as causes of one another (ἀλλήλων αἴτια), but rather as actions of interacting but autonomous subjects.

It is more likely, however, that the example serves a different purpose than to illustrate the thesis proposed in 99,13. Introduced by Clement himself, it rather seems to indicate how the discussion about reciprocal causation in his philosophical source can be used in the exegesis of the biblical passages mentioned in 99,25: more particularly, perhaps, in a polemic against those who, like Marcion, take such passages as an occasion to blame the Mosaic law for relishing in the "mutual exercise of injury".[454]

---

spite of meting out and receiving in turn the heaviest of blows, has neither killed anyone else, nor died himself." Here Galen pokes at Erasistratus' apparent denial of the efficacy of procatarctic causes. So the point is different than in our passage, but the underlying example seems to be the same.

454 Cf. Tertullian, *Adv. Marc.* II 18, 1 (trans. Evans): "None of the good things of the law do I find it more natural to defend than those which heresy has sought to break down. One of these is that law of equivalent retribution which demands an eye for an eye, a tooth for a tooth, and a bruise for a bruise (*talionis definitionem, oculum pro oculo, dentera pro dente, et livorem pro livore repetentis*). Its intention is not to give licence for the mutual exercise of injury: rather has it in view the total restraint of violence (*non enim iniuriae mutuo exercendae licentiam sapit, sed in totum cohibendae violentiae prospicit*)." Cf. Sebastian Moll, *Arch-Heretic Marcion* (Tübingen: Mohr Siebeck, 2010), p. 62.

100,2f. οὐκ ἀλλήλων τὰ αἴτια [ὡς αἴτια] κτλ. A summary of the discussion started above, 99,13. ὡς αἴτια is best explained as a scribal gloss (cf. above, on 84,15f.). If Clement or his source had wishes to speak more plainly than in 99,13, they could have added αἴτια after ἀλλήλων.

31, 1–32, 3

100,4–19 This section deals with the question of multiple causation. It is listed as SVF II 348, but in fact only the first half (31, 1–3) seems to reflect Stoic views, the second half (31, 4–5) reflecting the views of our author. Its introduction (ἔτι ζητεῖται κτλ.) suggests that our text registers, and comments upon, other people's views about causes, following a traditional division of questions pertaining to this subject. Cf. above, 26, 1.4.

100,4f. εἰ πολλὰ κατὰ σύνοδον ἑνὸς αἴτια γίνεται πολλά. Stählin adds αἴτια after γίνεται, thus taking the first αἴτια as a part of the subject and complementing the predicate. But this is unnecessary, as the phrase πολλὰ κατὰ σύνοδον can function as a subject (κατὰ σύνοδον having the same attributive role as συνελθόντες in 100,5) and ἑνὸς αἴτια πολλά as a predicate. The question is whether many factors coming together are (a) many causes of one effect or (b) one cause consisting of many factors. If the latter, none of these factors, taken separately, could be described as a cause (see the next note).

100,5–7 οὐ [L: οἱ] γὰρ ἄνθρωποι συνελθόντες αἴτιοί εἰσι τοῦ καθέλκεσθαι τὴν ναῦν, ἀλλὰ σὺν τοῖς ἄλλοις. As preserved in L, this sentence does not make sense in its context. Von Arnim thinks that something must have been lost between ναῦν and ἀλλά, and Stählin agrees.[455] However, we may solve the problem by changing one letter and reading οὐ instead of οἱ in 100,5.[456] If this is correct, the sentence says that "men coming together are *not* the causes of the ship having been drawn to the sea, *except* with the others (ἀλλὰ σὺν τοῖς ἄλλοις)". The point, apparently, is that the men drawing the ship are only causes of this effect along with other men doing the same thing, and, for this reason, none of them can be separately described as a cause. This solution closely agrees with a parallel in *Strom.* I (20) 97, 1, noted by Stählin: ὡς δὲ οἱ πολλοὶ ἄνθρωποι οἱ καθέλκοντες τὴν ναῦν οὐ πολλὰ αἴτια λέγοιντ' ἄν, ἀλλ' ἐκ πολλῶν αἴτιον ἕν (*οὐκ ἔστι γὰρ αἴτιος* ἕκαστος τοῦ καθέλκεσθαι τὴν ναῦν, *ἀλλὰ σὺν τοῖς ἄλλοις*), οὕτω καὶ ἡ φιλοσοφία (…) *οὐκ αἰτία*

---

455 Von Arnim (*ad* SVF II 348): ⟨οὐκ αὐτὸς ἕκαστος αἴτιος ὤν⟩; Stählin: ⟨ἀλλ' οὐχ ἕκαστος κατ' ἰδίαν αἴτιος,⟩.

456 I proposed this in my "Textual Notes on *Stromata* VIII," *Mnemosyne* 66/4–5 (2013), pp. 761–768, here 766.

οὖσα καταλήψεως, σὺν δὲ τοῖς ἄλλοις αἰτία καὶ συνεργός. On the relation between the two passages, cf. above, pp. 59–62.

100,7 εἰ μή τι καὶ τὸ συναίτιον αἴτιον. For the meaning of συναίτια ('joint-causes'), cf. below, 33, 8/102,4–6. For the force of εἰ μή τι, cf. *Strom.* I (20) 97, 1: τάχα δὲ καὶ τὸ συναίτιον αἴτιον.

100,7f. ἄλλοι δέ φασιν· εἰ πολλὰ αἴτια κτλ. Here speaks another party, possibly sharing the same view about the case in question, but adducing examples of multiple causal factors that do qualify individually as causes. The protasis εἰ πολλὰ αἴτια seems to echo the question raised in 100,4f. (εἰ πολλά ... ἑνὸς αἴτια γίνεται πολλά), laying down a general condition for an affirmative answer: "If many [factors are] causes [of one effect], then each [of them] is separately a cause of one effect."[457] This implies the same notion of a cause as the reasoning of the previous party, namely the notion of a cause as that which produces its effect in virtue of itself (cf. below, 33, 2/101,19–21).

100,8–13 There is a close parallel to this section in *Strom.* I (20) 97, 2–3; see the following notes.

100,8–10 τοῦ γοῦν εὐδαιμονεῖν ... καὶ τοῦ θερμαίνεσθαι καὶ τοῦ ἀλγεῖν ὁμοίως πολλὰ τὰ αἴτια. Here the second party provides some examples of multiple causal factors that qualify as causes: virtues (with respect to happiness) and things that bring about warmth and pain. In *Strom.* I (20) 97, 2–3, Clement is more specific, listing four cardinal virtues and such causes of warmth as the sun, fire, bath, and clothing (he does not mention pain). For the idea of all virtues having the same end, cf. e.g. Panaetius, fr. 109 van Straaten (LS 63G), SVF III 280 (LS 61D).

100,10–13 αἱ πολλαὶ ἀρεταὶ μία ἐστὶ κατὰ δύναμιν κτλ. The idea of the essential unity of virtues goes back (at least) to the Socratic philosopher Menedemus of Eretria. Among the Stoics, it is mainly associated with Aristo; cf. Plutarch, *Virt. Mor.* 2, 440e–441a (LS 61B); Galen, *PHP* VII 2, 1–3 (CMG V,4,1,2: 434,31–436,7/V,595f. K.). Cf. also Seneca, *Ep.* 113, 7–8 and the commentary of Inwood, *Seneca*, pp. 279f. Our text paraphrases this view by saying that many virtues are one *potentially*, thus implying that the plurality of virtues arises in the course of the actualization of the same potentiality in specific contexts; cf. above, on 96,6.

---

457 Thus the full form of the protasis would be: εἰ πολλὰ ἑνὸς αἴτια γίνεται κτλ.

Again, there is a close parallel in Clement, *Strom.* I (20) 97, 3: εἰ γοῦν σκοποῖμεν, *μία κατὰ δύναμίν ἐστιν ἡ ἀρετή*, ταύτην δὲ συμβέβηκεν τούτοις μὲν τοῖς πράγμασιν ἐγγενομένην λέγεσθαι φρόνησιν, ἐν τούτοις δὲ σωφροσύνην, ἐν τούτοις δὲ ἀνδρείαν ἢ δικαιοσύνην.[458]

100,11 καὶ τὰ θερμαίνοντα καὶ τὰ ἀλγοῦντα. These words are not satisfactorily integrated into the syntax of the sentence. Following Heyse, we might assume that a connecting element was omitted in the course of the textual transmission (e.g. ὁμοίως, which also appears on the previous line), but it is more likely that the words entered the text as a marginal gloss.[459]

100,12 f. καὶ τὸ πλῆθος δὲ τῶν ἀρετῶν κατὰ γένος ἓν τυγχάνον, ἑνὸς αἴτιον γίνεται κτλ. The view that all virtues are potentially one is now rephrased so that they are one *with respect to genus*. In *Strom.* I (20), 98, 1, Clement expresses the same idea by saying that "every virtue … is called [virtue] synonymously (συνωνύμως καλουμένη)". This probably means that every virtue is called 'virtue' on account of the same definition (or essence) as all the others.[460] Interestingly, in the same passage, Clement also makes a distinction between the effects of particular virtues (ἑκάστη … τοῦ καθ' ἑαυτὴν ἀποτελέσματός ἐστιν αἰτία) and the effect of virtues "when they are applied together" (κατὰ σύγχρησιν τούτων), describing the latter as 'living happily' (τὸ εὐδαιμόνως ζῆν). Thus, according to this account, virtues (*qua* virtues) are multiple causes of one effect, while, at the same time, each of them being a cause of its own effect (*qua* wisdom, temperance, etc.).

100,13–19 τῷ ὄντι δὲ κτλ. Unlike the second party, which constructs the unity of multiple causes as the unity of their genus (τὸ πλῆθος … κατὰ γένος ἕν), our author now proposes a different view, according to which this unity should be constructed with regard to the genus of their effects (αἴτια ἑνὸς … κατὰ γένος). This enables him to distinguish further between more and less general degrees of unity, one corresponding to the genus, and one to the species of the effects.

---

458 Cf. above, pp. 59–62. Von Arnim lists *Strom.* I (20) 97, 3 and 98, 1–2 as Aristo, fr. 376 (SVF I, p. 86). But it is unlikely that Clement draws directly from Aristo; rather, in the section I (20) 97, 1–98, 2, he uses a philosophical source dealing with the issue of multiple causation which incorporates, among other things, the Aristonean view of virtues.

459 As noted by von Arnim, ἀλγοῦντα is probably a corruption of ἀλγύνοντα.

460 On synonymy in this sense, cf. Aristotle, *Cat.* 1, 1a6–12. Cf. Alexander, *In Top.* VI 13, 150b19 (CAG II/2: 490,6–8): οἷον ἀρετή ἐστιν ἀνδρία, δικαιοσύνη καὶ φρόνησις· ἕκαστον γὰρ τούτων ἀρετὴ λέγεται συνωνύμως τῷ ὅλῳ.

Such a distinction, in turn, helps him to draw another difference between the procatarctic and synectic causes (cf. above, 28, 7/98,19–24): In the case of the procatarctic causes, there may be many causes of one effect according to genus (e.g. disease), but also according to species (e.g. fever). The synectic causes, on the other hand, may be multiple causes of one effect according to genus (e.g. scent), but not according to species, because each of these causes (e.g. different kinds of fragrance) produces its specific effect. For details, see below.

100,14–16 κατὰ γένος μὲν τοῦ νοσεῖν ὁπωσοῦν, οἷον ψῦξις, ἔγκαυσις, κόπος, ἀπεψία, μέθη, κατ' εἶδος δὲ τοῦ πυρετοῦ. Cold, heat, fatigue, etc. are stock examples of procatarctic causes of disease used in the medical literature.[461] Cf. esp. Galen, *Sect. Int.* 7 (16,16 f. Helmreich/I,84 f. K.): ψύξεις τε καὶ ἐγκαύσεις καὶ μέθας καὶ ἀπεψίας καὶ … κόπους κτλ. On the procatarctic causes of fever, cf. e.g. Galen, *Diff. Feb.* I 3/VII,279–282 K. (and the rest of book I); *MM* X 2 (X,666,2–11 K.).

100,17–19 τοῦ γὰρ εὐωδιάζεσθαι κατὰ γένος ἑνὸς ὄντος πολλὰ τὰ αἴτια κατ' εἶδος, οἷον λιβανωτός … μύρον. Here we have other examples of synectic causes: various kinds of fragrances (cf. above, on 98,19–22). Their effect is one according to genus, but not according to species, because each fragrance is of a different kind. It is slightly disturbing that, unlike in the previous sentence, the qualification κατ' εἶδος is attached to the causes rather than the effects; for this reason, von Wedel suggests to delete it.[462] But the point is explained in 100,19: "For rose certainly does not scent in the same way as myrrh."

For the examples of fragrances, cf. Galen, *Comp. Med. Loc.* IV 5 (XII,718,17–719,12 K.), where rose, frankincense, saffron, and myrrh (ῥόδα, λιβανωτός, κρόκος, σμύρνα) are mentioned as cures for ophthalmic diseases, which, despite their different capacities, nevertheless belong to the same genus (ἐκ τοῦ αὐτοῦ γένους). Cf. also the examples of drugs ibid. II 1 (XII,561,10–15 K.): σμύρνα, λίβανος … κρόκος … στύραξ.

100,20–101,3 Another consequence of the relativity of causes is that one and the same factor can produce opposite results, depending either on the strength of the agent or on the suitability of the patient. This is the end of the thematic section starting above, 29, 1–2/98,25–30. The passage is listed as SVF II 350, but the notion of suitability is not attested in Stoic sources (cf. above, on 98,9 f.).

---

461 For (mainly Galenic) references, cf. Hankinson, "Galen's Theory of Causation," p. 1761 and n. 23.

462 *Symbola*, p. 35. Stählin agrees.

100,20–22 παρὰ τὸ μέγεθος τοῦ αἰτίου ... παρὰ τὴν ἐπιτηδειότητα τοῦ πάσχοντος. In *Caus. Morb.* 2 (VII,8,4–9,14 K.), in the context of a discussion about the causes of fever, Galen mentions three factors, without which, even causes as efficient as a fire or a knife, cannot produce an effect: time, strength (ῥώμη, μέγεθος), and a suitable condition (τὸ ἐπιτηδείως ἔχον) of the patient. Cf. also *CP* 8, 101 (CMG Suppl. II: 24,24–34). Of these three factors, two (strength and suitability) are mentioned in our passage; whereas the third (time) is mentioned above, 28, 7/98,19–24.

100,24–101,3 τὸ μέλι ... ὁ οἶνος ... ὁ ἥλιος κτλ. For the example of honey, cf. Sextus, *PH* I 213 and II 63, referring to Democritus; for the effects of wine, cf. Philo, *Som.* II 164–168; for the effects of heat on wax as opposed to mud, cf. Sextus, *M.* VIII 192.194; IX 247.250. Cf. already Plato, *Tht.* 159c–160b.

32, 4–6

101,4–12 This section divides causes as objects of knowledge into two groups, the 'clear' (1) und the 'unclear' (2). Among the 'unclear' variety, two groups are distinguished, namely those unclear 'for the moment' (2.1) and those unclear 'by nature' (2.2); the latter group is further analysed into graspable causes (2.2.1) and those that are not graspable (2.2.2). The two major groups of this division correspond to two ways of reasoning: epilogism and analogism, respectively. There is a close parallel to this division in Sextus, *PH* II 97–99 and *M.* VIII 144–148; there, however, it does not pertain to causes but to 'things' (πράγματα) in general. Sextus divides things into 'clear' (πρόδηλα) and 'unclear' (ἄδηλα), further distinguishing the unclear variety into three groups, those unclear 'for the moment' (= 2.1 above), those unclear 'once and for all' (= 2.2.2 above), and those unclear 'by nature' (= 2.2.1 above).[463] Another variant, pertaining to causes, is found in Ps.Galenic *Medical Definitions* 161–163 (XIX,394 K.); here, three groups of causes are distinguished: the 'clear' (= 1 above), the 'not clear' (οὐ πρόδηλα, presumably corresponding to 2.2.1 above), and the 'unclear' (= 2.2.2 above). A fourth group consists of causes unclear 'for the moment' (= 2.1 above).[464] Neither Sextus nor Ps.Galen mentions the distinction between epilogism and analogism in this connection. For details, see the following notes.

---

463 For the two accounts in Sextus, their philosophical (probably medical) background, and their role within the context of Sextus' polemic, see esp. Allen, *Inference*, pp. 87–146.

464 As printed by Kühn, the description of the fourth group (*Def. Med.* 163) seems to be marred by textual problems. Students of this difficult text will have to wait for the edition being prepared by Jutta Kollesch for CMG.

101,4f. τὰ ἐπιλογισμῷ λαμβανόμενα ... τὰ ἀναλογισμῷ. According to Galen, the distinction between epilogism and analogism had been used by the Empiricist doctors in the context of a controversy concerning "the discovery of things which are not manifest" (ἐν ταῖς περὶ τῆς εὑρέσεως τῶν ἀφανῶν ζητήσεσι).[465] They described epilogism as a "reasoning in terms of what is apparent", claiming that it was useful "in the discovery of things which are unclear for the moment (τῶν προσκαίρων ἀδήλων)".[466] With this kind of reasoning, acceptable from the Empiricist point of view, they contrasted analogism, which "starts from what is apparent but then proceeds to matters that are entirely unclear (ἐπὶ τὰ διὰ παντὸς ἄδηλα)".[467] Galen occasionally adopts the distinction himself; cf. e.g. *Hipp. Prog.* I 6 (XVIIIb,26 K./CMG V,9,2: 211,10–17).[468]

101,5–7 τὰ μὲν πρὸς καιρὸν ἄδηλα κτλ. Sextus' example of items 'unclear for the moment' is "the city of Athens for us now" (*PH* II 98; *M.* VIII 145). This can hardly apply to causes. However, when discussing the distinction between the indicative and the recollective signs, Sextus says that the recollective sign "is thought to be useful mainly in the case of things that are unclear for the moment", citing examples that include effects as signs of their causes: smoke with regard to fire, a scar with regard to a wound (*M.* VIII 156–157).[469]

101,7 τὰ δὲ φύσει ἄδηλα. Unlike the 'dogmatists', the Empiricists claimed that these items (2.2 above, on 101,4–12) could not be known at all. Cf. Galen, *Sect. Int.* 5 (10,24f. Helmreich/I,77 K.), on the Empiricists: "... nor is there a sign of anything unclear by nature" (μηδ' εἶναί τι σημεῖον ἀδήλου φύσει πράγματος οὐδενός). Cf. Sextus, *PH* II 102; *M.* VIII 156–158.

---

465 *Sect. Int.* 5 (10,16f. and 11,8–20 Helmreich/I,77f. K.); trans. Michael Frede in Galen, *Three Treatises On the Nature of Science* (Indianapolis: Hackett, 1984), p. 12.

466 *Sect. Int.* 5 (11,8–10 Helmreich/I,78 K.): ὁ δ' ἐπιλογισμός, ὃν δὴ ⟨τῶν⟩ φαινομένων λόγον εἶναί φασι, χρήσιμος ... εἰς εὕρεσιν τῶν προσκαίρων ἀδήλων.

467 *Sect. Int.* 5 (11,16–18 Helmreich/I,78 K.).

468 For the history of the concept of *epilogismos* (going back to the Epicureans), cf. Malcolm Schofield, "*Epilogismos*: An Appraisal," in M. Frede and G. Striker (eds.), *Rationality in Greek Thought* (Oxford: Oxford University Press, 1996), pp. 221–237.

469 How does the example of a (past) wound inferred from a (present) scar tally with the description of πρὸς καιρὸν ἄδηλα? An inference that if there is a scar there must have been a wound can hardly be tested by an observation of that particular wound. Nevertheless, it can be tested by other observations showing that scars are preceded by wounds.

101,9 ἅπερ οὐκ ἄδηλά τινες ἐκάλουν. People referred to here presumably took the word ἄδηλος in the sense of 'inapprehensible' (cf. 101,10–12), thus describing the 'apprehensible' (but not 'clear') items as 'not unclear'. In *Def. Med.* 162, the name οὐ πρόδηλα seems to have the same function. Sextus says that his 'dogmatists' call these (not clear but apprehensible) items (2.2.1 above, on 101,4–12) φύσει ἄδηλα (*PH* II 98; *M.* VIII 145–146).

101,9 f. διὰ σημείων ἀναλογιστικῶς λαμβανόμενα. Both the items unclear 'for the moment' and those unclear 'by nature' (but apprehensible) are graspable by signs. Cf. Sextus, *PH* II 99; *M.* VIII 150; *Def. Med.* 163. The crucial difference is that, whereas the former can also be grasped by direct observation, the latter are grasped by signs *only*. Herein lies the distinction between epilogistic and analogistic inference (cf. above, on 101,4 f.). Sextus describes the signs by which the items 'unclear by nature' are grasped as 'indicative' (ἐνδεικτικά, *scil.* σημεῖα); cf. *PH* II 99.101; *M.* VIII 151.154–155. Cf. above, on 82,27 f.

101,10 καθάπερ ἡ συμμετρία τῶν λόγῳ θεωρητῶν πόρων. 'Intelligible pores' (οἱ νοητοὶ πόροι) are mentioned, in the same connection, by Sextus, who points out that "these are never apparent of themselves but would be deemed to be apprehended, if at all, by way of something else, e.g. by sweating or something similar" (*PH* II 98, trans. Annas and Barnes); cf. ibid. 140.142; *M.* VIII 146; DL IX 89. In *Def. Med.* 18 (XIX,353 f. K.), pores are an example of an item whose existence is inferred by ἀναλογισμός. However, in our passage, the issue at hand is not the existence of pores, but their symmetry. The concept of the symmetry of pores appears in reports on Empedoclean physics and the atomist theories of perception; cf. Plato, *Men.* 76c7–d5; Theophrastus, *Sens.* 12 (*Dox. Gr.* 502,28 and 503,3 f.); Plutarch, *Quaest. Conv.* 649d3–4; Ps.Plutarch, *Plac.* 899f9–10; DL X 107, etc. Later it is associated with the medical diagnostics of Asclepiades of Bithynia (2nd–1st cent. B.C.); cf. Galen, *PHP* V 3, 18 (CMG V 4,1,2: 308,27–29/V,449 K.); *MM* IV 4 (X,268,10 f. K.: ἐν συμμετρίᾳ μέν τινι πόρων τὸ ὑγιαίνειν ... ἐν ἀμετρίᾳ δὲ τὸ νοσεῖν), etc. For the expression 'pores discernible by reason', cf. also *Anon. Lond.* 39,31 (Manetti 94); Galen, *MM* II 4 (X,101,18 K.: λόγῳ θεωρητὰ ἀραιώματα); XIII 2 (X,876,11 K.); XIII 16 (X,919,16 K.: λόγῳ θεωρητοὶ πόροι); Sextus, *M.* III 5; Caelius Aurelianus, *Cel. Pass.* I 14, 106 (CML VI,1: 82,3 f.).[470]

470 For the last two references, cf. David Leith, "Pores and Void in Asclepiades' Physical Theory," *Phronesis* 57/2 (2012), pp. 164–191, here 179 f.

101,10–12 τὰ δὲ ἀκατάληπτα … ἃ δὴ καὶ ἄδηλα ἐν τῷ καθάπαξ λέγεται. Sextus' examples of τὰ καθάπαξ ἄδηλα include such questions as "whether there is an even or an odd number of stars" or statements "that there is such-and-such number of grains of sand in Libya" (*M.* VIII 147). These and similar questions belong among things about which even a wise man, according to the Stoics, withholds assent; cf. Alexander, *In Top.* I 11, 104b1 (CAG II/2: 76,4–6); Sextus, *M.* VII 243; XI 59; *PH* II 90–91.97.231; III 177; Clement, *Strom.* V (1) 5, 3, etc. None of these sources, however, gives an example of an 'inapprehensible' cause.[471]

32, 7–33, 9

101,13 f. καὶ τὰ μὲν προκαταρκτικά, τὰ δὲ συνεκτικά, τὰ δὲ συναίτια, τὰ δὲ συνεργά. Having divided causes as objects of knowledge, the text now divides them in view of their contribution to the causal process (both views are combined below, 33, 5/101,25 f.). The classification is different from the one drafted above, 25, 1/95,27 f., as it adds 'joint-causes' (first introduced in 31, 1/100,4–7) and omits the 'prerequisites' (cf. above, on 96,5 and 98,5–9). This sentence and its continuation in 33, 1–9 (101,17–102,12) is listed as SVF II 351 (= LS 55I, leaving out the first sentence). Close parallels to the classification are found in Sextus (*PH* III 15–16) and in medical sources, esp. in the Pseudo-Dioscorides and Pseudo-Galenic writings *History of Philosophy* and *Medical Definitions*. For details, see below, on 101,17–19 and the following.

101,14–16 τὰ μὲν τοῦ κατὰ φύσιν, τὰ δὲ τοῦ παρὰ φύσιν, τὰ δὲ νόσου, καὶ κατὰ τὸ ὑποβεβηκὸς τὰ μὲν παθῶν, τὰ δὲ τοῦ μεγέθους αὐτῶν, τὰ δὲ χρόνων καὶ καιρῶν. This excerpt or gloss introduces another division of causes, this time from the perspective of their effects. Six groups of causes are distinguished according to the following effects: (1) things according to nature, (2) things contrary to nature, (3) disease, (4) affections, (5) their magnitude, and (6) temporal intervals of some sort. Some of these groups are described as being 'lower' (κατὰ τὸ ὑποβεβηκός) than others, probably in the sense of being more specific. There are two ways of constructing this relation: (a) either it is a relation between the last three groups (4–6) and the first three (1–3), or (b) it is a relation of the last three groups to the third one only. Option (a) could still be

471 Of course, according to the sceptics (and the 'empiricist' doctors), all causes are 'inapprehensible'; cf. Sextus, *PH* III 17–29.38 (τὸ … αἴτιον ἀκατάληπτόν ἐστι); Ps.Dioscorides, *De iis quae virus ejaculantur*, *Praef.* (MGO XXVI/2 Sprengel: 49 f.). For the latter passage, referring to Herophilus' rejection of the 'empiricist' view of causes as 'inapprehensible' (ἀκατάληπτα), cf. Heinrich von Staden, *Herophilus: The Art of Medicine of Early Alexandria* (Cambridge: Cambridge University Press, 1989), p. 123.

taken in more than one way, but option (b) has the merit of providing a more coherent interpretation of the passage.

We shall start with the last bit of the puzzle, which appears to be the most perplexing of all. At first glance, the expression χρόνοι καὶ καιροί seems to refer to 'times and seasons' of the year. But it is hard to see why the causes of cosmic intervals (which presumably must be the heavenly bodies)[472] should be treated on the same level as the causes of affections and their magnitudes; or, why they are subordinate to one or more of the preceding groups. On the other hand, a plausible connection between affections, their magnitudes, and temporal intervals emerges as we think of them in relation to disease. In the medical literature, the word καιρός is used, among other things, to designate developmental stages of disease, differing from other stages by the strength (τὸ μέγεθος) of a particular affection (e.g. heat in the case of fever).[473] Each of these stages, of course, lasts for a certain amount of time, and thus it makes sense to describe them (by way of a hendiadys) as χρόνοι καὶ καιροί of disease.[474] If this is what happens in our passage, we may explain the last three items in the classification (4–6 above) as referring to three sorts of causes of disease, namely those bringing about a particular affection of the body (4), those responsible for the strengthening of the affection (5), and those responsible for the duration of its stages (6).

How shall we construe the rest of the classification? Here, the most difficult question is the extension of the pair natural/unnatural and the relation of the third effect ('disease') to it. Without any context, the first part of the question cannot be settled. Possibly, the pair extends to 'things' in general; if so, then 'disease' is likely introduced as a particular case of unnatural effects.[475]

---

472 Cf. e.g. Porphyry, *Agalm.* 8,42–44: χρόνων ἐστὶ ποιητικὸς καὶ καιρῶν ὁ ἥλιος.

473 Cf. especially Galen's treatises *De morborum temporibus* ('Stages of Diseases') and *De totius morbi temporibus* ('Stages of a Disease as a Whole'), *passim*. For the meaning of καιρός, cf. e.g. *Morb. Temp.* 1 (VII,406,3–9 K.), where Galen compares καιρός in this sense to a stage of life of an animal (ἡλικία). For the decisive role of μέγεθος in distinguishing different stages of the same disease cf. ibid. 1 (VII,408,3–6 K.). For the example of fever, cf. ibid. 1 (VII,409,2–8 K.).

474 For the duration (χρόνος) of the stages, cf. *Morb. Temp.* 2 (VII,411,13–18 K.). However, sometimes the words χρόνος and χρόνοι (counted as πρῶτος, δεύτερος, etc.) are used in a sense that is hardly distinguishable from that of καιρός/καιροί. Cf. ibid. 3 (VII,414,15–415,16 K.).

475 Cf. Galen, *Morb. Diff.* 2 (VI,838,5–9 K.): "In this case too there is an agreed starting-point, namely that things according to nature are proportionate (σύμμετρον μὲν εἶναι τὸ κατὰ φύσιν) not only in animals, but also in plants and seeds and every organ, whereas things contrary to nature are disproportionate (τὸ δὲ αὖ παρὰ φύσιν ἄμετρον). Health, then, appears to be some sort of proportion, disease disproportion."

101,17–19 τῶν μὲν ... προκαταρκτικῶν ... αἱρομένων μένει τὸ ἀποτέλεσμα, συνεκτικὸν δέ ... οὗ παρόντος μένει τὸ ἀποτέλεσμα καὶ αἱρομένου αἴρεται. Here, again, the contrast is drawn between the procatarctic and synectic causes; cf. above, 28, 7/98,19–24; 31, 4–5/100,13–19. The contrast is this: the effect of the procatarctic cause lasts even when the cause is removed, whereas the effect of the synectic cause lasts as long as the cause is present. This is a traditional description, attested by Sextus[476] and by medical sources. Cf. Ps.Galen, *Def. Med.* 155 (XIX,392 K.): "Procatarctic is that which brings about an effect and then separates itself (ὃ ποιῆσαν τὸ ἀποτέλεσμα κεχώρισται), as a biting dog or a stinging scorpion or a heat-stroke from the sun producing fever."[477] Ibid. 157 (XIX,393 K.): "Synectic cause is that which is such that when present it preserves a present disease and when removed it removes it (ὃ παρὸν μὲν παροῦσαν φυλάττει τὴν νόσον, ἀναιρούμενον δὲ ἀναιρεῖ), e.g. a stone in the bladder, hydatid disease, pterygium, swelling of the lacrimal caruncle, etc."[478] Similar definitions are found in Pseudo-Dioscorides' treatise *On Venomous Animals*, where they are said to be used by the Methodist doctors.[479] Cf. also Ps.Galen,

476 *PH* III 15 (trans. Annas and Barnes, slightly modified): "[Causes are] synectic if when they are present the effect is present and when they are removed it is removed and when they are diminished it is diminished (they say that in this way the placing of a noose is a cause of strangulation) [συνεκτικὰ μὲν ὑπάρχειν ὧν παρόντων πάρεστι τὸ ἀποτέλεσμα καὶ αἱρομένων αἴρεται καὶ μειουμένων μειοῦται (οὕτω γὰρ τὴν περίθεσιν τῆς στραγγάλης αἴτιον εἶναί φασι τοῦ πνιγμοῦ)]." Sextus describes this division as held by the majority of 'dogmatists'; cf. *PH* III 15: τούτων δὲ τῶν αἰτίων οἱ μὲν πλείους ("the majority of them", i.e. the dogmatists) ἡγοῦνται τὰ μὲν συνεκτικὰ εἶναι, τὰ δὲ συναίτια, τὰ δὲ συνεργά, κτλ.

477 For the last example, cf. also Sextus, *PH* III 16. Cf. also above, on 100,14–16 (with references).

478 Cf. also *Def. Med.* 156 (XIX,393 K.): "[X] is such that when present the effect is present, when increased the effect is increased, when diminished the effect is diminished and when removed the effect is removed (οὗ παρόντος πάρεστι τὸ ἀποτέλεσμα καὶ αὐξομένου αὔξεται καὶ μειουμένου μειοῦται καὶ αἱρουμένου αἱρεῖται)." Surprisingly, X is not the synectic cause here, but the preceding cause (τὸ προηγούμενον αἴτιον); cf. Kollesch, *Untersuchungen*, p. 123.

479 Ps.Dioscorides, *De iis quae virus ejaculantur*, praef. (MGO XXVI/2 Sprengel: 51 f.): "At this point they [i.e. the Methodists] turn to the classification of causes and say that of causes some are procatarctic, viz. those that bring about an affection and then separate themselves, e.g. fatigue, cold, heat-stroke, and so on, and some are synectic, because they bring about an affection and remain, i.e. when they are present the effect is present, when they increase the effect increases, when they are diminished the effect diminishes, when they cease, the effect ceases" (μεταβαίνουσι δὲ ἐνταῦθα ἐπὶ τὰς τῶν αἰτίων διαφοράς, καὶ λέγουσιν, ὅτι τῶν αἰτίων τινὰ μέν ἐστι τὰ προκατάρξαντα, ἃ ποιήσαντα πάθος χωρίζεται· οἷον κόπος καὶ ψύξις, ἔγκαυσις, καὶ τὰ παραπλήσια· τὰ δὲ συνεκτικά, διὰ τὸ ποιεῖν τὸ πάθος καὶ παραμένειν, τουτέστιν, ὧν μὲν παρόντων τὰ ἀποτελέσματα πάρεστιν, αὐξανομένων δὲ αὔξεται, καὶ μειουμένων μειοῦται, καὶ παυσαμένων παύεται).

Hist. Phil. 19.[480] For the synectic cause, cf. also Galen, *Adv. Jul.* 8, 22, quoted above, p. 270 and n. 387.

101,19 f. τὸ δὲ συνεκτικὸν συνωνύμως καὶ αὐτοτελὲς καλοῦσιν κτλ. Cf. above, 25, 3/95,31–96,2. Apart from these passages, I am not aware of any other source that would claim the identity of the synectic and the self-complete causes. The closest parallel is found in Ps.Dioscorides who says that the synectic causes "can also be self-complete".[481] Ps.Galen, *Def. Med.* 158, also includes the self-complete causes in his classifications, but does not identify them with the synectic.[482] The identification is also in conflict with other passages in our chapter; cf. esp. above, 28, 3–6 and notes on 98,5–18. But then our author does not say that he agrees with this identification. Rather, in our passage, he seems to report on other people's views, possibly because they are influential among his readers and/or because they provide a starting-point to his arguments.

101,21–102,12 Having explained the difference between the procatarctic and synectic causes, the author now turns to the next two groups mentioned above (32, 7/101,13 f.), namely the auxiliary causes and joint-causes. The aim of this (last) section is to show how these two groups of causes variously relate to the activity of the self-complete cause (described as 'synectic' or simply as 'the cause').

101,21 εἰ δὲ τὸ αἴτιον αὐτοτελοῦς ἐνεργείας ἐστὶ δηλωτικόν. This protasis proposes a view, according to which the word 'cause' indicates a self-complete activity.[483] This view is at the background to the following discussion; cf. esp. 33, 7–

---

480 *Dox. Gr.* 611,8–11: "Four causes have been postulated: the procatarctic cause, which brings about an effect and remains, [the synectic cause,] which is such that when present the effect is present and when it ceases the effect departs and when it increases the effect continues to grow and when it diminishes the effect diminishes" (αἴτια δὲ ὑπέθεντο τέσσαρα· προκαταρκτικόν, ὃ πεποιηκὸς παραμεμένηκεν, καὶ ⟨συνεκτικόν add. Diels⟩, οὗ παρόντος τὸ ἀποτέλεσμα πάρεστι καὶ πεπαυμένου παρενήλλακται καὶ αὐξομένου εἰς ἐπίδοσιν πρόεισι καὶ μειουμένου ἐλαττοῦται).

481 *De iis quae virus ejaculantur, Praef.* (MGO XXVI/2 Sprengel: 52): καὶ αὐτοτελῆ δύναται εἶναι [*scil.* τὰ συνεκτικά] καθ' ἑαυτὰ παραγινόμενα τῶν ἀποτελεσμάτων. I do not understand the meaning of παραγινόμενα with the genitive and suspect a textual corruption here. But the sentence surely says something to the effect that synectic causes, insofar as they produce their effects by themselves (καθ' ἑαυτά), can also be self-complete.

482 XIX,393,12 f. K.: αὐτοτελὲς αἴτιόν ἐστι τὸ αὐτὸ καθ' αὑτὸ ποιοῦν τέλος. As noted above, on 95,31–96,2, Stoic sources know of a self-sufficient cause, too, but never describe it as synectic.

483 For the phrase ἐνεργείας ἐστὶ δηλωτικόν, cf. above, 97,9 f.

9, where the synectic cause is described as τὸ κατ' ἰδίαν ποιοῦν (102,9 and, by implication, 102,4–6) and identified simply as τὸ αἴτιον (102,11 f.).[484] It is true that, in what follows, both the auxiliary factors and joint-causes are also called 'causes' (101,24.27; 102,1.6); however, in both cases the author qualifies (or defends) this description against the background of the aforementioned view. Cf. above, 28, 3/98,7 f., on the suitability of the patient: κατὰ ἀνάγκην ἐστὶν αἴτιον, αἴτιον δὲ οὐχ ἁπλῶς· οὐ γάρ ἐστι συνεκτικὸν κτλ.[485]

101,21 f. τὸ συνεργὸν ὑπηρεσίαν σημαίνει. Unlike above, 31, 5/98,12 f., the help provided by the auxiliary factor is not necessary for the outcome. See below, on 102,1 f.

101,23 f. τούτου πάντως γίνεται αἴτιον οὗ καὶ παρέχεται. Cf. above, 28, 3/98,7 f., where κατὰ ἀνάγκην has the same force as πάντως in this passage. Here, as there, this emphasis seems to be directed against the opinion that auxiliary factors do not count as causes at all.

101,25 οὗ παρόντος ἐγίνετο τὸ ἀποτέλεσμα. The auxiliary factor is a cause, because it has contributed to the effect under consideration by its presence. Cf. above, on 97,23–98,2.

101,26 προδήλῳ μὲν οὖν παρόντος ⟨προδήλου⟩, ἀδήλῳ δὲ ἀδήλου. I follow Stählin in reading προδήλῳ ... ἀδήλῳ, as it stands in the manuscript.[486] Long and Sedley (55I3) accept Potter's conjecture προδήλως ... ἀδήλως and translate the whole sentence as follows: "So an auxiliary cause is one which was present while its effect was coming about. When it is pre-evidently present, its effect is pre-evident, whereas when it is non-evidently present, its effect is non-evident." But the manuscript reading is probably correct. We have seen above (33, 3/101,21 f.) that the word 'auxiliary' "signifies assistance and service alongside something else (σὺν ἑτέρῳ λειτουργίαν)". Then we are told that the auxiliary factor con-

---

484 The last reference shows, in my view, that it is unnecessary to add αὐτοτελὲς before αἴτιον in 101,21, as Stählin suggests.

485 The similarity between the two passages consists in the following: both in 98,7 f. and in 101,21 (together with 102,11 f.) causes described as synectic are also regarded as causes without qualification. But there is a difference between the two sections: in the first, the synectic cause is not self-complete, as it depends on the suitability of the patient (cf. esp. 28, 4/98,9 f.); in the second, the synectic cause is described as self-complete and the contribution of the patient is not taken into account at all.

486 Apart from that, of course, I accept Potter's addition of προδήλου.

tributes to an effect by its *presence* (οὗ παρόντος κτλ. in 101,25), the item to which it is present presumably being τὸ ἕτερον, to which it provides assistance with regard to that effect. This is explicated in our passage, where the datives refer to this item, i.e. to the self-complete cause. Next our author applies the distinction between clear and unclear causes (cf. above, 32, 4–6/101,4–12) to point out that the epistemological status of the auxiliary cause is determined by the status of the main causal factor.

101,26–102,1 καὶ τὸ συναίτιον κτλ. The concept of joint-causes is explained below, 33, 8/102,4–6. As we have already seen, their causal status is controversial (31, 1/100,4–7). Here our author seems to present his own view of the matter, namely that joint-causes do belong among causes, just as a fellow soldier belongs among soldiers, etc. This is a common-sense argument, but it is slightly misleading. Later the view will be further qualified in the sense that a joint-cause is a cause only when joined with another cause (αἴτιον ὂν σὺν αἰτίῳ in 102,6), a qualification which, apparently, does not apply for soldiers.

102,1 f. τὸ μὲν οὖν συνεργὸν αἴτιον τῷ συνεκτικῷ πρὸς τὴν ἐπίτασιν βοηθεῖ κτλ. We have seen that the auxiliary cause provides assistance to the self-complete/synectic one (cf. above, 33, 3/101,21 f.). Here it is further specified that it helps the synectic cause "so as to intensify what comes about through the latter"; this characteristic is confirmed below, 102,8–10. For a different description of (what appears to be) the same notion, cf. Sextus, *PH* III 15 (trans. Annas and Barnes): "[Causes] are auxiliary if they supply a force which is slight and which contributes to the easy existence of the effect (e.g. when two men are lifting a weight with difficulty and a third one comes along and helps to lighten it)."[487] Cf. also Ps.Galen, *Hist. Phil.* 19;[488] *Med. Def.* 160.[489] The notion is such that it

---

487 *PH* III 15: συνεργὸν δὲ ὃ βραχεῖαν εἰσφέρεται δύναμιν καὶ πρὸς τὸ μετὰ ῥᾳστώνης ὑπάρχειν τὸ ἀποτέλεσμα, οἷον ὅταν δυοῖν βάρος τι βασταζόντων μόλις τρίτος τις προσελθὼν συγκουφίσῃ τοῦτο.

488 *Dox. Gr.* 611,13–15: "Auxiliary cause appears to be that which contributes a little force to the effect, e.g. when two people carry a beam with difficulty and they will carry it more easily if a third one joins them to make it lighter" (συνεργὸν δὲ δοκεῖ τὸ μικρὰν ῥοπὴν πρὸς τὸ ἀποτέλεσμα εἰσάγον ὡς ἐπὶ δυοῖν δοκὸν ἅμα φερόντων δυσκόλως καὶ ῥᾷον οἰσόντων, εἰ τρίτος ἐπιγίνοιτο συνδιακουφίζων).

489 XIX,393,16–18 K.: "Auxiliary cause is such that it produces an effect, but with difficulty, and contributes to its coming about more easily, not being able of doing anything on its own" (συνεργόν ἐστιν αἴτιον ὃ ποιοῦν ἀποτέλεσμα, δυσχερῶς δέ, συλλαμβάνον πρὸς τὸ ῥᾷον αὐτὸ

does not contradict the view of the main causal factor as a self-complete cause. Contrast the use of *συνεργόν* above, 28, 3–5 (with note on 98,5–9).

102,4f. νοεῖται γὰρ σὺν ἑτέρῳ τὸ συναίτιον οὐδ' αὐτῷ δυναμένῳ κατ' ἰδίαν ποιῆσαι τὸ ἀποτέλεσμα κτλ. The difference between auxiliary causes and joint-causes is constituted by their relations to that which produces an effect 'on its own' (*κατ' ἰδίαν*). This, of course, is the self-complete cause. The difference is this: For the auxiliary cause to be effective, it must join another, independently effective cause. In contrast, the joint-cause produces an effect along with another joint-cause, neither of which are independently effective. For the notion of joint-causes, cf. above, 31, 1/100,4–7; 33, 6/101,26–102,1. For parallels, cf. Sextus, *PH* III 15 (who adduces the example of oxen drawing a plough);[490] Ps.Galen, *Hist. Phil.* 19 (with the same example).[491] For the expression *κατ' ἰδίαν*, cf. Ps.Galen, *Def. Med.* 159: "Joint-cause is that which produces an effect along with something else while having the same force and being incapable of producing the effect just on its own."[492] Cf. also Clement, *Strom.* I (20), 99, 2: "That which produces [an effect] alongside something else (μεθ' ἑτέρου ποιεῖ), being incapable of acting by itself (ἀτελὲς ὂν καθ' αὑτὸ ἐνεργεῖν), is called *συνεργόν* and *συναίτιον*."

102,10 πρὸς τὸ σφοδρότερον γίνεσθαι τὸ ἀποτέλεσμα. Cf. above, 33, 7/102,2. Long and Sedley (55I5 in LS, Vol. II) follow L in reading *σφοδρότατον*, but the superlative does not seem to make any sense here and von Arnim's correction is preferable.[493]

102,10–12 μάλιστα δὲ τὸ ἐκ προκαταρκτικοῦ συνεργὸν γεγονέναι τὴν τοῦ αἰτίου διατείνειν δύναμιν παρίστησιν. The phrasing of this sentence is admittedly awk-

---

γενέσθαι, κατ' ἰδίαν τι ποιεῖν οὐ δυνάμενον). As is often the case in this source, the text seems to be muddled, but the core message is clear.

490 *PH* III 15: "[Joint-causes] supply a force equal to another joint-cause towards the existence of the effect (they say that in this way each of the oxen drawing the plough is a cause of the drawing of the plough)" [συναίτιον δὲ ὃ τὴν ἴσην εἰσφέρεται δύναμιν ἑτέρῳ συναιτίῳ πρὸς τὸ εἶναι τὸ ἀποτέλεσμα (οὕτω ⟨γὰρ⟩ ἕκαστον τῶν ἑλκόντων τὸ ἄροτρον βοῶν αἴτιον εἶναί φασι τῆς ὁλκῆς τοῦ ἀρότρου)]; trans. Annas and Barnes, slightly modified.

491 *Dox. Gr.* 611,11–13: ⟨τὸ δὲ⟩ συναίτιον ὑπάρχειν ἔφασαν, ὃ τὴν ἴσην δύναμιν συμβάλλεται πρὸς τὸ ἀποτέλεσμα ὡς ἐπὶ τῶν ἀροτικῶν βοῶν.

492 XIX,393,13–15 K.: συναίτιόν ἐστιν ὃ σὺν ἑτέρῳ δύναμιν ἴσην ἔχον ποιοῦν τὸ ἀποτέλεσμα, αὐτὸ δὲ κατ' ἰδίαν μόνον οὐ δυνάμενον ποιῆσαι.

493 It is telling that the manuscript reading is not reflected in Long and Sedley's translation ("... so that the effect is intensified").

ward, but (*pace* von Arnim and Stählin) the text needs no correction.[494] We have seen that the auxiliary cause intensifies the effect of the synectic cause (cf. 33, 7/102,1f.: πρὸς τὴν ἐπίτασιν βοηθεῖ). Here it is said what makes this function of the auxiliary cause especially apparent (μάλιστα παρίστησιν [τὸ συνεργόν *acc.*] τὴν δύναμιν τοῦ αἰτίου διατείνειν): when a procatarctic cause turns into an auxiliary one (τὸ … γεγονέναι). To take an example from the beginning of our chapter: The sight of Jason's beauty brings about an erotic condition in Medea, which then develops into her falling in love with him. In its capacity to provide the first impulse to love, the sight of Jason's beauty is a procatarctic cause. However, if Medea continues to see him even after she has already fallen in love, the same sight will turn into an auxiliary cause, making her "effort to produce friendship" with him stronger. Cf. above, on 25, 2/95,28–31.[495]

The idea that the procatarctic causes may turn into the auxiliary seems to be reflected in Ps.Galen, *Def. Med.* 156, where we read of the 'preceding' cause (προηγούμενον αἴτιον) as a cause "that is either constituted or supported (συνεργούμενον) by the procatarctic cause and precedes the synectic one".[496] A close association between the procatarctic and auxiliary causes seems to go back to Chrysippus; cf. Cicero, *Fat.* 41,5, where *adiuvantes* is the likely translation of συνεργά.[497]

---

494 Cf. LS 55I5.

495 For other examples, cf. LS I, p. 342.

496 XIX,392,17–393,1 K.: προηγούμενον αἴτιόν ἐστι τὸ ὑπὸ τοῦ προκαταρκτικοῦ ἤτοι κατασκευαζόμενον ἢ συνεργούμενον καὶ προηγούμενον τοῦ συνεκτικοῦ. The concept of the preceding cause is not mentioned in our text, but the condition described as ἐρωτικὴ διάθεσις in the example above (25, 2/95,30) seems to play a similar role in two respects: (a) it is constituted by a procatarctic cause; (b) it is not a synectic cause. Presumably, if the synectic cause of the intemperate person's behaviour corresponds to her ἔρως, the 'erotic condition' may also be taken as something preceding this cause. Galen too speaks of preceding causes as 'conditions' (διαθέσεις) of the body brought about by the procatarctic causes and having influence on the synectic causes; Cf. *Caus. Puls.* I 1 (IX,2,11–15 K.): "Speaking generally, things which are external to a body and alter it in some way are called procatarctic causes, because they precede the conditions of the body. Whenever these conditions influence synectic causes, they are preceding causes of them" (καθόλου γὰρ εἰπεῖν ὅσα μὲν ἔξωθεν ὄντα τοῦ σώματος ἀλλοιοῖ τι τῶν κατ' αὐτὸ προκαταρκτικὰ λέγεται, προκατάρχοντά γε δὴ τῶν ἐν τῷ σώματι διαθέσεων. αὗται δὲ αἱ διαθέσεις ὅταν τὰ συνεκτικὰ τρέπωσι, προηγούμεναι γίνονται αὐτῶν αἰτίαι). Cf. Hankinson, *Cause and Explanation*, pp. 378f.

497 For a possible relation between the two kinds of causes in Chrysippus, cf. Hankinson, *Cause and Explanation*, pp. 248f.

Bibliography

A 'Stromateis VIII': Editions and Translations

a *Editions*

Piero Vettori (Petrus Victorius), ed., ΚΛΗΜΕΝΤΟΣ ΑΛΕΞΑΝΔΡΕΩΣ ΤΑ ΕΥΡΙΣΚΟΜΕΝΑ ΑΠΑΝΤΑ, Florence: Laurentius Torrentinus, 1550

Friedrich Sylburg, ed., *Clementis Alexandrini opera quae exstant*, Heidelberg: Ex typographeio Hieronymi Commelini, 1592

Daniel Heinse, ed., *Clementis Alexandrini opera Graece et Latine quae extant*, Leiden: Ex bibliopolio Commeliniano, 1616

John Potter, ed., *Clementis Alexandrini opera quae extant*, Oxford: E theatro Sheldoniano, 1715

Reinhold Klotz, ed., *Titi Flaui Clementis Alexandrini opera omnia*, vol. III: continens *Stromatorum* libros V–VIII, libellum *Quis diues salvetur*, Leipzig: Schwickert, 1832

Christian Karl Josias von Bunsen, ed., *Clementis Hypotyposeon librorum octo quae supersunt* in idem, *Analecta Ante-Nicaena*, vol. I: *Reliquiae literariae* (London: Longman, Brown, Green, and Longmans, 1854), 159–340

Wilhelm Dindorf, ed., *Clementis Alexandrini opera*, vol. III: *Stromatum* V–VIII, scripta minora, fragmenta, Oxford: E typographeo Clarendoniano, 1869

Otto Stählin, ed., *Clemens Alexandrinus*, Bd. III: *Stromata* VII–VIII, *Excerpta ex Theodoto, Eclogae propheticae, Quis dives salvetur, Fragmente*, GCS 17; Leipzig: Hinrich 1909; 2nd edition, revised by Ludwig Früchtel and Ursula Treu, Berlin: Akademie-Verlag, 1970

b *Translations*

Gentien Hervet: *Clementis Alexandrini omnia quae quidem extant opera, nunc primum è tenebris eruta Latinitateque donata, Gentiano Herueto Aurelio interprete*, Florence: Laurentius Torrentinus, 1551

William Wilson: *The Writings of Clement of Alexandria*, Vol. II (*Ante-Nicene Christian Library* 12; Edinburgh: T&T Clark, 1869), 490–514

Janina Niemirska-Pliszczyńska: Klemens Aleksandryjski, *Kobierce zapisków filozoficznych dotyczących prawdziwej wiedzy*, II (Warszawa: PAX/ATK, 1994), 311–336

Marcelo Merino Rodríguez: Clemente de Alejandría, *Stromata VI–VIII. Vida intelectual y religiosa del cristiano*, Madrid: Ciudad Nueva, 2005

Annalisa Zanotti Fregonara: Clemente di Alessandria, *Gli Stromati. Note di vera filosofia* (Milan: Paoline, 2006), 855–882

Manabu Akiyama: "Arekusandoria no Kuremensu, *'Sutromateisu'* (*'Tsuzureori'*) 8," in

Bungei gengo kenkyū. Bungei hen [*Studies in Language and Literature. Literature*] 66 (2014), 87–115

B Cited Editions of Ancient Sources

a *Collections of Texts*

von Arnim, Hans, ed., *Stoicorum veterum fragmenta*, Vols. I–III, Leipzig: Teubner, 1903–1905

Diels, Hermann, ed., *Doxographi Graeci*, Berlin: Reimer, 1879

Diels, Hermann, and Kranz, Walter, eds., *Die Fragmente der Vorsokratiker*, Vols. I–II, 16th edition, Dublin/Zurich: Weidman, 1972

Dindorf, Wilhelm, ed., *Scholia Graeca in Homeri Odysseam*, Vol. I, Oxford: E typographeo academico, 1855

Hilgard, Alfred, ed., *Scholia in Dionysii Thracis Artem grammaticam, Grammatici Graeci* 3; Leipzig: Teubner, 1901

Hülser, Karlheinz, ed., *Fragmente zur Dialektik der Stoiker*, Vols. I–IV, Stuttgart/Bad Cannstatt: frommann-holzboog, 1987–1988

Long, Anthony A., and Sedley, David N., *The Hellenistic Philosophers*, Vol. II, Cambridge: Cambridge University Press, 1987

Rabe, Hugo, ed., *Rhetores Graeci XIV: Prolegomenon Sylloge*, Leipzig: Teubner, 1931

b *Ancient Authors*

Alcinous, *Didascalicus*, ed. J. Whittaker: *Alcinoos: Enseignement des doctrines de Platon*, Les Belles Lettres: Paris 1990

Alexander of Aphrodisias, *In Aristotelis Analyticorum priorum librum I commentarium*, ed. M. Wallies, CAG II/1; Berlin: Reimer, 1883

Alexander of Aphrodisias, *In Aristotelis Analytica posteriora commentariorum fragmenta*, ed. M. Wallies (CAG II/1; Berlin: Reimer, 1883), xix-xxii

Alexander of Aphrodisias, *De anima*, ed. I. Bruns: *Alexandri Aphrodisiensis praeter commentaria, scripta minora* (*De anima liber cum Mantissa*), Supplementum Aristotelicum II/1; Berlin: Reimer, 1887

Alexander of Aphrodisias, *De fato*, ed. I. Bruns: *Alexandri Aphrodisiensis praeter commentaria, scripta minora* (*Quaestiones, De fato, De mixtione*), Supplementum Aristotelicum II/2; Berlin: Reimer, 1892

Alexander of Aphrodisias, *Mantissa*, ed. I. Bruns: *Alexandri Aphrodisiensis praeter commentaria, scripta minora* (*De anima liber cum Mantissa*), Supplementum Aristotelicum II/1; Berlin: Reimer, 1887

Alexander of Aphrodisias, *In Aristotelis Metaphysica commentaria*, ed. M. Hayduck, CAG I; Berlin: Reimer, 1891

Alexander of Aphrodisias, *In Aristotelis Topicorum libros octo commentaria*, ed. M. Wallies, CAG II/2; Berlin: Reimer, 1891

Ammonius, *In Aristotelis Analyticorum priorum librum I commentarium*, ed. M. Wallies, CAG IV/6; Berlin: Reimer, 1899

Ammonius, *In Aristotelis Categorias commentarium*, ed. A. Busse, CAG IV/4; Berlin: Reimer, 1895

Anonymus Londinensis, *De medicina*, ed. D. Manetti, Berlin/New York: de Gruyter, 2011

Anonymi *In Theaetetum*, ed. G. Bastianini and D.N. Sedley, "Commentarium in Platonis *Theaetetum*," *Corpus dei papyri filosofici graeci et latini*, Vol. III (Florence: Leo S. Olschki, 1995), 227–562

Aristocles of Messene, ed. M.L. Chiesara, *Aristocles of Messene: Testimonies and Fragments*, Oxford: Oxford University Press, 2001

Aristotle, *Analytica priora, Analytica posteriora*, ed. W.D. Ross, *Aristotle's Prior and Posterior Analytics*, a revised text with introduction and commentary, Oxford: Clarendon Press, 1957

Aristotle, *De anima*, ed. W.D. Ross, *Aristotelis De anima*, OCT; Oxford: Clarendon Press, 1956

Aristotle, *Categoriae, De interpretatione*, ed. L. Minio-Paluello, *Aristotelis Categoriae et liber De interpretatione*, OCT; Oxford: Clarendon Press, 1949

Aristotle, *Ethica Eudemia*, ed. R.R. Walzer and J.M. Mingay, *Aristotelis Ethica Eudemia*, OCT; Oxford: Clarendon Press, 1991

Aristotle, *Ethica Nicomachea*, ed. I. Bywater, *Aristotelis Ethica Nicomachea*, OCT; Oxford: Clarendon Press, 1894

Aristotle, *De generatione animalium*, ed. H.J. Drossaart Lulofs, *Aristotelis De generatione animalium*, OCT; Oxford: Clarendon Press, 1965

Aristotle, *De memoria*, ed. W.D. Ross, *Aristotle: Parva naturalia*, Oxford: Clarendon Press, 1955

Aristotle, *Metaphysica*, ed. W.D. Ross, *Aristotle's Metaphysics*, a revised text with introduction and commentary, Vols. I–II, Oxford: Clarendon Press, 1924

Aristotle, *De motu animalium*, ed. M. Nussbaum, *Aristotle's De Motu Animalium*, Princeton, NJ: Princeton University Press, 2nd edition, 1985

Aristotle, *De partibus animalium*, ed. A.L. Peck, *Aristotle: Parts of Animals, Movement of Animals, Progression of Animals*, LCL 323; Cambridge, MA/London: Harvard University Press/Heinemann, 1961

Aristotle, *Physica*, ed. W.D. Ross, *Aristotle's Physics*, a revised text with introduction and commentary, Oxford: Clarendon Press, 1936

Aristotle, *Politica*, ed. W.D. Ross, *Aristotelis Politica*, OCT; Oxford: Clarendon Press, 1957

Aristotle, *Rhetorica*, ed. W.D. Ross, *Aristotelis Ars Rhetorica*, OCT; Oxford: Clarendon Press, 1959

Aristotle, *Topica, Sophistici elenchi*, ed. W.D. Ross, *Aristotelis Topica et Sophistici elenchi*, OCT; Oxford: Clarendon Press, 1958

[Aristotle], *Magna moralia*, ed. G. Cyril Armstrong, *Aristotle: Metaphysics Books 10–14, Oeconomica, Magna Moralia*, LCL 287; Cambridge, MA/London: Harvard University Press/Heinemann, 1947

Artemidorus, *Onirocriticon*, ed. R.A. Pack, *Artemidori Daldiani Onirocriticon libri V*, Leipzig: Teubner, 1963

Aspasius, *In Ethica Nicomachea quae supersunt commentaria*, ed. G. Heylbut, CAG XIX/1; Berlin: Reimer, 1889

Athenaeus, *Deipnosophistae*, ed. G. Kaibel, *Athenaei Naucratitae Deipnosophistarum libri XV*, Vols. I–III, Leipzig: Teubner, 1887–1890

Aulus Gellius, *Noctes Atticae*, ed. C. Hosius, *A. Gellii Noctium Atticarum libri XX*, Vols. I–II, Leipzig: Teubner, 1903

Boethius, *De divisione*, PL 64 (Garnier Brothers: Paris, 1891), 875–892

Boethius, *In Aristotelis De interpretatione librum commentaria*, ed. C. Meiser, *Anicii Manlii Severini Boethii commentarii in librum Aristotelis ΠΕΡΙ ΕΡΜΗΝΕΙΑΣ*, Vols. I–II, 2nd edition, Leipzig: Teubner, 1877–1880

Caelius Aurelianus, *Celeres passiones*, ed. G. Bendz, *Caelii Aureliani Celerum passionum libri III, Tardarum passionum libri V*, CML VI,1; Berlin: Akademie-Verlag, 1990

Cicero, *Academica*, ed. H. Rackham, *Cicero: On the Nature of the Gods, Academica*, LCL 268; Cambridge, MA/London: Harvard University Press/Heinemann, 1933

Cicero, *De fato*, ed. R.W. Sharples, *Cicero: On Fate (De Fato) and Boethius: The Consolation of Philosophy IV.5–7 (Philosophiae Consolationis)*, Warminster: Aris and Phillips, 1991

Cicero, *De inventione*, ed. H.M. Hubbell, *Cicero: On Invention, The Best Kind of Orator, Topics*, LCL 386; Cambridge, MA/London: Harvard University Press/Heinemann, 1949

Cicero, *De legibus*, ed. C.W. Keyes, *Cicero: On the Republic, On the Laws*, LCL 213; Cambridge, MA/London: Harvard University Press/Heinemann, 1928

Cicero, *Topica*, ed. T. Reinhardt, Oxford: Oxford University Press, 2003

Clement of Alexandria, *Protrepticus, Paedagogus*, ed. O. Stählin, *Clemens Alexandrinus*, Bd. I: *Protrepticus* und *Paedagogus*, GCS 12; Leipzig: Hinrich, 1905; 3rd edition, revised by Ursula Treu, Berlin: Akademie-Verlag, 1972

Clement of Alexandria, *Stromateis, Quis dives salvetur*, ed. O. Stählin, *Clemens Alexandrinus*, Bd. II: *Stromata I–VI*, GCS 15; Leipzig: Hinrich, 1906; 4th edition, revised by Ludwig Früchtel, Berlin: Akademie-Verlag, 1985; Bd. III: *Stromata* VII–VIII, *Excerpta ex Theodoto, Eclogae propheticae, Quis dives salvetur, Fragmente*, GCS 17; Leipzig: Hinrich 1909; 2nd edition, revised by Ludwig Früchtel and Ursula Treu, Berlin: Akademie-Verlag, 1970

Clement of Alexandria, *Excerpta ex Theodoto*, ed. F. Sagnard, *Clement d'Alexandrie: Extraits de Théodote*, SC 23; Paris: Éditions du Cerf, 1948

Clement of Alexandria, *Eclogae propheticae, Hypotyposeis*, ed. J. Plátová, *Klement Alexandrijský: Exegetické zlomky. Eclogae Propheticae—Hypotyposes*, Prague: ΟΙΚΟΥΜΕΝΗ, 2014

Dexippus, *In Aristotelis Categorias commentarium*, ed. A. Busse, CAG IV/2; Berlin: Reimer, 1888

Diodorus, *Bibliotheca historica* XII, ed. F. Vogel, *Diodori Bibliotheca historica*, Vol. II, Leipzig: Teubner, 1890; *Bibliotheca historica* XVI, ed. C.Th. Fischer, *Diodori Bibliotheca historica*, Vol. IV, Leipzig: Teubner, 1906

Diogenes Laertius, *Vitae philosophorum*, ed. T. Dorandi, *Diogenes Laertius: Lives of Eminent Philosophers*, CCTC 50; Cambridge: Cambridge University Press, 2013

Dionysius of Halicarnassus, *Antiquitates Romanae* VII, ed. C. Jacoby, *Dionysii Halicarnasei Antiquitatum Romanarum quae supersunt*, Vol. III, Stuttgart: Teubner, 1967

Dionysius Thrax, *Ars grammatica*, ed. G. Uhlig, *Dionysii Thracis Ars grammatica*, Leipzig: Teubner, 1883

[Dioscorides], *De iis quae virus ejaculantur* (Περὶ ἰοβόλων), ed. K. Sprengel, *Pedanii Dioscoridis Anazarbei tomus secundus: libri ΠΕΡΙ ΔΗΛΗΤΗΡΙΩΝ, ΙΟΒΟΛΩΝ ΚΑΙ ΕΥΠΟΡΙΣΤΩΝ spurii*, MGO XXVI; Leipzig: C. Knobloch, 1830

Epictetus, *Dissertationes*, ed. H. Schenkl, *Epicteti Dissertationes ab Arriano digestae*, Leipzig: Teubner, 1916

Epicurus, *Sententiae selectae*, ed. H. Usener, *Epicurea* (Leipzig: Teubner, 1887), 68–81

Eusebius of Caesarea, *Historia Ecclesiastica*, ed. E. Schwartz and Th. Mommsen, *Eusebius Werke*, Bd. II/1–2: *Die Kirchengeschichte*, GCS 9/1–2; Leipzig: Hichrichs, 1903–1908

Eusebius of Caesarea, *Praeparatio evangelica*, ed. K. Mras, *Eusebius Werke*, Bd. VIII/1–2: *Die Praeparatio evangelica*, GCS 43/1–2; Berlin: Akademie-Verlag, 1954–1956

Eustratius, *In Analyticorum posteriorum librum secundum commentarium*, ed. M. Hayduck CAG XXI/1; Berlin: Reimer, 1907

Galen, *Adversus eos, qui de typis scripserunt*, ed. K.G. Kühn, *Claudii Galeni opera omnia*, Vol. VII (Leipzig: Knobloch, 1824), 475–512

Galen, *Adversus Julianum, Adversus Lycum*, ed. E. Wenkebach, *Galeni Adversus Lycum et Adversus Iulianum libelli*, ed. E. Wenkebach, CMG V,10,3; Berlin, Akademie-Verlag, 1951

Galen, *De alimentorum facultatibus*, ed. K.G. Kühn, *Claudii Galeni opera omnia*, Vol. VI (Leipzig: Knobloch, 1823), 453–748

Galen, *De anatomicis administrationibus*, ed. K.G. Kühn, *Claudii Galeni opera omnia*, Vol. II (Leipzig: Knobloch, 1821), 215–731

Galen, *De animi cuiuslibet affectuum dignotione et curatione/De animi cuiuslibet peccatorum dignotione et curatione*, ed. W. De Boer, *Galeni De propriorum animi cuiuslibet affectuum dignotione et curatione, De animi cuiuslibet peccatorum dignotione et curatione, De atra bile* (CMG V,4,1,1; Leipzig/Berlin: Akademie-Verlag, 1937), 3–68

Galen, *Ars medica*, ed. V. Boudon, *Galien*, tome II: *Exhortation à l'Étude de la Médecine, Art Médical*, Paris: Les Belles Lettres, 2000

Galen, *An in arteriis natura sanguis contineatur*, ed. K.G. Kühn, *Claudii Galeni opera omnia*, Vol. IV (Leipzig: Knobloch, 1822), 703–736

Galen, *De atra bile*, ed. W. De Boer, *Galeni De propriorum animi cuiuslibet affectuum dignotione et curatione, De animi cuiuslibet peccatorum dignotione et curatione, De atra bile* (CMG V,4,1,1; Leipzig/Berlin: Akademie-Verlag, 1937), 71–93

Galen, *De causis contentivis*, ed. M. Lyons, CMG Suppl. Or. II; Berlin: Akademie-Verlag, 1969

Galen, *De causis morborum*, ed. K.G. Kühn, *Claudii Galeni opera omnia*, Vol. VII (Leipzig: Knobloch, 1824), 1–41Galen, *De causis pulsuum*, ed. K.G. Kühn, *Claudii Galeni opera omnia*, Vol. IX (Leipzig: Knobloch, 1825), 1–204

Galen, *De causis procatarcticis libellus*, a Nicolao Regino in sermonem Latinum translatus, recensuit et in Graecum sermonem retro vertit K. Bardong, CMG Suppl. II; Leipzig/Berlin: Akademie-Verlag, 1937

Galen, *De causis pulsuum*, ed. K.G. Kühn, *Claudii Galeni opera omnia*, Vol. IX (Leipzig: Knobloch, 1825), 1–204

Galen, *De compositione medicamentorum per genera*, ed. K.G. Kühn, *Claudii Galeni opera omnia*, Vol. XIII (Leipzig: Knobloch, 1827), 362–1058

Galen, *De compositione medicamentorum secundum locos*, ed. K.G. Kühn, *Claudii Galeni opera omnia*, Vol. XII (Leipzig: Knobloch, 1826), 378–1007

Galen, *De constitutione artis medicae ad Patrophilum*, ed. St. Fortuna, CMG V,1,3; Berlin: Akademie-Verlag, 1997

Galen, *De consuetudinibus*, ed. J.M. Schmutte, CMG Suppl. III; Leipzig/Berlin: Akademie-Verlag, 1941

Galen, *De crisibus*, ed. B. Alexanderson, Göteborg: Elanders, 1967

Galen, *De diebus decretoriis*, ed. K.G. Kühn, *Claudii Galeni opera omnia*, Vol. IX (Leipzig: Knobloch, 1825), 769–941

Galen, *De differentiis febrium*, ed. K.G. Kühn, *Claudii Galeni opera omnia*, Vol. VII (Leipzig: Knobloch, 1824), 273–405

Galen, *De differentia pulsuum*, ed. K.G. Kühn, *Claudii Galeni opera omnia*, Vol. VIII (Leipzig: Knobloch, 1824), 493–765

Galen, *De dignoscendis pulsibus*, ed. K.G. Kühn, *Claudii Galeni opera omnia*, Vol. VIII (Leipzig: Knobloch, 1824), 766–961

Galen, *De elementis ex Hippocratis sententia*, ed. Ph. De Lacy, CMG V,1,2; Berlin: Akademie-Verlag, 1996

Galen, *De foetuum formatione*, ed. D. Nickel, CMG V,3,3; Berlin: Akademie-Verlag, 2001

Galen, *Hippocratis Aphorismi et Galeni in eos commentarii*, ed. K.G. Kühn, *Claudii Galeni opera omnia*, Vol. XVIIb (Leipzig: Knobloch, 1829), 345–887

Galen, *In Hippocratis De natura hominis commentaria III*, CMG V,9,1; Leipzig/Berlin: Akademie-Verlag, 1914

Galen, *In Hippocratis Epidemiarum librum III commentaria III*, ed. E. Wenkebach, CMG V,10,2,1; Leipzig/Berlin: Akademie-Verlag, 1936

Galen, *In Hippocratis Epidemiarum librum VI commentaria I–VIII*, edd. E. Wenkebach et F. Pfaff, CMG V,10,2,2; Berlin: Akademie-Verlag, 1956

Galen, *In Hippocratis De officina medici librum commentarii*, ed. K.G. Kühn, *Claudii Galeni opera omnia*, Vol. XVIIIb (Leipzig: Knobloch, 1830), 629–925

Galen, *In Hippocratis Prorrheticum I commentaria III*, ed. H. Diels; *De comate secundum Hippocratem*, ed. J. Mewaldt; *In Hippocratis Prognosticum commentaria III*, ed. J. Heeg, CMG V,9,2; Leipzig/Berlin: Akademie-Verlag, 1915

Galen, *Institutio logica*, ed. K. Kalbfleisch, Leipzig: Teubner, 1896

Galen, *De instrumento odoratus*, ed. J. Kollesch, CMG Suppl. V; Berlin: Akademie Verlag, 1964

Galen, *De libris propriis*, ed. V. Boudon-Millot, *Galien*, tome I: *Introduction générale. Sur l'ordre de ses propres livres, Sur ses propres livres, Que l'excellent médecin est aussi philosophe*, Paris: Les Belles Lettres, 2007

Galen, *De locis affectis*, ed. K.G. Kühn, *Claudii Galeni opera omnia*, Vol. VIII (Leipzig: Knobloch, 1824), 1–452

Galen, *De medendi methodo ad Glauconem*, ed. K.G. Kühn, *Claudii Galeni opera omnia*, Vol. XI (Leipzig: Knobloch, 1826), 1–146

Galen, *De methodo medendi*, ed. K.G. Kühn, *Claudii Galeni opera omnia*, Vol. X (Leipzig: Knobloch, 1825), 1–1021

Galen, *De morborum differentiis*, ed. K.G. Kühn, *Claudii Galeni opera omnia*, Vol. VI (Leipzig: Knobloch, 1823), 836–880

Galen, *De morborum temporibus*, ed. K.G. Kühn, *Claudii Galeni opera omnia*, Vol. VII (Leipzig: Knobloch, 1824), 406–439

Galen, *De motu musculorum*, ed. K.G. Kühn, *Claudii Galeni opera omnia*, Vol. IV (Leipzig: Knobloch, 1822), 367–464

Galen, *De naturalibus facultatibus*, ed. G. Helmreich, *Claudii Galeni Pergameni scripta minora*, Vol. III (Leipzig: Teubner, 1893), 101–257

Galen, *De optima doctrina*, ed. A. Barigazzi, *Galeni De optimo docendi genere, Exhortatio ad medicinam* (*Protrepticus*) (CMG V,1,1; Berlin: Akademie-Verlag, 1991), 92–108

Galen, *De optima nostri corporis constitutione*, ed. K.G. Kühn, *Claudii Galeni opera omnia*, Vol. IV (Leipzig: Knobloch, 1822), 737–749

Galen, *De ordine librorum suorum ad Eugenianum*, ed. V. Boudon-Millot, *Galien*, tome I: *Introduction générale. Sur l'ordre de ses propres livres, Sur ses propres livres, Que l'excellent médecin est aussi philosophe*, Paris: Les Belles Lettres, 2007

Galen, *De ossibus ad tirones*, ed. K.G. Kühn, *Claudii Galeni opera omnia*, Vol. II (Leipzig: Knobloch, 1821), 732–778

Galen, *De placitis Hippocratis et Platonis*, ed. Ph. De Lacy, Vols. I–III, CMG V,4,1,2; Berlin: Akademie-Verlag, 1978–1984

Galen, *De plenitudine*, ed. K.G. Kühn, *Claudii Galeni opera omnia*, Vol. VII (Leipzig: Knobloch, 1824), 513–583

Galen, *De propriis placitis*, ed. V. Nutton, CMG V,3,2; Berlin: Akademie Verlag, 1999

Galen, *Protrepticus*, ed. A. Barigazzi, *Galeni De optimo docendi genere, Exhortatio ad medicinam (Protrepticus)* (CMG V,1,1; Berlin: Akademie-Verlag, 1991), 114–151

Galen, *Quod animi mores corporis temperamenta sequantur*, ed. I. Müller, *Claudii Galeni Pergameni scripta minora*, Vol. II (Leipzig: Teubner, 1891), 32–79

Galen, *Quod optimus medicus sit quoque philosophus*, ed. V. Boudon-Millot, *Galien*, tome I: *Introduction générale. Sur l'ordre de ses propres livres, Sur ses propres livres, Que l'excellent médecin est aussi philosophe*, Paris: Les Belles Lettres, 2007

Galen, *De sanitate tuenda*, ed. K. Koch, CMG V,4,2; Leipzig/Berlin: Akademie-Verlag, 1923

Galen, *De sectis ad eos, qui introducuntur*, ed. G. Helmreich, *Claudii Galeni Pergameni scripta minora*, Vol. III (Leipzig: Teubner, 1893), 1–32

Galen, *De semine*, ed. Ph. De Lacy, CMG V,3,1; Berlin: Akademie-Verlag, 1992

Galen, *De simplicium medicamentorum temperamentis et facultatibus*, ed. K.G. Kühn, *Claudii Galeni opera omnia*, Vols. XI–XII (Leipzig: Knobloch, 1826), XI: 379–892; XII: 1–377

Galen, *De sophismatis seu captionibus penes dictionem*, ed. K.G. Kühn, *Claudii Galeni opera omnia*, Vol. XIV (Leipzig: Knobloch, 1827), 582–598

Galen, *Subfiguratio empirica*, ed. K. Deichgräber, *Die griechische Empirikerschule. Sammlung der Fragmente und Darstellung der Lehre* (2nd ed.; Berlin/Zurich: Weidmann, 1965), 43–90

Galen, *De symptomatum differentiis*, ed. B. Gundert, CMG V,5,1; Berlin: Akademie-Verlag, 2009

Galen, *Synopsis librorum suorum sedecim De pulsibus*, ed. K.G. Kühn, *Claudii Galeni opera omnia*, Vol. IX (Leipzig: Knobloch, 1825), 431–549

Galen, *De temperamentis*, ed. G. Helmreich, Leipzig: Teubner, 1904

Galen, *Ad Thrasybulum*, ed. G. Helmreich, *Claudii Galeni Pergameni scripta minora*, Vol. III (Leipzig: Teubner, 1893), 33–100

Galen, *De tumoribus praeter naturam*, ed. K.G. Kühn, *Claudii Galeni opera omnia*, Vol. VII (Leipzig: Knobloch, 1824), 705–732

Galen, *De usu partium*, ed. G. Helmreich, Bd. I–II, Leipzig: Teubner, 1907–1909

Galen, *De utilita respirationis*, ed. K.G. Kühn, *Claudii Galeni opera omnia*, Vol. IV (Leipzig: Knobloch, 1822), 470–511

Galen, *De venae sectione adversus Erasistrateos Romae degentes*, ed. K.G. Kühn, *Claudii Galeni opera omnia*, Vol. XI (Leipzig: Knobloch, 1826), 187–249

[Galen,] *An animal sit quod est in utero*, ed. K.G. Kühn, *Claudii Galeni opera omnia*, Vol. XIX (Leipzig: Knobloch, 1830), 158–181

[Galen,] *Definitiones medicae*, ed. K.G. Kühn, *Claudii Galeni opera omnia*, Vol. XIX (Leipzig: Knobloch, 1830), 346–462

[Galen,] *Ad Gaurum*, ed. K. Kalbfleisch, *Die neuplatonische, fälschlich dem Galen zugeschriebene Schrift Πρὸς Γαῦρον περὶ τοῦ πῶς ἐμψυχοῦται τὰ ἔμβρυα*, Abhandlungen der königlichen Akademie der Wissenschaften zu Berlin; Berlin: Reimer, 1895

[Galen,] *In Hippocratis De alimento librum commentarii*, ed. K.G. Kühn, *Claudii Galeni opera omnia*, Vol. XV (Leipzig: Knobloch, 1828), 224–417

[Galen,] *De historia philosophica*, ed. H. Diels, *Doxographi Graeci* (Berlin: Reimer, 1879), 595–648

Hermogenes of Tarsus, *Hermogenis opera*, ed. H. Rabe, Leipzig: Teubner, 1913

Hero of Alexandria, *Pneumatica*, ed. W. Schmidt, *Heronis Alexandrini opera quae supersunt omnia*, Vol. I: *Pneumatica et Automata*, Leipzig: Teubner, 1899

Hippocrates, *De alimento*, ed. W.H.S. Jones, *Hippocrates*, Vol. 1 (LCL 147; Cambridge, MA: Harvard University Press, 1923), 342–361

Hippolytus of Rome, *Refutatio omnium haeresium*, ed. M. Marcovich, Patristische Texte und Studien 25; Berlin/New York: de Gruyter, 1986

Isocrates, *Oratio* 7 (*Areopagiticus*), ed. G.E. Benseler, F. Blass, *Isocratis Orationes* (Leipzig: Teubner, 1907), 152–171

John of Damascus, *Sacra parallela*, ed. K. Holl, *Fragmente vornicänischer Kirchenväter aus den Sacra parallela*, Leipzig: Hinrichs, 1899

Lucianus, *Opera*, ed. M.D. Macleod, Vols. I–IV, Oxford: Oxford University Press, 1972–1987

Marcus Aurelius, *Ad se ipsum*, ed. J. Dalfen, *Marci Aurelii Ad se ipsum libri XII*, Leipzig: Teubner, 1987

Menander, ed. A. Koerte, *Menandri quae supersunt*, Vol. I, Leipzig: Teubner, 1957

Menander, *Sententiae*, ed. C. Pernigotti, *Menandri Sententiae*, Florence: Leo S. Olschki, 2008

Nicomachus, *Harmonikon encheiridion*, ed. K. von Jan, *Musici scriptores Graeci* (Leipzig: Teubner, 1899), 235–265

Numenius of Apamea, ed. É. des Places, *Numénius: Fragments*, Paris: Les Belles Lettres, 1973

Oribasius, ed. J. Raeder, *Oribasii Collectionum medicarum reliquiae, libri XLIX–L, libri incerti, eclogae medicamentorum*, CMG VI,2,2; Leipzig/Berlin: Teubner, 1933

Origen, *Contra Celsum* V–VI, ed. M. Borret, Origène, *Contre Celse*, Vol. III, SC 147; Paris: Éditions du Cerf, 1969

Panaetius, ed. M. van Straaten, *Panaetii Rhodii Fragmenta*, Leiden: Brill, 1952

Philo of Alexandria, ed. L. Cohn and P. Wendland/S. Reiter, *Philonis Alexandrini opera quae supersunt*, Vols. I–VI, Berlin: Reimer, 1896–1915

Philoponus, *In Aristotelis Analytica Posteriora commentaria*, ed. M. Wallies, CAG XIII/3; Berlin: Reimer, 1909

Philoponus, *In Aristotelis De anima libro commentaria*, ed. M. Hayduck, CAG XV; Berlin: Reimer, 1897

Philostorgius, *Historia ecclesiastica*, ed. J. Bidez; GCS 21; Leipzig: Hinrichs, 1913

Philostratus, *Vitae sophistarum*, ed. K.L. Kayser, *Flavii Philostrati opera*, Vol. II (Leipzig: Teubner, 1871), 1–127

Photius, *Bibliotheca*, ed. I. Bekker, Berlin: Reimer, 1824

Plato, ed. J. Burnet, *Platonis opera*, Vols. I–V, OCT; Oxford: Clarendon Press, 1900–1907

Pliny the Elder, *Naturalis historia*, ed. K.F.T. Mayhoff, *C. Plini Secundi Naturalis historiae libri XXXVII*, Vols. I–V, Leipzig: Teubner, 1892–1909

Plutarch, *Adversus Colotem*, ed. B. Einarson and Ph. De Lacy, *Plutarch: Moralia*, Vol. XIV (LCL 428; Cambridge, MA/London: Harvard University Press/Heinemann, 1967), 151–315

Plutarch, *Cato Minor*, ed. B. Perrin, *Plutarch: Lives*, Vol. VIII (LCL 100; London/New York: Heinemann/G.P. Putnam's Sons, 1919), 235–411

Plutarch, *Cicero*, ed. B. Perrin, *Plutarch: Lives*, Vol. VII (LCL 99; London/New York: Heinemann/G.P. Putnam's Sons, 1919), 81–209

Plutarch, *De communibus notitiis adversus Stoicos*, ed. H. Cherniss, Plutarch's *Moralia*, Vol. XIII/2 (LCL 470; Cambridge, MA/London: Harvard University Press/Heinemann, 1976), 660–873

Plutarch, *Platonicae quaestiones*, ed. H. Cherniss, *Plutarch's Moralia*, Vol. XIII/1 (LCL 427; Cambridge, MA/London: Harvard University Press/Heinemann, 1976), 1–129

Plutarch, *Quaestiones convivales*, ed. C. Hubert, *Plutarchi Moralia*, Vol. IV, Leipzig: Teubner, 1938

Plutarch, *De Stoicorum repugnantiis*, ed. H. Cherniss, *Plutarch's Moralia*, Vol. XIII/2 (LCL 470; Cambridge, MA/London: Harvard University Press/Heinemann, 1976), 412–603

Plutarch, *De virtute morali*, ed. W.C. Helmbold, *Plutarch's Moralia*, Vol. VI (LCL 337; Cambridge, MA/London: Harvard University Press/Heinemann, 1939) 15–87

[Plutarch], *Placita philosophorum*, ed. J. Mau, *Plutarchi Moralia*, Vol. V, fasc. 2, pars 1 (Leipzig: Teubner, 1971) 50–153

Porphyry, Περὶ ἀγαλμάτων, ed. A. Smith, *Porphyrii philosophi fragmenta* (Leipzig: Teubner, 1993) 407–434

Porphyry, *In Aristotelis Categorias expositio per interrogationem et responsionem*, ed. A. Busse, *Porphyrii Isagoge et In Aristotelis Categorias commentarium* (CAG IV/1; Berlin: Reimer, 1887) 55–142

Porphyry, *Isagoge*, ed. A. Busse, *Porphyrii Isagoge et In Aristotelis Categorias commentarium* (CAG IV/1; Berlin: Reimer, 1887) 1–22

Posidonius, ed. L. Edelstein and I.G. Kidd, *Posidonius*, Vols. I–III, CCTC 13, 14A–B, and 36; Cambridge: Cambridge University Press, 1972–1999

Proclus, *In Platonis Timaeum commentaria*, ed. E. Diehl, Vols. I–III, Leipzig: Teubner, 1903–1906

Quintilian, *Institutio oratoria*, ed. D.A. Russell, *The Orator's Education*, Vols. I–V, LCL 124–127 and 494; Cambridge, MA: Harvard University Press, 2002

Seneca, *Epistulae*, ed. L.D. Reynolds, *L. Annaei Senecae Ad Lucilium epistulae morales*, Vols. I–II, OCT; Oxford: Clarendon Press, 1965

Sextus Empiricus, *Adversus dogmaticos* (= *Adversus mathematicos* VII–XI), ed. H. Mutschmann, *Sexti Empirici Opera*, Vol. II, Leipzig: Teubner, 1914

Sextus Empiricus, Πυρρωνείων ὑποτυπώσεων ΑΒΓ, ed. H. Mutschmann and J. Mau, *Sexti Empirici Opera*, Vol. I, Leipzig: Teubner, 1958

Simplicius, *In Aristotelis De caelo commentaria*, ed. J.L. Heiberg, CAG VII; Berlin: Reimer, 1894

Simplicius, *In Aristotelis Categorias commentarium*, ed. K. Kalbfleisch, CAG VIII; Berlin: Reimer, 1907

Simplicius, *In Aristotelis Physicorum libros quattuor priores commentaria*, ed. H. Diels, CAG IX; Berlin: Reimer, 1882

Stobaeus, *Anthologium*, ed. K. Wachsmuth and O. Hense, *Ioannis Stobaei Anthologium*, Vols. I–V, Berlin: Weidmann, 1884–1912

Tertullian, *Adversus Marcionem*, ed. E. Evans, Oxford: Clarendon Press, 1972

Themistius, *Analyticorum posteriorum paraphrasis*, ed. M. Wallies, CAG V/1; Berlin: Reimer, 1900

Theophrastus, *fragmenta*, ed. W.W. Fortenbaugh et al., *Theophrastus of Eresus: Sources for his Life, Writings, Thought & Influence*, edited and translated by W.W. Fortenbaugh, P. Huby, R. Sharples and D. Gutas, PhA 54; Leiden/New York/Cologne: Brill, 1992

Theophrastus, Περὶ ἀνέμων, ed. F. Wimmer, *Theophrasti Eresii opera quae supersunt omnia*, Vol. III: *Fragmenta* (Leipzig: Teubner, 1862), 94–115

Theophrastus, *De lapidibus*, ed. D.E. Eichholz, Oxford: Clarendon Press, 1965

Troilus the Sophist, *Prolegomena in Hermogenis Artem rhetoricam*, ed. C. Walz, *Rhetores Graeci*, Vol. VI (Stuttgart/Tübingen: Cotta, 1834), 42–55

Varro, *De lingua Latina* X, ed. D.J. Taylor, *Varro: De Lingua Latina X*, a new critical text and English translation with prolegomena and commentary, Amsterdam/Philadelphia: John Benjamins Publishing Company, 1996

C Cited Translations

Ackrill, John Lloyd, *Aristotle's Categories and De Interpretatione*, translated with notes, Oxford: Clarendon Press, 1963

Baltzly, Dirk, *Proclus: Commentary on Plato's Timaeus*, vol. IV, translated with notes, Cambridge: Cambridge University Press, 2009

Barnes, Jonathan, et al., *Alexander of Aphrodisias: On Aristotle's Prior Analytics 1.1–7*, translation and notes by J. Barnes, S. Bobzien, K. Flannery, and K. Ierodiakonou, Ithaca/New York: Cornell University Press, 1991

Barnes, Jonathan, *Aristotle: Posterior Analytics*, translated with a commentary, 2nd edition, Oxford: Clarendon Press, 1993

Barnes, Jonathan, and Annas, Julia, *Sextus Empiricus: Outlines of Scepticism*, translated with notes, Cambridge: Cambridge University Press, 2000

Barnes, Jonathan, *Porphyry: Introduction*, translated with a commentary, Oxford: Clarendon Press, 2003

Bett, Richard, *Sextus Empiricus: Against the Logicians*, translated with notes, Cambridge: Cambridge University Press, 2005

Boys-Stones, George, *Numenius: Fragments*, draft translation, academia.edu, 2014

Brittain, Charles, *Cicero: On Academic Scepticism*, translation of Cicero's *Academica* with notes, Indianapolis: Hackett, 2006

Butterworth, G.W., *Clement of Alexandria: The Exhortation to the Greeks, The Rich Man's Salvation, To the Newly Baptized*, Loeb Classical Library 92; Cambridge, MA: Harvard University Press, 1919

Chadwick, Henry, *Origen: Contra Celsum*, translated with notes, Cambridge: Cambridge University Press, 1953

Dillon, John, *Alcinous: The Handbook of Platonism*, translated with a commentary, Oxford: Clarendon Press, 1993

Evans, Ernest, *Tertullian: Adversus Marcionem*, Oxford: Clarendon Press, 1972

Ferguson, John, *Clement of Alexandria: Stromateis, Books 1–3*, Fathers of the Church 85; Washington, D.C.: Catholic University of America Press, 1991

Frede, Michael, *Galen: Three Treatises on the Nature of Science*, translation of Galen's *On the Sects for Beginners, An Outline of Empiricism*, and *On Medical Experience* by M. Frede and R. Waltzer, Indianapolis: Hackett, 1984

Furley, David J., and Wilkie, James S., *Galen on Respiration and the Arteries*, Princeton, NJ: Princeton University Press, 1984

Hankinson, Robert James, *Galen: On the Therapeutic Method*, translated with a commentary, Oxford: Oxford University Press, 1991

Hankinson, Robert James, *Galen: On Antecedent Causes*, CCTC 35; Cambridge: Cambridge University Press, 1998

Heath, Malcolm, *Hermogenes: On Issues. Strategies of Argument in Early Greek Rhetoric*, translated with a commentary, Oxford: Clarendon Press, 1995

Hort, Fenton John Anthony, and Mayor, Joseph B., *Clement* of *Alexandria: Miscellanies* Book VII, London/New York: Macmillan&Co., 1902

Inwood, Brad, *Seneca: Selected Philosophical Letters*, translated with a commentary, Oxford: Oxford University Press, 2007

Inwood, Brad, and Gerson, Lloyd P., *Hellenistic Philosophy: Introductory Readings*, 2nd edition, Indianapolis: Hackett, 1997

Johnston, Ian, *Galen on Diseases and Symptoms*, Cambridge: Cambridge University Press, 2006

Johnston, Ian, and Horsley, G.H.R., *Galen: Method of Medicine*, Vols. I–III, Loeb Classical Library 516–518; Cambridge, MA: Harvard University Press, 2011

De Lacy, Phillip, *Galen: On the Doctrines of Hippocrates and Plato*, edited with a translation and notes, Vols. I–III, CMG V,4,1,2; Berlin: Akademie-Verlag, 1978–1984

Lamb, W.R.M., *Plato in Twelve Volumes*, Vol. VIII: *The Statesman, Philebus, Ion*, LCL 164; Cambridge, MA/London: Harvard University Press/Heinemann, 1955

Lattimore, Richmond, *The Iliad of Homer*, Chicago: University of Chicago Press, 1951

Long, Anthony A., and Sedley, David N., *The Hellenistic Philosophers*, I: Translations of the principal sources, with philosophical commentary, Cambridge: Cambridge University Press, 1987

Lyons, Malcolm, *Galen: On the Parts of Medicine, On Cohesive Causes, On Regimen in Acute Diseases*, editions of the Arabic versions with English translation, CMG Suppl. Or. II; Berlin: Akademie-Verlag, 1965

May, Margaret Tallmadge, *Galen: On the Usefulness of the Parts of the Body*, translated with notes, Ithaca: Cornell University Press, 1968

Menn, Stephen, *Galen: On the Best Kind of Teaching*, draft translation, unpublished

Mueller, Ian, *Alexander of Aphrodisias: On Aristotle Prior Analytics 1.23–31*, translated with notes, London/New York: Bloomsbury, 2006

Nussbaum, Martha Craven, *Aristotle's De Motu Animalium*, edited with a translation and commentary, Princeton, NJ: Princeton University Press, 2nd edition, 1985

Pickard-Cambridge, Arthur Wallace, *Aristotle: Topics* in *The Complete Works of Aristotle: The Revised Oxford Translation*, vol. I, ed. J. Barnes (Princeton, NJ: Princeton University Press, 1984), 167–277

Rackham, Harris, *Cicero: On the Orator Book 3, On Fate, Stoic Paradoxes, Divisions of Oratory*, Loeb Classical Library 349; Cambridge, MA: Harvard University Press, 1942

Reinhardt, Tobias, *Cicero's Topica*, edited with a translation and commentary, Oxford: Oxford University Press, 2003

Ross, William David, *Aristotle: Metaphysics* in *The Complete Works of Aristotle: The Revised Oxford Translation*, vol. II, ed. J. Barnes (Princeton, NJ: Princeton University Press, 1991), 1552–1728

Runia, David T., *Philo of Alexandria: On the Creation of the Cosmos according to Moses*, translated with a commentary, Leiden/Boston/Cologne: Brill, 2001

Sedley, David N., and Long, Anthony A., *The Hellenistic Philosophers*, see Long

Sharples, Robert W., *Cicero: On Fate (De Fato) and Boethius: The Consolation of Philosophy IV.5–7 (Philosophiae Consolationis)*, edited with a translation and commentary, Warminster: Aris and Phillips, 1991

Singer, Peter N., *Galen: Selected Works*, Oxford: Oxford University Press, 1997

Smith, Robin, *Aristotle's Topics, Books I and VIII*, translated with a commentary, Oxford: Clarendon Press, 1997

Striker, Gisela, *Aristotle's Prior Analytics: Book I*, translated with a commentary, Oxford: Clarendon Press, 2009

Taylor, Daniel J., *Varro: De Lingua Latina X*, edited with a translation and commentary, Amsterdam/Philadelphia: John Benjamins Publishing Company, 1996

Walzer, Richard, see Michael Frede, *Galen: Three Treatises on the Nature of Science*

D Modern Authors

Acerbi, Fabio, "The Language of the 'Givens': Its Forms and Its Use as a Deductive Tool in Greek Mathematics," *Archive for History of Exact Sciences* 65/2 (2011), 119–153

Acerbi, Fabio, "The Geometry of Burning Mirrors in Classical Antiquity: Analysis, Heuristic, Projections, Lemmatic Fragmentation," *Archive for History of Exact Sciences* 65/5 (2011), 471–497

Alessandrelli, Michele, *Il problema del λεκτόν nello stoicismo antico: Origine e statuto di una nozione controversa*, Florence: Leo S. Olschki Editore, 2013

Allen, James, *Inference from Signs*, Oxford: Oxford University Press, 2001

Annas, Julia, and Barnes, Jonathan, *The Modes of Scepticism: Ancient Texts and Modern Interpretations*, Cambridge: Cambridge University Press, 1985

von Arnim, Hans, *De octavo Clementis Stromateorum libro*, Rostock: Adler, 1894

von Arnim, Hans, *Quellenstudien zu Philo von Alexandria*, Berlin: Weidmann, 1888

Aubry, Gwenaëlle, "Capacité et convenance: la notion d'epitêdeiotês dans la théorie porphyrienne de l'embryon," in L. Brisson, M.-H. Congourdeau, and J.-L. Solère (eds.), *L'embryon: formation et animation* (Paris: Vrin, 2008), 139–155

Barnes, Jonathan, "Proof Destroyed," in M. Schofield, M. Burnyeat, and J. Barnes (eds.), *Doubt and Dogmatism* (Oxford: Clarendon Press, 1980), 161–181

Barnes, Jonathan, "Galen on Logic and Therapy," in F. Kudlien and R.J. Durling (eds.), *Galen's Method of Healing* (Leiden: Brill, 1991), 50–102

Barnes, Jonathan, "Diogenes Laertius IX 61–116: The Philosophy of Pyrrhonism," in *Aufstieg und Niedergang der römischen Welt*, II: Principat, Band 36.6, ed. W. Haase (Berlin/New York: de Gruyter, 1992), 4241–301

Barnes, Jonathan, *Aristotle: Posterior Analytics*, translated with a commentary, 2nd edition, Oxford: Clarendon Press, 1993

Barnes, Jonathan, "Galen on the Utility of Logic," in J. Kollesch and D. Nickel (eds.), *Galen und das hellenistische Erbe* (Stuttgart: Steiner, 1993), 33–52

Barnes, Jonathan, "The Beliefs of a Pyrrhonist," in M. Burnyeat and M. Frede (eds.), *The Original Sceptics: A Controversy* (Indianapolis: Hackett, 1997), 58–91

Barnes, Jonathan, "Logic I: The Peripatetics" and "Language I.2: Meaning," in K. Algra,

J. Barnes, J. Mansfeld, and M. Schofield (eds.), *The Cambridge History of Hellenistic Philosophy* (Cambridge: Cambridge University Press, 1999), 77–83 and 193–213

Barnes, Jonathan, *Porphyry: Introduction*, translated with a commentary, Oxford: Clarendon Press, 2003

Barnes, Jonathan, *Truth, etc.: Six Lectures on Ancient Logic*, Oxford: Clarendon Press, 2007

Barnes, Jonathan, "Galen, Christians, Logic," in *Logical Matters: Essays in Ancient Philosophy* II (Oxford: Clarendon Press, 2012), 1–21

Barnes, Jonathan, and Annas, Julia, *The Modes of Scepticism*, see Annas

Bergjan, Silke-Petra, "Logic and Theology in Clement of Alexandria: The Purpose of the 8th Book of the *Stromata*," *Zeitschrift für antikes Christentum* 12 (2009), 396–413

Berti, Enrico, "Amicizia e «focal meaning»," in A. Alberti and R.W. Sharples (eds.), *Aspasius: The Earliest Extant Commentary on Aristotle's Ethics* (Berlin/New York: de Gruyter, 1999), 176–190

Biester, Björn, *Der innere Beruf zur Wissenschaft: Paul Ruben (1866–1943). Studien zur deutsch-jüdischen Wissenschaftsgeschichte*, Berlin: Dietrich Reimer Verlag, 2001

Biriukov, Dmitry, "О платонизме в философско-богословской системе Климента Александрийского" (On Platonism in the Theological and Philosophical Doctrine of Clement of Alexandria) in *Plato Collection* II (Supplement to the Bulletin of the Russian Christian Academy for the Humanities 14; Moscow/Saint Petersburg: RGGU-RHGA, 2013), 276–483

Bobzien, Susanne, "Logic III: The Stoics," in K. Algra, J. Barnes, J. Mansfeld and M. Schofield (eds.), *The Cambridge History of Hellenistic Philosophy* (Cambridge: Cambridge University Press, 1999), 92–176

Bobzien, Susanne, "Chrysippus' Theory of Causes," in K. Ierodiakonou (ed.), *Topics in Stoic Philosophy* (Oxford: Clarendon Press, 1999), 196–242

Bobzien, Susanne, *Determinism and Freedom in Stoic Philosophy*, Oxford: Clarendon Press, 1998

Bolton, Robert, "Aristotle's Method in Natural Science: Physics I," in L. Judson (ed.), *Aristotle's Physics: A Collection of Essays* (Oxford: Clarendon Press, 1991), 1–29

Boudon-Millot, Véronique, and Pietrobelli, Antoine, "Galien ressuscité: Édition princeps du texte grec du *De propriis placitis*," *Revue des études grecques* 118 (2005), 168–213

Boudon-Millot, Véronique, "La naissance de la vie dans la théorie médicale et philosophique de Galien," in L. Brisson, M.-H. Congourdeau, and J.-L. Solère (eds.), *L'embryon: formation et animation* (Paris: Vrin, 2008), 79–94

Le Boulluec, Alain, *Clément d'Alexandrie: Les Stromates. Stromate V*, Tome II (commentaire), SC 279; Paris: Éditions du Cerf, 1981

Le Boulluec, Alain, *La notion d'hérésie dans la littérature grecque IIe–IIIe siècles*, II: *Clément d'Alexandrie et Origène*, Paris: Institut d'Études Augustiniennes, 1985

Le Boulluec, Alain, "La rencontre de l'hellénisme et de la « philosophie barbare » selon Clément d'Alexandrie," in idem, *Alexandrie antique et chrétienne: Clément et Origène* (Paris: Institut d'Études Augustiniennes, 2006), 81–93

Le Boulluec, Alain, "Extraits d'œuvres de Clement d'Alexandrie: La transmission et le sens de leur titres," in idem, *Alexandrie antique et chrétienne: Clément et Origène* (Paris: Institut d'Études Augustiniennes, 2006), pp. 109–122

Le Boulluec, Alain, "L'identité chrétienne en autodéfinition chez Clément d'Alexandrie," in N. Belayche and S. Mimouni (eds.), *Les identités religieuses dans l'Antiquité grecque et romaine, Collection de la Revue des Études Juives* (Leuven: Peeters, 2009), 437–458

Le Boulluec, Alain, "Comment Clément applique t-il dans le *Stromate* VII, à l'intention des philosophes, la méthode définie dans le prologue (1–3)?" in M. Havrda, V. Hušek, and J. Plátová (eds.), *The Seventh Book of the Stromateis: Proceedings of the Colloquium on Clement of Alexandria (Olomouc, October 21–23, 2010)* (Suppl. Vig. Chr. 117; Leiden/Boston: Brill, 2012), 39–62

Bousset, Wilhelm, *Jüdisch-christlicher Schulbetrieb in Alexandria und Rom: Literarische Untersuchungen zu Philo und Clemens von Alexandria, Justin und Irenäus*, Göttingen: Vandenhoeck & Ruprecht, 1915

Boys-Stones, George, *Post-Hellenistic Philosophy*, Oxford: Oxford University Press, 2001

Brunschwig, Jacques, "Proof Defined," in M. Schofield, M. Burnyeat, and J. Barnes (eds.), *Doubt and Dogmatism* (Oxford: Clarendon Press, 1980), 125–160

Bucur, Bogdan Gabriel, *Angelomorphic Pneumatology. Clement of Alexandria and Other Early Christian Witnesses*, Suppl. Vig. Chr. 95; Leiden/Boston: Brill, 2009

Cambe, Michel, *Avenir solaire et angélique des justes: le psaume 19 (18) commenté par Clément d'Alexandrie*, Cahiers de la Biblia patristica 10; Strasbourg: Brepols, 2009

Carriker, Andrew, *The Library of Eusebius of Caesarea*, Suppl. Vig. Chr. 67; Leiden/Boston: Brill, 2003

Castagnoli, Luca, *Ancient Self-Refutation: The Logic and History of the Self-Refutation Argument from Democritus to Augustine*, Cambridge: Cambridge University Press, 2010

Chiaradonna, Riccardo, "Le traité de Galien *Sur la démonstration* et sa postérité tardoantique," in R. Chiaradonna and F. Trabattoni (eds.), *Physics and Philosophy of Nature in Greek Neoplatonism* (PhA 115; Leiden/Boston: Brill, 2008), 43–77

Chiaradonna, Riccardo, "Autour d'Eudore: Les débuts de l'exégèse des *Catégories* dans le Moyen Platonisme," in M. Bonazzi and J. Opsomer, *The Origins of the Platonic System: Platonisms of the Early Empire and Their Philosophical Contexts* (Leuven: Peeters, 2009), 89–111

Chiaradonna, Riccardo, "Galen and Middle Platonism," in C. Gill, T. Whitmarsh, and J. Wilkins (eds.), *Galen and the World of Knowledge* (Cambridge: Cambridge University Press, 2009), 243–260

Chiesara, Maria Lorenza, *Aristocles of Messene: Testimonies and Fragments*, edited with translation and commentary, Oxford: Oxford University Press, 2001

Congourdeau, Marie-Hélène, *L'embryon et son âme dans les sources grecques (VIe siècle av. J.-C.–Ve siècle apr. J.-C.)*, Paris: Association des amis du Centre d'histoire et civilisation de Byzance, 2007

Cook, Brad L., "The Essential Philip of Macedon: A Byzantine Epitome of His Life," *Greek, Roman and Byzantine Studies* 45/2 (2005), 189–211

Deichgräber, Karl, *Pseudhippokrates: Über die Nährung*, Mainz: Akademie der Wissenschaften und der Literatur, 1973

Denniston, John Dewar, *The Greek Particles*, Oxford: Clarendon Press, 1953

Descourtieux, Patrick, *Clément d'Alexandrie: Les Stromates. Stromate* VI, introduction, texte critique, traduction et notes; SC 446, Paris: Éditions du Cerf, 1999

Dillon, John, *Alcinous: The Handbook of Platonism*, translated with an introduction and commentary, Oxford: Clarendon Press, 1993

Dorandi, Tiziano, *Nell' officina dei classici: Come lavoravano gli autori antichi*, Rome: Carocci, 2007

Duhot, Jean-Joël, *La conception stoicienne de la causalité*, Paris: Vrin, 1989

Dyson, Henry, *Prolepsis and Ennoia in the Early Stoa*, Berlin/New York: de Gruyter, 2009

Edlow, Robert Blair, *Galen on Language and Ambiguity*, PhA 31; Leiden: Brill, 1977

van der Eijk, Philip J., "Therapeutics," in R.J. Hankinson (ed.), *The Cambridge Companion to Galen* (Cambridge: Cambridge University Press, 2008), 283–303

van der Eijk, Philip J., "Galen and the Scientific Treatise: a Case Study of *Mixtures*," in M. Asper (ed.), *Writing Science: Medical and Mathematical Authorship in Ancient Greece* (Berlin/Boston: de Gruyter, 2013), 145–176

Ernst, Wilhelm, *De Clementis Alexandrini Stromatum libro VIII. qui fertur*, Göttingen: Officina Hubertiana, 1910

Frede, Michael, "The Original Notion of a Cause," in idem, *Essays in Ancient Philosophy* (Minneapolis: University of Minnesota Press, 1987), 125–150

Frede, Michael, "The Method of the So-Called Methodical School of Medicine," in idem, *Essays in Ancient Philosophy* (Minneapolis: University of Minnesota Press, 1987), 261–278

Frede, Michael, "The Stoic Notion of a Grammatical Case," *Bulletin of the Institute of Classical Studies* 39/1 (1994), 13–24

Frede, Michael, "The Stoic Notion of *lekton*," in S. Everson (ed.), *Language* (Companions to Ancient Thought 3; Cambridge: Cambridge University Press, 1994), 109–128

Frede, Michael, "Stoic Epistemology," in K. Algra, J. Barnes, J. Mansfeld, and M. Schofield (eds.), *The Cambridge History of Hellenistic Philosophy* (Cambridge: Cambridge University Press, 1999), 295–322

Frede, Michael, "Les *Catégories* d'Aristote et les Pères de l'Église grecs," in O. Bruun and L. Corti (eds.), *Les Catégories et leur histoire* (Paris: Vrin, 2005), 135–173

Gamble, Harry Y., *Books and Readers in the Early Church: A History of Early Christian Texts*, New Haven/London: Yale University Press, 1995

Gaskin, Richard, "The Stoics on Cases, Predicates, and the Unity of Proposition," in R. Sorabji (ed.), *Aristotle and After* (London: Institute of Classical Studies, School of Advanced Study, University of London, 1997), 91–108

Garofalo, Ivan, "Il falso commento di Galeno al *De humoribus* e un saggio di edizione del vero," in S. Fortuna, I. Garofalo, A. Lami, and A. Roselli (eds.), *Sulla tradizione indiretta dei testi medici greci* (Pisa/Rome: Fabrizio Serra, 2009), 201–218

Garofalo, Ivan, "Il commento di Galeno al *De alimento* e gli estratti di Ibn Riḍwān," *Galenos* 6 (2012), 123–164

Gill, Christopher, "Did Chrysippus Understand Medea?" *Phronesis* 28/2 (1983), 136–149

Gill, Christopher, "Stoic Erôs—Is There Such a Thing?" in E. Sanders, C. Thumiger, C. Carey, and N.J. Lowe (eds.), *Erôs in Ancient Greece* (Oxford: Oxford University Press, 2013), 143–161

Good, John Mason, *The Study of Medicine*, II, London: Baldwin, Cradock, and Joy, 1822

Görler, Woldemar, "Les «évidences» dans la Philosophie hellenistique," in C. Lévy and L. Pernot, *Dire l'évidence* (*philosophie et rhétorique antiques*) (Paris/Montréal: L'Harmattan, 1997), 131–143

Gourinat, Jean-Baptiste, *La dialectique des stoïciens*, Paris: Vrin, 2000

Graeser, Andreas, "The Stoic Theory of Meaning," in J. Rist (ed.), *The Stoics* (Berkeley/Los Angeles: University of California Press, 1978), 77–100

Grant, Robert M., "Paul, Galen, and Origen," in *Journal of Theological Studies* 34/2 (1983), 533–536

Graverini, Luca, "A *lepidus susurrus*. Apuleius and the Fascination of Poetry," in R.R. Nauta, (ed.), *Desultoria Scientia. Genre in Apuleius' Metamorphoses and Related Texts* (Leuven/Paris/Dudley, MA: Peeters, 2006), 1–18

Griffin, Michael, *Aristotle's* Categories *in the Early Roman Empire*, Oxford: Oxford University Press, 2015

Hadot, Ilsetraut (ed.), *Simplicius: Commentaire sur les Catégories*, III: Préambule aux *Catégories*/Commentaire aux premier chapitre des *Catégories*; PhA 51; Leiden: Brill, 1990

Hadot, Pierre, "Sur divers sens du mot *pragma* dans la tradition philosophique grecque," in P. Aubenque (ed.), *Concepts et catégories dans la pensée antique* (Paris: Vrin, 1980), 309–319

Hankinson, Robert James, *Galen: On the Therapeutic Method*, Oxford: Oxford University Press, 1991

Hankinson, Robert James, "Galen's Theory of Causation," in *Aufstieg und Niedergang*

der römischen Welt, II: Principat, Band 37.2, ed. W. Haase (Berlin/New York: de Gruyter, 1994), 1757–1774

Hankinson, Robert James, *The Skeptics: Arguments of the Philosophers*, London/New York: Routledge, 1995

Hankinson, Robert James, *Galen: On Antecedent Causes*, Cambridge: Cambridge University Press, 1998

Hankinson, Robert James, "Causation in Galen," in *Galien et la philosophie* (Entretiens Hardt 49; Genève: Vandœuvre, 2002), 31–72

Hankinson, Robert James, "Stoic Epistemology," in B. Inwood (ed.), *The Cambridge Companion to the Stoics* (Cambridge: Cambridge University Press, 2003), 59–84

Hankinson, Robert James, "Epistemology," in idem (ed.), *The Cambridge Companion to Galen* (Cambridge: Cambridge University Press, 2008), 157–183

Hankinson, Robert James, "Galen on the Limitations of Knowledge," in C. Gill, T. Whitmarch, and J. Wilkins (eds.), *Galen and the World of Knowledge* (Cambridge: Cambridge University Press, 2009), 206–242

Hanson, Ann Ellis, "Galen: Author and Critic," in G.W. Most (ed.), *Editing Texts/Texte Edieren. Aporemata: Kritische Studien zur Philologiegeschichte*, II (Göttingen: Vandenhoeck & Ruprecht, 1998), 22–53

von Harnack, Adolf, *Geschichte der altchristlichen Litteratur bis Eusebius*, II/2, Leipzig: Hinrich, 1904

Havrda, Matyáš, "Some Observations on Clement of Alexandria, *Stromata*, Book Five," *Vigiliae Christianae* 64/1 (2010), 1–30

Havrda, Matyáš, "Galenus Christianus? The Doctrine of Demonstration in *Stromata* VIII and the Question of its Source," *Vigiliae Christianae* 65/4 (2011), 343–375

Havrda, Matyáš, "Demonstrative Method in *Stromateis* VII: Context, Principles, and Purpose," in M. Havrda, V. Hušek, and J. Plátová (eds.), *The Seventh Book of the Stromateis: Proceedings of the Colloquium on Clement of Alexandria (Olomouc, October 21–23, 2010)* (Suppl. Vig. Chr. 117; Leiden/Boston: Brill, 2012), 261–275

Havrda, Matyáš, "Categories in *Stromata* VIII," *Elenchos* 33/2 (2012), 197–225

Havrda, Matyáš, "Textual Notes on *Stromata* VIII," *Mnemosyne* 66/4–5 (2013), 761–768

Havrda, Matyáš, "The Purpose of Galen's Treatise *On Demonstration*," *Early Science and Medicine* 20/3 (2015), 265–287

Havrda, Matyáš, "Clement's Exegetical Interests in *Stromateis* VIII," in V. Černušková, J.L. Kovacs, and J. Plátová (eds.), *Clement's Biblical Exegesis. Proceedings of the Second Colloquium on Clement of Alexandria (Olomouc, May 29–31, 2014)* (Suppl. Vig. Chr. 139; Leiden/Boston: Brill, 2016), 162–178

Heinse, Daniel, "Annotationes in Clementem Alexandrinum," attached to D. Heinse, ed., *Clementis Alexandrini opera Graece et Latine quae extant* (Leiden: Ex bibliopolio Commeliniano, 1616); contains the notes of F. Sylburg (pp. 1–38) and D. Heinse (pp. 39–50)

Hervet, Gentien, *Clementis Alexandrini presbyteri, et ecclesiasticae scholae magistri, qui Pantaeni quidem martyris fuit discipulus, praeceptor vero Origenis, opera omnia ante annos quadraginta e Graeco in Latinum conversa, nun vero recognita interpretatione, amplissimis commentariis illustrata*, Paris: Sebastianus Nivellius, 1590

van den Hoek, Annewies, *Clement of Alexandria and His Use of Philo in the* Stromateis: *An Early Christian Reshaping of a Jewish Model*, Suppl. VigChr 3; Leiden: Brill, 1988

van den Hoek, Annewies, "Techniques of Quotation in Clement of Alexandria: A View of Ancient Literary Working Methods," *Vigiliae Christianae* 50/3 (1996), 223–243

van den Hoek, Annewies, *Clement d'Alexandrie: Les Stromates. Stromate IV*, introduction, texte critique, traduction et notes, SC 463; Paris: Éditions du Cerf, 2001

van den Hoek, Annewies, "Apologetic and Protreptic Discourse in Clement of Alexandria," in *L'apologétique chrétienne gréco-latine à l'époque prénicénienne*, ed. A. Wlosok and F. Paschoud (Entretiens Hardt 51; Genève: Vandœuvre, 2005), 69–102

van den Hoek, Annewies, "*Stromateis* Book VII in Light of Recent Scholarship: Approaches and Perspectives," in M. Havrda, V. Hušek, and J. Plátová (eds.), *The Seventh Book of the Stromateis: Proceedings of the Colloquium on Clement of Alexandria (Olomouc, October 21–23, 2010)* (Suppl. Vig. Chr. 117; Leiden/Boston: Brill, 2012), 3–36

Hoffmann, Petra, *Weibliche Arbeitswelten in der Wissenschaft: Frauen an der Preußischen Akademie der Wissenschaften zu Berlin 1890–1945*, Berlin: Humboldt-Universität, diss., 2008

Hood, Jane, "Galen's Aristotelian Definitions," in D. Charles (ed.), *Definition in Greek Philosophy* (Oxford: Oxford University Press, 2010), 450–466

Huby, Pamela, *Theophrastus of Eresus: Sources for his Life, Writings, Thought and Influence, Commentary, IV: Psychology*, Leiden: Brill, 1999

Ierodiakonou, Katerina, "The notion of *enargeia* in Hellenistic philosophy," in K. Ierodiakonou and B. Morison (eds.), *Episteme, Etc.: Essays in Honour of Jonathan Barnes* (Oxford: Oxford University Press, 2011), 60–73

Inwood, Brad, *Ethics and Human Action in Early Stoicism*, Oxford: Clarendon Press, 1985.

Itter, Andrew C., *Esoteric Teaching in the Stromateis of Clement of Alexandria*, Suppl. Vig. Chr. 97; Leiden/Boston: Brill, 2009

Janáček, Karel, *Studien zu Sextus Empiricus, Diogenes Laertius und zur pyrrhonischen Skepsis*, edited by J. Janda and F. Karfík, Berlin/New York: de Gruyter, 2008

Janáček, Karel, "Zur Interpretation des Photiosabschnittes über Ainesidemos" (1976), in idem, *Studien*, 216–224

Janáček, Karel, "Ainesidemos und Sextos Empeirikos" (1980), in idem, *Studien*, 251–264

Janáček, Karel, "Philon von Alexandreia und skeptische Tropen" (1982), in idem, *Studien*, 265–282

Jouanna, Jacques, "Médecine et philosophie: la reception de la science aristotélicienne chez Galien," in Y. Lehman (ed.), *Aristoteles Romanus. La réception de la science aristotélicienne dans l'Empire gréco-romain* (Turnhout: Brepols, 2013), 159–181

Karamanolis, George, *Plato and Aristotle in Agreement? Platonists on Aristotle from Antiochus to Porphyry*, Oxford: Oxford University Press, 2006

Karamanolis, George, *The Philosophy of Early Christianity*, Durham: Acumen, 2013

Kollesch, Jutta, *Untersuchungen zu den pseudogalenischen Definitiones medicae*, Berlin: Akademie-Verlag, 1973

Krämer, Hans Joachim, *Platonismus und hellenistische Philosophie*, Berlin/New York: de Gruyter, 1971

Kudlien, Fridolf, "'Endeixis' as a Scientific Term: (A) Galen's Usage of the Word in Medicine and Logic," in F. Kudlien and R.J. Durling (eds.), *Galen's Method of Healing* (Leiden: Brill, 1991), 103–111

Kühner, Raphael, *Ausführliche Grammatik der griechischen Sprache*, 3rd edition, Hannover/Leipzig: Hahnsche Buchhandlung, 1904

De Lacy, Phillip, "Galen's Response to Skepticism," in M. Marcovich (ed.), *Illinois Classical Studies* 16/1–2 (1991), 283–306

Lampe, Geoffrey W.H., *A Patristic Greek Lexicon*, Oxford: Oxford University Press, 1969

Leith, David, "Pores and Void in Asclepiades' Physical Theory," *Phronesis* 57/2 (2012), 164–191

Lennox, James G., "Aristotelian Problems," in idem, *Aristotle's Philosophy of Biology: Studies in the Origins of Life Science* (Cambridge: Cambridge University Press, 2001), 72–97

Lennox, James G., *Aristotle: On the Parts of Animals* I–IV, translated with a commentary, Oxford: Clarendon Press, 2001

Lévy, Carlos, *Cicero Academicus: Recherches sur les* Académiques *et sur la philosophie cicéronienne*, Rome: École française de Rome, 1992

Lévy, Carlos, "Deux problèmes doxographiques chez Philon d'Alexandrie: Posidonius et Énésidème," in A. Brancacci (ed.), *Philosophy and Doxography in the Imperial Age* (Florence: Leo S. Olschki Editore, 2005), 79–102

Lévy, Carlos, "Philon d'Alexandrie est-il inutilisable pour connaître Énésidème? Étude méthodologique," *Philosophie Antique* 15 (2015), 5–26

Lilla, Salvatore R.C., *Clement of Alexandria: A Study in Christian Platonism and Gnosticism*, Oxford: Oxford University Press, 1971

Lloyd, Anthony C., "*Nosce Te ipsum and Conscientia*," *Archiv für Geschichte der Philosophie* 46 (1964), 188–200

Lloyd, Geoffrey E.R., "The Theories and Practices of Demonstration," in idem, *Aristotelian Explorations* (Cambridge: Cambridge University Press, 1996), 7–37

Long, Anthony A., "Language and Thought in Stoicism," in idem (ed.), *Problems in Stoicism* (London: The Athlone Press, 1996), 75–113

Mansfeld, Jaap, "Doxography and Dialectic. The *Sitz im Leben* of the 'Placita'," in *Aufstieg und Niedergang der römischen Welt*, II: Principat, Band 36.3, ed. W. Haase (Berlin/New York: de Gruyter, 1990), 3056–3229

Mansfeld, Jaap, *Heresiography in Context: Hippolytus'* Elenchos *as a Source for Greek Philosophy*, PhA 56; Leiden/New York/Cologne: Brill, 1992

Mansfeld Jaap, and Runia, David T., *Aëtiana: The Method and Intellectual Context of a Doxographer*, II/1, PhA 114; Leiden/Boston: Brill, 2009

Menn, Stephen, "The Origins of Aristotle's Concept of *Energeia*: *Energeia* and *Dynamis*," *Ancient Philosophy* 14/1 (1994), 73–114

Menn, Stephen, "Plato and the Method of Analysis," in *Phronesis* 47/3 (2002), 193–223

Merino Rodríguez, Marcelo, "El *Stromata* VIII de Clemente de Alejandría," *Scripta Theologica* 37/1 (2005), 13–51

Moll, Sebastian, *Arch-Heretic Marcion*, Tübingen: Mohr Siebeck, 2010

Morison, Benjamin, "Logic," in R.J. Hankinson (ed.), *The Cambridge Companion to Galen* (Cambridge: Cambridge University Press, 2008), 66–115

Morison, Benjamin, "Language," in R.J. Hankinson (ed.), *The Cambridge Companion to Galen* (Cambridge: Cambridge University Press, 2008), 116–156

Morison, Benjamin, "Les *Catégories* d'Aristote comme introduction à la logique," in O. Bruun and L. Corti (eds.), *Les Catégories et leur histoire* (Paris: Vrin, 2005), 103–119

von Müller, Iwan, "Über Galens Werk vom wissenschaftlichen Beweis," in *Abhandlungen der königlich-bayerischen Akademie der Wissenschaften, phil.-philol. Kl.* 20 (1897), 403–478

Le Nain de Tillemont, Louis-Sébastien, *Mémoires pour servir à l'histoire ecclesiastique des six premiers siècles*, III, Paris: Robustel, 1695

Nardi, Carlo, "Socratismo evangelico nell' *Ottavo Stromateus* (cap. I) di Clemente Alessandrino," *Annali del Dipartimento di Filosofia dell'Università di Firenze* 5 (1989), 23–36

Nardi, Carlo, *Clemente Alessandrino: Estratti profetici*, Florence: Nardini, 1985

Nautin, Pierre, "La fin des *Stromates* et les *Hypotyposes* de Clément d'Alexandrie," *Vigiliae Christianae* 30/4 (1976), 268–302

Le Nourry, Nicolas, "De omnibus Clementis Alexandrini operibus. Dissertatio tertia," in idem, *Apparatus ad bibliothecam maximam veterum patrum et antiquorum scriptorum ecclesiasticorum*, III, Paris: Anisson, 1703

Opsomer, Jan, *In Search of the Truth: Academic Tendencies in Middle Platonism*, Brussels: Koninklijke Academie voor Wetenschappen, 1998

Orbe, Antonio, *Parábolas evangélicas en San Ireneo*, I, Madrid: La Editorial Católica, 1972

Pépin, Jean, "Clément d'Alexandrie, les *Catégories* d'Aristote et le fragment 60 d'Héraclite," in idem, *De la philosophie ancienne à la théologie patristique* (London: Variorum, 1986), IV, 271–284

Plátová, Jana, "Bemerkungen zu den Hypotyposen-Fragmenten des Clemens Alexandrinus," in *Studia Patristica* 46, ed. J. Baun, A. Cameron, M. Edwards, and M. Vinzent (Leuven/Paris/Walpole, MA: Peeters, 2010), 181–187

Plátová, Jana, "Die Fragmente des Clemens Alexandrinus in den griechischen und arabischen Katenen," in *Studia Patristica* 54, ed. M. Vinzent, L. Mellerin, and H.A.G. Houghton (Leuven/Paris/Walpole, MA: Peeters, 2013), 3–9

Primavesi, Oliver, "Dionysios der Dialektiker und Aristoteles über die Definition des Lebens," *Rheinisches Museum für Philologie* 135 (1992), 246–261

Raiola, Tommaso, "Alle origini di un falso galenico: Il *Commento* al *De alimento* e una citazione di Sabino," *Aion* 32 (2010), 101–110

Ramelli, Ilaria, "Origen, Greek Philosophy, and the Birth of the Trinitarian Meaning of Hypostasis," *Harvard Theological Review* 105/3 (2012), 302–350

Reed, Charles M., *Maritime Traders in the Ancient Greek World*, Cambridge: Cambridge University Press, 2003

Reinhardt, Tobias, *Cicero's Topica*, edited with an introduction, translation, and commentary, Oxford: Oxford University Press, 2003

Reinkens, Joseph Hubert, *De Clemente presbytero Alexandrino, homine, scriptore, philosopho, theologo liber*, Wrocław: Aderholz, 1851

Renan, Ernst, *Marc Aurèle et la fin du monde antique*, Paris: Calmann-Lévy, 1882

Rizzerio, Laura, "La nozione di ἀκολουθία come logica della verità in Clemente di Alessandria," *Rivista di Filosofia Neo-Scolastica* 79 (1987), 175–195

Rizzerio, Laura, "Foi, Gnosis, Dialectique, Logique: notes à propos de *Stromates* VIII de Clément d'Alexandrie," in *Studia Patristica* 31, ed. E.A. Livingstone (Leuven: Peeters, 1997), 522–529

Rizzi, Marco, "The End of *Stromateis* VII and Clement's Literary Project," in M. Havrda, V. Hušek, and J. Plátová (eds.), *The Seventh Book of the Stromateis: Proceedings of the Colloquium on Clement of Alexandria (Olomouc, October 21–23, 2010)* (Suppl. Vig. Chr. 117; Leiden/Boston: Brill, 2012), 299–311

Ross, William David, *Aristotle's Metaphysics*, a revised text with introduction and commentary, Vols. I–II, Oxford: Clarendon Press, 1924

Ross, William David, *Aristotle's Prior and Posterior Analytics*, a revised text with introduction and commentary, Oxford: Clarendon Press, 1957

Ruben, Paul, *Clementis Alexandrini Excerpta ex Theodoto*, Leipzig: Teubner, 1892

Runia, David T., and Mansfeld, Jaap, *Aëtiana: The Method and Intellectual Context of a Doxographer*, see Mansfeld

Scherer, Jean, *Extraits des livres I et II du* Contre Celse *d'Origène d'après la papyrus no. 88747 du Musée du Caire*, Cairo: Institute français d'archéologie orientale, 1956

Schofield, Malcolm, "*Epilogismos*: An Appraisal," in M. Frede and G. Striker (eds.), *Rationality in Greek Thought* (Oxford: Oxford University Press, 1996), 221–237

Schubert, Andreas, *Untersuchungen zur stoischen Bedeutungslehre*, Vandenhoeck & Ruprecht: Göttingen, 1994

Schumrick, Arnold, *Observationes ad rem librariam pertinentes. De syntaxis, syntagma, pragmateia, hypomnēma vocabulis*, Marburg diss. 1909

Sedley, David N., "On Signs," in J. Barnes, J. Brunschwig, M. Burnyeat, and M. Schofield (eds.), *Science and Speculation* (Cambridge: Cambridge University Press, 1982), 239–272

Servino, Antonio, "Clemente Alessandrino: il problema di Stromata VIII," *Quaderni del Dipartimento di filologia, linguistica e tradizione classica «Augusto Rostagni»* 17 (2001), 97–104

Smyth, Herbert Weir, *Greek Grammar*, revised by G.M. Messing, Cambridge, MA: Harvard University Press, 1956

Solmsen, Friedrich, "Early Christian Interest in the Theory of Demonstration," in W. den Boer, P.G. van der Nat, C.M.J. Sicking, and J.C.M. van Winden (eds.), *Romanitas et Christianitas: Studia J.H. Waszink Oblata* (Amsterdam: North-Holland, 1973), 281–291

von Staden, Heinrich, *Herophilus: The Art of Medicine of Early Alexandria*, edition, translation, and essays, Cambridge: Cambridge University Press, 1989

Steneker, Henricus, *ΠΕΙΘΟΥΣ ΔΗΜΙΟΥΡΓΙΑ. Observation sur la fonction du style dans le Protreptique de Clément d'Alexandrie*, Nijmegen/Utrecht: Dekker and Van de Weckt, 1967

Thomassen, Einar, *The Spiritual Seed: The Church of the 'Valentinians'*, Nag Hammadi & Manichaean Studies 60; Leiden/Boston: Brill, 2006

Tieleman, Teun, *Galen and Chrysippus On the Soul: Argument and Refutation in* De placitis, *Books II–III*, PhA 68; Leiden/New York/Cologne: Brill, 1996

Tieleman, Teun, *Chrysippus' On Affections: Reconstruction and Interpretation*, PhA 94; Leiden/Boston: Brill, 2003

Tieleman, Teun, "Methodology," in R.J. Hankinson (ed.), *The Cambridge Companion to Galen* (Cambridge: Cambridge University Press, 2008), 49–65

Todd, Robert B., "*Epitedeiotes* in Philosophical Literature: Towards an Analysis," in *Acta Classica* 15 (1972), 25–35

von Wedel, Christiane, *Symbola ad Clementis Alexandrini stromatum librum VIII. interpretandum*, Weimar: R. Wagner iun., 1905

Witt, Reginald E., *Albinus and the History of Middle Platonism*, Cambridge: Cambridge University Press, 1937

Wright, William, *Catalogue of the Syriac Manuscripts in the British Museum*, Vol. II, London: Trustees of the British Museum, 1871

Wyrwa, Dietmar, *Die christliche Platonaneignung in den* Stromateis *des Clemens von Alexandrien*, Berlin/New York: de Gruyter, 1983

von Zahn, Theodor, *Forschungen zur Geschichte des neutestamentichen Kanon und der altkirchlichen Literatur*, III: *Supplementum Clementinum*, Erlangen: Deichert, 1884

Zeller, Eduard G., *Die Philosophie der Griechen*, III/1, 5th edition, Leipzig: Reisland, 1923

Zhyrkova, Anna, "Reconstructing Clement of Alexandria's Doctrine of the Categories," in J.J. Finamore and R.M. Berchman (eds.), *Conversations Platonic and Neoplatonic: Intellect, Soul, and Nature. Papers from the 6th annual conference of the International Society for Platonic Studies* (Sankt Augustin: Academia Verlag, 2010), 145–154

Index of Modern Authors[1]

1 The index does not include translators and editors, unless a particular solution of theirs is the point at issue. Page numbers in italics refer to the apparatus of the Greek text.

Index of Subjects and Names[1]

1 For the names of ancient authors, see also Index of sources. Page numbers in italics refer to the translation of '*Strom.* VIII'. Numbers in bold refer to particularly enlightening passages.

Index of Sources

1 NB: Titles of Galen's works are abbreviated according to R.J. Hankinson (ed.), *The Cambridge Companion to Galen* (Cambridge: Cambridge University Press, 2008), pp. 391–397. References in the Index include (1) the book, chapter, and section numbers (if applicable) of the cited work according to the edition used in this book (see Bibliography) and (2) volume/page numbers according to the Kühn edition, if the work is contained therein.

www.ingramcontent.com/pod-product-compliance
Lightning Source LLC
LaVergne TN
LVHW010542300926
849162LV00002B/209